Decentralized Frameworks for Future Power Systems

Decentralized Frameworks for Future Power Systems

Operation, Planning, and Control Perspectives

Edited by

Mohsen Parsa Moghaddam
Faculty of Electrical and Computer Engineering, Tarbiat Modares University, Tehran, Iran

Reza Zamani
Faculty of Electrical and Computer Engineering, Tarbiat Modares University, Tehran, Iran

Hassan Haes Alhelou
Department of Electrical and Computer Systems Engineering, Monash University, Melbourne, Australia

Pierluigi Siano
Department of Management and Innovation Systems, University of Salerno, Fisciano, Italy

ELSEVIER

ACADEMIC PRESS
An imprint of Elsevier

Academic Press is an imprint of Elsevier
125 London Wall, London EC2Y 5AS, United Kingdom
525 B Street, Suite 1650, San Diego, CA 92101, United States
50 Hampshire Street, 5th Floor, Cambridge, MA 02139, United States
The Boulevard, Langford Lane, Kidlington, Oxford OX5 1GB, United Kingdom

ISBN 978-0-323-91698-1

For information on all Academic Press publications
visit our website at https://www.elsevier.com/books-and-journals

Publisher: Charlotte Cockle
Acquisitions Editor: Graham Nisbet
Editorial Project Manager: Mica Ella Ortega
Production Project Manager: Poulouse Joseph
Cover Designer: Matthew Limbert

Typeset by STRAIVE, India

Contents

Contributors

Amir Abdollahi Department of Electrical Engineering, Shahid Bahonar University of Kerman, Kerman, Iran

Shahab Afshar ConnectSmart Research Laboratory, Department of Electrical Engineering, University of Tennessee at Chattanooga, Chattanooga, TN, United States

Muhammad Afzal Sichuan Provincial Key Lab of Power System Wide-Area Measurement and Control, School of Mechanical and Electrical Engineering, University of Electronic Science and Technology of China, Chengdu, China

Yaser Al Mtaw Department of Applied Computer Science, The University of Winnipeg, Winnipeg, MB, Canada

Waqas Amin Department of Electronics & Power Engineering, PN Engineering College, National University of Sciences & Technology, Karachi, Pakistan

F. Conte Department of Electrical, Electronic, Telecommunication Engineering and Naval Architecture, University of Genova, Genova, Italy

F. D'Agostino Department of Electrical, Electronic, Telecommunication Engineering and Naval Architecture, University of Genova, Genova, Italy

Thusitha Thilina Dayaratne Faculty of IT, Monash University, Clayton, VIC, Australia

Hamed Delkhosh Faculty of Electrical and Computer Engineering, Tarbiat Modares University, Tehran, Iran

Jianguo Ding Department of Computer Science, Blekinge Institute of Technology, Karlskrona; School of Informatics, University of Skövde, Skövde, Sweden

Vahid R. Disfani ConnectSmart Research Laboratory, Department of Electrical Engineering, University of Tennessee at Chattanooga, Chattanooga, TN, United States

Daniel Eghbal Manager Future Network Strategy, Energy Queensland, Brisbane, QLD, Australia

Sajjad Fattaheian-Dehkordi Aalto University, Espoo, Finland; Sharif University of Technology, Tehran, Iran

Yi Shyh Foo Eddy School of Electrical and Electronic Engineering, Nanyang Technological University, Singapore, Singapore

Mahmud Fotuhi-Firuzabad Sharif University of Technology, Tehran, Iran

Gevork B. Gharehpetian Department of Electrical Engineering, Amirkabir University of Technology, Tehran, Iran

Hoay Beng Gooi School of Electrical and Electronic Engineering, Nanyang Technological University, Singapore, Singapore

James Griffin ConnectSmart Research Laboratory, Department of Electrical Engineering, University of Tennessee at Chattanooga, Chattanooga, TN, United States

Anwar Haqu Department of Computer Science, Western University, London, ON, Canada

Miguel Heleno Lawrence Berkeley National Laboratory, Berkeley, CA, United States

Qi Huang Sichuan Provincial Key Lab of Power System Wide-Area Measurement and Control, School of Mechanical and Electrical Engineering, University of Electronic Science and Technology of China, Chengdu, China

Li Jain Sichuan Provincial Key Lab of Power System Wide-Area Measurement and Control, School of Mechanical and Electrical Engineering, University of Electronic Science and Technology of China, Chengdu, China

Anssi Jäntti School of Technology and Innovations, University of Vaasa, Vaasa, Finland

Maarit Jäntti School of Technology and Innovations, University of Vaasa, Vaasa, Finland

Mohsen Jorjani Faculty of Electrical and Computer Engineering, Tarbiat Modares University, Tehran, Iran

Ali Karimi Electrical and Computer Engineering, University of Kashan, Kashan, Iran

Ariel Liebman Faculty of IT, Monash University, Clayton, VIC, Australia

Pablo Macedo ConnectSmart Research Laboratory, Department of Electrical Engineering, University of Tennessee at Chattanooga, Chattanooga, TN, United States

Luciana Marques Universidade Federal de Minas Gerais, Belo Horizonte, Brazil; Flemish Institute for Technological Research (VITO), Mol, Belgium

Mohsen Parsa Moghaddam Faculty of Electrical and Computer Engineering, Tarbiat Modares University, Tehran, Iran

Davis Montenegro Power Delivery and Utilization, EPRI, Knoxville, TN, United States

Vahid Naserinia School of Informatics, University of Skövde, Skövde, Sweden

Saeed Nasiri Faculty of Electrical and Computer Engineering, Tarbiat Modares University, Tehran, Iran

Amirhossein Nasri Department of Electrical Engineering, Shahid Bahonar University of Kerman, Kerman, Iran

Yousef Noorizadeh Electrical and Computer Engineering, University of Kashan, Kashan, Iran

Wei Peng Faculty of Engineering and Applied Science, University of Regina, Regina, SK, Canada

Masoud Rashidinejad Department of Electrical Engineering, Shahid Bahonar University of Kerman, Kerman, Iran

Carsten Rudolph Faculty of IT, Monash University, Clayton, VIC, Australia

Mahsa Salehi Faculty of IT, Monash University, Clayton, VIC, Australia

Miadreza Shafie-khah School of Technology and Innovations, University of Vaasa, Vaasa, Finland

Pierluigi Siano Department of Management and Innovation Systems, University of Salerno, Fisciano (SA), Italy

F. Silvestro Department of Electrical, Electronic, Telecommunication Engineering and Naval Architecture, University of Genova, Genova, Italy

Fereidoon P. Sioshansi Menlo Energy Economics Energy Consultant, Walnut Creek, CA, United States

Nader Tarashandeh Electrical and Computer Engineering, University of Kashan, Kashan, Iran

Jason Taylor Power Delivery and Utilization, EPRI, Knoxville, TN, United States

Mahyar Tofighi-Milani Sharif University of Technology, Tehran, Iran

Khalid Umer Sichuan Provincial Key Lab of Power System Wide-Area Measurement and Control, School of Mechanical and Electrical Engineering, University of Electronic Science and Technology of China, Chengdu, China

Wadaed Uturbey Universidade Federal de Minas Gerais, Belo Horizonte, Brazil

Fei Wang North China Electric Power University, Beijing, China

Shailesh Wasti ConnectSmart Research Laboratory, Department of Electrical Engineering, University of Tennessee at Chattanooga, Chattanooga, TN, United States

Morteza Yousefian Faculty of Electrical and Computer Engineering, Tarbiat Modares University, Tehran, Iran

Reza Zamani Faculty of Electrical and Computer Engineering, Tarbiat Modares University, Tehran, Iran

Mahdi Zolfaghari Power System Secure Operation Research Centre, Amirkabir University of Technology, Tehran, Iran

Preface

Societies are shifting toward new equilibria characterized by decentralized, digitalized, decarbonized, deregulated, and democratized attributes known as 5D Giga trends. These global multidestination attributes are of varying depth and speed in different societies around the world with heterogeneous impacts on the legacy infrastructures along with sophisticated intertwining with different aspects of human life.

With this in mind, the power system is among those infrastructures that will face transformation waves driven by 5D Giga trends and emergence of disruptive technologies. Decentralization of electricity generation has already begun in several countries and is predominantly driven by high penetration of renewable energy resources and, in particular, high uptake of distributed energy resources.

Transformation edge utilities have already started the digitalization journey by supporting digital grade loads through expanding smart grids and deploying customer-centric activities along with big data-driven services. The status quo of carbon footprints of electricity generation is still far from global decarbonization targets; therefore, higher uptake of renewable power generation is inevitable.

Deregulation in the power industry is mainly driven by the economics around the power delivery value chain resulting in a widespread competition of emerging entities in the marketplace. The emergence of peer-to-peer (P2P) and peer-to-x (P2X) transactions and prosumerization could create a paradigm shift in the deregulation reform that started almost three decades ago.

A democratic grid concept will play a key role in enabling the energy transition driven by the 5D megatrends. Such a grid provides a democratized connectivity in the network between all active participants in the electric power ecosystem for mutual benefits.

This book addresses various aspects of the decentralized frameworks in future power systems where challenges including great complexity, dimensionality, uncertainties, and the curse of big data will be created. These challenges have been the basis and motivation for preparing the book aiming at graduate students and the power industry engineers and experts intending to become familiar and/or update their knowledge and skills related to energy industry transformation.

It is a pleasure to acknowledge all the contributors of the book chapters for sharing their knowledge and experience through the creation of this edition.

Mohsen Parsa Moghaddam
Reza Zamani
Hassan Haes Alhelou
Pierluigi Siano

Energy transformation and decentralization in future power systems

1

Fereidoon P. Sioshansi[a], *Reza Zamani*[b], *and Mohsen Parsa Moghaddam*[b]
[a]Menlo Energy Economics Energy Consultant, Walnut Creek, CA, United States, [b]Faculty of Electrical and Computer Engineering, Tarbiat Modares University, Tehran, Iran

1 Introduction

Energy transformation is addressed as a global trend to more sustainable world and future. Zero-emission targets pave the way to reach this goal with brilliant presence of renewable energy resources. Obviously, transforming the structure of the energy industries including power systems to decentralized frameworks is inevitable. Challenges and issues in future power systems can be met by managing the evolution of the grid as a system of systems (SOS) [1]. The growing complexity of the future power system and also decentralized nature of the data and information in this environment urge moving toward decentralized decision-making (DDM) in the future power system that can be tackled as decentralized framework approaches. In addition, the conventional decision-making approaches in which the problems are solved in a centralized manner can fail particularly for large-scale systems with problems of large size. Modeling and solving the large-scale problems are often an arduous work which requires intensive data. However, extensive data communications lead to security concerns and the privacy as well. In such cases, the problem should be solved based on decentralized algorithms by several dispersed decision makers in which their communication and computational resources are limited but coordinated. The main challenge is defining decision polices with a specified structure using distributed algorithms to solve large-scale power system problems. This approach can be applied for several aspects of power system engineering including operation, planning, and control problems. Disruption in future power system due to the presence of new technologies supported by big data issues and security of cyber physical systems stimulates researches to tackle the raised challenges in a decentralized manner. The DDM have recently attracted considerable interest from researchers in the field of power system.

The main purpose of this chapter is to address the energy transformation trend and comprehensive issues of the decentralized approach for decision-making in the future power systems. Also, the book covers very important topics in power system studies, which are of decentralized nature. One of the main issues that arise with the integration of distributed energy resources (DERs) into the grid is the increase in the information and complexity of the network. Information privacy is another challenge in

Decentralized Frameworks for Future Power Systems. https://doi.org/10.1016/B978-0-323-91698-1.00009-1

smart grids of the future in which the entities try to preserve their privacy, and this situation makes the decision-making process more difficult. Accessing and analyzing the required information is a time-consuming work that should be performed in a central unit and at the same time in the centralized decision-making process. All of the aforementioned issues justify the necessity of using DDM in future power system environment. In this chapter and remaining parts of the chapter, decentralized decision-making frameworks applied to the future power system in order to provide a network of more efficient, secure, and reliable attributes are discussed. The main topics of this chapter include: (1) how energy transition can change the world and consequently power system, (2) decentralized decision-making process and requirements in future power system, (3) necessity of applying decentralized decision-making in future power system, and (4) the application of decentralized decision-making in future power systems.

2 Energy transformation

What is a **decentralized, decarbonized, digitalized** future energy system likely to look like and what will be the central roles and functions of the future electric power system at its core? These are timely questions to ask as the world is finally transitioning to a more sustainable, low-carbon future, and these are among the questions addressed in this collected volume appropriately titled **Decentralized Frameworks for Future Power Systems**.

The starting point, of course, is to ask why we are *transitioning* our existing energy system and replacing our existing *infrastructure*. The answer is climate change, or rather, the unsustainability of the business-as-usual (BAU) approach, which has got us to our present predicament. As explained in the latest report by the Intergovernmental Panel on Climate Change (IPCC), *Climate Change 2021: the Physical Science Basis* [2] in August 2021, to limit global warming to a maximum of 1.5°C, or even 2°C, requires urgent reduction in greenhouse gas emissions on a grand scale. According to the global climate change trend, the observed temperature rise in 2020 was quite alarming [2].

In short, the *energy transition* that everybody is talking about means that we must dramatically reduce our dependence on fossil fuels—among other things—and do it in a relatively short order. Achieving this requires a herculean effort sustained roughly over the next three decades across the globe. It requires massive investments and global cooperation and collaboration on a scale and speed never attempted before.

According to a study by the Bloomberg New Energy Finance (BNEF) up to $173 trillion of investments will be needed to achieve net zero emissions by 2050 [3]. As a point of reference, US GDP in 2020 of $21 trillion was used. That is, globally we must invest over eight times the size of the US economy to reach zero carbon by 2050.

Assuming we can get there, and in time, what will that future energy system look like? There are many projections and many paths we can potentially take. The International Energy Agency (IEA) [4] and DNV [5], for example, have published reports that suggest feasible pathways to carbon neutrality by 2050.

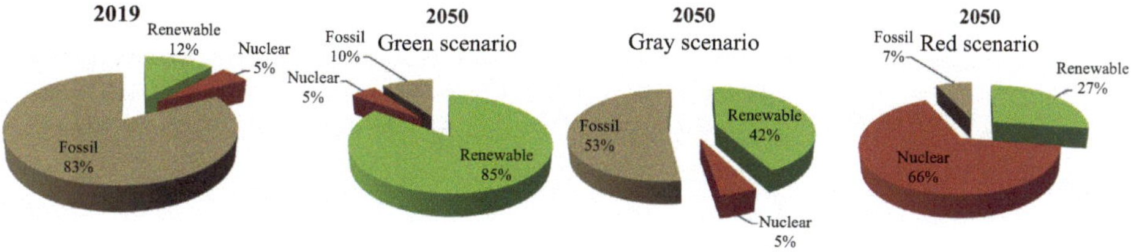

Fig. 1.1 Pathways to net zero by 2050.

In all cases, there are options or paths we can follow, each with its pros and cons. The BNEF, for example, examines three scenarios to achieve net zero by 2050, each presenting major transformations to the primary energy supply as illustrated in Fig. 1.1 [3].

- The **Green Scenario** prioritizes **clean electricity** and **green hydrogen**, with solar and wind production growing to 70% of primary global energy in 2050 and other renewables comprising a further 15%.
- The **Red Scenario** prioritizes **nuclear energy** for **hydrogen** production, causing nuclear to account for 66% of primary energy by 2050, a significant rise from its current 5%.
- The **Gray Scenario** depicts widespread use of **carbon capture and storage** (CCS), allowing for continued use of coal and gas. In this scenario, fossil fuel consumption could be reduced by 2% a year to 53% by 2050, with only oil seeing a significant decline.

As noted, there are many ways to get to zero carbon by 2050, all presenting significant challenges. Clearly, there is no simple, no single right or wrong way, and no silver bullet. But most analysts believe that we must try because the alternative—failure to act decisively—is likely to be far more expensive and potentially catastrophic.

It should be equally clear that the transition toward a more sustainable energy future not only impacts those directly involved in the fossil fuel business but also virtually every business that uses fossil fuels, which include everything from banking to baking, from investing to transportation and beyond.

After procrastinating for decades, many governments are now driving the transformation from the conventional fuels to renewables. Different societies have set ambitious targets to turn their economies into carbon-neutral by years 2045–50 such as European Commission, United Kingdom, and California (United States). The United Kingdom has announced a plan to ban the sale of the internal combustion engines (ICEs) by 2035. Even China, a developing economy, has pledged to peak its emissions by 2030 and reach zero carbon state by 2060 [6]. Under the Biden Administration, the United States is poised to lead efforts to make definitive progress.

Critically, the investment community, long resistant to change, is now embracing environmental, social, and governance (ESG) principles in deciding long-term investment strategies due to the relentless pressure from environmental activists, such as Greta Thunberg and her followers. An increasing number of global and local businesses are adopting strategies to reduce their carbon footprint while shifting from fossil fuels to renewables. Such a massive transformation would have seemed improbable a mere decade ago.

At the same time, virtually all European oil majors have announced plans to become net zero carbon by 2050 as have a number of airlines such as United Airlines, British Airways, and Qantas and the world's biggest container shipping company, Maersk. Major car makers including GM have said that they will stop building internal combustion engines (ICEs) as they rapidly shift toward electric vehicles (EVs). The pace and audacity of these moves are simply stunning and are changing the energy landscape.

This leads to a host of questions as follows:

- How is the world going to get by with less or—perhaps someday—no fossil fuels?
- Can we continue with economic growth, prosperity, and high living standards without the fossil fuels? Or in fact can we?

Many "traditional" energy experts have come to gradually—sometimes grudgingly—accept the emerging view that fossil fuels are no longer the *future* of energy. They have accepted that we must—and will—gradually replace fossil fuels with increasing amounts of renewable sources, notably wind and solar.

Both solar and wind technologies have experienced declining costs and improved performance and are now broadly recognized as the cheapest source of electricity generation. But neither resource is "dispatchable" as are the conventional plants. They are plentiful at certain times and scarce or nonexistent at others, such as when there is no wind or when the sun sets at the end of the day. This presents a major challenge for the electric grid operators who must balance supply and demand at all times.[a] The skeptics legitimately ask how can this inherent variability be satisfactorily and economically resolved? Energy storage and flexible demands offer partial solutions; however, more solutions and suggestions are needed.

This, needless to say, is not a trivial issue especially as the transportation, heating, and industry are being electrified. How can entire modern economies depend on electricity when it is totally dependent on variable renewable generation resources?

But ultimately, most observers agree that the future energy system will be mostly if not exclusively *electrified* with most if not all electricity generated from low—such as nuclear—or noncarbon—such as renewable—resources. This means radical new ways to generate, transmit, distribute, and consume energy.

At the same time, it is recognized that the electric power business, traditionally dominated by a few big players generating and transmitting undifferentiated commodity in kilowatt-hour (kWh) to millions of passive consumers in a one-way flow across a vast delivery network, is changing—slowly in some places and rather fast in others—with increasing focus on the behind-the-meter (BTM) space. The transition of consumers to prosumers, who generate much of what they consume via rooftop solar PVs, is already underway. This is likely be followed by *prosumagers* [7], who are *prosumers* who invest in BTM storage, including EVs—nothing but massive batteries on wheels. What used to be an academic curiosity is increasingly the future of electricity business.

[a] Power outages in mid-August 2020 in California were blamed on the difficulty of filling the gap created by lack of solar generation after sunset.

That, however, is not the end of the likely developments of BTM storage. In the future, there will be increased opportunities for trading and transactions among and between consumers, *prosumers* and *prosumagers*, in many cases assisted and/or enabled by intelligent aggregators and intermediaries. But how will transactions to BTM become part of the future? According to Glachant and Rossetto [8].

> ***Peer-to-peer** (P2P) and **peer-to-x** (P2X) open up a new world of transactions in the electricity sector. We have already seen in the past **business-to-business** (B2B) with the wholesale markets, opening around 1990, and **business-to-consumer** (B2C) with the retail markets, opening around 2000.*

For now, no one has all the answers or the perfect crystal ball. But the fundamental challenge of variability of renewables can be addressed through a combination of cures, some more advanced, cost-effective, and practical than others.

Moreover, solving the problem of the variability of renewables [9], as daunting as it may appear, is *not* significantly more difficult than a number of other challenges that have already been resolved by good engineering and design.

This, or for that matter, any book about the future energy systems must begin with an acknowledgement that the *energy transition* has already started and will gain momentum over the next couple of decades. The future of life on Earth depends on its successful execution.

3 Decentralized decision-making

Decision-making is a cognitive process that leads to the selection of an action among several alternatives based on the estimation of values and advantages. The decision-making process is done by identifying the decision, gathering information, and evaluating all aspects. Using the decision-making process, we can achieve the best possible solution by organizing the relevant information and defining alternatives. The decision-making process can be divided into seven steps, the objective of which is to increase the probability of reaching the desired answer:

- Step 1: *Identify the decision.* The need for decision-making must be determined and introduced. In this step the nature of the decision is identified.
- Step 2: *Gather information.* Complete information must be reviewed before making a decision, that is, the information that is needed, the best sources of information, and how to get it. Some are internal information and belong to the same system that intends to make decisions, and others are external information that includes information from other systems but is effective in the decision-making process.
- Step 3: *Identify alternatives.* Once the information has been collected, all possible possibilities or different solutions are identified. Existing information can be used to build alternatives or even predict future information. In this step, all possible and desirable options are listed.
- Step 4: *Evaluate the evidence.* Based on the available information and knowledge, the outcome of each of the available options should be reviewed and evaluated. In this step, it is examined whether the needs identified in step 1 will be met using these alternatives.

Therefore, at this stage, some options that seem to have a high potential to achieve the desired goal are supported. Finally, alternatives are prioritized based on predetermined values.

- Step 5: *Choose from alternatives*. Once all the evidence have been evaluated, we are ready to choose the option that seems best. A combination of options may even be selected in the process. The item selected at this stage can most likely be the same or similar to the one at the top of the list at the end of step 4.
- Step 6: *Execute the decision*. In this step, we are ready to take the necessary steps and start the implementation of the alternative selected in step 5.
- Step 7: *Review the decision and its consequences*. In this step, the results of the decision are evaluated to determine whether the needs identified in step 1 have been met. If the decision does not meet the identified needs, certain steps of the process may be repeated to make a new decision. For example, more accurate or different information may replace existing information and be redecided with existing sources.

Therefore, decision-making can be a difficult process, and the accuracy of the answer depends on the accuracy of all the steps. For example, if the sources of information in the decision-making process are wrong or even large, the decision-making will be difficult. Traditionally, the decision-making process is considered centralized and is performed at the same time. A specific central unit is responsible for carrying out this process from the beginning to the end. Therefore, all decision-making steps are done by this center independently. Therefore, if there is a disturbance in this center, the whole decision-making process is stopped, and if a decision is implemented, the decision cannot be accepted. Gathering information in a central unit will also raise security concerns. Although several algorithms have been proposed to increase the security of this approach, the existence of these algorithms themselves increases the complexity of the decision-making process and make it difficult and time consuming to achieve the desired answer.

On the other hand, if there are several decision makers, the decision-making process becomes very difficult and it will be difficult or even impossible to make a decision that is acceptable to all of these decision makers. Given the above, a centralized decision-making approach cannot be an appropriate option in complex systems such as future power systems to solve problems of large size and large amount of information. Therefore, increased size of the problem, high volume of information, maintaining information security, the existence of multiple decision makers, and the complexity of the under studied system are among the weaknesses of integrated (centralized) decision-making. In the following, we will describe the DDM approach in order to solve these problems.

3.1 Concepts of decentralized decision-making

Modern societies faced with a wide variety of interests and evolving complexities can no longer understand and use the decision-making process centrally. Under these circumstances, instead of following the centralized approach, DDM becomes the dominant methodology in managing complex systems. Also, democratic structures tend to annihilate decision-making power to those sections of society that are actually affected. It is desired that the corporations become separated with different profits, which result in facing complex centralized decision-making problems by dividing them. As a result, complex centralized decision-making problems are solved by dividing them into several components. In fact, predefined relationships, especially those with complex hierarchical natures, are obsolete and replaced by free and transparent activities.

Thus, DDM is a vital, fundamental, and rapidly growing issue in decision-making theory. This includes a variety of areas such as multilevel optimization, multistage stochastic programming, hierarchical programming, multiagent system (MAS), key factor theory, supply chain management, contract theory, auction theory, etc. In most cases, these areas are part of various disciplines such as operations research, computer science, economics, game theory, managerial accounting, organizational theory, psychology, sociology, and so on. From the perspective of decision-making theory, these areas range from DDM (such as multilevel optimization) to multifaceted situations (such as principal factor theory). If a human is tried to make a decision, DDM approach can be used to better understand or simplify complex decision-making situations. This approach is especially important for dynamic systems and when decisions need to be made over time and in a distributed manner, as well as when new information is being obtained (updated). In a situation which several persons (multiple subsystems) in interaction with each other want to conduct a decision-making process, the need to apply the DDM is clearly greater. Apart from the variety of information that may be available to decision makers, decision rights, in particular, should be specified and decision-making authority and the type of relationship between decision-making units should be considered [10].

Therefore, the DDM has considerable applications, especially when the issues of thought and its coordination are fully disseminated among the various decision makers, and all of them participate in matters of mutual interest. Therefore, according to these considerations, distributed decisions can be used as a strategy in interconnected decisions. Most of these distributed decision-making issues do not have the same rank, which in many cases create a kind of hierarchical dependence on one-way relationships. Accordingly, many theories adopt an asymmetric description that examines DDM from a top party perspective. In fact, decisions that do not have some hierarchical features are no exception. For example, decisions made at different points and at the same time, or made with different levels of power, are the prime examples of asymmetric (hierarchical) dependency. In microeconomics, this type of dependence is often referred to as the Stackelberg model [11,12].

3.2 Application of DDM in engineering

Due to the increasing complexity and growth of data in various engineering sciences, the DDM approach has been considered. Among these applications, we can mention the importance of using decentralized decision-making in coordinating several aircrafts simultaneously [13]. Spatially and temporally interconnected systems (including vehicle fleets, service providers' communication with shared information and telecommunications resources) have a distinct structural feature in which different decision makers coordinate with centralized and limited information. When it is not possible to have a centralized coordinated plan, independent decision makers are forced to work together to achieve common or independent goals while having both local and communication constraints. In Ref. [13] a decentralized optimization method is introduced to solve the problem of coordination of several nonlinear dynamic systems with several decision makers. In Ref. [14] the authors present a sensor network for magnetic resonance imaging with a decentralized decision approach.

While the success of modern machine learning models is the foundation of many intelligent services, the performance of a complex model is often limited by data access. On the other hand, in most applications, a large amount of useful information may be generated and stored by several individuals (departments). Gathering such information leads to the existence of a central reference for training, additional management and business-related costs, privacy concerns, or even litigation. As a solution, a number of distributed machine learning techniques have been proposed to create a harmonized learning model that allows each section to update local models and exchange local calculations [15] or model parameters [16] with the central server. A model is given to improve accuracy. Decentralized machine learning has been extensively studied in references to enhance machine learning model training with increasing data [17]. In Ref. [18] distributed machine learning has been studied where information about teaching similar examples is inherently decentralized and is in the hands of different people (different sections). In Ref. [19] a framework for extensive training in kernel-based statistical models based on the combination of distributed convex optimization with stochastic techniques is proposed.

Machine learning also plays an important role in big data systems, which is very efficient due to its ability to discover and extract valuable knowledge and hidden information from data. In most cases, big data in systems such as the health-care system or financial systems may exist with multiple organizations that may have different privacy policies or may not explicitly share their data publicly. In order to process and analyze the system, shared data may be required. Thus, sharing big data between distributed data processors has become a challenging issue due to privacy concerns. Conventional methods of privacy include cryptographic tools or random data transmission. These methods may be inadequate for some of the emerging complex data systems, because these methods are mainly designed for small-scale data and traditional systems. In recent years, many methods have been proposed to maintain security and privacy by using decentralized and distributed data learning [20]. Using this approach, in addition to increase the speed of decision-making, data encryption methods can be used due to the reduction in data in different subsystems between which the data is divided.

Image reconstruction is one of the most important tasks in image processing. In many imaging applications, images inevitably include unrealistic and abnormal noise, such as impact noise, Poisson noise, and Cauchy noise. At the same time, images may be blurred due to the use of noise canceling functions. In order to revive a blurry image with noise as well as to preserve the edges of the image, the total variation (TV) optimization method is often used. Cauchy noise, which is often used in engineering applications, is a kind of impact and non-Gaussian noise. On the other hand, Cauchy noise can be detected and eliminated by solving a nonconvex TV optimization problem, which is very difficult to solve due to its nonconvexity. In Ref. [21] a DDM method is proposed to solve this problem.

Another example of DDM is programming issues for multiple robots. For example, guiding robots to specific goals, each with a limited communication range, is one of the most common programming problems in robotics. Programming the common space of all robots with complex constraints can be very difficult and practically impossible. Multirobot route planning suffers from the inherent complexity of the need to configure and move in Cartesian space [22]. Solving the robot's continuous path planning problem is very difficult using centralized decision-making [23], unless the problem is solved separately and asynchronously for each robot [24], in which each robot has to wait for the previous robot to complete its movement and a lot of time is lost. Decentralized approaches are used to solve multirobot path optimization problems, task allocation, and other robot requirements to optimize a general objective function. The goal of these approaches is to solve a series of small optimization problems for each robot while sharing information among the robots. In this process, the general optimization problem is solved, if this method is not used, the problem will be an unsolvable problem [24].

4 Implementation of DDM in future power systems

One common fact in DDM approach is that components made up of limited communication and computational capabilities must act collectively to communicate in order to perform a complex task [25]. In other words, there are intelligent interactions between interconnected components that make the overall system intelligent. One way to manage these interconnected systems is to manage them intuitively or by trial and error. This approach has been widely used in the past decades and its applications are in computer networks, wireless sensor networks, and MAS [26]. However, this approach seems to be hampered by the proliferation of more complex interactive systems. To fully empower the potential of modern systems, we need regular techniques for designing mechanisms that coordinate interconnected components. Ideally, it should be ensured that the components approach the optimum point quickly, and this is done with energy savings and minimal information exchange. In some applications, the final implementation must be performed correctly and securely, and the system must adhere to user privacy and data security. All of these issues, as well as others, indicate the need to take advantage of DDM in complex future systems.

Optimization theory provides an attractive framework for solving multiple decision problems. This method provides a method for modeling and formulating an engineering problem and its operational limitations mathematically, and then pursues the best solution. Therefore, an optimization problem can be formulated as follows:

$$\begin{aligned} &\text{minimize } f(x) \\ &\text{subject to } x \in \chi \end{aligned} \tag{1.1}$$

where $x \in \mathbb{R}^n$ is the optimization variable (expressing the decision variable), $f(x)$: $\mathbb{R}^n \rightarrow \mathbb{R}$ is the objective function, and χ is a set of decision constraints that must be met. Classical optimization algorithms typically run on a central computer where the objective function and constraint set are defined and described in closed form. The decentralized algorithm, on the other hand, divides the problem into several parts and breaks down separate processors or agents that solve the problem by interacting with each other. Due to the limited capacity of agents or subsystems, simple computations and collaboration mechanisms are often necessary for agents to perform local computations and communicate with neighbors. An example of a distributed decision problem is shown in Fig. 1.2. This network consists of four agents that work together to solve a decision problem.

As aforementioned, future power systems are so complex and extensive that it is no longer possible to address issues through a centralized decision-making approach, and the DDM approach should be used to address these issues. The application of DDM in power systems can be studied in its planning and operation. This section provides an overview of the application of the DDM approach in power systems. Fig. 1.3 shows this classification. In the following, this classification and existing researches are reviewed and investigated.

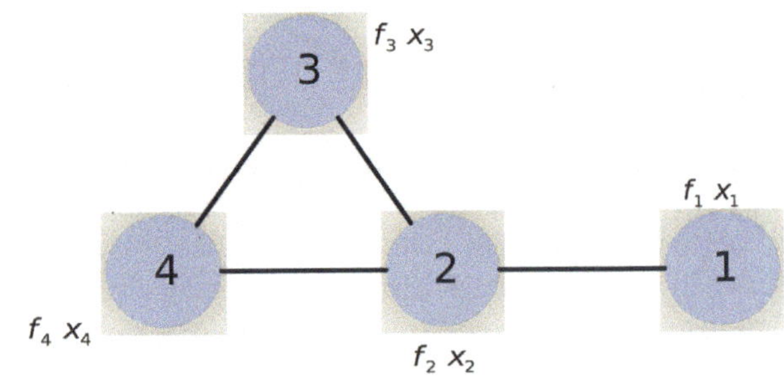

Fig. 1.2 Example of DDM approach in problem solving.

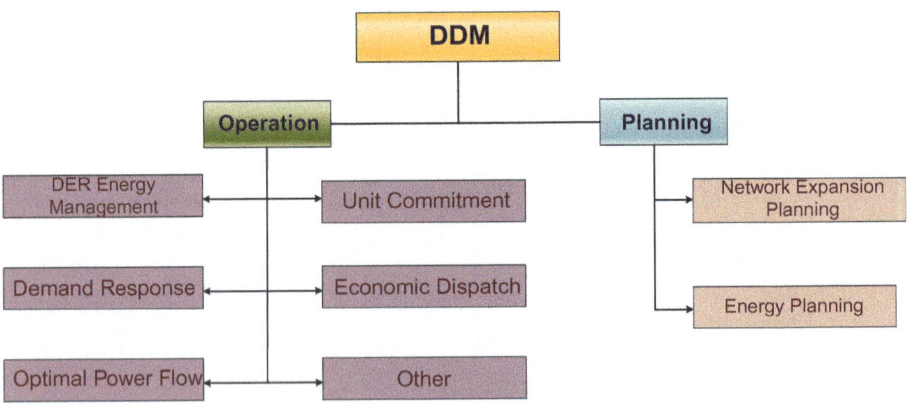

Fig. 1.3 Main application of DDM in power system issues.

4.1 DDM based on MAS

In the MAS-based DDM, each agent (or node) has a local objective function, and all agents are related to each other in order to reach the final decision. To achieve this goal, agents must exchange information in a network, which normally should be limited to adjacent agents.

4.2 Big data and decentralized data analytics

The pervasive sensor networks in future power systems are responsible for collecting a lot of information at any given time, as a result a huge flow of raw data in different formats across the network and between different parts of the system (generation, transmission, distribution, and consumption) are exchanged. Data mining extracts the desired useful information from this big data, as a result many benefits can be gained. Intelligent agents are a good infrastructure for DDM process and extraction of information from big data. Data analytics can be done seamlessly or using distributed MAS. Intelligent and autonomous agents are the basis for solving decentralized decision-making problems. Information extracted from big data has several applications in power systems such as load forecasting, load behavior extraction, load anomaly detection, wind product forecasting, monitoring all system components using real-time data with high accuracy, system operating condition monitoring, modeling of various system components (particularly load modeling), and so on. Information from distributed big data in a DDM approach can be extracted in three ways:

- The first method is to collect data in a central database. In this method, information is extracted using all the data stored in this center.
- The second method is the exchange of information between agents, in which data analytics is performed using local data.
- The third method is data analytics using local data and sharing its results with other agents, which ultimately corrects (updates) the results after receiving answers from other agents.

The purpose of decentralized data mining is to extract a useful model from distributed databases and use them with distributed knowledge and apply them in the decision-making process. In modern and complex systems, the use of decentralized data mining has become a dominant methodology. In future power systems, due to the emergence of big data and also the distributed nature of data, the use of decentralized data mining approach should be considered.

5 Application of DDM in future power system planning

Numerous resources of uncertainty can be considered for future power systems that influence the planning process dramatically. Conventional methods for managing decision-making uncertainties in the field of planning include stochastic and robust planning [27]. In stochastic programming, it is assumed that the probabilistic parameters follow a certain probabilistic distribution, and thus generate scenarios in which the decision-making problem is studied. This approach applies to many planning issues. In robust planning, only the worst-case scenario is known in terms of imposing the most impact on the decision-making process and the decision is taken by considering the occurrence of this scenario. With the introduced uncertainties that we will face in future power systems, as well as the existence of common uncertainties such as climate change uncertainties, fuel and equipment prices, customer consumption and the production of renewable resources, the decision-making process in future power systems is difficult and it will be challenging.

On the other hand, security and privacy concerns have caused the administrators of the network infrastructure and generation resources or exploiters not to make their information available to other entities, and as a result, the planning process will face the challenge of not having accurate information. Therefore, the planning in future power systems will face the problem of large size, excessive uncertainties, and insufficient information, which will be very difficult and time consuming to solve using centralized and conventional methods. Therefore, the DDM approach can be a suitable solution to solve planning problems in future power systems in order to achieve the desired solution, maintain security and privacy in the network and take into account all the uncertainties.

5.1 Decentralized network expansion planning

Transmission line expansion planning is traditionally done with a centralized approach. In other words, the planning of power systems is done in a centralized manner with the objective function of reducing investment and operating costs in order to maximize social welfare. As mentioned, the existence of new challenges such as many uncertainties, increased size of the decision makers, increased data, and the security and privacy of agents, has changed the issue of decision-making. Also, to provide electricity to customers economically with high reliability, planners must consider not only plans for the generation, development, and expansion of transmission lines, but also the entire wholesale electricity system, including transmission and purification of the market by independent system operator (ISO), to ensure that there is enough energy to meet future loads.

On the other hand, investors' decisions have a great impact on market results. Due to the restructuring of the electricity industry, some research works based on planning in competitive spaces have been proposed. Due to the complexity of this issue and some problems in policies related to the relationship between the long-term planning horizon and the day-to-day operation of the power system in a deregulated environment, the developed models so far cannot be adapted to the needs of planners and policy makers. In the new paradigm, the development of the power system must have a mechanism for negotiating between all actors with the objective function of maximizing the distributed profit for each actor in the system.

Uncertainties, efficiency, and productivity can be considered for all agents including independent generations, transmission system owners, ISOs, and consumers in decision-making to choose to invest and exploit the power system. As a result, future power network planning has to deal with decentralized multiobjective planning in a general framework, and the mechanism of this planning must be consistent across different planning horizons.

5.2 Decentralized energy planning

Studies in the field of energy planning include finding a set of energy sources and means of energy conversion in order to meet the needs of energy and load in a completely optimal way. This activity can be done using both centralized and decentralized approaches. The current pattern of energy trade development, especially in the field of fossil fuels and electricity, has been concentrated, leading to neglect of external issues and environmental degradation. For example, a large portion of the rural population and poor urban dwellers continue to depend on inadequate, low-quality energy sources and inefficient devices, leading to a decline in their quality of life. The current situation is mainly the result of the application of centralized energy planning (CEP) methods that ignore the energy needs of villages and people with disabilities and also lead to environmental degradation due to fossil fuel consumption, deforestation and coal mining. Also, centralized energy planning cannot pay attention to changes in socioeconomic and environmental factors of an area, which directly affects energy resources and the level of social welfare of society as a whole. The decentralized energy planning (DEP) approach can be used to improve the efficiency of energy resources.

6 Power system operation issues based on DDM

Independent system operators and regional transmission organizations (RTOs) use centralized optimization methods for optimal operation of power systems, which, in this process, gather all the necessary information and make centralized operation decisions in the central control center [28]. As the size of power systems expands and flexible, demand-driven resources enter power systems, such a centralized framework increases computational and communication problems. Therefore, problems will be difficult to solve or even unsolvable. The DDM is considered as an alternative way to solve the challenges of centralized decision-making mechanism in future power systems.

So far, the centralized decision-making framework has been widely used in power system operation, in which all the necessary information is collected and decisions are made centrally by the control center. However, this centralized framework faces significant challenges in future power systems, as follows:

- The first challenge is complex communications and computation burden for large-scale power systems. Given the significant size of future power systems, the need for extensive communication to gather relevant information and computational complexity to solve large-scale decision-making problems (such as security constraints problems) increases significantly. This problem will be exacerbated by increase in the size of the power system and increase in the flexible and distributed resources on the demand side.
- The second challenge is the challenge of political and technical issues for the coordination of several areas. Planning the connection of regional electricity infrastructure and aggregating large-scale renewable energy demand and planning to coordinate them to achieve overall economic reliability and efficiency will be a very difficult task in future power systems.

Given these challenges, the DDM as an alternative approach has attracted more attention. In this framework, a large-scale power system problem is divided into small-scale subsystems and subproblems for small areas, and effectively the desired decision relating to the overall system is made. In addition, local subproblems can be solved in parallel with the help of high-performance computational techniques that can increase the computational performance of the DDM. In fact, the DDM provides a structure for analyzing a decision-making problem to increase computational efficiency across multiple problems. In the DDM process, local subproblems can be distributed directly to separate data processing units.

6.1 DER energy management

One goal of future power systems is to create smarter and more secure distribution systems by integrating a two-way communication infrastructure. Data exchange provides distribution network operators (DNOs) with sophisticated management techniques for complex analysis and automated operations [29–31]. At the same time, various incentives from distribution companies have accelerated the use of renewable generators for smart grids. Power system analysis and control are the major challenges for distribution network operators. Uncertainty in demand and renewable generations can lead to unforeseen problems such as voltage drops or swells. Violation of the privacy of stakeholders, such as subscriber information and the cost of generators, is another problem. In addition, system analysis and calculations can be very difficult, especially when the number of decision variables increases with the participation of loads and renewable generators in the energy market. Hence, this problem can be tackled using DDM approach more conveniently.

6.2 Decentralized demand side management

One of the main goals of sustainable and efficient operation of the electricity network is to balance generation and consumption. In today's power system, most of the demand is met by large power plants, which is the task of the ISOs to determine the appropriate amount of generation to maintain the optimal performance of the power grid. In future power systems, with the deployment of intelligent equipment and communication infrastructure in the smart grid, demand-side resources can play an active role in energy management to balance generation and consumption. In particular, price-based demand response (DR) programs can well encourage consumers to change their consumption patterns in response to market price fluctuations. Also, the response of different consumers in a DR program depends mainly on the flexibility of the demand. Centralized management and decision-making for demand response programs in future power systems will be difficult and complex with a centralized decision-making approach due to the large amount of information and security issues and concerns. Therefore, a decentralized decision-making approach can be a good solution for managing load response programs in future power systems.

6.3 Decentralized optimal power flow

Optimal power flow problem determines the minimum operating costs of power systems by dispatching generation resources to supply system loads and to take into account generation units constraints. The DC power flow is an approximation of

AC power flow to obtain the amount of active power dispatch of the system. The emergence of new technologies in future power systems is attracting increasing attention of power markets, including owners of renewable electricity generation, as well as active customers with their own generation resources and advanced metering infrastructure. All players tend to automatically maximize their profits while they may not want to disclose their true financial information to system operators or other players. As mentioned, DR programs are one of the most important applications of active consumer engagement with the network to reduce the need to build new power plants or lines to deliver peak load. But the centralized approach does not meet the necessary speed in decision-making on these issues, and the increasing use of the programs such as DR, as well as the expansion of DERs, requires the use of DDM algorithms in future power systems. The advantages of optimal power flow using a decentralized approach are as follows:

(1) Centralized traditional optimal power flow requires a lot of bandwidth to gather all the information in the control centers, while using a decentralized approach, only limited information is transferred during the decision-making process.
(2) In the case of optimal power flow using the decentralized approach, there is no need to transfer the confidential financial information of the subsystems to other subsystems or control centers.
(3) The method of decentralized optimal power flow is more flexible than changes in the centralized system, especially in cases of changing the topology of the power system or telecommunications infrastructure, which has significant dynamics.
(4) Decentralized optimal power flow is more robust than centralized one. In other words, in centralized optimal power flow, the decision-making process is disrupted throughout the system by losing and disconnecting from the control center. While in DDM approach, if one of the subsystems is lost, the other subsystems continue to operate and the accuracy of the decision can be increased by returning the lost subsystem.

6.4 Decentralized economic dispatch

The economic dispatch problem distributes all energy consumed among power plant units by minimizing operating costs and meeting generators and system constraints. Economic dispatch is the basic mechanism for determining the best cost effective point of operation for all controllable devices connected to the power system according to their economic efficiency in real time. In its traditional form, economic load dispatch consists mainly of conventional generators, known as renewable generations, and loads that can be well approximated to a definite problem that usually covers a short period of time. However, this approach should be changed when unforeseen loads are used, such as when the grid uses energy purchased from electric vehicles or storage devices whose price depends on future energy prices and is typically determined by large transmission units at the transmission level.

However, some operators use economic dispatch mechanisms that take into account more than 2 h ahead, which is insufficient to plan these generations. On the other hand, lack of coordination between demand response decisions and generation units scheduling can increase fluctuations in energy price. Also, the increase in these productions puts a lot of pressure on the distribution network infrastructure. Conventional optimization methods such as Landa iteration method, gradient method, linear programming, and Newton methods are used to solve the problem of economic dispatch and one of their weaknesses is that they are sensitive to the initial point and converge to the local optimal point. As the power system transforms from its traditional structure with a classical centralized structure to an intelligent network of scattered nature, a similar paradigm shift occurs in decision-making areas. As mentioned, centralized algorithms typically require a high-bandwidth communication infrastructure to send aggregated data, a high level of communication, and concerns about system reliability due to their susceptibility to modeling error. Decentralized algorithms can dispatch a much larger number of generation units than centralized methods. Also, in decentralized methods, smaller volumes of data and information are exchanged between the participating units in the economic dispatch problem.

6.5 Decentralized unit commitment

With the growth of power systems, a new operating framework will be needed. One solution is to integrate system operations such as economic load distribution, congestion analysis, and unit planning (UC) problem. The problem of unit commitment also faces the challenges expressed in economic load dispatch as well as optimal power flow which solves and sends its injection power information to a high level. Each coordinator communicates with other neighboring coordinators at a higher level and updates their local pricing.

7 Conclusions

This chapter discussed the energy transformation outlook, zero-emission target for sustainable world, comparison of the concepts of centralized and decentralized decision-making, and the need to apply decentralized decision-making in modern systems as well as future power systems. Weaknesses as well as shortcomings of the centralized decision-making approach for applying in future power systems were addressed and discussed in this chapter. Applications of decentralized decision-making and the need to use this approach in various engineering sciences were explained and some applications were described. The applications of decentralized decision-making and the reasons for applying it in future power systems were studied. A comprehensive new classification of these applications, which includes application of decentralized decision-making in power system planning and operation, was also presented in this chapter. Applications of this approach in power system planning were divided into two parts: network expansion planning in power systems and energy planning. Also, application of this approach in power system operation has been investigated in different problems including DER management, demand response program, optimal power flow, economic dispatch, and unit commitment problems.

References

[1] R. Zamani, M.P. Moghaddam, M.R. Haghifam, Evaluating the impact of connectivity on transactive energy in smart grid, IEEE Trans. Smart Grid 1 (2021) 1–4, https://doi.org/10.1109/TSG.2021.3136776.

[2] J.-Y. Lee, et al., Climate Change 2021: The Physical Science Basis. Contribution of Working Group I to the Sixth Assessment Report of the Intergovernmental Panel on Climate Change, Cambridge University Press, 2021.

[3] BloombergNEF, New Energy Outlook 2021, 2021, [Online]. Available from: https://about.bnef.com/new-energy-outlook/.

[4] IEA, Net Zero by 2050: A Roadmap for the Global Energy Sector, International Energy Agency, 2021.

[5] DNV-GL, Rising the Challenge of a Hydrogen Economy, DNV, 2021.

[6] China pledges to become carbon neutral before 2060. *Guardian Article*. [Online]. Available from: https://www.theguardian.com/environment/2020/sep/22/china-pledges-to-reach-carbon-neutrality-before-2060.

[7] F. Sioshansi, Consumer, Prosumer, Prosumager: How Service Innovations Will Disrupt the Utility Business Model, Elsevier, 2019.

[8] J.M. Glachant, N. Rossetto, New transactions in electricity: peer-to-peer and peer-to-X, Econ. Energy Environ. Policy 10 (2021) 1–2.

[9] F. Sioshansi, Variable Generation, Flexible Demand, Elsevier, 2021.

[10] C. Schneeweiss, Distributed Decision Making, Springer Science & Business Media, 2012.

[11] A. Belgana, B.P. Rimal, M. Maier, Open energy market strategies in microgrids: a stackelberg game approach based on a hybrid multiobjective evolutionary algorithm, IEEE Trans. Smart Grid 6 (2015) 1243–1252.

[12] J. Wang, A Stackelberg differential game for defence and economy, Optim. Lett. 12 (2018) 375–386.

[13] G. Inalhan, D.M. Stipanović, C.J. Tomlin, Decentralized optimization, with application to multiple aircraft coordination, in: Proceedings of the IEEE Conference on Decision and Control, 2002.

[14] Y. Yang, J. Sun, H. Li, Z. Xu, Deep ADMM-Net for compressive sensing MRI, in: Advances in Neural Information Processing Systems, 2016.

[15] R. Shokri, V. Shmatikov, Privacy-preserving deep learning, in: 2015 53rd Annual Allerton Conference on Communication, Control, and Computing, Allerton 2015, 2016.

[16] H. Brendan McMahan, E. Moore, D. Ramage, S. Hampson, B. Agüera y Arcas, Communication-efficient learning of deep networks from decentralized data, in: Proceedings of the 20th International Conference on Artificial Intelligence and Statistics, AISTATS 2017, 2017.

[17] A. Nathan, D. Klabjan, Optimization for large-scale machine learning with distributed features and observations, in: Lecture Notes in Computer Science (Including Subseries Lecture Notes in Artificial Intelligence and Lecture Notes in Bioinformatics), 2017.

[18] Y. Hu, D. Niu, J. Yang, S. Zhou, FDML: a collaborative machine learning framework for distributed features, in: Proceedings of the ACM SIGKDD International Conference on Knowledge Discovery and Data Mining, 2019.

[19] H. Avron, V. Sindhwani, High-performance kernel machines with implicit distributed optimization and randomization, Technometrics 38 (2016) 341–349.

[20] C. Li, P. Zhou, L. Xiong, Q. Wang, T. Wang, Differentially private distributed online learning, IEEE Trans. Knowl. Data Eng. 30 (2018) 1440–1453.

[21] J.J. Mei, Y. Dong, T.Z. Huang, W. Yin, Cauchy noise removal by nonconvex ADMM with convergence guarantees, J. Sci. Comput. 74 (2018) 743–766.
[22] J.P. Desai, J.P. Ostrowski, V. Kumar, Modeling and control of formations of nonholonomic mobile robots, IEEE Trans. Robot. Autom. 17 (2001) 905–908.
[23] S.G. Loizou, K.J. Kyriakopoulos, Closed loop navigation for multiple non-holonomic vehicles, in: Proceedings—IEEE International Conference on Robotics and Automation, 2003.
[24] B. Penin, P.R. Giordano, F. Chaumette, Minimum-time trajectory planning under intermittent measurements, IEEE Robot. Autom. Lett. 4 (2019) 153–160.
[25] R. Zamani, M.P. Moghaddam, M. Imani, H.H. Alhelou, M.E.H. Golshan, P. Siano, A novel improved Hilbert-Huang transform technique for implementation of power system local oscillation monitoring, in: 2019 IEEE Milan PowerTech, PowerTech 2019, 2019.
[26] D.E. Boyle, D.C. Yates, E.M. Yeatman, Urban sensor data streams: London 2013, IEEE Internet Comput. 17 (2013) 12–20.
[27] H. Haes Alhelou, S.J. Mirjalili, R. Zamani, P. Siano, Assessing the optimal generation technology mix determination considering demand response and EVs, Int. J. Electr. Power Energy Syst. 119 (2020) 1–9.
[28] R. Zamani, M.E.H. Golshan, H. Haes Alhelou, N. Hatziargyriou, A novel synchronous DGs islanding detection method based on online dynamic features extraction, Electr. Power Syst. Res. 195 (2021) 1–12.
[29] R. Zamani, M.E.H. Golshan, Islanding detection of synchronous machine-based distributed generators using signal trajectory pattern recognition, in: 2018 6th Int. Istanbul Smart Grids Cities Congr. Fair, 2018, pp. 91–95.
[30] A. Abyaz, et al., An effective passive islanding detection algorithm for distributed generations, Energies 12 (16) (2019) 1–19.
[31] R. Zamani, M.P. Moghaddam, H. Panahi, M. Sanaye-Pasand, Fast islanding detection of nested grids including multiple resources based on phase criteria, IEEE Trans. Smart Grid 12 (6) (2021) 4962–4970.

5D Giga Trends in future power systems

2

Mohsen Parsa Moghaddam, Saeed Nasiri, and Morteza Yousefian
Faculty of Electrical and Computer Engineering, Tarbiat Modares University, Tehran, Iran

1 Introduction

Fundamental changes that have taken place in different societies are often subject to the trends that are originated from the foundations of that society. For example, the "industrial revolution" in Europe was caused by the process of "mechanization" of systems with the invention of the "steam engine" [1]. Scientific advances in the early 20th century were also a function of the "militarization" of the First and Second World Wars [2]. Technological advances in the late 20th and early 21st centuries can also be attributed to the process of "digitalization" due to the invention of the "transistor" [3]. Therefore, by looking at these examples, we can understand the importance of trends in various developments.

In general, trends are of such importance that if they are identified and properly understood, it will be possible to predict the future of societies and systems to an acceptable level. For example, the trend of "global warming" indicates an increase in greenhouse gas emissions and the need to take precautionary measures to prevent this disaster. These measures could include the increasing use of "carbon capture and storage" equipment, reducing the use of fossil fuels, increasing the use of renewable energy sources, and so forth. Accordingly, considering the future predicted through the trends, there will be necessary conditions in order to provide the necessary ground actions to lead the communities in the right direction.

According to the above explanations, trends can be considered as predictors that determine the future of societies with appropriate estimates, and therefore it is necessary to identify the needs of implementing strategic planning in order to guide the societies in the right direction. It should be noted that neglecting these trends can cause problems for the development of societies or even have irreversible consequences in the long run. In the following, the 5D Giga Trends are briefly introduced.

Decentralization has already begun in every aspect of human life. It may have been referred differently in various contexts, but the most inherent concept is that: the centralized approaches, utilized so far, cannot meet the future demands. The main challenge is to maintain the system optimized, predictable, and under control, as much as the centralized version where each participant behaves according to its local objectives. Moving toward decentralized decision-making in power systems will be inevitable because of the decentralized structure of the future power systems.

Decentralized Frameworks for Future Power Systems. https://doi.org/10.1016/B978-0-323-91698-1.00015-7

Decarbonization in the power system atmosphere is a vital decision taken by international organizations in response to the overwhelming demand to clean human life. Climate changes have forced governments to take proper actions regarding carbon emissions. For example, most European countries have started to reduce their overall emissions by 20% since 1990 and have plans to make it negligible by 2050. Such plans mean that a huge amount of electricity must be generated by renewables and low carbon resources, and considerable impacts on the future power systems.

A comparison of the largest technology market caps vs top countries' GDPs indicates how fast the **digitalization** age is being spread; a plain sign that human beings have already arrived at the digital era. Also, the nature of the load is rapidly changing. Digital grade loads such as data centers, which demand high reliability, are increasing, meaning that the future power systems will be heavily affected from both the demanding nature and technical aspects.

Besides the main goal of **deregulation** which resulted in the creation of competitive markets for participants, the future power systems cannot be made upon old regulatory regimes, as many features and entities are not recognized. New regulatory practices must be defined for energy trading and appropriate incentive mechanisms must be introduced to encourage investors and costumers to be involved in different aspects of the market and network management. Tangible and intangible asset valuation mechanisms must be designed in order to support new business models that deal with data. For example, cybersecurity providers, situational awareness services, big data analytics for behavioral science purposes, flexibility providers, etc.

All the Ds mentioned so far support the fifth D, i.e., **democratization**. From a deep point of view, it is not the lack of generation or transmission facilities that makes power shortages, it is the lack of demand-side responsibilities that make energy crisis, which is also a direct consequence of lack of consumer's engagement in the network activities. Democratization in the future power systems means to empower the consumers and, in turn, to make the demand responsible. In such an atmosphere, complex business models will be born, offering flexible reliability, energy trading, data processing applications, sophisticated protection schemes, cybersecurity initiatives, etc.

2 What are the 5D Giga Trends?

When we refer to 5D Giga Trends, we aim to characterize a system such as social or economic systems so that only by identifying such trends, the overall direction can be predicted. These trends overshadow all aspects of society, from politics, economics, and trade to culture and education. These 5D Giga Trends that we aim to discuss fall into five major branches, as shown in Fig. 2.1. Since each of them begins with the letter "D," we call them 5D Giga Trends.

2.1 Decentralization

One of the important problems in many political, economic, and social institutions is the excessive focus on a specific sector. This has led to the idea of decentralization, initially manifested by the formation of decentralized communities and gradually appeared in different parts of each society.

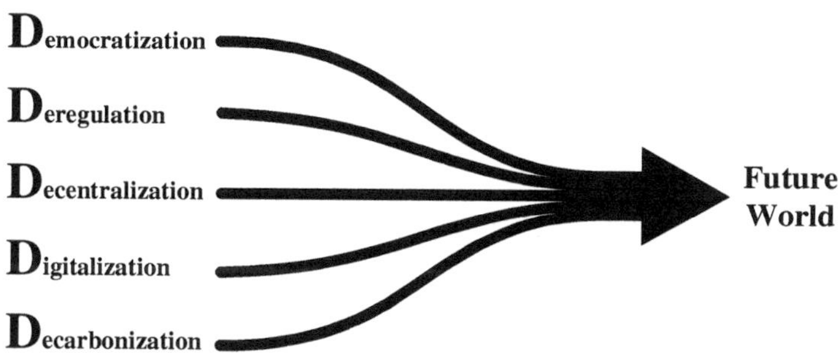

Fig. 2.1 5D Giga Trends.

Although new paradigm changes call for changing centralized structures and turning them into distributed ones, doing so is a significant challenge. Decentralization means decomposing a system into several systems while there is no particular central agent. In completely decentralized systems, the decentralized agents interact directly with each other (P2P) in order to achieve a specific goal. On the contrary, in a fully centralized structure, all procedures and processes pass through a single centralized agent. However, it is possible to define systems between the two, which are known as "semicentralized" structures [4]. The term "decentralization" was first used in the late 18th century by the "Association of French Operators" after the French Revolution. This concept also had manifestations in the 19th century, but it became especially popular in the 1970s of the 20th century. During this period, many European societies tried to abandon the centralized management structures that had previously resorted to decentralized management methods [5,6].

The most important problems of centralized structures include the following:

- One of the most important problems of centralized structures is the high overhead costs of the system. In fact, in centralized structures, additional costs are always imposed on different sectors. For example, some organizations in a country may have a branch concentrated only in the capital, which makes it difficult and expensive for people to access this organization. In the power system, for example, the costs of sending and receiving information, central processing, data storage, etc. are among the overhead costs faced by centralized networks.
- Another significant challenge of centralized systems compared to decentralized and distributed systems is the lack of agility. Due to the centrality of decision-making in centralized systems, a lot of information must be transferred to the central body, which significantly reduces the response speed of the system. However, in distributed systems, decisions in each part are made according to local data and exchanged data with the neighbors, thus increasing the agility of the system to a considerable extent.
- High volume of information required to be relocated in centralized structures requires a lot of processing and storage equipment, which in addition to increasing costs, may also make decisions out of optimality. The extensive information flow in these structures can cause problems in sending and receiving information and may lead to improper data transfer.
- In centralized structures, it is possible that the decisions made for each sector do not fully conform to the conditions of that sector. In other words, the restriction of information transfer often prevents the transfer of detailed local information to the core of the system, which prevents the specific conditions associated with each situation.

- Another important problem in central structures is lower reliability. In these structures, due to the urgent need for information transfer between different departments and the core, the possibility of interrupting this information transfer due to natural events or physical and cyber sabotages greatly increases. However, moving toward distributed structures reduces the possibility of such problems due to the removal of the central body.

The problems mentioned for centralized structures clearly show the need for communities to move in the direction of decentralization in all dimensions. Therefore, "decentralization" is one of the 5D Giga Trends in today's societies that has received much attention. In the meantime, it is necessary to mention that the movement toward decentralization requires special platforms, the most important of which are "deregulation," "application of distributed decision-making methods" and "democratization."

2.2 Deregulation

The existence of a large number of strict rules based on vertical structures has caused many sectors to not have the opportunity to participate in the movement of institutions and systems toward optimality and efficiency. An example of this can be seen in the pursuit of "privatization" policies in each country which by reducing the existing complex regulations, investors are allowed to operate freely in order to create new values and businesses.

At first glance, deregulation may be confused with the concept of "eliminating rules" and "creating an atmosphere without law" whereas, the main goal of deregulation is to transform "strict governance laws" into laws that allow more freedom of action to "all sections of a society." This will make it possible for micro and macroactivists to participate in various affairs in various fields. Thus, what is known as "deregulation" is a modified term for "reregulation" that may even increase the number of existing laws, but these laws will not impede the participation of different actors. The first signs of "deregulation" can be seen in the reduction of "administrative bureaucracy" and the reduction of barriers to public activity in society. In the final ideal situation, this leads to a completely free space for interaction between actors in different fields, which is called "perfect competition" [7].

"Increasing the competitive environment," "reducing rents," and "distributing power among all small and large actors" can be considered as the most important benefits of "deregulation." On the other hand, "increasing the possibility of collusion between actors" and "severely affecting different areas of society" can be considered as two main problems of "deregulation" which need appropriate solutions to be developed.

The first signs of deregulation date back to the 1970s. At that time, economic thoughts about the inefficiency of government regulation and the dangers of controlling the prices of products and equipment by regulated industries were formed in developed countries. It was shown that this issue causes serious harm to consumers. Thus, some countries began to reduce regulations in order to increase "level of competition," "productivity," "efficiency," and "price reduction" [8]. The most important problems of "highly regulated" sectors can be expressed as follows:

- Perhaps the main problem with the old regulatory schemes is the risk transfer to weak actors and often consumers. In centralized and highly regulated environments, powerful institutions such as the "government" can change market conditions arbitrarily (in order to make more profit), and because this process does not follow economic patterns, most micro-investors suffer the consequences.
- Existence of severe information rent is another problem with the highly regulated schemes. The ability to make decisions about market conditions by powerful institutions makes this preperformance information so valuable that only certain actors have access to it or other actors need to pay a lot of money to achieve it. However, in deregulated environments, economic interactions between actors determine market conditions instead of relying on the information rent.
- Administrative bureaucracy is another problem of highly regulated structures. In such structures, the process of registering a company, obtaining a license, registering a brand, and so forth, are complex since most micro-actors can hardly walk in this direction. In fact, in these structures, powerful actors always have priority, and small actors act as employees of major ones. One of the main goals of deregulation is to reduce administrative bureaucracy so that all actors, both micro and macro, have the opportunity to operate freely in the economy which leads to increasing market dynamics and reducing the power of macro actors and information rents. In New Zealand, for example, it takes less than a day to establish a firm. Ref. [9] provides a list of the open countries with the fastest time to start the economic activity.
- The possibility of tax evasion may be another problem with regulated systems. In these structures, according to the existing laws, the possibility of tax evasion is often higher for large actors. In this way, institutions that can have a significant impact on government funding can escape the process, and only micro-actors and micro-consumers, in general, will be forced to pay taxes. This not only prevents the government from properly financing through taxes, but it also leads to public dissatisfaction with the nonpayment of taxes by various social groups.

As a result, the "deregulation" Giga Trend can be considered as a necessary platform to move in the direction of the "decentralization" Giga Trend. In other words, the departure from centralized structures requires that the necessary regulatory reforms be made first and different segments of society, especially micro actors, be able to actively participate in various activities in order to move toward distributed structures. The absence of such a platform will prevent the implementation of decentralization policies.

2.3 Digitalization

"Digitization" literally means the conversion of analog data to digital data [10], which simplifies data storage and makes their transfer more reliable than analog data [11]. This has led to the foundation of the "information age," the "computer age," or the "digital age." As shown in Fig. 2.2, the process of digitizing or storing information digitally began in the late 1980s (c.1986) and has evolved exponentially [12]. Since 2014, almost all of the information used is stored digitally [13]. The explanations provided so far have emphasized the literal meaning of the term "digitization," while the Giga Trend of "digitization" has a meaning beyond the conversion of analog to digital systems. In fact, the digitalization Giga Trend refers to a major change in the industry

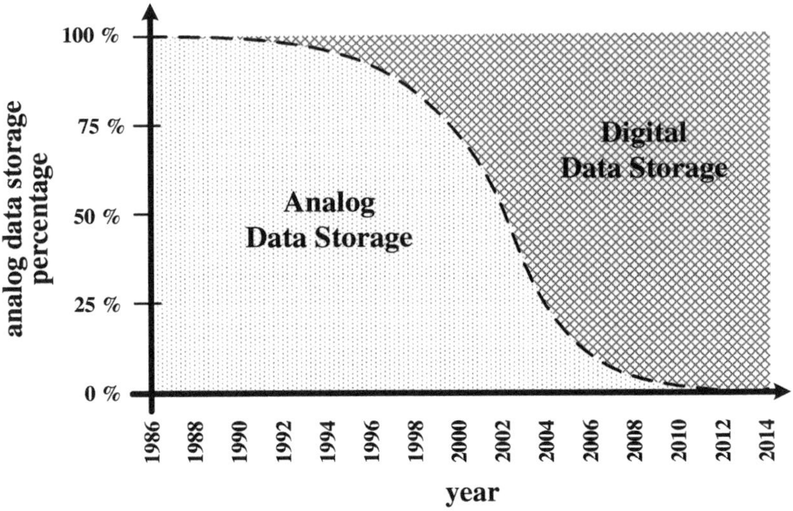

Fig. 2.2 The comparison of analog and digital data storage systems over years.

structure, based on digital technologies and computers. With the proliferation of digital technologies in the domains of "storage," "processing," "transfer," and "display" of information, the world has undergone a dramatic change, which is constantly evolving with the emergence of new technologies. This path, which brings about dramatic changes in different parts of every industry, is known as the "digitalization" Giga Trend.

The increasing development of computer-based technologies has necessitated the development of digital tools in today's industries more than ever before. This has also led to such amount of information produced in recent years that are thousands of times more than the past centuries. Accordingly, the development of information and communication technologies, digital and intelligent tools, and the move toward e-learning is among the basic needs of today's societies.

Another indication to show that how fast digitalization grows is to compare total value of the technology giants with the GDP of those countries. Wealth production in countries is usually measured by the "GDP" index. According to some reports, the total value of the four companies "Google," "Amazon," "Microsoft," and "Apple" in June 2021 was about $8.5 trillion, which is more than the total GDP of many other countries [14]. This also clearly shows the impact of digitalization on wealth creation and increasing social welfare.

2.4 Decarbonization

The word "decarbonization" means a reduction in carbon levels and refers to a set of methods that reduce the amount of "carbon dioxide" in order to prevent global warming. Since carbon dioxide is the most important greenhouse gas in global warming, its reduction is known as "decarbonization." "Electricity generation" and "transportation" are the two main reasons for using fossil fuels. Since the burning of fossil fuels also produces a set of polluting gases such as Sachs, Knox, carbon monoxide, and "very tiny solid particles," the word "decarbonization" is synonymous with the word "pollution reduction."

Some studies show that global climate change has always existed, but after the Industrial Revolution, and especially since the middle of the 20th century, human actions in the indiscriminate use of fossil fuels have had a significant impact on global warming [15]. Earth temperature measurements have been in place since 1880, and according to Fig. 2.3, the results show sharp temperature changes in recent years [16]. Cutting trees and excessive use of fossil fuels (in which more than 90% of those are composed of "carbon dioxide" and "methane") are the most important factors in global warming in recent years [14]. Rising global temperatures have irreparable consequences such as the melting of polar ice caps, rising ocean water levels, declining ice and snow reflecting sunlight, increasing sea water evaporation and consequently increasing greenhouse gases and creating storms, and displacement or extinction of animal species. In ecosystems, climate change threatens social life, which may lead to drought, food shortages, floods, infectious diseases, and so on [17]. This global crisis has been a major factor in initiating the Giga Trend of "decarbonization" that is affecting all sections of societies.

The Kyoto Protocol may be the first serious global action to reduce greenhouse gas emissions. The Kyoto Protocol, a complement to the Rio Convention, was established in 1997 within the framework of the United Nations, under which the industrialized nations pledged to reduce their greenhouse gas emissions by 5% within the next 10 years.

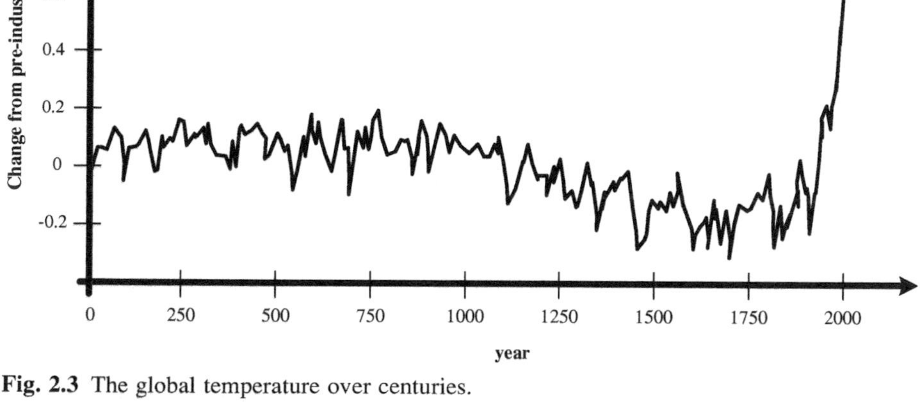

Fig. 2.3 The global temperature over centuries.

Industrialized countries also pledged to provide financial assistance to developing countries to increase the penetration of renewable energy sources such as solar and wind energy [18].

Following the failures of the Kyoto Protocol, the Paris Agreement under the United Nations Framework Convention on Climate Change (UNFCCC) to reduce greenhouse gas emissions, according to the specific conditions of each region and taking into account the specific financial conditions of countries, by representatives of 195 countries at the conference Climate 2015 UN, was negotiated and approved in Paris. The aim of the agreement is to keep the global average temperature rise much lower than 2 °C above preindustrial levels. This agreement has been in execution since 2020 [19].

Measures such as "constructing large power plants in outside and far away from cities," "requiring power plants to use storage and carbon capture equipment," "moving towards the use of clean energy to generate electricity," "monitoring the amount of pollution made by vehicles and requiring them to obtain nonpollution certificates," "encouraging the use of electric or hybrid vehicles," "encouraging the use of public transport instead of private vehicles," "using artificial rain," "arboriculture and forest development," and so forth, have been implemented in many countries around the world in order to overcome the climate change. All the abovementioned measures show the movement toward the "decarbonization" Giga Trend.

2.5 Democratization

"Democratization" literally means the transition to a society in which a higher level of importance is given to the people [20]. To examine the extent of democracy in different societies, data from a project called "Polity" was first used in 1960 by Ted Robert Gurr and then was followed by Monty G. Marshall, one of his students. This project has been repeated five times so far, the last one was in 2017. Fig. 2.4 shows the results of this project in the last period of its implementation. In this project, according to different parameters, a score between 0 and 10 is considered for the degree of "democratic" and "authoritarian" societies. As can be seen, the number of authoritarian societies in the world has been decreasing every year since 1990 and the number of democratic societies has been increasing exponentially, which underscores the great importance of democracy and the movement of societies [21].

Going beyond the literal meaning of "democratization," the word as a Giga Trend means "giving the people the right to choose," "creating free competition and eliminating monopoly," and "fair distribution of facilities among different classes." This issue manifests itself in different ways in different infrastructures. For example, in the political domain of societies, it shows the possibility of the activity of different parties, free elections, the right to vote for all sections of society, holding referendums, and consulting the people on various issues and the like. As it turns out, the Giga Trend of "democratization" expresses the concepts that provide the necessary basis for the formation of other 5D Giga Trends. In other words, if the role and position of the people as the most important building block of society is not well defined,

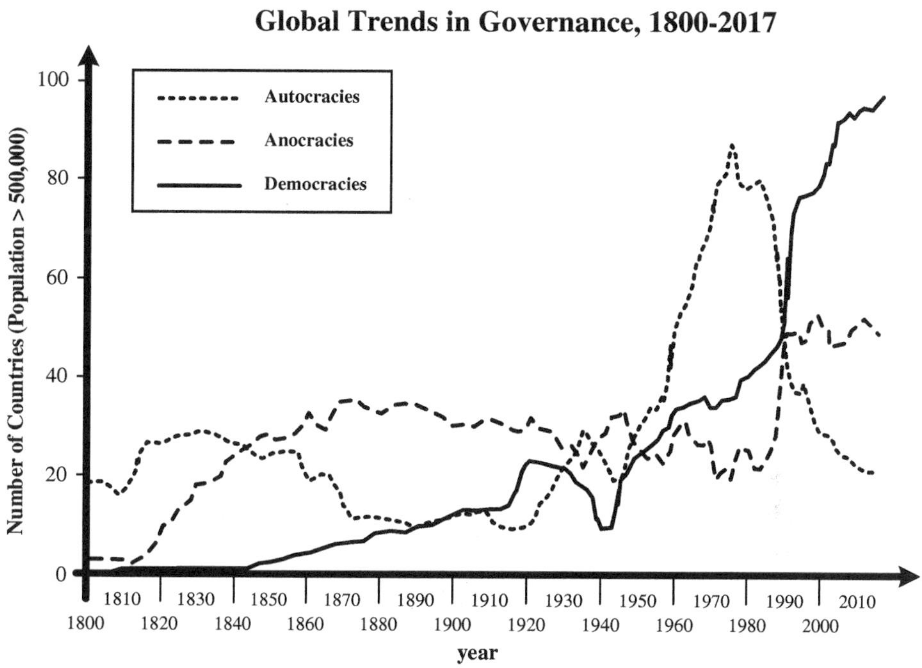

Fig. 2.4 The trend of the countries governing forms from 1800 to 2017.

the creation of distributed structures and the transition from centralism will not be possible. Also, it will not be possible to move toward deregulation to create a free and competitive environment. On the other hand, creating the necessary conditions for achieving digital and carbon-free societies has been possible with the cooperation of people, which requires the implementation of "democratization" contexts in societies.

According to the explanations provided, the main problems of societies that have not moved in the direction of "democratization" can be mentioned as follows:

- Neglecting the role of people in society has hindered their free activity in various matters, which will clearly prevent the dynamism and mobility in different parts including political, economic, cultural, and social. Ignoring the activities of the private sector in the management of industrial companies, for example, can result in corruption, less efficiency, and also low-quality and expensive products for consumers.
- Lack of right to choose people is another important problem in dictatorial societies. In these societies or systems, people are subject only to the views and opinions of the rulers and are not given any choice. In nondemocratic power systems, for example, consumers only have to use electricity provided by upstream institutions, and they will not be able to choose electricity of different qualities.
- Not paying attention to the Giga Trend of democratization can lead to an unfair distribution of resources among different sections of society. An example of this is the different access levels to facilities for different classes of society, which creates class discrimination.

3 The existing power systems issues

In the previous section, a general view of 5D Giga Trends was provided. In this section and before investigating the impact of these 5D Giga Trends on power systems, it is necessary to review the issues and challenges faced by power systems. The main issues of the existing power systems can be summarized as follows:

Centralized production: Centralized power generation causes a large amount of pollution in the production area. On the other hand, this method of production is unreliable from the security point of view. Centralized power generation also requires the development of a vast transmission network to provide the consumers with electrical energy, incurring huge transmission expansion costs.

Power losses: According to the studies, the amount of power losses in transmission and distribution networks is about 5%–10% of overall load [22]. This amount increases to 20% in some problematic networks. This amount of loss is highly related to the centralized production, imposing considerable costs on the network, annually.

Monitoring issues: Due to the size of the networks, integrated monitoring of all sectors has many technical problems and costs. Also, the need for monitoring large number of equipment puts the system at risk because of the need for transmitting and processing a huge amount of data. Moreover, the existing power systems suffer from a lack of fast communication and processing facilities. Furthermore, as power systems are facing big data issues, integrated monitoring methods may be no longer efficient.

The nature of the load: Due to technological advances, the nature of the load has changed completely and the amount of nonlinear loads with high harmonic injection (such as inverter-based loads) on one hand, and on the other hand, sensitive digital loads (always-on loads) in the network have increased. This phenomenon requires a higher level of power quality and reliability, urging to move toward restructured power systems.

Monopolistic competition: The monopoly in production, transmission, and distribution sectors of power systems leads to major problems such as nonoptimal operation, risk transfer to the weak participants, lack of demand response, nonoptimal expansions of generation, transmission, and distribution systems, the free-riding issues, and so forth. All these will consequently provide low quality and expensive electricity.

Reliability issues: Increasing the amount of sensitive and always-on loads in the network requires a high level of reliability to be continuously provided, which cannot be provided through equipment such as uninterruptible power supplies (UPS).

Limited service to the consumers: In traditional networks, only limited services can be provided to the consumers and there are no platforms for creating new businesses using new platforms such as the Internet of Things (IoT).

Inactive role of the consumers: In traditional power networks, consumers are only end users of electricity and have no role in generating power or operating the network.

Limited information exchange: The existence of limited one-way communication links in the transmission and distribution networks has led to a condition that almost no real-time information from consumers is available to network operators.

The environmental issues: Fossil fuel power plants release large amounts of pollutants into the environment, which in addition, lead to global warming and cause various diseases for living organisms. According to statistics, coal, natural gas, and fuel oil power plants in the United States produce 1002, 413 g/kWh of carbon dioxide [23].

Low supply chain efficiency: The power supply chain consisting of "energy resource," "energy conversion," "energy transfer," "energy distribution," "energy consumption," and "energy storage" has low productivity, due to the abovementioned issues of traditional power systems.

Operation in stressful situations: The aging of many power grid equipment (often over 50 years) puts the operation of the network at risk and reduces the overall reliability.

Not being prepared for technological changes: Rapid technological advances require the power systems to accept the necessary platforms in the network, while, for example, the current networks are not ready to host a large number of electric vehicles owing to both electrical and telecommunications problems.

Inappropriate asset management: Due to the limitations of information flow in the existing networks, it is not possible to properly monitor assets, which, therefore, results in high failure rates. As such, the preventive and corrective decisions cannot be optimally and timely made. Moreover, the employees of a system are one of the most important assets in centralized systems. The human resources may not be well managed because of the issues such systems inherently have, as discussed in the previous sections about centralized systems problems.

Limited business models: Due to the centralized and monopolistic nature of traditional power grids, the opportunity for creating new businesses such as "aggregators," "retailers," "electric vehicle parking lots," "private operators of distribution networks," "virtual power plant owners," and so on, is limited. Also, there is no necessary platforms for consumers to participate in the process of power generation and implementation of peer-to-peer energy exchanges between them.

Unfavorable level of passive defense: As the power grid is a geographically vast cyber-physical system, in the case of centralized management, there is a high probability of physical-cyberattacks against the grid, and also the occurrence of natural disasters such as floods or storms can pose serious risks to its operation.

Limited use of renewable energy: Each power grid has a limited hosting capacity to use renewable energy resources due to some limitations (such as grid inertia and over-voltage issues). However, the share of renewables in most power grids in the world is not adequate. In fact, this level should be close to hosting capacity constraints.

Network agility problems: Owing to the limited use of modern communication systems in existing power networks, it is not possible to react quickly to different situations, and therefore these networks have little agility. However, the presence of sensitive loads in the network requires high agility when there are needs for fast response to manage the network in situations such as restoration conditions.

Not being self-healing: Increased sensitivity often necessitates a high level of network reliability due to technological advances and the constant need for energy. This requires the network to be able to respond appropriately to different situations and take actions to resolve the error. Current networks do not have satisfactory abilities to self-heal the issues, as they need human-made factors in the decision-making process.

Connectivity issues: Increasing the amount of intelligent equipment in power networks has created a large amount of information that needs to be transferred. This calls for sufficient telecommunications infrastructure, which may not be available in the existing power grids. Also, the electrical connectivity in the downstream power networks should be in such a way that the opportunity to take part is provided to all the actors and there must be as little discrimination between them as possible.

4 The impacts of 5D Giga Trends on future power systems

Each of the 5D Giga Trends introduced so far affects the power grid as well as other sectors of many political, social, and economic systems. Therefore, the power grids, as one of the most critical human-made infrastructures, will face significant changes in their structures. Having comprehensive knowledge about the impacts of 5D Giga Trends on power girds is a vital prerequisite for future developments to respond to the changes.

4.1 Decentralization in power systems

The first steps toward decentralization in power networks were observed with developments called "restructuring." The concept of "restructuring" in the power network infrastructure has evolved around the 1970s and gradually led to the transformation of vertical power networks into horizontal networks, from a managerial point of view. In the traditional power networks, the management of all parts of the network, including production, transmission, and distribution, is in the hands of a central government institution, and all these parts follow centralized and integrated management. However, in restructured networks, the system allows private sectors to own generating units and, in turn, compete to take the market.

As shown in Fig. 2.5, moving in the direction of the decentralization Giga Trend in the power system will lead to a point that centralized networks with single management gradually become decentralized networks with multiple management sectors. In decentralized networks, different areas are managed by area operators who may be monitored by the main network operator. This is not the end of the story, and as this process progresses, power grids become "distributed" networks with local management structures. Being distributed causes the operation of different parts of the network to be granular and partial. The concepts such as microgrids and local networks are signs of moving toward decentralization in the power grids, and this

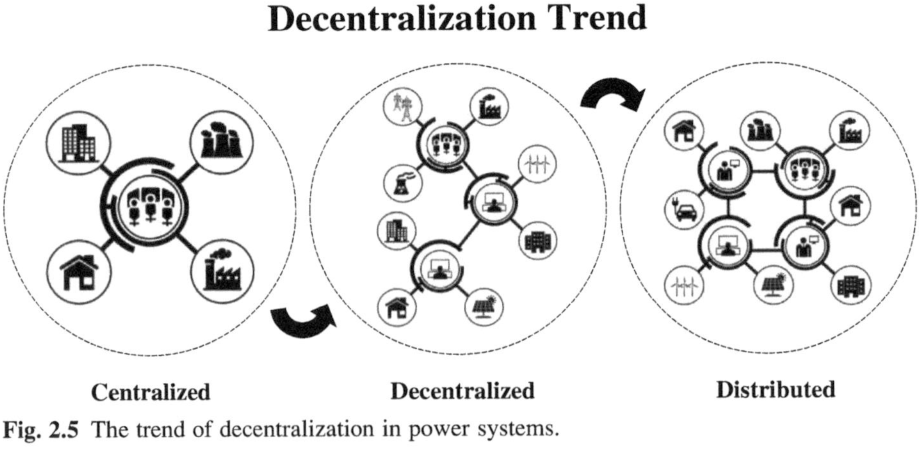

Fig. 2.5 The trend of decentralization in power systems.

trend will impose itself at such a level that each tiny facility with a roof-top PV can be a part of the operation of the network.

Moving toward the decentralization Giga Trend in power grids can solve the problems such as "centralized production," "power losses," "low efficiency of the supply chain," "operating under stress," and "undesirable passive defense," directly, and many other problems, indirectly. For example, decentralization will reduce the need for power transmission networks due to decentralized generation, and therefore concerns about the costs imposed on the network will be reduced.

As stated, the Giga Trend of decentralization in power networks allows "local networks" to play a significant role in future power networks. This paves the path for the development of "microgrids" and "virtual power plants," and in turn, strengthens another dimension of the power networks, namely, "smart grids," which is of great importance. Note that extensive use of distributed generations resources such as electric vehicles requires the provision of the necessary electrical and telecommunication platforms on the distribution network. The construction of electric parking lots, the existence of two-way telecommunication platforms, the use of smart meters, facilities for buying and selling power at the distribution network level, and so on, are some of the platforms that must be provided in response to the decentralization Giga Trend.

4.2 Deregulation in power systems

Deregulation in power systems took place with the development of economic thoughts about the inefficiency of government regulation in the operation of power networks, as well as the risks of controlling the purchase and sale of electricity on both the production and consumption sides. Those studies led to a conclusion that the way the regulation was done transferred almost all the risk to the end user side. At the same time, as the consumers are not involved in any production and operational issues of power grids, they are not responsible for their consumption level, environmental issues, etc. Finally, the deregulation Giga Trend persuades the decision-makers to reduce and change regulations in order to increase "competition," "productivity," and "efficiency."

In particular, deregulation in power systems was a part of the National Energy Law in the United States in the Public Service Oversight Policy Act on November 9, 1978. The law states that cogenerators and independent power producers (IPPs) can sell their generated electricity to local regulated investor-owned utilities (IOUs). This changed in 1992, 1994, and 1996, until finally in 2002, the existing laws allow all consumers, regardless of size, to purchase the power they need from any company or independent power producer. It is worth mentioning that the concept of "electricity market" was also introduced in this period and was formed following the deregulation in the electricity industry.

Deregulation and the creation of a market for buying and selling power have resulted in changes in the structure of the electricity industry, known as "restructuring." This broke the governmental structure of power networks in all three sectors of "production," "transmission," and "consumption" and opened the competition door for private investors. This, as shown in Fig. 2.6, has created "generation

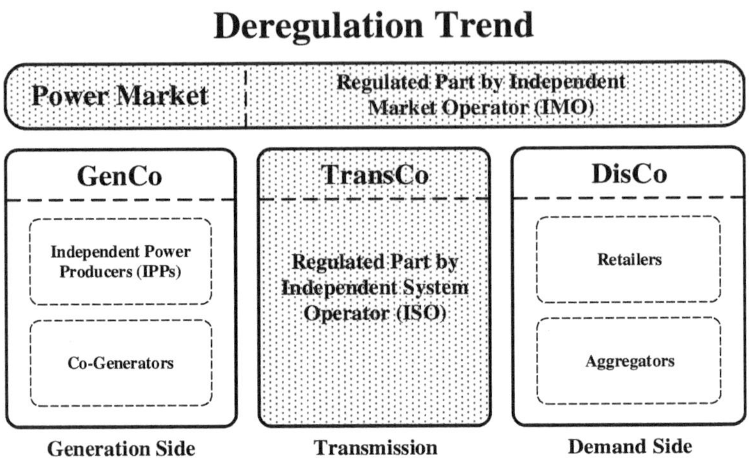

Fig. 2.6 The recent deregulation impact on power systems.

companies," "transmission companies," and "distribution companies," and new actors such as "independent power producers," "joint power companies" on the production side, and "retailers" and "aggregators" on the consumption side.

The two sections of "transmission network" and "electricity market" are regulated in nature due to the need for free access of all network actors to these two sections. In the meantime, although the transmission network can be owned by private companies, it is still necessary to follow certain rules set by the "independent system operator." The electricity market process is also carried out under the supervision of an "independent market operator" and it is necessary to strictly follow the rules developed by this institution (the electricity market includes both "energy" and "ancillary" markets).

The remarkable point about Fig. 2.6 is that although the aggregators are on the consumption side, these actors are actually power providers and offer a price to sell electricity in the electricity market. In general, aggregators can be divided into three categories: "demand response aggregators," "renewable energy aggregators," and "electric vehicle aggregators (power stored in their batteries)," respectively. In other words, deregulation Giga Trend provides the basis for the free competition of actors in the electricity market environment and allows the creation of business models such as "retailers" or "aggregators." It should be noted that deregulation Giga Trend is the bedrock of other developments in future power networks.

4.3 Digitalization in power systems

From the power system point of view, digitalization means the use of digital equipment and increasing reliance on them in order to monitor and control power grids. This action can play an effective role in reducing costs and improving the speed and accuracy of operating the power system. It brings some new business models and opportunities for new actors such as data analytics experts, cybersecurity initiatives, and so on.

As power systems are becoming intelligent day by day, moving in the direction of digitalization is one of the most important needs of this system. The smart grid, as the most important dimension of the future power grids, is closely related to this Giga Trend. The development of information and communication technologies, the increasing use of computers, and the development of control systems can be considered as the basic needs of intelligent networks. One of the most important features of smart grids is the ability to instantly monitor all parts of the network, which requires the ability to store, process, transmit, and display large-scale information which we call big data.

Moving toward the digitalization in power networks can pave the path for solving many problems of future power networks, such as "partial observation," "limited information exchange," "inappropriate asset management," "lack of agility," and "not being self-healing." In this way, electrical transportation as one of the important dimensions in future power networks, due to its high dependence on communication networks and control systems, greatly needs advanced digitalized processes. The need for the use of smart controllers and two-way online communication between consumers and regional control centers is one of the major aspects of digitalization in power grids. Increasing the active participation of consumers in the operation of the power networks and expanding the role of the distribution side in power generation require digitalized platforms. Implementing demand response programs, for example, would not be possible without adequate telecommunication and digital platforms. Not to mention that the aforementioned digital platforms consist of hardware and software technologies that require a multidisciplinary view into the future power system.

Moreover, it can be stated that the use of renewable energy sources requires control systems based on digital equipment. The use of these resources, whether at the household level or at a large scale farm level, requires control and telecommunications technologies in order to accurately monitor and control them.

4.4 Decarbonization in power systems

The power grid, as one of the main systems producing carbon pollutants, is strongly influenced by the decarbonization Giga Trend. Fig. 2.7 shows the strategies used to move in the direction of "decarbonization," from the perspective of the power system. As can be seen, various renewable sources such as "bioenergy," "geothermal energy," "power generation through solar energy (solar panels)," "wind turbines," "power generation through sea waves." "Fuel cells" and "hydropower plants" are the most important ways to reduce the production of carbon pollutants. There are also other solutions of which "power generation using high-efficiency generators (such as combined cycle power plants and generators of simultaneous production of electricity, heat, and even refrigeration)," "increasing the efficiency of existing power plants (for example, by placing steam cycle power plants in conditions 'Supercritical')," "use of carbon capture and storage equipment" and "use of methods to increase efficiency (such as energy audits and implementation of demand response programs)" can be mentioned.

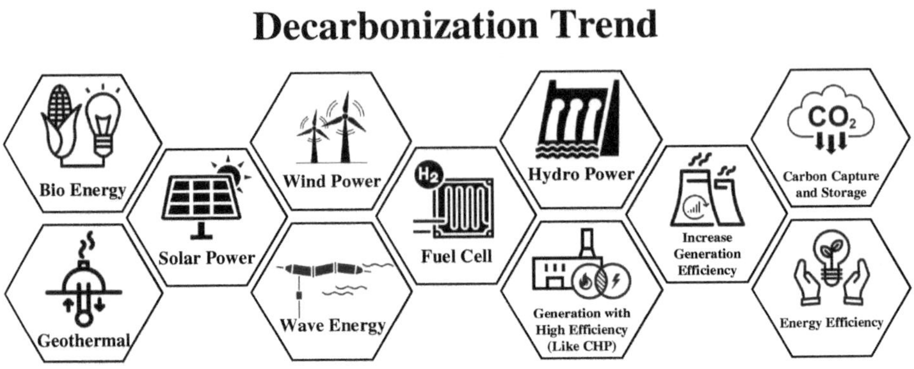

Fig. 2.7 Decarbonization resources in power systems.

Extensive use of electric vehicles can also be considered as a low carbon power source which is already a target motivated by the decarbonization Giga Trend. It is worth mentioning that the high use of electric vehicles in networks will generally increase the amount of carbon pollutants, taking into account the efficiency of traditional power plants as well as the efficiency of electric vehicles. Hence, power systems should be benefited from the energy stored in the batteries of electric vehicles, when they are not in use for transportation purposes. This can be a great help to solve network problems during peak times. Therefore, in the absence of the necessary infrastructures in power networks, the use of electric vehicles is not sufficiently justified.

As a result, moving toward the decarbonization Giga Trend solves the two basic problems of "environmental issues" and "limited use of renewable energy" in power systems [24,25]. It should be noted that the full implementation of this Giga Trend in power networks is highly dependent on other 5D Giga Trends.

4.5 Democratization in power systems

The "democratization" Giga Trend, which can be called the most important future-making Giga Trend, is the final target that imposes fundamental changes in different systems, institutions, and organizations. The conversion of consumers who have previously been the only power end users to active consumers who, in addition to power consumption, generate energy and play an important role in the operation of the network, is the beginning of thinking about the formation of smart grids and local area networks. This is manifested in two ways: increasing the installation and utilization of household power generation resources (micro-turbines, wind turbines, roof-top solar panels, etc.) and the participation of consumers in demand response programs. It should be noted that participation in demand response programs, as one of the "energy efficiency" means, reducing the need for power generation and is referred to as the sixth fuel.

Consumers can play a key role in local networks by allowing the power storage resources to be controlled by the networks operators. Doing so means that the consumers are now actively engaged in balancing network demand-supply conditions.

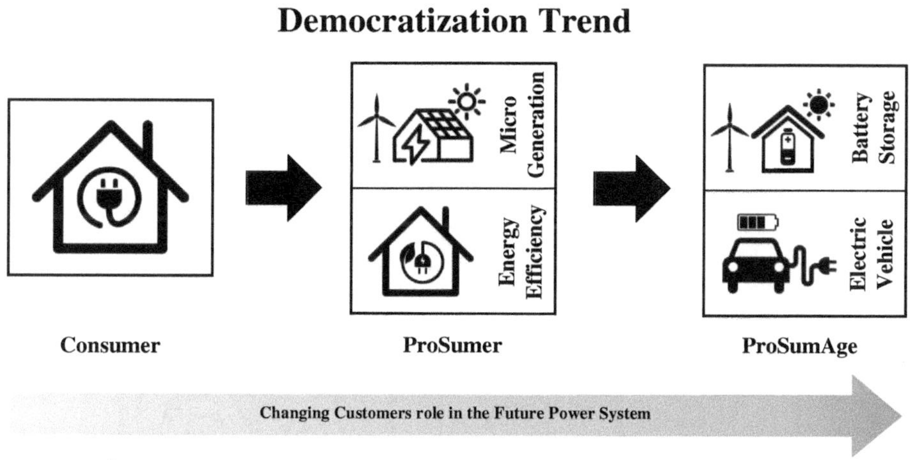

Fig. 2.8 The impact of democratization trend in changing consumers into ProSumAge.

In such a situation, the consumers become "ProSumAge"s, which produce, consume, and store electrical energy depending on their optimal benefits. As shown in Fig. 2.8, by providing consumers with the right to choose, a fundamental paradigm change will happen so that one can deduce the power grids that are much more democratized compared to the traditional power grids. The future democratized power systems should be decentralized, decarbonized, deregulated, and digitalized in such a way that, on one hand, the demand side is an active entity with the power of choice and, on the other hand, the power systems major issues are tackled using vast digitalized solutions.

In such an environment, a high level of connectivity in power networks is one of the most important achievements in moving toward democracy. Given that, in a democratic power system, it is necessary to distribute the facilities fairly among all participants, and there needs a high level of physical connectivity with massive communications links. In other words, the free and fair activity of the actors requires that no actor has any restrictions on the transmission of power or the sending and receiving of information. "Transactive energy" and P2P energy trade are two major concepts that seek optimal democratized energy exchanges while the benefits of the demand side and the network constraints are met.

5 Future power systems affected by 5D Giga Trends

In the previous section, the footprints of the 5D Giga Trends in power systems were discussed. In the following, the relationship between the 5D Giga Trends and four major infrastructures of the future power systems are presented. As shown in Fig. 2.9, smart grids, local energy networks, low-carbon generation, and electric transportation are the four main infrastructures of future power systems [26].

Smart grid is a concept that encompasses many other fundamental and technical concepts including operational, managerial, telecommunication, and control theories, and

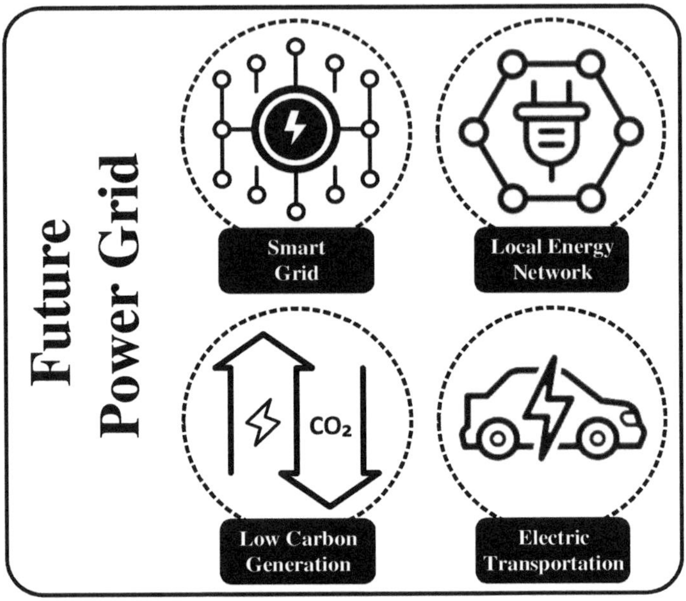

Fig. 2.9 The main infrastructures of the future power systems.

implementations in order to respond to the future demands of power systems. One of the most important goals for smart grids is to achieve a high level of characteristics of power grids, namely "Security," "Quality," "Reliability," and "Availability." These characteristics are known as "SQRA" [26]. These attributes were completed by adding two more attributes, namely, "Efficiency," and "Connectivity" in [27,28], which is abbreviated SQRAEC. This requires many changes in various dimensions of power networks. For example, to increase the security of networks, it is necessary to use sufficient monitoring equipment to control the grid in a real-time manner. On the other hand, in order to improve the quality of delivered electricity, it is necessary to use dynamic compensation equipment such as FACTS devices. The movement of power structures from a vertical structure to horizontal networks with distributed structure is also one of the important factors in increasing reliability. Also, intelligent maintenance methods in order to prevent damage to the network equipment can increase their availability to an acceptable level. Meanwhile, the transfer of production to distribution networks and the conversion of end users who have only been power consumers to active entities with the ability to generate power reduce network losses and increase efficiency.

The concept of local area networks has been around since the beginning of the power system, but in recent years, it has become more important under the influence of the 5D Giga Trends. The local area networks are independent, flexible, intelligent, and high-performance networks that allow consumers to increase participation in the operation of the power grid. In such networks, due to the strong telecommunication platforms that exist, it will be possible to provide electrical power with the characteristics desired by each actor. These networks, which consist of several microgrids, are managed independently or by power distribution companies and can be under operation in grid-connected mode or islanded-mode [29]. In general, local area networks can be "self-sensor," "secure," "self-correcting," and "self-healing" networks with the ability to meet the different needs of consumers at reasonable prices, to use fewer resources, and not to damage the environment.

In the case of low-carbon generation, various solutions have been devised to generate low-carbon power in future power grids. The most important way is to increase the use of "renewable power sources." These sources, which use renewable fuels to generate electricity, greatly reduce carbon emissions. It should be noted that for these resources, a small percentage of carbon production can be considered, which is due to the use of various machinery and transportation needed for operation and repair. Also, the increasing use of renewable power generation resources requires technological advances to reduce the cost of construction and operation of renewable power plants.

Another way to generate low-carbon power in future power systems is to use "carbon capture and storage equipment" that can convert the smoke from power plants into "synthetic oil." "Using high-efficiency power plants" or "using methods to increase the efficiency of power plants" are other methods of generating low-carbon power, which was mentioned earlier.

Electric transport is one of the infrastructures that will revolutionize the future of power networks and will bring many social benefits. "Reducing urban carbon emissions," "reducing noise," and "ease of transport" can be considered as the main benefits of electric transport from a social point of view.

The main advantage of electric transportation from the perspective of future power networks is the use of these vehicles as mobile storage devices in the network. Batteries used in electric vehicles have the potential to increase the flexibility of power grids by using renewable sources with fluctuating production capacity, as many electric vehicles are housed in electric car parks for a long time during the day and night. These car parks can act as a small power plant and balance the supply-demand in the network by considering the charging status of cars to discharge or vice versa. It should be noted that electric vehicles can have other benefits such as peak reduction. In fact, the energy stored in the batteries of electric vehicles can provide the energy needed at the time of the peak without the need to build a power plant, and thus can save a lot of costs imposed on the network [30]. This dimension of future power networks requires the provision of many infrastructures such as "development of charging technologies," "development of communication and control platforms," "increase of electric parking lots," and so on, in power networks.

According to the explanations provided by now, it is clear that the basic perspectives of the future power grids are based on the 5D Giga Trends. To illustrate this point, Fig. 2.10 shows the connections between the 5D Giga Trends and the basic perspectives of future power systems. In this figure, direct connections with solid lines and indirect connections with dashed lines are shown.

As can be seen in this figure, the Giga Trend of "democratization" as the future-making trend of the world affects all dimensions of the future power systems. The "decentralization" and "digitalization" are also important trends in power networks that have a significant impact on the future of these networks. The "deregulation" and "decarbonization" also jointly affect all four dimensions of future power networks.

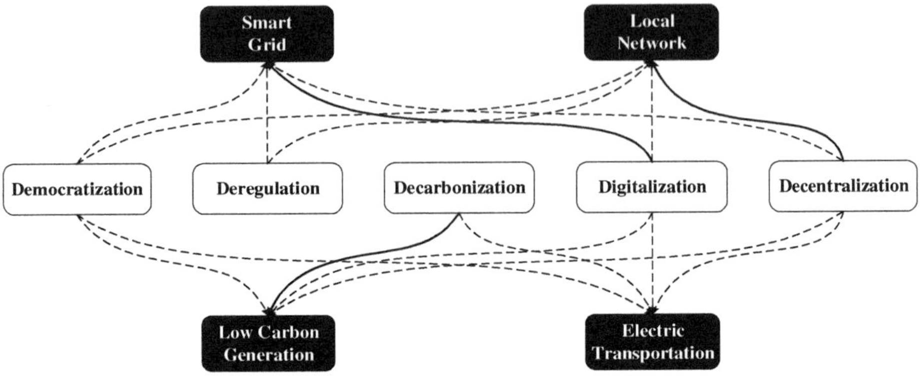

Fig. 2.10 Direct and indirect interactions among the 5D Giga Trends and the main infrastructures of future power systems.

6 Opportunities, challenges, and new issues of the future power systems under 5D Giga Trends

After the introduction of the 5D Giga Trends and how they are defined in the power system context, in this section, particular new challenges which the future power systems will face are investigated. Meanwhile, there will be new opportunities based on which new business models will be born. In the following, both the challenges and opportunities are introduced separately for each Giga Trend.

6.1 Opportunities and challenges of decentralization Giga Trend in power systems

Decentralization has a significant impact on future power systems. The most important advantages of this Giga Trend in power networks are as follows:

Improved passive defense: Reducing centralization in the power grid and increasing decentralization cause network hotspots to be distributed across the grid, and in turn, prevents the reliance on some limited central entities which can be prone to different cyber or physical attacks. Hence, having decentralized structures leads to a higher level of reliability.

Increased number of stakeholders: Decentralization requires that the management of different parts of the network be delegated to companies (mostly private) with expertise in that field with special measures (such as regular repairs, use of specialized personnel, etc.) to reduce operation costs. In addition to the increasing efficiency of the network, this will create the necessary competitiveness in the industry so that the end users can be final beneficiaries.

Activated demand-side resources: One of the important consequences of decentralization is to increase the power of end users and their active cooperation in the operation of the network. This active cooperation manifests itself in two ways: power generation by the end users (through the household renewable resources) and participation in demand management programs (through participation in demand response and energy efficiency programs) [31,32].

Reduced expansion costs: Decentralization allows a significant portion of the grid's power to be provided through demand-side. This makes it much less necessary to expand the network and build new transmission lines, substations, and power plants to meet the increase in electricity consumption. In an ideal condition, there may even be a need for de-expansion plans in order to optimally dismantle the old facilities.

Reduction of transmission losses: Reducing the need for a power transmission network, in accordance with the previous paragraph, causes the transmission losses (which amount to around 3%–4% of the network consumption), to be significantly reduced.

Efficient asset management: Decentralization by distributing the managerial, operational, and planning sectors among different companies makes it possible to better manage assets such as the specialized workforce in the field.

Increased efficiency of the power supply chain: The power supply chain includes six dimensions of "energy resource," "energy conversion," "energy transmission," "energy distribution," "energy consumption," and "energy storage." Decentralization increases the productivity of this chain by increasing the use of demand-side resources. Decentralization increases competition in the supply chain by improving the two areas of "asset management" and "business environment."

New business models: Decentralization and the increased importance of the role of end users, create new entities such as "virtual power plants" and "aggregators" in the power system.

Possibility of peer-to-peer interactions: Direct buying and selling of electricity among end users without any intermediaries can change the structure of the power network dramatically. Such a concept requires the elimination of centralized structures and the move toward distributed structures. It should be noted that decentralization is the most basic need to implement this concept.

Decentralization in power networks, in addition to its benefits, also creates some problems and challenges. The most significant concern of decentralized systems is to optimize the performance of systems so that they behave as optimal as their centralized version, while the participants in these systems try to act according to personal objectives and constraints. Therefore, in order to make the power networks decentralized, it is necessary to conduct comprehensive changes in various areas of the power system. Some of the most important requirements are listed below:

Decentralized decision-making: With the increasing participation of distributed generation in the future power grids, it is very important to make appropriate decisions that fit the conditions of decentralized management. In fact, in future power networks, power generation is dispersed, while network performance should remain consistent and predictable. In fact, the purpose of "decentralized decision-making" is to create appropriate local decisions so that the overall goals of the network are achieved. However, this needs special mechanisms to be designed to make timely consensual decisions. To achieve such a goal, mathematical models are needed along with fast software and hardware technologies.

Decentralized power supply models: Changing the network structure from vertically concentrated networks to horizontally distributed ones requires making different operational methods to supply load. For example, with the increase of distributed generation resources, the need for developing the transmission network has decreased while distribution networks should be expanded to increase highly connected distribution networks with much more capacity to host distributed generations. Also, load-centric grid management and using

demand response programs require bidirection fast data exchanges among entities such as grid operators, prosumers, aggregators, virtual power plants, etc.
Decentralized energy storage: In the future distributed and intelligent power networks, many entities have the potential capacity in power storage and the network operator can use this potential to balance the supply-demand in real-time. Refrigerators, for example, can act as storage devices by injecting negative power into the grid by increasing the operating temperature during peak periods (e.g., −28°C to −23°C). This also applies to electric vehicles in parking lots. There must be appropriate platforms and mechanisms to do so.
Decentralized embedded energy sources: Sources such as water heaters, air conditioners, and fans that have the ability to cut off power in critical conditions have the potential to play an energy-saving role in some ways. Utilization of these resources requires the development of communication platforms in the network.

All the abovementioned opportunities and challenges call for developing new models and computational tools. Some of them are decentralized state estimation, decentralized optimal power flow, decentralized voltage, and frequency control, decentralized energy management, decentralized energy exchange platforms, and settlement mechanisms, coordinated behaviors of distribution and transmission operators, etc. [33].

6.2 Opportunities and challenges of deregulation Giga Trend in power systems

The main purpose of deregulation in the power networks is to create competitive markets for market players. In other words, many of the goals defined for future power networks require a competitive open environment for different actors, which is prevented by the existing rules in traditional networks. Four of the most important regulatory issues which should be addressed are:

Peer-to-peer interactions: This is inherently based on competition between micro-prosumers and therefore requires a new regulatory regime.
Development of new business models: When it comes to business models, it entails a concept of fair competition among actors and how to settle the market financially. In the future power grid which will be dramatically affected by the 5D Giga Trends, there must be new regulatory regimes encompassing new business models such as virtual power plants, aggregators, retailers, local markets, etc.
Democratic electricity services: In future power grids, it must be possible for each consumer to purchase electricity according to its desired reliability and cost. In this situation, the problem of free-riding arises, meaning, due to technical constraints, it is not possible to separate two neighbors' electricity service qualities and one of them can benefit from the other one service for free. Such issues need to be addressed using technical or financial mechanisms.
Development of valuation mechanisms: In the future load-centric power grids, there exist various stakeholders and entities each of which provides a particular service to the network. However, these services have different characteristics such as service time-frame, level of criticality for the network security, and so on. On the other hand, the beneficiaries of these services must be distinguished from the value they get from the services. Also, there may be some service makers that may not be able to deliver their services only due to network

constraints (lack of democratization) or, on the other hand, there may be some service takers which are free-riders, again due to technical constraints. Thus, some of the most important potential challenges for the future power grids that need to be addressed in light of the Deregulation Giga Trend are:

- asset valuation methods based on the position of the player in the electricity market;
- determining the position of emerging and competitive business models in the electricity network;
- the importance of cybersecurity service providers of power networks;
- the importance of situation awareness services (such as data analytic based services) in the electricity market environment;
- the importance of flexibility providers in the network and their role in maintaining the security of power grids;
- issues related to the economic participation of actors and determining the market equilibrium point.

6.3 Opportunities and challenges of digitalization Giga Trend in power systems

Perhaps the most fundamental indication of the digitalization in the electricity industry can be seen in the "change in the nature of loads." Nowadays, there are many loads that are highly sensitive and need to be always on, so-called "always on loads" or "digitally charged loads." Some studies have shown that power outages of critical loads such as "data centers," "information exchange centers," "stock exchanges," and "central banks" can cause million-dollar-scale damages (based on Ref. [34], 16% of data center outages in the United States cost more than $1 million). Therefore, the power supply of these loads has a different standard from the other network loads. This raises the question of "the possibility of selecting the characteristics of the received power" in future power grids, which is under consideration by some researchers as "flexible reliability."

Another important aspect of the future power networks that align with the Giga Trend of "digitalization" is the issue of "big data analytics." In order to extract valuable information from a very large amount of raw data, data mining needs to go through the following three basic steps:

Data analysis: Converting a lot of raw data into valuable information (such as charts, tables, etc.).
Knowledge generation: Transformation of acquired information into applied knowledge (e.g., knowledge of urban transportation according to traffic information on different days).
Insights generation: Transforming knowledge extracted from information into insights of the subject (e.g., converting urban transportation knowledge into urban transportation management insights).

Clearly, this set of actions can well reveal the significance of raw data, which seemed sporadic and hard-to-use. It should be noted that one of the main functions of data processing in future power networks is "extracting information from beyond the meters." Using this measure, without interfering with the field and privacy of the consumer and only according to the information collected by smart meters, the consumption pattern and even the type of energy consumption of consumers can be determined. This concept is so-called "Nonintrusive Load Monitoring (NILM)."

Other dimensions of moving in the direction of the "digitalization" Giga Trend in the power networks is the change of the equipment and communication links used in the power network from analog to digital. An example of this can be seen in the change of many protection relays used in different parts of the power grid. Also, data exchange between meters located in different parts of the network and control centers has gone beyond analog modes based on "power line carriers (PLC)" and has been converted into optical fibers or digital wireless communications. All of these measures pave the way for the implementation of "smart grid," which is mainly manifested by "digitalization."

One of the important issues that can be considered as the result of digitalization in the electricity industry is the emergence of "Internet of Things." "Industrial Internet of Things" allows equipment around life (washing machine, refrigerator, etc.) that can be coordinately monitored and controlled, with the ability to connect to the internet through smartphones or personal computers. This concept can be as simple as connecting a smartphone to a TV or as complex as monitoring urban and traffic infrastructure [35]. Meanwhile, by increasing the intelligence of power networks and expanding the use of telecommunication equipment, the exchange of information between smart devices with each other and with local control centers, with the aim of increasing network efficiency, is a crucial task. The Internet of Things is one of the concepts that can play a key role in this context. For example, the IoT can help aggregate beyond the meter's resources and overreach the observability on the network to the home appliances.

As a result, the more information received from the power network and sent to the control center, the more it will be possible to monitor the network more accurately. This can be done due to the existence of digital measurement equipment at the network level, the existence of high-bandwidth digital data transmission platforms, as well as the implementation of data analytics methods to extract valuable information from raw data. Therefore, in this situation, the network operator has the opportunity to intelligently monitor the various parts of the network and make the right decisions, accordingly.

Strong connections between smart grid equipment and the ability to process this information quickly can turn future power grids into "self-healing" networks that, without any external action, detect incidents and take the necessary control measures to completely eliminate the problem. This issue, which can be classified as a subset of a broad topic in future power grids as "outage management," plays an important role in network resilience [36].

The concept of "cloud computing" is another issue that has been shaped by global movements in the direction of "digitalization" Giga Trend. Cloud computing is one of the best measures that the age of digitalization has provided in order to solve the problem of running heavy processing without the need for expensive hardware and equipment. Using cloud computing platforms, it is possible to upload the desired information in the cloud environment and with the help of powerful hardware equipment connected to this cloud environment anywhere in the world, the desired

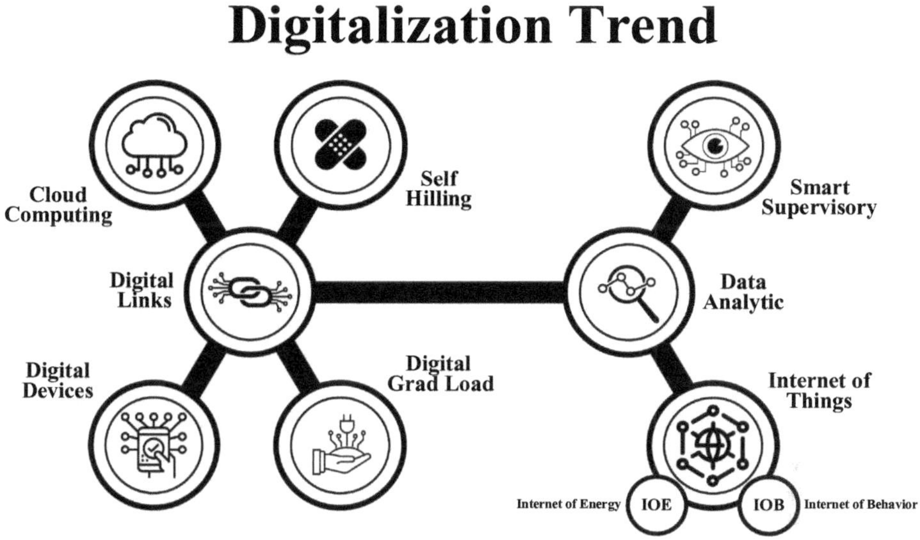

Fig. 2.11 Main digitalization impacts on the future power systems.

processing is performed and the results are received [37]. This feature is vital in future power grids because it has made it possible to increase the number of local control centers without any need for expensive hardware equipment and thus turn centralized processing into distributed one [38]. According to the explanations provided, Fig. 2.11 shows the most important dimensions and major effects of the digitalization process on power networks.

Moving power networks toward digitalization, in addition to the many benefits it brings, also creates problems, some of the most important are:

✓With the spread of digitally sensitive loads in the network that needs to be always on, it will no longer be possible to use alternative and limited solutions such as the use of "uninterruptible power supplies" to provide these loads, and a higher quality and reliability level of electricity is needed.

✓In order to enable peer-to-peer interactions in future power networks, it is necessary to use highly reliable encrypted digital communication platforms such as blockchain-based networks. In these networks, information about each transaction is placed in a block and encrypted. These blocks are connected to each other to form a chain of blocks and each node has a copy of the information so that the transaction can be performed securely. Given that there is no intermediary in this structure, it is necessary that the transactions be confirmed by all or some of the nodes based on a consensus mechanism [39]. However, similar to every distributed system, there are challenges in reaching consensus considering time limitations and cybersecurity issues, as some of the nodes can act maliciously. In such systems, anomaly detection mechanisms and the defense methods against cyberattacks are crucial tasks that should be taken into account.

✓Extensive use of digital equipment in power networks in order to manage, control, and protect different sectors, requires appropriate security measures to prevent intruders from conducting cyber-threats. The issue of "cybersecurity" is one of the topics that has received a lot of attention due to the movement of power networks in the direction of "digitalization" and has been the subject of many studies. In this way, privacy is also a vital issue that should be considered.

✓Although big data is a paradigm shift and brings many unique benefits for the power systems, unique challenges also emerge. One of the main challenges is the well-known curse of big data. Roughly speaking, when the dimensions of data are big, some of the statistical and practical inferences about the data are misleading. In fact, when dealing with big data, although the null hypothesis can be rejected, some noninteresting or insignificant correlations and effects may be recognized significant while there is only a statistical significance due to a larger amount of data not having a practically significant effect. Moreover, the future power systems will deal with huge volumes of big data as well as dimensions of data that need particular measures.

6.4 Opportunities and challenges of decarbonization Giga Trend in power systems

As it is clear, reducing carbon emissions has many benefits from a national and international perspective, but moving in this direction from the perspective of the power grid faces several challenges. Some of the most important ones are:

Lack of grid inertia: Increased use of renewable energy resources can leave power grids with a lack of inertia, which poses major challenges to grid stability. Therefore, it is necessary to think of appropriate solutions for this issue. Two of the most important strategies include the use of some technologies (such as the synchronous condenser, flywheel, compressed air storage, pumped-storage power plant, and supercapacitor) and the creation of virtual inertia through control systems.

Limitation of hosting capacity: Due to the intermittent nature of power generation by renewable resources and their dependencies to geographical location, the power network may not host any amount of renewables penetration level. Limitation in transmission network capacity, overvoltage issues within the time of off-peak load profile, lack of inertia issues are some of the technical constraints that limit the hosting capacity of the power grids for renewables. Not to mention that one of the hot topics in this area is to develop methods to increase hosting capacity. Also, the real-time approach for monitoring the hosting capacity of the network so that the operator can decide proper actions is another subject to work on.

The need for proper communication in the network: Increasing the use of demand-side resources (including renewable resources with positive power generation and demand response programs with negative power generation) requires appropriate communication and control platforms, especially in the distribution networks. Uncertainty of power generation by renewable sources on one hand and the need for accurate control and management of responsive loads, on the other hand, requires proper communication between these sources and local control centers.

Changing the method of power generation scheduling: The intermittent nature of power generation by renewable energy sources introduces many uncertainties into power generation planning and therefore requires different methods than the current ones to balance the load and production in the network. It should be noted that the use of production capacity with fluctuations of renewable sources, along with the use of responsive loads and also the management of storage devices such as electric vehicles, provides the possibility of covering the existing uncertainties, but there is a need for particular grid operational scheduling [40].

Location-dependent power plants: Renewable energy sources, due to the use of renewable energies as fuel, are inherently location-dependent resources. In other words, it is possible that using these resources in some networks is not economically viable and requires sufficient field studies to determine this issue.

Power quality issues: Due to the fact that many renewable resources require power electronic devices to connect to the network, the switching done on these devices produces a lot of the harmonics, but if appropriate methods are not considered to remove these harmonics, it results in major power quality issues. Therefore, determining the amount and type of harmonics produced by renewable resources and finding out appropriate solutions for this problem is a must-to-do task in the networks with high penetration of renewables [41].

Increased network operating costs: The use of renewable resources causes the network operator to turn off and on the traditional power generation sources frequently due to the intermittent power generation. The same goes for storage devices and flexible loads. All these high rate switching issues incur "cycling cost" to the network.

6.5 Opportunities and challenges of democratization Giga Trend in power systems

A comprehensive look at the infrastructure of the electricity industry shows that the main challenge of this industry which faces the network with electricity crisis is the lack of response from the load. In fact, the consumer is not involved in different characteristics of the network. Consequently, the consumers do not take responsibility for what is going on with the network. The solution for this problem is undoubtedly to democratize the electricity industry in order to create the necessary conditions for consumers to play an active role. In other words, when consumers become aware of their valuable position in future power networks and receive the necessary incentives to increase their activities, they can play a significant role in improving network performance and reducing the expansion and operational cost of the network.

In fact, the Giga Trend of democratization in the power networks can be defined in parallel with informing, valuing, and motivating consumers to become aware of their key role in the electricity industry, in order to actively participate in providing "energy efficiency" services or P2P energy trade. As mentioned earlier, energy efficiency which is the sixth fuel resource in the world can be divided into two main parts: "energy auditing" in the long run and "demand response programs" in the short term.

It has already been stated that one of the most important goals of future power networks is to achieve the six attributes, namely "security," "quality," "reliability," "availability," "efficiency," and "connectivity." Meanwhile, "connection" is a topic that has recently been added to this collection. Connection means prosumers communications at different levels of the network, especially on the distribution network, to operate in optimal conditions. This issue is closely related to moving toward "democratization" in the future power grids. Increasing the participation and dynamism of consumers in future power grids and increasing energy exchanges between them, which is widely discussed today as "exchanged energy," requires complete communication between them and democratic electrical connections in the network. Therefore, one of the basic factors in the realization of the Giga Trend of democratization in the electricity industry can be realized by developing the concept of highly connected electricity networks in which all participants can exchange energy in P2P or local market platforms.

The movement of future power networks in the direction of activating end users is a gradual and step-by-step process. As mentioned, the first step in this direction is to provide the necessary awareness to the consumers about their high position and value in the power network. The next steps can be "creating the necessary requirements for the participation of prosumers," "proposing incentive partnership projects," and "helping them to make them prosumers instead of only being consumers (such as loans to buy renewable equipment)."

Finally, the main issues that need to be addressed in the process of achieving democracy in the future power systems are development of new business models, providing flexible network reliability, preparing the groundwork for peer-to-peer energy transactions, development of cybersecurity programs, and so forth.

7 Life cycle of 5D Giga Trends

In this study, the direction of future power networks has been manifested, the basic dimensions of future power networks have been discussed, the importance of the 5D Giga Trends in this direction, and the impact they will have on the movement of power networks have been explored. However, there are some questions to be answered: "Does the realization of these 5D Giga Trends takes precedence over each other?" "What is the path to implementing future power networks?"

We need to know that the 5D Giga Trends have two aspects when it comes to answering the above questions. The 5D Giga Trends have a historical route based on which one can know which trend took precedence historically. However, the most important aspect here is to determine the causal relations among the trends, as the realization of the 5D Giga Trends is different from a historical point of view. In the former, we usually face a history about each trend. However, in the latter aspect, we usually face a cycle of trends. Generally speaking, the idea of democratization dates back to the ancient Greek philosophers who established the concept of a government in which people somehow have the right of free choice, making important decisions, ending the conflictions, etc. Since then, that idea has manifested itself differently in different contexts. The other four 5D Giga Trends are a kind of newcomers. Decentralization, digitalization, decarbonization, and deregulation have received special attention in recent decades.

However, in the casual-effect analysis, we want to reason that there is a life-cycle relationship between the trends. In fact, each time that this life cycle completes a round, one can find more democratization, digitalization, decentralization, deregulation, and decarbonization when compared to the previous round. Usually, the starting point is a kind of decentralization so that the aim is to outsource some of the services that were previously done by a central mechanism. For example, consider the banking system or educational system, which used to be governmental institutions with limited services, similar quality for all, centralized management, etc. However, the first stages of decentralization took place in such systems with limited decentralized managerial processes such as building several branches of public schools and banks. Note that, although it is a kind of decentralization, there is usually no considerable flexibility in the services the end users receive. In the case of power systems, as mentioned earlier, departing the generation side from the transmission and distribution parts can be considered as the mentioned stage of decentralization.

After an extent of decentralization, there is usually an urgent need for changing regulatory regimes and obsoleting the old ones, namely, deregulation. In this stage, some new (though limited) categorized services are offered to the end users to make a competitive environment. Also, some mechanisms are designed to address financial matters and legal conflictions. For example, the schooling system may be categorized into a few subsystems each of which proposes different quality of education that provides freedom of choice for the end users to some extent. Note that the level of deregulation is extremely constrained to the level of decentralization in the corresponding industry.

The next stage in this life cycle is digitalization. In fact, the more the lifestyle of society is digitalized, there is more capability to make a system decentralized. However, before society moves to extensive use of digitalized platforms, there should be a sufficient deregulated atmosphere in which private sectors have motivations to make new services and values. Thus, after each stage of deregulation, we expect to see new creative technologies. For example, by growing computers and internet technologies, distance learning and home-schooling systems were realized, although the idea had existed for a long time. As well, the banking system, for example, was able to outsource some duties such as transactions between individuals to people while such a duty was previously an inherent task of bankers.

In the next stage, when a system is decentralized, deregulated, and digitalized to an extent, the decarbonization stage begins, as many works can be done without lots of transportations. In this stage, there are also efficient ways to monitor the systems, to consume less energy in the supply chain to do the task, to produce clean energy, etc., which lead to a more decarbonized environment.

Finally, after all the stages regarding decentralization, deregulation, digitalization, and decarbonization, the main goal of the 5D Giga Trends (i.e., democratization) emerges. In this stage, depending on to what extent the previous stages have grown, the end users and all the participants are more engaged in the system's management and decision-making process. They have more power of choice and take the responsibility of their acts and the impact their decisions can have on the environment. In such a democratized atmosphere, there exists a different quality of services that can be chosen by the users. On the other hand, there are motivations to make creative businesses and new values.

Fig. 2.12 shows the life cycle of the 5D Giga Trends. As mentioned, irrelevant to the historical path each trend has passed, from a causal-effect point of view, there is round-based relation between the trends. Each time that this life cycle takes a round, all aspects of a system, namely, political, educational, technical, economical, etc. alter. For example, in the cinema industry, this life cycle can be traced back from a few cinemas across each country to social-media-based movies where all trends are on a level that every single individual can make a movie and release it on its YouTube channel or the other platforms. Such a revolutionary change has been built on many rounds of decentralization, deregulation, digitalization, decarbonization (in

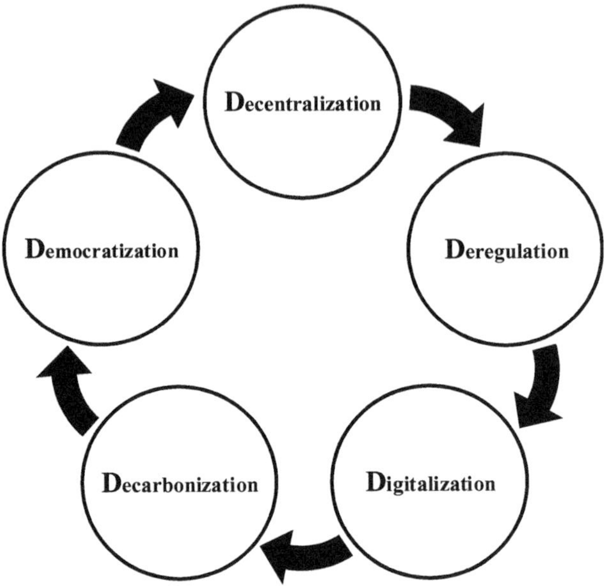

Fig. 2.12 The life cycle of 5D Giga Trends.

some cases such as this example, this trend usually is indirect. What would happen to decarbonization and democratization if everything was produced, transferred, and consumed in the same way as it was going on the 16th century), and democratization.

Similarly, the power system industry has experienced the same evolution pathway. After some rounds of being affected with the 5D Giga Trends life cycle, the power systems are growing based on the 5D Giga Trends as follows. They are moving toward a level of decentralization so that the tiny power generations of roof-top PVs, batteries of electric vehicles, embedded storages, etc. get involved in the supply chain of the power systems. Newborn technologies such as blockchain, digital twin, machine learning, and artificial intelligence have prepared the playground for a high level of decentralization. The combination of these technologies can lead to a democratic autonomous power system in which thousands of energy transactions can take place within a short period of time and many local energy markets can be settled without involving central entities. In such a situation, as much as people have the power of choice in their energy productions and consumptions, at the same time, they take responsibility for environmental issues, network costs, operational costs, and so forth.

References

[1] N. Crafts, The first industrial revolution: resolving the slow growth/rapid industrialization paradox, J. Eur. Econ. Assoc. 3 (2–3) (2005) 525–534.
[2] E. Schatzberg, Eleven. Science and technology between the world wars, in: Technology, University of Chicago Press, 2018, pp. 174–193.
[3] A.S.G. Andrae, Prediction studies of electricity use of global computing in 2030, Int. J. Sci. Eng. Investig. 8 (86) (2019) 64–79.

[4] A.K. Sharma, G. Tjandraatmadja, S. Cook, T. Gardner, Decentralised systems—definition and drivers in the current context, Water Sci. Technol. 67 (9) (2013) 2091–2101.
[5] V.A. Schmidt, Democratizing France: The Political and Administrative History of Decentralization, Cambridge University Press, 2007.
[6] A. Forte, V. Larco, A. Bruckman, Decentralization in Wikipedia governance, J. Manage. Inf. Syst. 26 (1) (2009) 49–72.
[7] N. Castree, Neoliberalising nature: the logics of deregulation and reregulation, Environ. Plan. A 40 (1) (2008) 131–152.
[8] H. Daly, R. Goodland, An ecological-economic assessment of deregulation of international commerce under GATT Part I, Popul. Environ. 15 (5) (1994) 395–427.
[9] https://www.inc.com/peter-economy/top-10-fastest-countries-to-start-up-a-new-business.html.
[10] M. Guha, Digitization/digitisation, in: Collins English Dictionary (Reference Edition), Reference Reviews, 2016.
[11] M. Maguire, F. Murphy, Neoliberalism, securitization and racialization in the Irish taxi industry, Eur. J. Cult. Stud. 17 (3) (2014) 282–297.
[12] K.A. Zimmermann, History of Computers: A Brief Timeline, Live Science, 2017.
[13] M. Hilbert, P. López, The world's technological capacity to store, communicate, and compute information, Science 332 (6025) (2011) 60–65.
[14] Global Annual Mean Surface Air Temperature Change, NASA, 2020. Retrieved 23 February 2020.
[15] https://www.visualcapitalist.com/the-tech-giants-worth-compared-economies-countries/. https://companiesmarketcap.com/.
[16] J. Rogelj, Summary for policymakers, Climate Change 2013: The Physical Science Basis, Cambridge University Press, 2013.
[17] G. Janssens-Maenhout, M. Crippa, D. Guizzardi, M. Muntean, E. Schaaf, F. Dentener, P. Bergamaschi, et al., EDGAR v4. 3.2 Global Atlas of the three major greenhouse gas emissions for the period 1970–2012, Earth Syst. Sci. Data 11 (3) (2019) 959–1002.
[18] H. Welzer, Climate Wars: What People Will Be Killed for in the 21st Century, John Wiley & Sons, 2015.
[19] C. Breidenich, D. Magraw, A. Rowley, J.W. Rubin, The Kyoto protocol to the United Nations framework convention on climate change, Am. J. Int. Law 92 (2) (1998) 315–331.
[20] Agreement, Paris, Paris agreement, in: Report of the Conference of the Parties to the United Nations Framework Convention on Climate Change (21st Session, 2015, Paris), vol. 4, 2015, p. 2017. Retrieved December.
[21] B. Geddes, What do we know about democratization after twenty years? Annu. Rev. Polit. Sci. 2 (1) (1999) 115–144.
[22] https://www.systemicpeace.org/polityproject.html.
[23] https://www.eia.gov/tools/faqs/faq.php?id=74&t=11.
[24] R.A. Mehrabadi, M.P. Moghaddam, M.K. Sheikh-El-Eslami, Generation expansion planning in multi electricity markets considering environmental impacts, J. Clean. Prod. 243 (2020) 118611.
[25] R.A. Mehrabadi, M.P. Moghaddam, M.K. Sheikh-El-Eslami, Regulatory-intervented sustainable generation expansion planning in multi-electricity markets, Sustain. Cities Soc. 52 (2020) 101794.
[26] C.W. Gellings, The Smart Grid: Enabling Energy Efficiency and Demand Response, CRC Press, 2020.
[27] M.P. Moghaddam, Lecture Notes of Smart Grids Course, Department of Electrical Engineering, Tarbiat Modares University, 2020.

[28] R. Zamani, M.P. Moghaddam, M.-R. Haghifam, Evaluating the impact of connectivity on transactive energy in smart grid, IEEE Trans. Smart Grid (2021), https://doi.org/10.1109/TSG.2021.3136776.
[29] R. Zamani, M.P. Moghaddam, H. Panahi, M. Sanaye-Pasand, Fast islanding detection of nested grids including multiple resources based on phase criteria, IEEE Trans. Smart Grid 12 (6) (2021).
[30] S. Nasiri, A. Ghadiri, A. Shahmirzaee, M. Parsa, Probabilistic and fuzzy modeling of charging and discharging of electric vehicles in presence of uncertainties applied to unit commitment problem, in: Power System Conference (PSC), 2014.
[31] H.A. Aalami, M. Parsa Moghaddam, G.R. Yousefi, Demand response modeling considering interruptible/curtailable loads and capacity market programs, Appl. Energy 87 (1) (2010) 243–250.
[32] A. Mansoori, A.S. Fini, M.P. Moghaddam, Power system robust day-ahead scheduling with the presence of fast-response resources both on generation and demand sides under high penetration of wind generation units, Int. J. Electr. Power Energy Syst. 131 (2021) 107149.
[33] M. Khodadadi, M.E.H. Golshan, M.P. Moghaddam, Non-cooperative operation of transmission and distribution systems, IEEE Trans. Ind. Inf. 18 (1) (2020) 153–162.
[34] A. Lawrence, Annual Outage Analysis 2021—The Causes and Impacts of Data Center Outages, Uptime Institute of Intelligence, 2021.
[35] B. Dorsemaine, J.-P. Gaulier, J.-P. Wary, N. Kheir, P. Urien, Internet of things: a definition & taxonomy, in: 2015 9th International Conference on Next Generation Mobile Applications, Services and Technologies, IEEE, 2015, pp. 72–77.
[36] M. Mahzarnia, M.P. Moghaddam, P.T. Baboli, P. Siano, A review of the measures to enhance power systems resilience, IEEE Syst. J. 14 (3) (2020) 4059–4070.
[37] P. Mell, T. Grance, The NIST Definition of Cloud Computing, National Institute of Standards and Technology, 2011.
[38] Y. Allahvirdizadeh, M.P. Moghaddam, H. Shayanfar, A survey on cloud computing in energy management of the smart grids, Int. J. Electr. Power Energy Syst. 29 (10) (2019), e12094.
[39] N. Wang, X. Zhou, X. Lu, Z. Guan, L. Wu, D. Xiaojiang, M. Guizani, When energy trading meets blockchain in electrical power system: the state of the art, Appl. Sci. 9 (8) (2019) 1561.
[40] R. Tahmasebifar, M.P. Moghaddam, M.K. Sheikh-El-Eslami, R. Kheirollahi, A new hybrid model for point and probabilistic forecasting of wind power, Energy 211 (2020) 119016.
[41] S. Nasiri, H. Seifi, Robust probabilistic optimal voltage sag monitoring in presence of uncertainties, IET Gener. Transm. Distrib. 10 (16) (2016) 4240–4248.

Grid transformation driven by high uptake of distributed energy resources—An Australian case study

3

Daniel Eghbal
Manager Future Network Strategy, Energy Queensland, Brisbane, QLD, Australia

1 Introduction

Distributed energy resources (DER) enable customers to lower their electricity bills, contribute toward a low-carbon future, and take more ownership in supplying their own energy as well as participate in energy trading opportunities and consequently gain a higher return on their investment in DER. This means customers and communities can trade energy locally and potentially provide services to the local network operator to manage the electricity network more efficiently at a lower overall cost to all customers. High penetration of DER introduces a paradigm shift in electricity demand and supply chain and requires significant changes in traditional planning and operation of the distribution networks. DER such as rooftop solar PV is a variable and nonschedulable generation resource. Efficient operation of the distribution network requires optimal coordination between flexible demand and variable generation resources connected to the distribution network.

This chapter consists of two main sections. The first section briefly explains the driving forces behind grid transformation in advanced electricity markets with high penetration of large-scale renewables and DER. While utility-scale wind and solar connected to transmission networks remain the dominant sources of renewable energy, unprecedented growth in the uptake of DER introduces a paradigm shift toward decentralized electricity generation that leads to significant changes in the role of traditional distribution network utilities. In the future, distribution network service providers need to optimally coordinate demand and supply on a localized basis based on local network constraints and maintain network reliability.

The second section gives an overview of grid transformation in Australia that has already embarked on the grid transformation journey. Today, one in every three houses in many suburbs in Australia (South Australia and Queensland among the highest) generates a significant portion of their daily energy from their rooftop solar system. This high uptake within only a few years was driven by factors such as a generous feed in tariffs, abundant availability of sunshine, and significant cost reduction in installing rooftop solar systems. The boom in the rooftop solar, along with expected

Decentralized Frameworks for Future Power Systems. https://doi.org/10.1016/B978-0-323-91698-1.00004-2

similar trends for electric vehicles (EVs) and battery storage in the near future, has created an opportunity to progressively transform the distribution network into an intelligent grid that supports a sustainable low-carbon future while benefiting all customers by maximizing the value from DER. Grid transformation at this scale requires major regulatory reforms that will be briefly discussed. This overview is based on the public information available at the time without specific recommendations toward a particular solution or a zero-emission target.

2 Energy transition

The energy system has experienced some significant changes over the last decade. As shown in Fig. 3.1, global installed capacity grew sevenfold for solar PV and approximately threefold for wind energy from 2011 to 2019 [1]. Improving cost competitiveness and operational efficiency were the key drivers behind this growth. Fig. 3.2 shows the global levelized cost of electricity from grid-scale renewable sources in 2010 and 2017 [2]. While most of the world's 10 largest economies have committed to achieve net-zero emissions by mid-century, only 13 of the 115 benchmarked countries have made consistent progress on their decarbonization journey over the past 10 years [3]. This demonstrates the difficulty in sustaining progress and the complexities of the energy transition.

Fig. 3.3 shows the percentage of world electricity generation mix by fuel from 1971 to 2019 [4]. It clearly shows that while there has been a considerable increase in

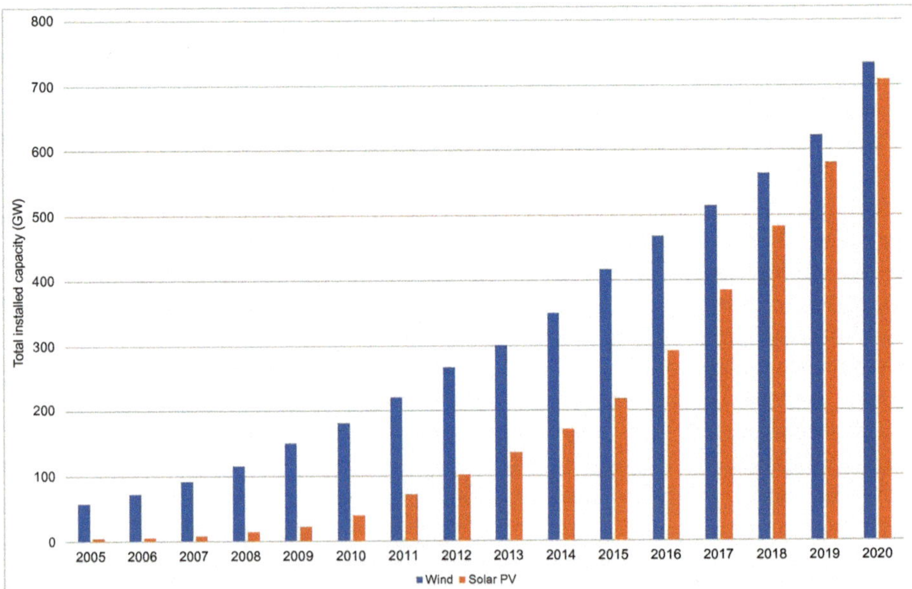

Fig. 3.1 Cumulative global installed capacity of wind and solar energy 2005–20 (GW). Credit: Renewable Energy, from https://ourworldindata.org/energy.

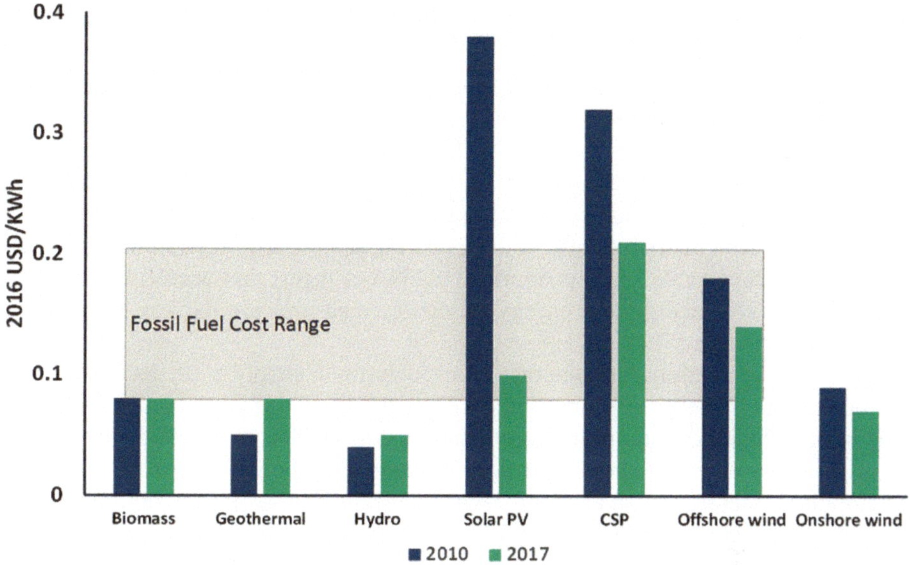

Fig. 3.2 Global levelized cost of electricity from grid-scale renewable sources, 2010–17. Credit: IEA, World Electricity Generation Mix by Fuel, 1971–2019, IEA, Paris, n.d.

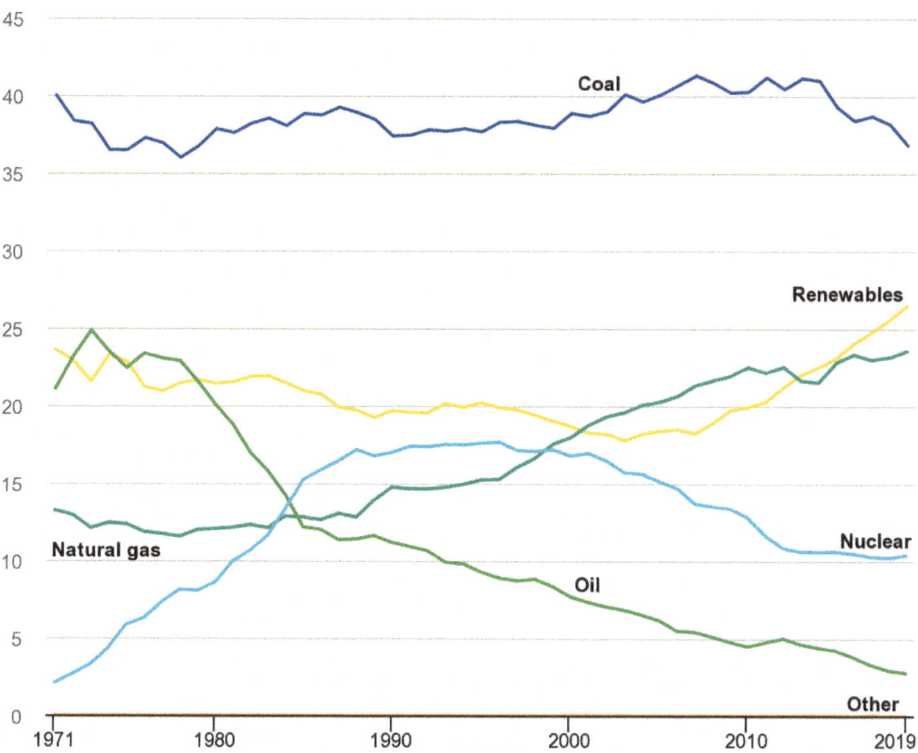

Fig. 3.3 World electricity generation mix by fuel—shown in percentage (1971–2019). Credit: IEA, World Electricity Generation Mix by Fuel, 1971–2019, IEA, Paris, n.d.

renewables over the last decade, fossil fuels still account for a majority of the fuel source for electricity generation. This makes the electricity industry a prime potential for reducing the dependency on fossil fuels and managing climate change.

International Energy Agency (IEA) findings suggest that an energy transition of exceptional scope, depth, and speed is required to limit the global mean temperature rise to below 2°C with a probability of 66%. In the same study, key findings from the International Renewable Energy Agency (IRENA) suggest that accelerated deployment of renewable energy and energy efficiency measures are the key elements of the energy transition [5].

While IEA projections indicate that fossil fuels could supply more than 70%–75% of global primary energy consumption by 2040, the share of renewables in electricity generation in major economies is significantly growing. The progress on energy transition varies by an enabling environment influenced by political support, economic development, security of investment, regulatory frameworks, incentives for innovation, and adoption of emerging technologies. This makes energy transition a different journey for each country.

3 Grid transformation

The global energy transformation is driven by a shift from centralized fossil fuel-based generation to more decentralized renewable energy sources over the next 10–30 years. This transformation is likely to see a dramatic change in the rapid adoption of emerging technologies, driven by falling production costs and global measures to combat climate change. The pace of such a significant transformation is highly influenced by measures such as economic development and commitment to climate change policies in each country.

Electricity networks, arguably the largest machines built by humans, were designed and built based on a top-down approach to deliver electricity from large-scale centralized power plants to consumers through long transmission and distribution power lines. This centralized architecture has enabled cost-effective and reliable supply of electricity to the customer. Colocation of large power plants and fuel sources such as coal and hydro outweighed the losses associated with long transmission and distribution lines. This architecture has proven to be effective even after the emergence of large-scale renewables. Large wind resources are often offshore or located in remote areas. Large solar farms are built far from major load centers due to land cost and availability. Hence, long transmission lines are still required for integrating large-scale renewables.

In some advanced electricity markets, grid-scale renewables already account for a major portion of the generation portfolio. This milestone was achieved following the previous major electricity reform: the introduction of electricity markets in the late 1990s. The power industry developed new technical solutions, utilities amended the grid codes, and regulatory bodies have changed the rules to facilitate large-scale integration of variable renewables. However, the centralized model of the power system and trading energy remained mostly the same. While wholesale energy markets enabled a degree of competition, customers still did not have an alternative source to procure their energy other than the centrally managed grid.

Levelized costs of small-scale renewables are higher than large-scale renewables, however; the enormous cost reductions in rooftop solar PV technology and government subsidies are attributed to the significant boost in uptake of DER. In countries such as Australia, more customers can now generate a portion of their own energy needs and even export some excess power to the grid at times. In areas with a high penetration of rooftop solar PV, the aggregated installed capacity of DER could exceed the size of the largest power plant in that area.

High uptake of DER in Australia is a significant step toward meeting net-zero emission goals, but it comes with a few caveats. Socially, vulnerable customers and those without roof space are disadvantaged and miss out on government subsidies and arguably pay more for their electricity, while rooftop solar PV owners pay less for energy and often get paid for excess solar generation during the day when wholesale energy prices could be even negative. Technically, high volumes of distributed solar generation without storage and lack of appropriate tariffs cause unbalanced demand and supply issues. The system operator has to deal with retirement of based load coal-fired power stations, system strength issues due to growing inverter-based generation capacity, and poor visibility of distributed generation connected to low voltage networks. This in turn ignited ideas for a mechanism to centrally shut down customers' rooftop solar PV as a last resort with a vision to dynamically manage rooftop solar PV exports in the future.

This is not an optimal outcome for customers and neither a cost-effective grid transformation; however, a significant paradigm shift occurs when distributed energy storage, EVs, and home energy management systems (HEMS) become cost competitive to alternative options. The ongoing technological advancement in battery chemistry is a catalyst for high uptake of battery energy storage and EVs. These DER technologies empower customers to become prosumers that can choose how to procure their electricity and also trade the excess energy from their DER with their peers, either through peer-to-peer trading platforms or via aggregators in the energy markets.

In countries such as Australia and the United Kingdom, the early impacts of this transformation are becoming increasingly visible and continue to stretch the current system to its technical limits and push regulatory boundaries to the extent that pragmatic and holistic reviews of regulatory frameworks are deemed to be more adequate than modest regulatory changes.

Fig. 3.4 illustrates the transformational change in the electricity supply chain and shows that while the physical aspects of the grid interconnections remain mostly unchanged, the supply and demand portfolios will change significantly. The future grid should be able to maintain a reliable and secure balance between large-scale renewables on the supply side and prosumers' DER on the demand side. A large amount of DER means that more electricity could be generated and consumed locally within the distribution networks and consequently bypass the upstream centralized generation and transmission networks.

Availability of DER to a majority of customers enables a democratic grid concept that merges the technological energy transition with customer empowerment and

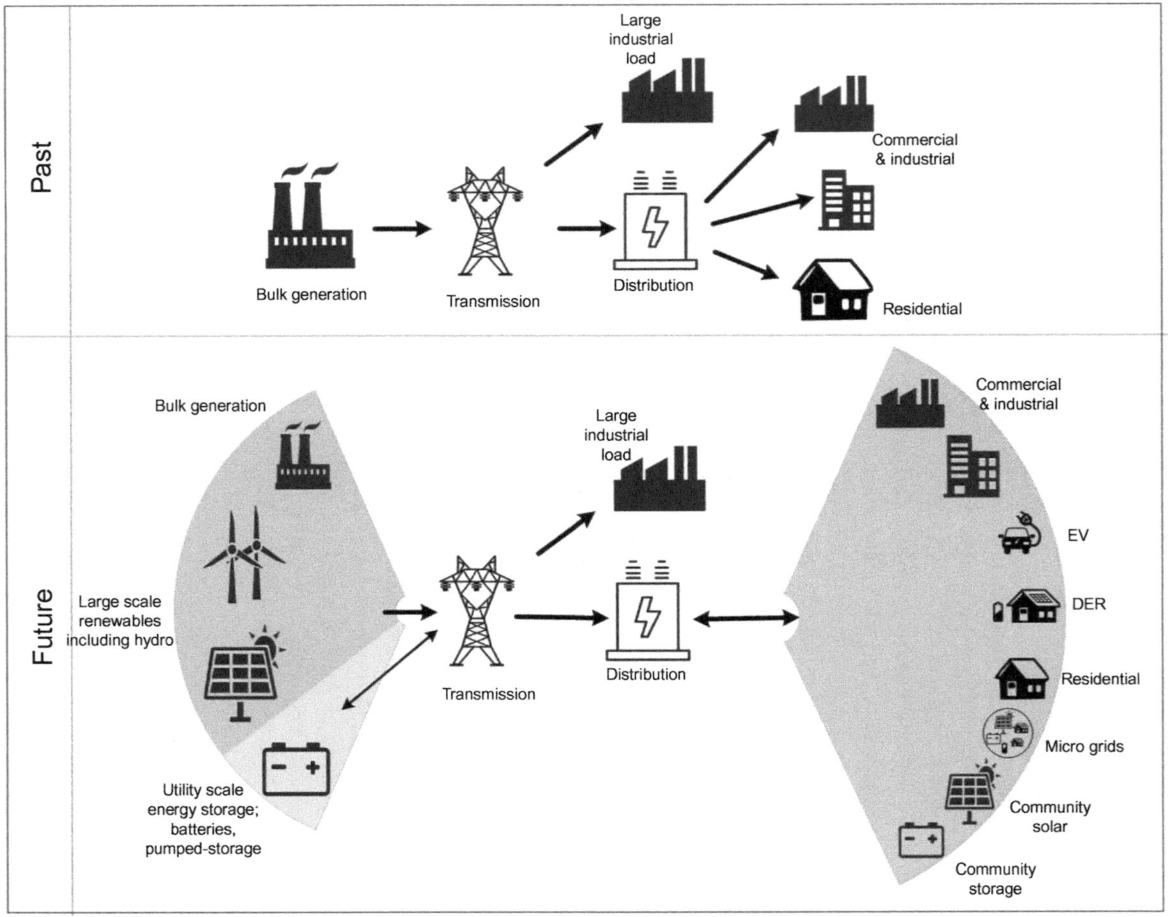

Fig. 3.4 Past and future supply chain diagram for power systems.
Credit: Author generated.

public participation. Once a wide range of DER technologies become affordable to communities, all customers can benefit from a more reliable, resilient, and clean energy regardless of their DER ownership. Some traditional network solutions (expansion of poles and wires of the distribution network) can be replaced by a more cost-effective nonnetwork solution offered by customers' DER. This in turn reduces the overall electricity price that benefits customers with and without DER. The DERtopia discussion paper [6] and the proposed model by Kristov et al. [7] explain the democratic grid concept in the Australian and US contexts, respectively. The latter also introduces the concept of Grand Central Optimization and Layered Decentralized Optimization that will be explained further in the next section.

Regardless of the future grid architecture model, which is explained further in the next section in this chapter, the evidence from countries with abundant renewable sources reveals that the impacts of this transition are becoming increasingly visible, and continue to stretch the system to its technical limits and push the regulatory boundaries to the extent that revolution of regulatory frameworks are required rather than modest changes to the regulatory frameworks.

Another key aspect of grid transformation is the key role of digital transformation. Digital transformation is a catalyst that technically enables active participation of prosumers in the electricity market in conjunction with regulatory reforms. To date, energy market participation is limited to large entities that can manage huge financial risks and afford large capital investments. The market transactions are mostly settled based on near real-time data from major connection points at high voltage transmission networks without accurate models of distribution networks. In future, a digital twin of the grid will be the core enabler for active interaction between grid elements and making cost-effective decisions in near real time. In simple terms, a digital twin of the grid is a digital version of the physical grid that enables a system operator to model the entire network and all connected assets in real time. The digital twin of an entire grid entails data collection and analysis from millions of connected assets owned by different parties that was not feasible before digital technological advancements such as Internet of Things (IoT), cloud-based digital services, and big data analytics.

In addition to advanced network modeling and vast data collection from multiple sources, interoperability between grid and DER assets is a fundamental capability that enables demand and supply orchestration. While digital metering infrastructures (also known as smart meters) already exist in some distribution networks, the data handling requirements of a modern grid with very high penetration of DER are of a truly unprecedented scale. Moreover, the social aspects of such a digital transformation—e.g., customer data privacy—should not be underestimated.

Cybersecurity is another key pillar of the grid transformation. Data exchange between multiple IoT platforms owned by grid operators, energy retailers, and aggregators enables optimal utilization of network assets and DER. However, the increased risk of new vectors of intrusion could impact the operation of critical grid infrastructure with possible breaches of data protection deteriorating customer trust. The cybersecurity risks are particularly high during early stages of grid transformation while rigorous standards and protocols are still being developed. Another risk is the skill gap within the power industry. With the growing number of digital assets in the power system and creation of the digital twin of the grid, cybersecurity should be treated similarly to physical asset safety and become an integral part of grid planning, design, and operation.

Grid transformation requires significant capital investment as it entails significant changes in all layers of the power system, from large-scale generation and wholesale market mechanisms, down to low voltage (LV) distribution networks. The capital cost of such transformation is expected to unlock new revenue streams for all stakeholders including customers who invest in DER. However, some of the benefits will be realized well into the future while capital investment would be required up front. Stranded investment is a major risk because of the rapid change in emerging technologies and uncertain political and economic environments. The key DER benefits and use cases include self-resiliency, customer choice, demand response, flexibility services, energy efficiency, and obviously emission reduction. Once multiple DER use cases are satisfied concurrently, the associated "value stacking" improves the economics and return on investment and enables a more democratic energy ecosystem. Fig. 3.5 shows a conceptual high-level value stack in the Australian electricity market context.

While uptake of DER is continuing to surge, large-scale renewables remain a key element of grid transformation. Technical barriers to full utilization of abundant renewable energy sources are their variable nature, lack of economic large-scale energy storage, and often establishment of additional transmission infrastructure with significant up-front costs. The nascent global focus on green hydrogen from renewables could potentially lift some of these barriers. Producing hydrogen is nothing new, but it is the source and process of producing hydrogen that is expected to cause a significant boost to large-scale renewable projects. The idea is to use renewables to supply an electrolyzer to break water into oxygen and hydrogen. Then the hydrogen can be stored and delivered to multiple end users including fuel cells to power homes or vehicles. This provides an alternative to building long transmission lines and even

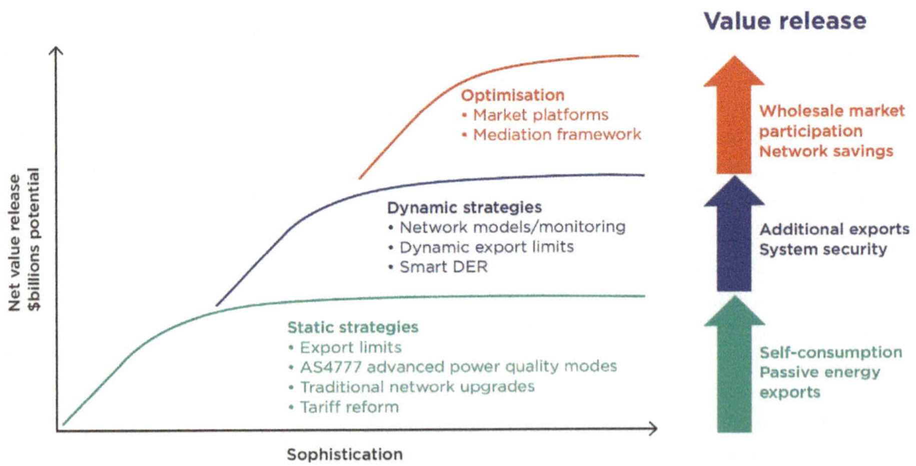

Fig. 3.5 Additional value release enabled by optimization of active DER [8].
Credit: ENA.

enables opportunities to export hydrogen internationally. Large-scale production of hydrogen from renewables is still at the infancy stage and fuel cells are not cost competitive. A comprehensive overview of hydrogen supply, distribution, end-use technologies, and cost comparison to incumbent technology pathways can be found in a recent report by Advisian [9]. While the report focuses on the Australian market, some of the proposed production and delivery models can be applied in other markets.

In summary, transition to net-zero emission will be of an unprecedented scale, the likes of which may never be repeated again. However, the pace of change will be influenced by national policies, economic growth, and technological advancement in different countries. While designing the future grid, it is important to adopt a customer-centric design that increases the interest of customers to participate in DER orchestration programs that could benefit all customers. While DER and other customer-initiated technologies may introduce technical challenges to traditional grid management, they also have the potential to provide new services and enable new business models that could stimulate increased medium and long-term business values.

Regardless of the end state, major regulatory and tariff reforms will be inevitable to accommodate high penetration of grid-scale renewables and DER. In regions with very high penetration of DER, the potential DER services could turn the grid pyramid upside down by enabling a totally decentralized grid in that region, which is fundamentally very different from what we see today. Hence, the next section is dedicated to an overview of decentralized versus centralized grid architectures.

4 Centralized versus decentralized

As outlined earlier, a major shift to renewable energies together with growing penetration of DER and electrification of transport is introducing alternative ways to generate, distribute, and consume electricity in many countries including Australia, the United States, and the United Kingdom. The early grid transformation is manifested in the ongoing rapid boost in rooftop solar PV. A significant portion of Australia's future energy production could be from DER—far in excess of any major economy in the world [10]. With the expected uptake of EVs and battery storage in addition to ongoing rooftop solar installations, policymakers started the discussion around future operating models of distribution networks and how roles and responsibilities of incumbent market participants should change to support a grid with high penetration of renewables and DER. While a consensus seems to exist among stakeholder with regard to the need for a reform and even a market redesign [11], there is no agreed view on roles and responsibilities of market participants and most of the proposed options are based on a centralized grid architecture.

Grid transformation at this scale requires alternative grid architecture models rather than developing incremental and separate technical and regulatory changes to integrate DER into the current model. The growing uptake of DER introduces a paradigm shift in the distribution networks as it enables more autonomy at the local level. In the future, a distribution network could be considered as the number of interconnected microgrids where each would have a different level of autonomy and the role of

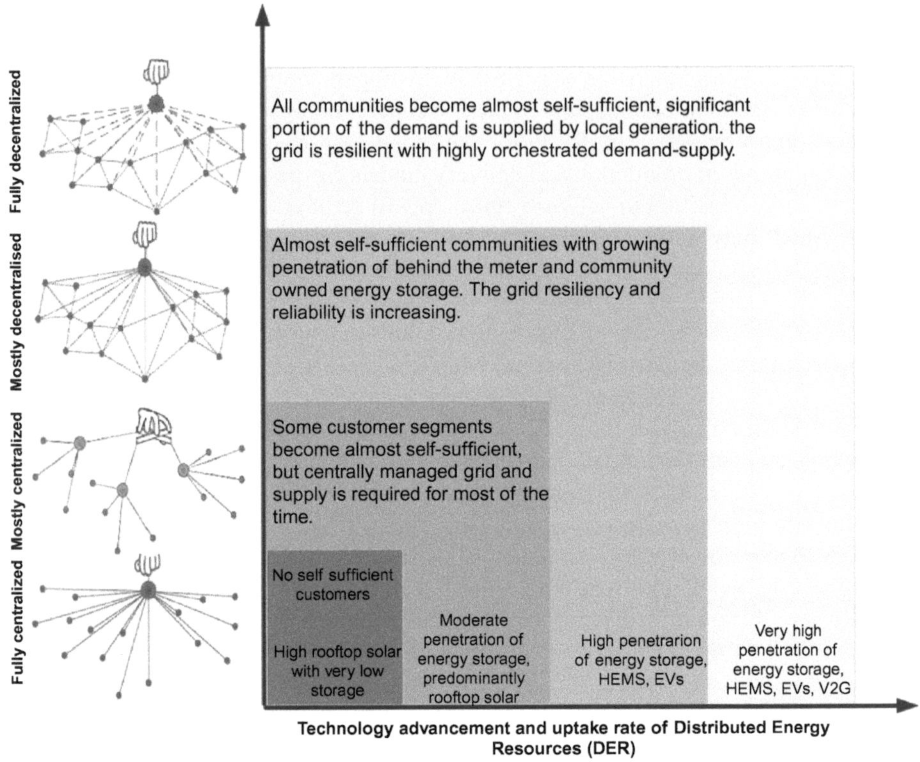

Fig. 3.6 Evolution of decentralized grids.
Credit: Author generated.

the grid operator would be to ensure a balance between these microgrids. This is a partially or fully decentralized model that is fundamentally very different from the existing grid and significantly changes all aspects of network planning, operation, and maintenance. Fig. 3.6 illustrates a conceptual evolution from a fully centralized to a fully decentralized model. Such an evolution cannot occur over a short period of time, and two key enablers are DER technological advancements and customer participations.

The last time the electricity system went through a major reform was with the introduction of electricity markets and privatization of government-owned energy companies in the late 1990s. However, the centralized model of the electricity system and trading energy remained mostly the same. This model was also suitable for integration of large-scale renewables, since major wind and solar resources are generally located in remote areas far from major load centers.

The dichotomy between centralized and decentralized grids is not novel, but often it is focused on technical and economic advantages and disadvantages. As discussed earlier, the nascent grid transformation requires holistic and pragmatic approaches. Grid architecture entails various aspects such as the location of generation and demand, flexibility resources, control mechanism, visibility, and data exchange

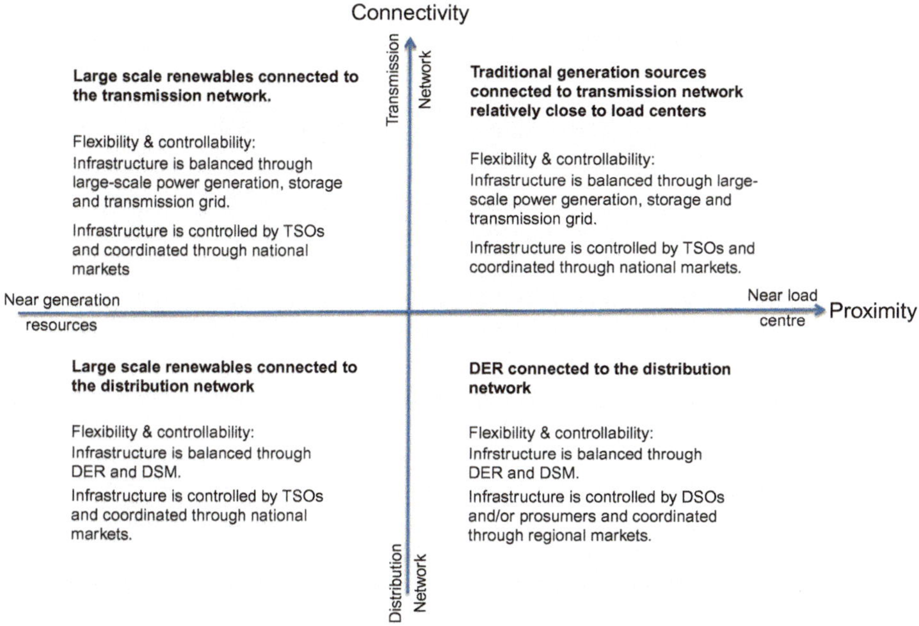

Fig. 3.7 Centralized and decentralized grid infrastructure dimensions.
Credit: Author generated.

options. Fig. 3.7 summarizes the infrastructure dimensions of the centralized and decentralized grid architecture based on the work presented by Simon Funcke [12]. This work is focused on Germany; however, the same principle can be applied to other markets.

One of the economic advantages of the decentralized grid is achieved through less capital investment in the grid infrastructure and more efficient utilization of all available resources including DER that customers have already invested in. In a grid with high DER penetration, the grid operator could procure network services such as flexibility services through aggregators or local markets. Innovative community storage business models are being considered as an option that could reduce the capital investment for grid operators and communities, and maximize the revenue streams for the community battery by enabling multiple use cases such as network services, energy arbitrage, and ancillary services [13–15]. The emergence of cloud-based services and the 4G/5G communication network facilitate the coordination among assets at different locations to achieve different objectives at different times. Zhang et al. [16] introduced a cloud-based platform to aggregate distributed energy storage to provide services across multiple users including flexibility services to the grid operator. Similar concepts utilize the available capacity from decentralized energy storage sources at a lower cost than by installing dedicated facilities.

The concept of community DER is one of the core elements of a democratic grid that could be developed gradually in either communities with a net-zero emission vision or remote communities as early adopters with possible deployment in other communities at a later stage. In a democratic grid, consumers are enabled to participate directly in the financing of much of the new grid infrastructure, gaining more autonomy and control while maximizing the return on their investment in DER. This could bring more competition and lower the overall system costs. Customers become increasingly responsible for their own supply, relying on the grid as backup. This is a significant change for customers and extreme care should be taken when designing such democratic grids. The majority of customers prefer a set and forget style setup for their electricity connection rather than being responsible for the daily operation and maintenance of their electricity supply (except off-grid customers). While it is envisaged that aggregators and demand response service providers will act on behalf of customers to maximize the value of DER, customers should not be disadvantaged or locked into a particular solution.

In recent years, there have been debates on whether governments and policymakers should subsidize large-scale renewables or small-scale DER projects. The operating model of a grid with a high penetration of large-scale renewables and bulk energy storage seems a logical upgrade of the incumbent grid with minimum fundamental changes in the demand-supply chain. On the other hand, a high penetration of DER means the majority of the energy is generated locally and consumed locally and favors a bottom-up approach, which is fundamentally different from the existing grid. Two key benefits of DER are customer empowerment and offering flexibility services at different levels. Even in a fully centralized grid architecture, DER can offer distribution flexibility services (e.g., real and reactive power and voltage control). DER services can be integrated into the network planning process to defer, reduce, or avoid network augmentation. This in turn improves the reliability of the grid and enables new business models that offer new value streams for prosumers. Aggregated flexibility resources from DER can also complement the flexibility resources on the upstream network required by large-scale renewables. As illustrated in Fig. 3.8, DER can support the decarbonization of the grid in two ways: directly through generating renewable energy and indirectly through providing flexibility services to large-scale renewables.

Today's grid transformation projects are aimed at developing pathways to include various aspects of the decentralized model in the incumbent centralized model.

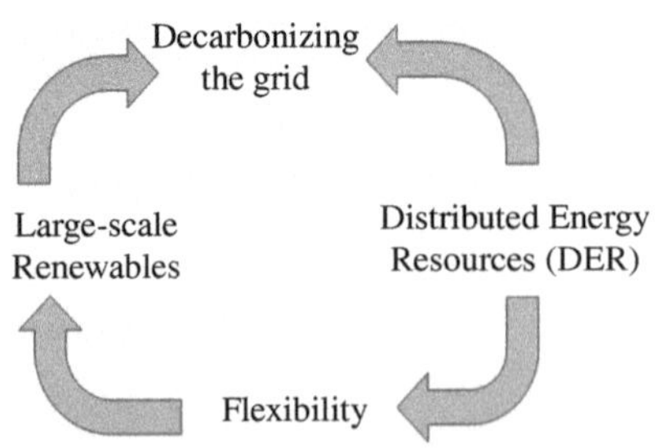

Fig. 3.8 Role of flexibility in supporting decarbonization of the grid.
Credit: Author generated.

The evolution of a fully decentralized grid at large scale appears to be a massive undertaking as it is significantly and fundamentally different from the incumbent grid that has been operating for decades. Such a dramatic shift requires cultural and skill set changes, strong community engagement and customer education, major regulatory reforms in conjunction with new business models for market participants.

Kristov et al. [7] proposed two distinct visions for the design and operation of a decentralized and transactive energy system with a large amount of DER. One vision, called Grand Central Optimization, is a straightforward extension of how demand response participates in the wholesale market. The market operator receives and dispatches bids from DER and aggregators and the Distribution system operator (DSO) coordinates DER to maintain the reliability of the network. The alternative and preferred vision—from a grid architecture perspective—is the layered decentralized optimization (LDO) that represents a substantial shift from today's wholesale market. The DSO aggregates all the DER within each local distribution network physically connected to a connection point to the upstream (transmission) network and submits a single bid to the wholesale market. The key advantage of this method is that it can be easily scaled and replicated at various levels because optimization at any given layer of the system only requires visibility to the interface points with the upstream and downstream networks instead of visibility and access to all layers.

Ongoing DER growth in Australia is pushing the grid more toward a decentralized energy ecosystem—where an increasing amount of electricity is generated at a smaller scale across vast geographical areas. This is a catalyst to a paradigm shift in the energy landscape where a bottom-up grid architecture (similar to the abovementioned layered decentralized optimization) would be more suitable.

The idea of decentralization of the Australian grid was first introduced in the 2015–16 Network Transformation Roadmap (NTR) by CSIRO and ENA [17]. The NTR concept paper proposed a Network Optimization Model as a technology neutral mechanism for procuring network optimization services. The road map also contended a growing number of opportunities, such as

- reduced power bills
- greater integration of renewables
- alternatives to large-scale generation
- potential for reduced network augmentation
- potential market opportunities in harnessing of aggregated residual DER generation to wholesale markets

Following rapid increases in DER uptake, ENA and AEMO initiated the Open Energy Network Project, mirroring the Open Network project in the United Kingdom, and proposed four operating models for managing the grid [8]. The key differences between the proposed models were around the roles and responsibilities of the wholesale energy market operator and distribution service operators. In essence, all proposed models were based on a centralized grid architecture in the sense that a single market platform procures aggregated services from a large number of DER. The virtual power plant (VPP) trial project in South Australia is a good example of attempts at integrating distributed flexibility into a centrally controlled grid [18].

A vision for a democratic grid for Australia, inspired by the layered decentralized optimization concept, was discussed in TEC [6]. The vision offers a bottom-up approach as an alternative to the incumbent top-down approach that could progressively enable a democratic grid. Distribution Trading Platforms were proposed as an alternative to a single Distribution Market Operator to facilitate local energy trading and also enable the distribution network operator to balance the network more locally, where possible.

Microgrids and off-grid communities are not within the scope of this chapter; nevertheless, the same architecture and technical capabilities could be applied to grid connected microgrids and communities. The grid infrastructures in remote or even isolated communities in countries such as Australia are usually maintained and operated by local Distribution Network Service Providers (DNSPs). Low demand, long distance from the national grid and road access issues make the extension of the centralized grid to these communities very expensive. With the falling costs of renewables and promising energy storage technological advancements, low carbon stand-alone power systems are becoming more cost competitive to the extent that networks and policymakers are considering the possibility of extending this capability to fringe of grid communities which have an inherent low reliability of supply or high maintenance cost. Stronger resiliency during natural disasters is another advantage of a totally decentralized grid. Without underestimating the efforts required to further advance and scale decentralized stand-alone grids, it is of strategic value to take them into consideration when developing the future grid architecture.

In summary, it is evident that first of all the transition to a fully decentralized (democratic) grid is not feasible in the near future and more likely various aspects of a decentralized grid architecture model will coexist with the incumbent grid model. A centralized national grid with smaller local decentralized grids in areas with very high penetration of DER would be a more practical scenario. The key point to consider is to ensure that regardless of the end state, the decision made and the platforms established during the transition period do not preclude the emergence of a decentralized model that enables a democratic grid with optimal outcomes for customers. Regardless of the end state, it is obvious that DNSPs will need additional capabilities to coordinate DER that entail improved network visibility and dynamic management of local constraints to maintain grid reliability. The next section summarizes the key required capabilities and a transition pathway from a DNSP model (asset manager) to a more modern distribution system operator (DSO) model (system coordination).

5 Distribution system operator

The way customers generate and consume electricity is changing. More customers install small variable generators at their own premises. Penetration of EVs is expected to grow significantly in the coming years as they become more cost competitive with internal combustion engine vehicles. A full charge of a typical EV requires a similar amount of energy as an average residential house consumes in a few days. EVs with vehicle to grid/building capability could export that amount of energy into the network. Consequently, LV networks require either significant augmentation to accommodate these changes or more active management of the energy flow and dynamic coordination between demand and generation at a local level. Major network augmentation comes with a significant risk of stranded assets (as customers become more self-sufficient) and upward pressure on electricity prices. The alternative is holistic integration of nonnetwork solutions (e.g., demand response) into planning and operation of the network. But traditional demand management and energy efficiency programs are usually passive programs designed to conserve energy during the peak with little incentive to customers.

Distribution networks incorporate a large number of assets (e.g., distribution transformers) and connection points with minimum visibility and limited real-time data from the field, to the extent that the existing wholesale market or transmission system operator (TSO) platforms cannot simply be extended to integrate millions of DER. This means either an independent DSO needs to be established or the distribution network utility acts as the DSO. The merits of each option are specific to each power system and out of the scope of this chapter. Instead, we focus on the key DSO functions and capabilities.

A key objective of the DSO is to optimally coordinate demand and supply on a localized basis based on local network constraints and maintain network reliability. Importantly, the DSO will provide a platform for energy suppliers and other parties to deliver new and improved services to customers within the physical limits of the local network. In areas with a high penetration of DER, optimal coordination of local demand and supply could reduce the reliance on the upstream transmission network and consequently minimize the variability and volatility impacts on the overall system. In addition to local flexibility services, DER can coordinate system wide services such as frequency control. Utilizing customer assets that have already been invested in reduces the overall system costs and also unlocks additional revenue for customers.

The transition to a DSO is a significant change and a key part of the grid transformation. Fig. 3.9 summarizes the key DSO enablers.

DER technology advancement: while growth in rooftop solar PV might have ignited grid transformation, high uptake of active DERs such as battery storage, EVs and HEMS are required before DSO services at scale become economically feasible.

Customer participation: customers who invest in DER are going to play a key role in the development of the DSO. Without active customer participation or enough incentive for customer engagement, the DSO will be unable to integrate nonnetwork solutions into the planning and operation process.

Digital transformation: emerging digital technologies enable secure and cost-effective interoperability between customer DER technologies and grid assets that is a fundamental capability for orchestrating a large number of load and generation resources.

Regulatory reform: DSO is a new business model that requires significant changes in tariff structures, commercial frameworks, new roles and responsibilities, and investment guidelines that support innovative solutions.

Fig. 3.9 Key DSO enablers.
Credit: Author generated.

While DSO functions could vary with grid architecture, the following DSO functions are common in the leading DSO initiatives worldwide and are considered no-regret investments. As summarized in Fig. 3.10, some of these functions are existing functions that need to be enhanced, and some are new functions.

Technical aspects of DSO functions are mostly common across different markets; however, regulatory aspects of new and enhanced DSO functions are region specific. The abovementioned key DSO functions are explained below.

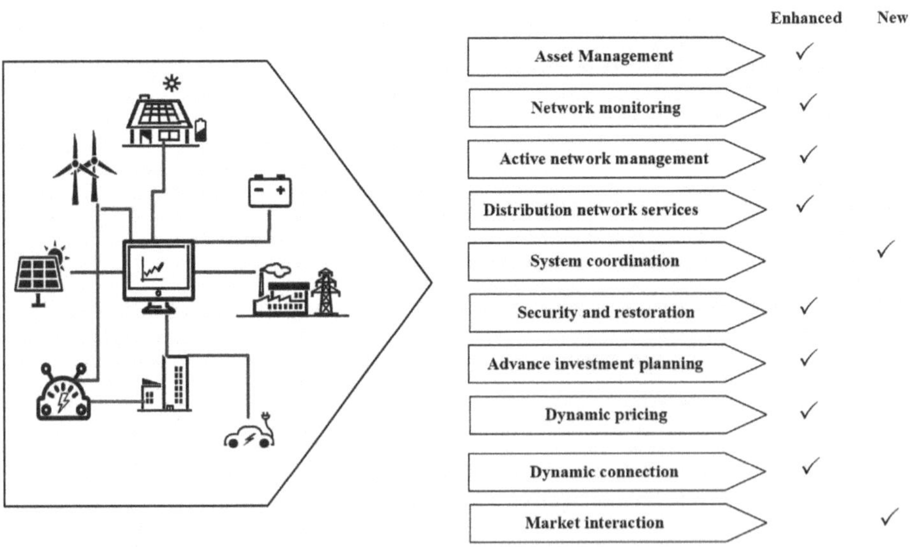

Fig. 3.10 Key DSO functions (new and enhanced).
Credit: Author generated.

Asset management: In cases where the DSO also owns the distribution network, asset management remains a key function. Planning, designing, and maintaining network assets form a significant part of the distribution utilities role. In future, availability of real-time asset monitoring and advanced data analytics can enhance asset life cycle management in distribution networks.

Network monitoring: Active coordination of large numbers of variable resources connected to the network requires comprehensive visibility of power flows across the network, together with the dynamic rating of network assets. In addition to monitoring access to network assets, the DSO would need some monitoring access to DER. There are different levels of monitoring and control access depending on the grid architecture and the governance model but monitoring access at the connection point is the common function.

Active network management: DER can participate either directly or through aggregators in different markets; however, energy transactions at any time should be within network limits. Safe, secure, and economic operation of the grid is one of the key responsibilities of the DSO. The DSO needs to leverage emerging technologies to manage the quality of supply as well as modern fit-for-purpose protection schemes for a high DER future. Direct load control is a common nonnetwork solution to manage network constraints (mainly during the peak hours). However, most of the current programs are either passive or not fit for the purpose with emerging DER technologies. Some transactive energy or market policy experts contend that appropriate price signals and market incentives can deliver the most cost-effective flexibility; however, based on control theory for large complex systems, there are certain situations where markets are not sufficiently effective to maintain reliable system operation [7]. An optimal mix of firm network and flexible nonnetwork resources will be required.

Distribution network services: The DSO will provide distribution network services such as voltage control and power quality through the optimal utilization of network and nonnetwork solutions. DER can provide flexibility services that assist the grid operator with generation intermittencies. While some of these services are in response to market signals and provided through aggregators, DER can also respond to local constraints/signals from the DSO.

System coordination: The DSO, as a layer between DER and the market, needs to set up and manage secure platforms to exchange data with the market operator and other relevant market participants (e.g., aggregators). In a high DER environment where DER could participate in the wholesale market, the DSO plays a key role in coordination between aggregators, transmission service providers, and market operators (either the wholesale market operator or the local distribution market operator).

Network security and restoration: The key responsibility of the DSO is to maintain the security of the network; hence, the DSO needs appropriate physical and cybersecurity standards as well as effective measures to restore power after major network contingencies or natural disasters.

Advanced investment planning: Dynamic calculation of local network DER hosting capacity as well as planning for adequate additional capacity is one of the key responsibilities of the DSO. Advanced planning tools along with data analytics enable the DSO to determine the optimal balance between network and nonnetwork solutions.

Dynamic pricing: Customers with and without DER use the grid in different ways. The DSO needs to provide cost reflective tariffs and fair access to all customers through efficient network investment to promote efficient bidirectional power flow across the network.

Dynamic connections: Customers with DER have more dynamic and sometimes complex load and generation connection requirements. The DSO needs to offer dynamic and fit-for-purpose network connections to customers that meet their individual needs within local network constraints.

Market interaction: The DSO is not a market operator, nor does it submit bids into the market; hence, it can provide market access platforms and services to customers and energy service providers without conflict of interest.

In order to deliver the aforementioned functions, the DSO needs to develop new and relevant capabilities. The minimum required capabilities depend on factors such as the penetration level of DER, type of DER, network topology, and the operating model. In order to avoid stranded assets or not having the functionality that network users need in place in a timely manner, it is important to have a long-term strategy aligned with the future grid architecture vision. It is important to ensure that all new DSO capabilities procured in the near term provide rich functionality, adhere to open standards, are extendable and flexible, and avoid vendor lock-in [19]. Below are three key capabilities regardless of the end state and market framework:

Network visibility: Visibility is one of the foundational capabilities that enables more sophisticated and active network management. However, traditionally, visibility across distribution networks is very poor/limited as SCADA and DMS systems are often limited to the HV and MV networks. Data from customers' smart meters is a great source for LV network visibility. The data from DER is another additional and important data source that can be procured through communication protocols such as IEEE 2030.5. The DSO needs flexible and cyber secure data analytics platforms to procure data from multiple sources and enhance network visibility by leveraging tools such as network state estimation for advanced monitoring and management of DER.

Network forecasting: The DSO needs accurate short-term and day ahead forecasts of the LV network with a variety of timeframes. Forecasts of individual customers are generally very difficult and only accurate with aggregation. Individual customer (behind the meter) forecasting would have more value for aggregators who manage customer DER.

Advanced data analytics: The DSO needs cyber secure and agile data analytics platforms to collect data from multiple sources and integrate with other platforms for forecasting, planning, and operation.

DER Management System (DERMS): Regardless of the DSO governance and market framework, the DSO needs to actively manage the network with high DER penetration. The control boundaries and contracts depend on the operating model, but DERMS platforms provide the required monitoring and control of DER for a wide range of value streams, functions, and use cases under a wide range of DSO operating models.

Advanced network modeling: The DSO needs advanced tools capable of using time series data to support efficient network planning, considering a range of network and nonnetwork solutions. The DSO needs to be able to model and calculate network DER hosting capacity for short time intervals to support dynamic DER connections. Calculating network dynamic operating envelopes enables better utilization of network assets and customers' DER.

6 Grid transformation in Australia

Australia has one of the highest concentrations of rooftop solar PV in the world, which accounts for about 24% of the total renewable energy generation in 2020. Fig. 3.11 shows the breakdown of annual renewable energy generation in 2020. While in the state of South Australia, 60.1% of the annual energy consumption was met by renewables, the average national energy consumption met by renewables was 27.7% in 2020 [20].

The growing uptake of rooftop solar PV, electrification of transport along with retirement of coal-fired power plants are the key drivers behind grid transformation in Australia in the next decade. Fig. 3.12 summarizes the existing and planned generation by fuel technology in Australia provided by Australian Energy Market Operator [21,22] that includes about 6 GW of coal-fired power plants already committed to retire by 2030.

Although more than 60% of the annual energy consumption was met by coal, 100% of electricity demand in South Australia came from solar for 1 h in October 2020. This is claimed to be the first time anywhere in the world for a region as large as South Australia. While the penetration of DER in Australia is rapidly growing, AEMO's projections indicate that the large-scale wind and solar generation capacity by 2040 will be double the rooftop solar PV generation capacity [23].

The growth in rooftop solar PV was originally attributed to the generous feed-in tariffs subsidized by state governments and increasing electricity prices. Even after the feed-in tariff subsidies were removed for new systems, affordable installation costs

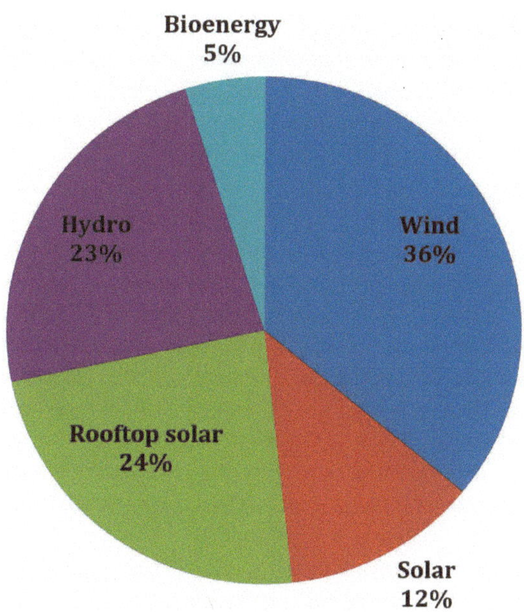

Fig. 3.11 Renewable generation by technology type in 2020.
Credit: Author generated from AEMO data.

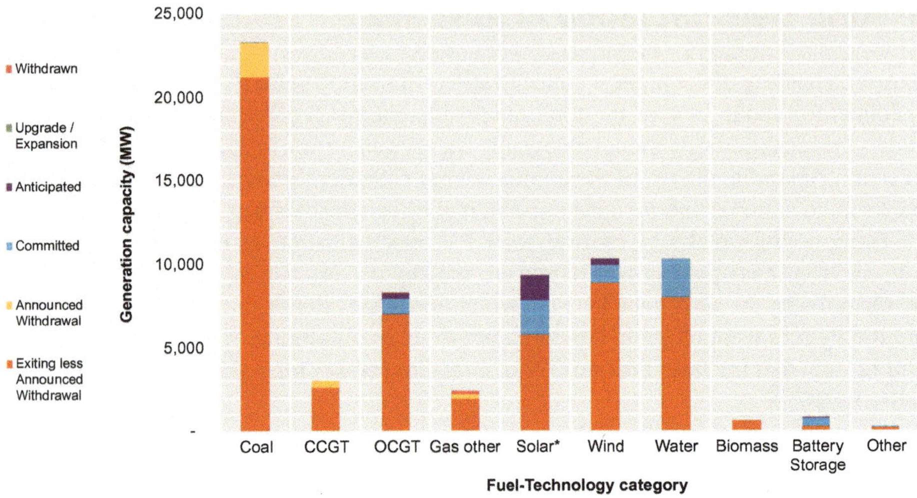

Fig. 3.12 Scheduled, semischeduled, and nonscheduled generation (MW)—existing and new developments by fuel technology category.
Credit: AEMO.

of reasonably sized rooftop solar PV contributed to the growing uptake in the residential sector in many states across Australia. The cost of residential rooftop solar PV systems decreased from 2.4 $/W in 2012 to 1.17 $/W in 2018 [24].

Unlike solar generation that is an attractive financial proposition, battery energy storage and home energy management systems are not economical for customers, because of the relatively high cost of battery energy storage systems coupled with flat electricity tariffs. This means there is not enough incentive for customers to store the surplus solar energy from their rooftop solar PV to offset their peak demand or even shift their load to the daytime. As the generation profile of solar is not matched to the underlying demand profile, the excess solar generation flows back to the grid. Fig. 3.13 shows that about 77% of minimum demand (sunny daytime during the off-peak season) would be met by rooftop solar PV by 2026. This means the minimum demand across the NEM (east coast of Australia excluding Tasmania) is expected to drop to a record low of 4–6 GW by 2025, down from 15 GW in 2019 [21,22].

In states with high penetration of rooftop solar PV, many distribution network feeders are already experiencing significant reverse power flow every day. Fig. 3.14 shows the load profile of a 500 kVA distribution transformer during the peak day in a residential area with over 600 kW of rooftop solar PV in Queensland.

Fig. 3.15 clearly shows the impact of growing excess rooftop solar PV on the load profile of a residential 33/11 kV substation over a period of 10 years in Queensland [25].

With the growing penetration of rooftop solar PV across the country, operational demand could reach zero in regions such as South Australia by 2023. This is believed to be the first GW scale grid in the world to approach zero operational demand due to the high penetration of unscheduled rooftop solar PV [26].

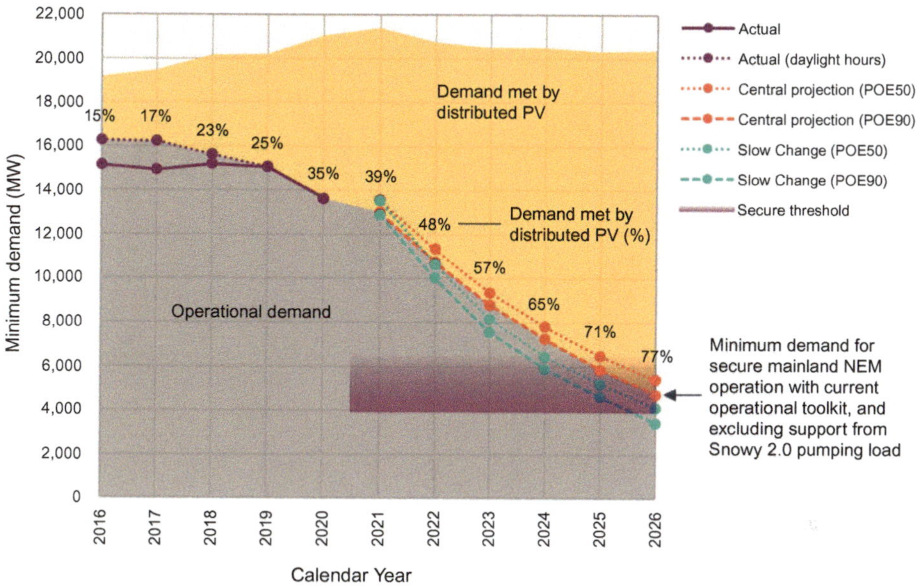

Fig. 3.13 Minimum demand on Australia's NEM (excluding Tasmania) [21,22]. Credit: AEMO.

Fig. 3.14 Peak load profile of an LV network with a very high penetration of rooftop PV. Credit: Author generated.

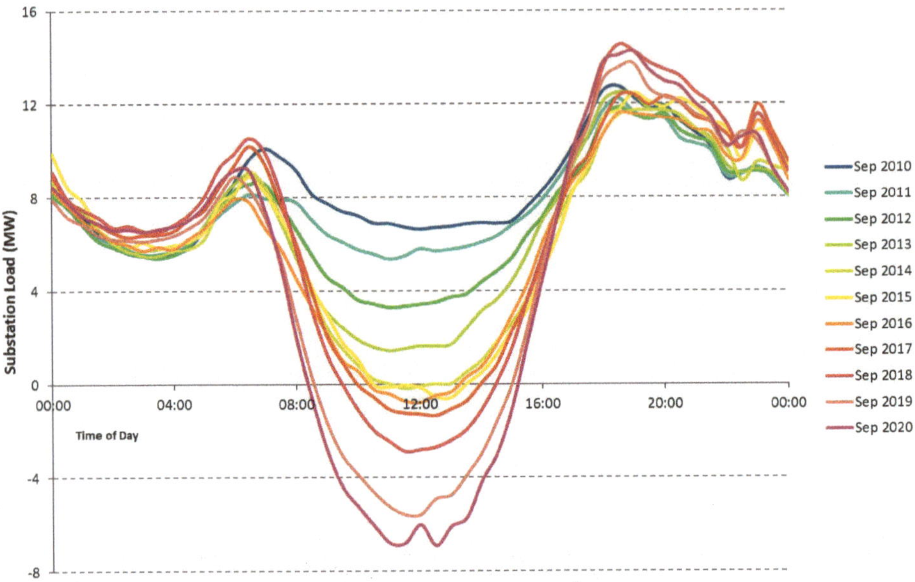

Fig. 3.15 Impact of growing uptake of rooftop solar PV on a 33/11 kV substation load profile (September 2010–20).
Credit: Author generated.

A large-scale trial project is set up to demonstrate the performance and verify the effectiveness of a centrally managed virtual power plant (VPP) to deliver benefits to households, energy retailers, and the local network. As part of the trial, 5 kW rooftop solar PV and 13.5 kWh battery energy storage systems will be installed at 3000 properties owned by Housing South Australia across Adelaide [27].

While there is a long-term vision for developing mechanisms for the daily operation of active two-way markets, engaging customers and unlocking value from DER assets, in the short term AEMO is proposing to shed generation from DER in abnormal system conditions, to maintain system security.

In anticipation of a very high penetration of DER, in 2016 Energy Networks Australia (ENA) partnered with Australia's national science agency, CSIRO, to develop a network transforming road map [17]. In 2020, ENA published findings of the Open Energy Network Project—mirroring the Open Network project in the United Kingdom—in which four frameworks for DER integration were reviewed [8].

The project explored distribution markets by developing new system architectures, including market designs and operational structures, then identified the minimum required capabilities that DNSPs would need regardless of the end state (either of the four proposed frameworks). These capabilities, referred to as least-regret action, were focused on improving network visibility, calculating network operating envelopes and protocols for communicating the operating envelopes [28].

All four frameworks entailed the creation of two new entities, called the DSO and the distribution market operator (DMO), and are defined as below [8]:

> *A Distribution System Operator (DSO), with visibility of power flows and DER on the network, will be required to manage the network within the technical constraints of the assets (otherwise known as "operating envelopes"), identify when network issues emerge and act to manage these issues. To do this, the DSO will need to see the flow of power across the distribution network in real-time. Where an issue on the network emerges, the DSO may obtain services to support the operation of the network from DER directly, or via aggregators, retailers and third parties and such services would be compensated. The DSO provides inputs to the DMO to ensure DER participation in markets does not compromise system security at the distribution level.*
>
> *DMO manages the distribution market, optimizing the provision of services and energy from DER within operating envelopes provided by the DSO. The DMO also provides information to AEMO to support the participation of DER in the wholesale market and ancillary service provision. At the distribution level, a DMO administers, operates, and manages platforms for aggregators, the DSO and AEMO to access flexibility services. The DMO might also administer, operate and manage platforms to support local market trading for energy and capacity.*

Details of the proposed frameworks together with the list of advantages and disadvantages can be found in the report [8]. All proposed models are based on a centralized grid architecture with variations on the integration of the distribution market operator. Some models simply suggest an extension of the current market platform to include the future distribution market, where others offer a separate market platform for distribution services.

The hybrid model, as shown in Fig. 3.16, is a conceptual framework that aims to ideally capture the merits of other proposed options. The cost benefits analysis findings suggest that any of the framework models could deliver positive net benefits under a very high DER uptake scenario. The challenge remains in defining the role and responsibilities of the market participants and the pace of adoption of DER (EVs and battery storage in addition to rooftop PV) by customers.

A number of projects were initiated to trial some of these frameworks and verify the effectiveness and complexity of calculating the dynamic operating envelopes and communicating to aggregators to ensure distribution market activities do not violate network constraints [29–32]. One common element of these projects is the development of dynamic operating envelopes (DOEs) and building interfaces to communicate these envelopes to DER either directly or via an aggregator. Operating envelopes basically refer to the maximum power limits for import and export at a connection point for a period of time that are issued by the DSO based on the available network capacity. Traditionally, DNSPs calculate a fixed limit for all customers connected to a transformer. This value is determined based on the rated capacity of the transformer, number of customers, diversity factor of the load and generation, and some rudimentary network modeling. On the other hand, the operating envelope concept determines the limit based on the available capacity at any point in time. And because of the dynamic nature of the load, the limit for each connection point changes over time.

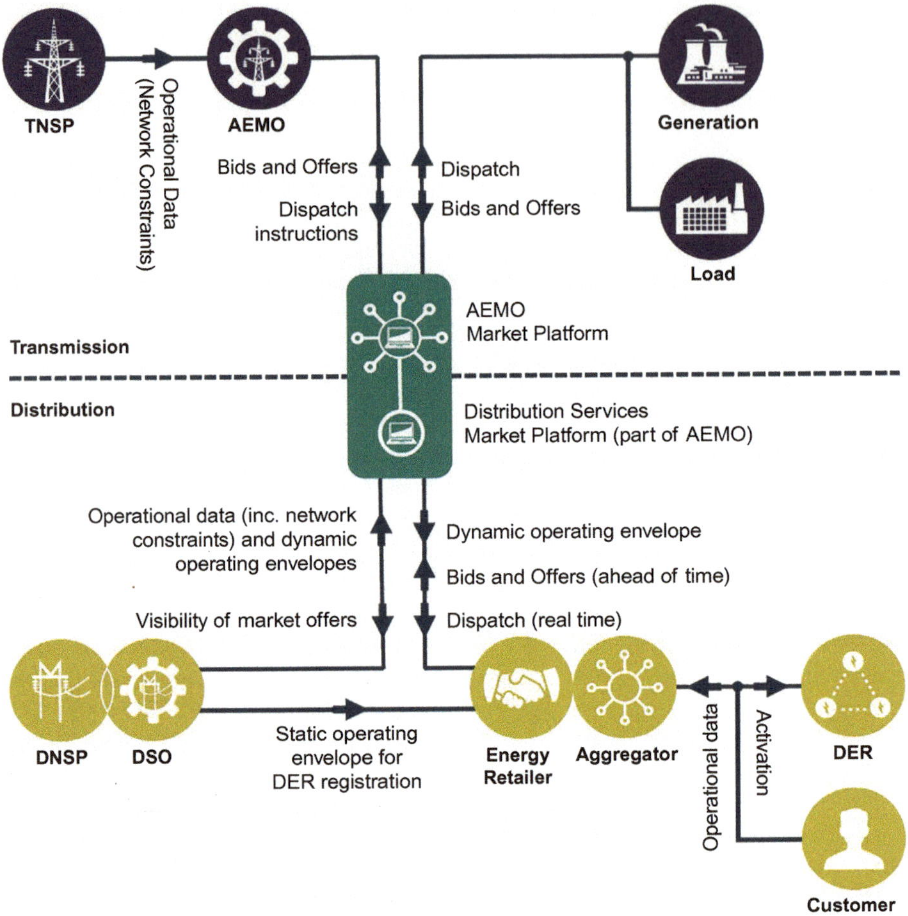

Fig. 3.16 A conceptual technical framework for DER integration—hybrid model. Credit: ENA.

Hence, these limits are also called Dynamic Operating Envelopes (DOEs). Fig. 3.17 illustrates the concept of DOE on a very simplified single line diagram of an LV feeder.

Obviously, more accurate and frequent DOEs (e.g., every 5 min) require more data from the field including real-time data. Calculating more accurate DOE would be more complex and require more data and deeper visibility. One of the parameters that could determine the length of the intervals of DOE is the type of service that DER is subscribed to. More details on DOEs can be found in [33], where the authors provide foundational insights on how DOEs are calculated and used to facilitate the provision flexibility and grid services.

Currently, all DER connections in Australia are based on a static limit; however, DNSPs are working toward the development of dynamic DER connections based on

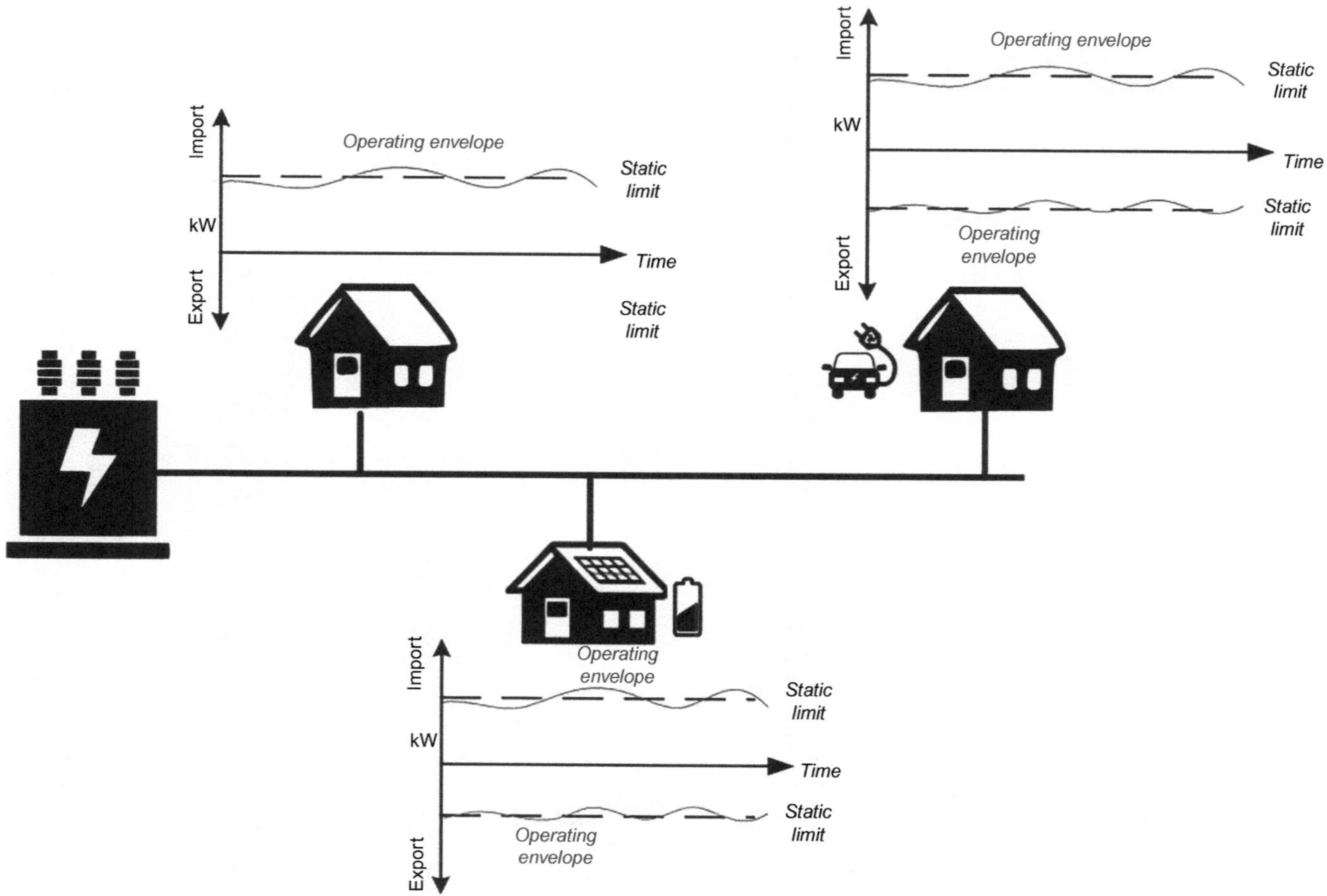

Fig. 3.17 Network dynamic operating envelopes (conceptual single line diagram). Credit: Author generated.

the DOE concept that supports higher penetration of DER. In addition to methods for calculating DOE, there are a number of other challenging technical issues that need to be solved such as a standard protocol to interface with the aggregator platforms, cyber-security, capacity allocation to DER, and generation of operational short-term forecasts for each LV network before implementation of dynamic DER connections.

In addition to developing new technical capabilities, regulatory frameworks need to be changed. The Australian Energy Market Commission (AEMC) also proposed a move to a two-sided market to streamline market rules and suggested replacing the existing diverse arrangements around retailers, generators, aggregators with two simple categories: those who use electricity and those who sell it on behalf of end users [34]. To meet the needs of the market transition up to and beyond 2025, the Energy Security Board (ESB) was tasked to develop advice on reforms to the National Electricity Market (NEM). Over a period of 2 years, the project team reviewed four key areas and set out a reform pathway based on the urgency of the identified issues in generation adequacy, transmission planning, essential services to maintain system security and DER integration [35]. Introducing technical standards for DER and development of emergency backstop measures to shut down rooftop PV were identified as immediate measures. The immediate reforms were aimed at addressing system security issues while broader reforms are planned in the near future. ESB developed a DER Implementation Plan to integrate the evolution of the roles and responsibilities of market participants into a suite of technical, market, and regulatory reforms leading to 2025.

While details of the role and responsibilities of market participants are yet to be defined, the ESB proposed that DNSPs assume the responsibilities of the DSO that include:

Publication of DOE and capacity allocation.

- Develop dynamic network tariffs to support automated response from DER and flexible loads.
- DER and load shedding under the direction of AEMO to maintain network security.

Development and implementation of DOE are no-regret technical capabilities that support the growing uptake of rooftop solar PV in the short term and in the future will be used by DSO to communicate local network constraints to the aggregators. This is expected to support higher penetration of DER connected to the distribution network without significant network augmentations to increase network capacity. It could also contribute toward maintaining system security during minimum operational demand periods. DNSPs with a high penetration of DER have already started trialing calculation of DOE with the target of deployment to all new DER within the next 1–2 years. Meanwhile, the growing uptake of rooftop solar PV and lack of flexible load during the daytime pose serious system security concerns for AEMO as the likelihood of minimum operational demand events is increasing [36].

Deployment of DOE is a fundamental capability for DSO; however, the solution to the minimum demand issue is strong incentives for customers to participate in demand response programs and leverage the flexibility services that DER can offer. Electrification of transport could introduce a significant amount of new and potentially flexible load that, if coordinated, could be used to address some of the emerging challenges. Immediate technical solutions need to be deployed to maintain the security and reliability of the grid, while the regulatory reform process facilitates issues such as the procurement of demand response from the market, enabling DER market participation, and ultimately establishing a two-sided market.

More than 50 years ago, a hot water load control program was designed to balance the cheap, abundant generation from coal-fired power stations in Queensland. The program allowed DNSPs to improve the utilization of the cheap base load power generation. Simplicity is the key feature of this successful program, which is still in operation. The program is using ripple control (Audio Frequency Load Control) to switch on and off the residential hot water systems linked to a discounted tariff. While the program was originally designed to shift the load to off-peak hours (10 p.m.–7 a. m.), it has been trialed to soak excess solar generation to manage voltage rise issues due to reverse power flow at the distribution network level as well as respond to system security issues during contingencies.

The growing uptake of residential battery storage and EVs driven by battery technological advancements along with regulatory reforms are expected to enable compelling business cases for aggregators to offer distribution network services to DSO. In the meantime, leveraging inherent network capabilities is the most cost-effective measure to address some of the emerging challenges.

Grid connected batteries (including community batteries) are another source of flexibility that can be charged during low demand and high solar export periods and discharged during peak hours. Grid connected batteries could also be used for energy arbitrage by charging when wholesale prices are negative and discharging during peak hours. Currently, DNSPs in Australia are not allowed to own and operate generation units (except a mobile generator for planned outages or after natural disasters). With possible changes to the ring-fencing guidelines, DNSPs would be able to deploy grid-connected batteries in areas with high penetration of rooftop solar PV to address local network issues, avoid network augmentation, and potentially use batteries to respond to system events. Also, DNSPs could potentially unlock more value streams through partnerships with retailers and trading the excess capacity in wholesale power and frequency control ancillary services (FCAS). Trial projects such as United Energy's pole-mounted batteries in Victoria are designed to demonstrate the values of operating 40 custom-built LV batteries (each at least 30 kW/66 kWh) as a virtual power plant (VPP) to provide demand management services and increase hosting capacity for rooftop solar PV. United Energy, the DNSP which owns and installs the batteries, has signed a deal with a retailer for the market trading rights of the VPP [29–31].

Western Power, a transmission and distribution company in Western Australia, included community batteries as part of their DER road map to address emerging technical issues in areas with high uptake of rooftop solar PV and constrained LV networks. In partnership with a retailer, the program allows customers to access a shared storage resource as an alternative to investing in their own battery, which reduces the up-front costs for households, while unlocking additional value to the network and potentially within the system [37].

Insights from these projects along with regulatory reforms could unlock new business models and enable more innovative solutions for integration of community battery energy storages in Australia. Community storage could play a significant role in building resilient and net-zero emission communities in the future.

In summary, Australia is one of the leading countries that have already embarked on the grid transformation journey driven by a shift from coal-fired power stations to renewables and high penetration of decentralized generation (DER) being embedded within the distribution networks. Customers with DER are empowered with greater choice, control, and autonomy while continuing to benefit from a secure and reliable grid connection. The enablement of DER integration in the local and wholesale energy markets means DER customers will be able to extract even more value from their investment. Without a pragmatic customer-centric approach, there is a risk that vulnerable customers or those who cannot install DER would be left behind. The rapid and high uptake of rooftop solar PV in Australia so far has shown that the paradigm shift in the operation of the distribution network can occur very quickly and the industry needs to develop new and innovative solutions to support a cost-effective transition.

During the early stages of grid transformation, not all customers can afford to install DER or own an EV. Hence, it is vital to develop holistic and pragmatic solutions and design an ecosystem where DER benefits are offered to all customers (even those without DER) and customers pay their fair share of using the network. The key elements of such a road map are pragmatic regulatory and tariff reforms, leveraging emerging technologies, taking the customer along on the grid transformation journey, and cross-industry collaboration.

References

[1] H. Ritchie, M. Roser, Renewable Energy, 2020, Retrieved 13 August 2021, from: *https://ourworldindata.org/energy*.

[2] IREA, 2018 Renewable Power Generation Costs in 2017, International Renewable Energy Agency, Abu Dhabi, United Arab Emirates, 2018.

[3] World Economic Forum, Fostering Effective Energy Transition—Insight Report April 2021, World Economic Forum, 2021. Retrieved from: *http://www3.weforum.org/docs/WEF_Fostering_Effective_Energy_Transition_2021.pdf*.

[4] IEA, World Electricity Generation Mix by Fuel, 1971–2019, IEA, Paris, 2021. Retrieved 13 August 2021, from: *https://www.iea.org/data-and-statistics/charts/world-electricity-generation-mix-by-fuel-1971-2019*.

[5] OECD/IEA and IRENA, [Executive Summary/Chapter [1/4]] of Perspectives for the Energy Transition—Investment Needs for a Low-Carbon Energy System, IEA and IRENA, 2017. Retrieved from: *https://www.irena.org/-/media/Files/IRENA/Agency/Publication/2017/Mar/Perspectives_for_the_Energy_Trans*.

[6] TEC, Designing DERtopia—How Do We Get to a Decentralised, Democtratic Grid From Here, Total Environment Centre for the Future, 2021.

[7] L. Kristov, P.D. Martini, J. Taft, A tale of two visions: designing a decentralized transactive electric system, IEEE Power Energy Mag. 14 (3) (2016) 63–69, https://doi.org/10.1109/MPE.2016.2524964.

[8] ENA, Open Energy Networks Project—Position Paper, Energy Network Australia, 2020. Retrieved 15 August 2021, from: *https://www.energynetworks.com.au/projects/open-energy-networks/*.
[9] Advisian, Australian Hydrogen Market Study—Sector Analysis Summary, Advisian Pty Ltd, 2021. *https://www.cefc.com.au/media/nhnhwlxu/australian-hydrogen-market-study.pdf*. (Accessed 5 January 2022).
[10] S. Henbest, 2017. *https://about.bnef.com/blog/henbest-energy-2040-faster-shift-clean-dynamic-distributed/*. Bloomberg New Energy FInance.
[11] ESB, Post 2025 Market Design Options—A Paper for Consultation, Energy Security Board, 2021. Retrieved from: *https://esb-post2025-market-design.aemc.gov.au/*.
[12] D.B. Simon Funcke, Typology of centralised and decentralised visions for electricity infrastructure, Util. Policy 40 (2016) 67–74.
[13] G. Fitzgerald, J. Mandel, J. Morris, H. Touti, The Economics of Battery Energy Storage, Rocky Mountain Institute, 2015.
[14] ITP, Business Models and Regulatory Considerations for Storage on the Distribution Network (for the ESB), Project No. A0350, ITP, 2020. Retrieved from: *https://itpau.com.au/publications/*.
[15] The Mckell Institute, Power to the People: Proposals to Increase the Rollout of Community Batteries, 2021, Retrieved from: *https://mckellinstitute.org.au/wp-content/uploads/Power-to-the-people.pdf-new.pdf*.
[16] N. Zhang, H. Jiang, Y. Li, P. Yong, M. Li, H. Zhu, S. Ci, C. Kang, Aggregating distributed energy storage, IEEE Power Energy Mag. 19 (4) (2021) 63–73.
[17] ENA, Electricity Network Transformation Roadmap (ENTR)—Final Report, Energy Network Australia, 2017. Retrieved 16 August 2021, from: *https://www.energynetworks.com.au/projects/electricity-network-transformation-roadmap/*.
[18] SAPN, Advanced Virtual Power Plant Grid Integration Trial, South Australia Power Networks, 2021. Retrieved 16 August 2021, from: *https://www.sapowernetworks.com.au/future-energy/projects-and-trials/advanced-virtual-power-plant-grid-integration-trial/*.
[19] B. Currie, Distributed Resource Energy Management Systems (DERMS)-White Paper, Smarter Grid Solutions, 2019. Retrieved from: *https://info.smartergridsolutions.com/whitepapers*.
[20] CEC, Clean Energy Australia 2021, Clean Energy Council (CEC), 2021. Retrieved, 16 August 2021, from: *https://assets.cleanenergycouncil.org.au/documents/resources/reports/clean-energy-australia/clean-energy-australia-report-2021.pdf*.
[21] AEMO, 2021 Electricity Statement of Opportunities, Australian Energy Market Operator, 2021. Retrieved from: *https://aemo.com.au/-/media/files/electricity/nem/planning_and_forecasting/nem_esoo/2021/2021-nem-esoo.pdf?la=en&hash=D53ED10E2E0D452C79F97812BDD926ED*.
[22] AEMO, Generation Information, Australian Energy Market Operator, 2021. Retrieved 16 August 2021, from: *https://aemo.com.au/en/energy-systems/electricity/national-electricity-market-nem/nem-forecasting-and-planning/forecasting-and-planning-data/generation-information*.
[23] AEMO, 2020 ISP Appendix 4 Energy Outlook, Australian Energy Market Operator, 2020. Retrieved 16 August 2021, from: *https://www.aemo.com.au/-/media/files/major-publications/isp/2020/appendix--4.pdf*.
[24] Solar Choice, Solar Choice, Residential Solar PV Price Index, 2012–2018, 2018, Retrieved 16 August 2021, from: *https://solarchoice.net.au/solar-power-system-prices*.

[25] Energex, Distribution Annual Planning Report (DAPR), Energex, Brisbane, 2020. Retrieved 16 August 2021, from: *https://www.energex.com.au/__data/assets/pdf_file/0008/867716/Distribution-Annual-Planning-Report-2020.pdf.*

[26] AEMO, Minimum Operational Demand Threshold in South Australia—Technical Report, Australian Energy Market Operator, 2020.

[27] ARENA, Australia's Largest Virtual Power Plant Ramps Up in South Australia, Australian Renewable Energy Agency (ARENA), 2020. Retrieved from: *https://arena.gov.au/news/australias-largest-virtual-power-plant-ramps-up-in-south-australia/.*

[28] ENA, Open Energy Networks-Interim Report, Required Capabilities and Recommended Actions, Energy Network Australia, 2019.

[29] ARENA, Victorian Distributed Energy Resources Marketplace Trial – Project EDGE (Energy Demand and Generation Exchange), Australian Renewable Energy Agency (ARENA), 2020. Retrieved 16 August 2021, from: *https://arena.gov.au/projects/victorian-distributed-energy-resources-marketplace-trial/.*

[30] ARENA, Evolve DER Project, Australian Renewable Energy Agency (ARENA), 2019. Retrieved 16 August 2021, from: *https://arena.gov.au/projects/evolve-der-project/.*

[31] ARENA, United Energy Low Voltage Battery Trial, Australia Renewable Energy Agency (ARENA), 2021. Retrieved 31 August 2021, from: *https://arena.gov.au/projects/united-energy-low-voltage-battery-trial/.*

[32] Government of Western Australia, Virtual Power Plants to Become a Reality in WA-First, 2021, Retrieved 31 August 2021, from: *https://www.mediastatements.wa.gov.au/Pages/McGowan/2021/02/Virtual-Power-Plants-to-become-a-reality-in-WA-first.aspx.*

[33] Z.M. Liu, L. Ochoa, S. Riaz, P. Mancarella, T. Ting, J. San, J. Theunissen, Grid and market services from the edge, IEEE Power Energy Mag. 19 (4) (2021) 52–62.

[34] AEMC, Energy Security Board's Two-Sided Market Paper, Australian Energy Market Commission, 2021. Retrieved 16 August 2021, from: *https://www.aemc.gov.au/news-centre/media-releases/consultation-open-energy-security-boards-two-sided-market-paper.*

[35] ESB, Post-2025 Market Design Final Advice to Energy Ministers—Part A, Energy Security Board, 2021. Retrieved 31 August 2021, from: *https://energyministers.gov.au/energy-security-board/post-2025.*

[36] Australian Energy Market Operator, Solar PV Curtailment Initiative by SA Government Supports the NEM, 2021, Retrieved 31 August 2021, from: *https://aemo.com.au/en/newsroom/media-release/solar-pv-curtailment-initiative-by-sa-government-supports-the-nem.*

[37] J.C. David Malcolm, Distributed Energy Resources Roadmap, Energy Transformation Taskforce, Perth, 2019. Retrieved from: *https://www.brighterenergyfuture.wa.gov.au/wp-content/uploads/2020/10/DER-Roadmap_April2020.pdf.*

Multidimensional method for assessing nonwires alternatives within distribution system planning

4

Davis Montenegro and Jason Taylor
Power Delivery and Utilization, EPRI, Knoxville, TN, United States

Abbreviations

ADAPT	automated distribution assessment planning tools
aOc	alternative occurrence
DBT	Doc. brown tree
DER	distributed energy resources
DG	distributed generation
ES	energy storage
NWA	nonwires alternatives
Oc	occurrence matrix
PV	photovoltaic panel array
QSTS	quasistatic-time-series
SoC	state of charge
tOc	time occurrence

1 Introduction

Planning studies are an essential component for ensuring cost-effective and timely distribution system upgrades and expansion. Traditionally, planning studies are based on load flow simulations representing the instantaneous peak demand projected for 1–5 years into the future. As the traditional alternatives were passive in nature and tended to provide a significant increase in capacity, short-range planning studies generally only considered a fixed planning horizon [1]. In some cases, the resulting expansion plan may be evaluated against long-range planning horizons; however, there was no need to evaluate whether a decision would be viable or cost-effective considering the years immediately following its implementation.

A significant change occurring across the industry is an increased focus on the emerging technologies that may be able to offset the need for expensive system upgrades and help meet other system objectives. These nontraditional solutions, referred to as nonwires alternatives (NWAs) [2], while not necessarily altering the

Decentralized Frameworks for Future Power Systems. https://doi.org/10.1016/B978-0-323-91698-1.00007-8

overall planning process, are expected to increase the complexity of the simulations and studies needed [3]. As most NWAs are active in nature—due to the reliance upon variable or constrained energy sources as well as advanced controls—they intrinsically introduce new dynamic characteristics that need to be accounted for within distribution planning studies [4]. In the context of decentralized frameworks, NWA comprehends distributed energy resources (DER), which require an adequate coordination for not only supplying energy to customers but also to maintain or improve the security, reliability, quality, and availability of the grid. These devices need to be coordinated considering their deployment across the power system, in many cases operating in a decentralized and autonomous fashion.

NWAs introduce dynamics that influence the planning time horizon deferring the need for subsequent upgrades for a year or more. In other cases, NWAs are modular in nature and can be incrementally added to match the system needs at a potentially lower capital cost after the initial investment.

The operational life span of NWA technologies tends to be shorter than most traditional solutions such as reconductoring and should be appropriately accounted for in the calculation of costs, if not directly through load-flow simulations, and other assessment techniques. The identification and evaluation of multiple potential future alternative deployment options, occurring in different years and comprising different alternative types, is necessary to fully assess and compare NWAs to traditional solutions.

For harmonizing new and existing technologies, power system planning tools and techniques require to consider the time-changing features of distribution systems. Several efforts have been conducted in this area such as the use of linear programming [5,6], AC optimal power flow, metaheuristics [7], and multiobjective programming just to name a few [8]. However, these techniques cover a limited spectrum of possibilities when integrating NWA into planning studies, particularly when considering the NWA dynamics and combinations across a given planning horizon [8].

This chapter introduces a multidimensional planning technique for NWA integration in power systems. With this technique, planning studies can integrate new technologies to visualize their effect on the current and future distribution systems. The multiple dimensions generated correspond to viable technical solutions, which can be sorted according to their economic cost to obtain the best investment plan for the years ahead. The analysis proposed considers the variability in time of NWAs, the dynamic load growth, and the harmonization of NWAs with traditional alternatives for relieving thermal and voltage violations.

The content is organized as follows: first, this chapter introduces the features of NWA and why they need to be addressed in planning studies. Second, an introduction to the multidimensional planning technique is presented. Then a case study and results are discussed to finalize with the conclusions.

2 Nonwires alternatives

NWAs are progressively being considered by distribution and transmission planners for economically addressing acknowledged grid limitations. Gaining popularity due to technological advancements, falling costs, and supportive regulatory orders, regulators in several jurisdictions are now mandating that utilities evaluate the feasibility of DER-based alternatives prior to conducting any major grid reinforcements [2].

But while NWAs present opportunities for distribution and transmission planners, they have a significant impact on the complexity of the planning process and pose new challenges to associated analytics, tools, and business processes [9].

Traditionally, utility electric power systems were not designed to accommodate NWAs such as distributed generation (DG) and energy storage (ES) at distribution level. Adding DG and energy storage among other NWAs changes the distribution system response in time as well as the operation of the control devices across it [10].

A major concern about integrating NWAs is due to their time variability. Compared to their predecessors, NWAs are not suitable for energy generation constantly since they depend on variable weather conditions and other specific features of the technology used.

A good example of this are photovoltaic panel arrays (PV), which are highly dependent on the amount of sunlight and temperature they are exposed to. Fig. 4.1 presents three different irradiance profiles for three different days. These represent the sunlight incidence on clear, variable, and overcast days for the daily operation of PV. As can be seen, not all days have the same irradiance, resulting in a variable energy production from PV.

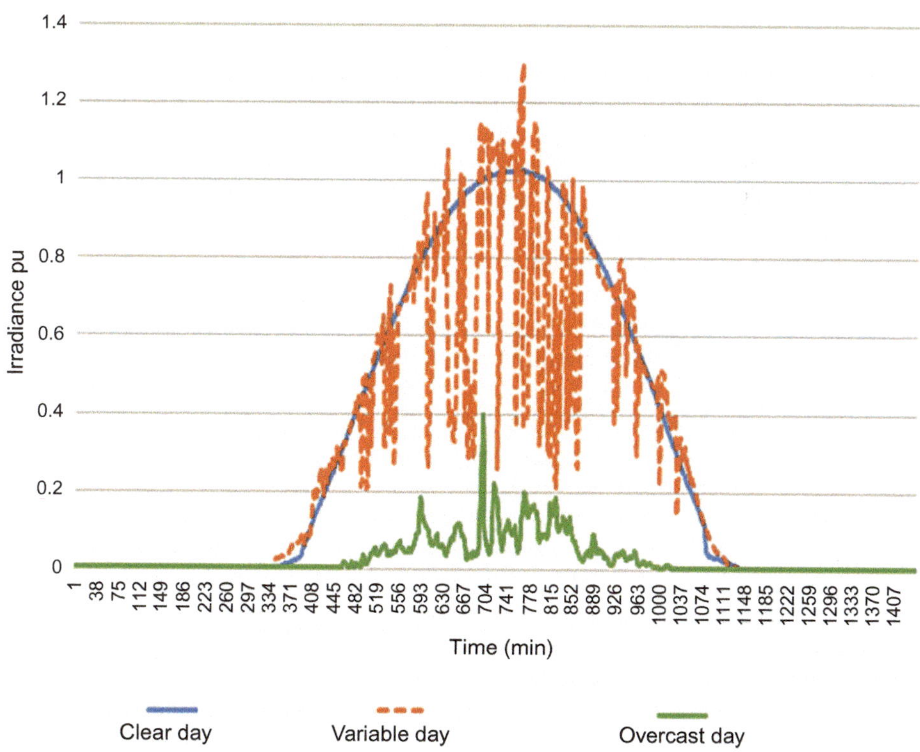

Fig. 4.1 Solar irradiance profiles for PV.
Credit: Authors.

Similarly, ES is an element that changes its electrical behavior depending on its state of charge (SoC). When the SoC is within an acceptable range, ES will act as DG if required.

But the energy stored in the ES is not unlimited, requiring ES to recharge contributing to increasing the power system loading during the charging period.

Analogous to the cases mentioned above, other NWAs variability will affect the traditional dynamic of the power distribution system at different points in time, requiring a deeper planning study to verify their effects in time.

Quasistatic-time-series (QSTS) simulation is a tool used for assessing the system response in time and has found wide acceptance across the industry. Furthermore, it can serve as the basis for designing and assessing the technical feasibility of proposed NWA implementations [11,12], advancing traditional planning studies to consider emerging technologies for reinforcing the distribution system when required.

QSTS simulations propose an accurate framework for evaluating the influence of NWAs in the distribution network, facilitating their incorporation when assessing planning problems. The NWA's variability is a feature that has not been considered and therefore accurately evaluated in traditional planning studies [13,14].

3 Multidimensional planning

Bringing together the technologies mentioned above and applying them in distribution planning is a challenge. This section describes the methodology used at EPRI for bringing these technologies together and for applying the final tool in real planning studies.

A planning horizon encompasses several years in which the power system loading varies representing new customers and the grid expansion. In many cases, this growth triggers thermal and voltage violations in time that need to be addressed year by year [13].

Conventionally, to mitigate these violations, distribution planners propose equipment upgrades and the addition/adjustment of passive and regulating devices across the power system. These tasks can be performed in one or several years ahead depending on the utility's investing interests. In the planning study, it is expected that the combination of different technical alternatives in time will reveal the best investment criteria for the years to come, which is validated through an economic analysis [15].

Adding NWAs to the equation increases the number of possible combinations possible for the planning study, highlighting the importance of a method to identify and map these scenarios efficiently during the study.

The method proposed in this paper is a graphical framework for describing the relationship, progression, and dependencies between the power distribution system state through time and the alternatives needed to solve the thermal and voltage violations found during the analysis period [16].

The proposed method, which the authors called the Doc. Brown Tree (DBT), agrees with all the characteristics of a directed graph. The cumulative nature of the deployed alternatives can be defined within the graph as the dependency between edges. This means that for evaluating an edge all its predecessors need to be evaluated first [17,18]. For example, if a graph S contains edges A, B, C, D as follows:

$$S=(A,B,C,D) \tag{4.1}$$

The transitive relations between the edges (link branches) are defined as

$$R=((A,B),(A,C),(B,D)) \tag{4.2}$$

It can be inferred that A depends on C and B; consequently, B needs to be calculated before A. C and D are given. This way, the dependency graph used to describe this case will be given by

$$G=(S,T) \tag{4.3}$$

where T is the transitive reduction of R. This definition is intended to describe the relationship between an edge and its predecessors when evaluating the possible solutions for a system violation in a year under study.

To explain this concept, let us assume a time horizon of 10 years for a planning study. The system is assessed on a yearly basis and any identified impacts are sufficiently mitigated using the correct alternative. These investments are assumed to be deployed in the system in the year associated with the identified need and are cumulative over the study period. An alternative implemented in the first year will remain in place in the analysis of all the subsequent years in the same timeline.

To create the reference timeline, the analysis will start at year one and try to solve all the violations within the 10-year period using any combination of methods. The goal is to create a timeline in which all the system needs have been identified and successfully mitigated. This will create a reference that can be described as a graph, where the edges are the years in which a violation was solved, and the link branches are the method used to solve the violation at the origin edge (year). The years where no violation was reported will not be included in the graph.

The new reference timeline can be used to return to years in which a relieving action was required and explore a different alternative, which will create a new branch in the graph located at the same edge and result in a new possible future path to year 10. All the new combinations coming at the critical points on each deployment path create a directed graph as illustrated in Fig. 4.2. The graph shown in Fig. 4.2 is the DBT mentioned above.

Graphing the timelines in this fashion provides the ability to easily track and return to critical points where additional alternatives may be considered, providing consistency between other deployment scenarios sharing a common set of historic system changes.

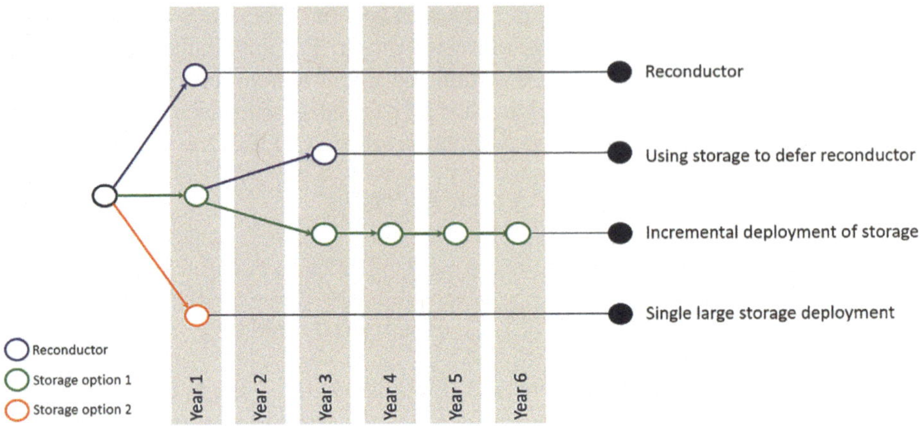

Fig. 4.2 Example of DBT.
Credit: Authors.

4 Case study

The DBT method is implemented within the automated distribution assessment planning tools (ADAPT), which is currently developed by EPRI for proposing the next generation of planning practices and tools. This tool uses OpenDSS for simulation purposes [19] and the DBT for navigating through the different timelines created during the study.

Consider the feeder shown in Fig. 4.3. The model of this feeder contains the forecasted load growth for 9 years, which is provided by the utility, as well as all the controls already deployed across the circuit such as regulators, capacitor banks, and existing DR. For the study, the utility also provided the seasonal ratings (winter and summer) for the lines.

After prescreening the feeder for the 9-year period, the thermal violations have been identified as shown in Fig. 4.4. As can be seen in Fig. 4.4, the feeder is a winter peaking feeder, revealing that most of the violations take place during the winter period.

The feeder has a peak demand of 4.57 MW and 1.3696 Mvar in year 1. The features and the alternatives to be used for the study are presented in Table 4.1. For this example, only four alternatives will be used: two traditional (reconductor and load transfer) and two NWAs (ES and PV).

The use of ES for relieving violations is limited to two sizes available for the utility. The number of sizes and the features of ES can be modified as needed for the study.

For adding PV during the study, a maximum limit of PV penetration has been established. This limit is expected to reflect the feeder's hosting capacity study performed previously.

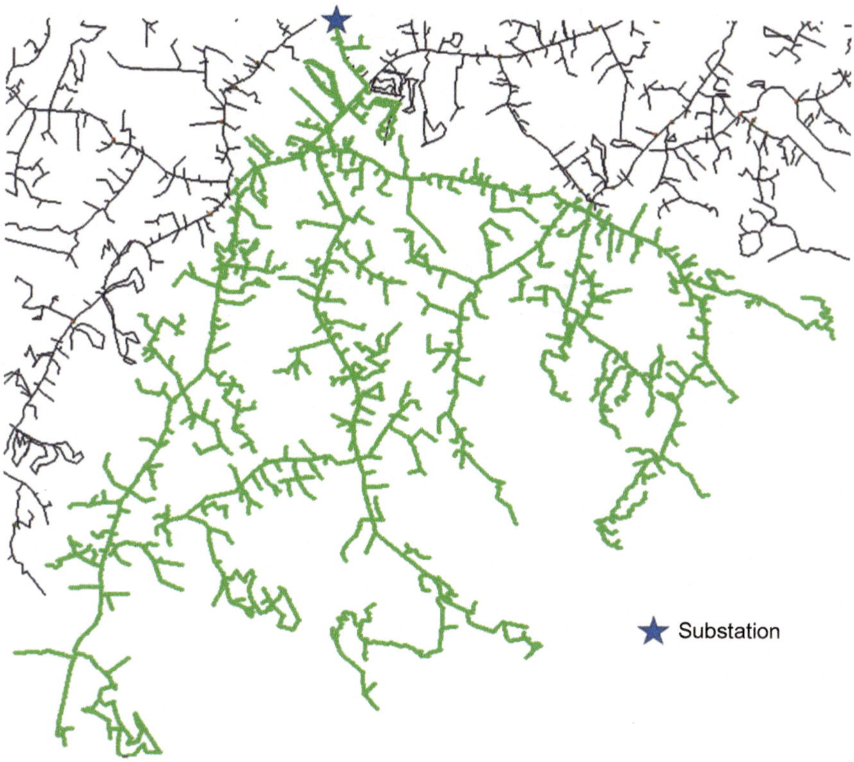

Fig. 4.3 Feeder under study (highlighted in *green*).
Credit: Authors.

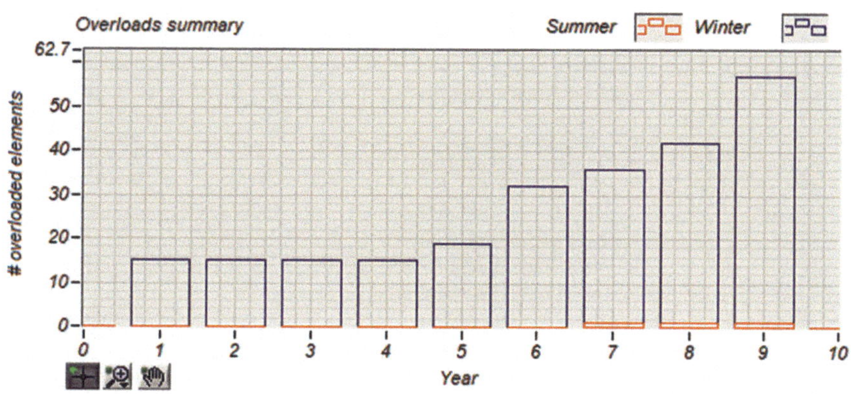

Fig. 4.4 Prescreening results for the proposed feeder.
Credit: Authors.

Table 4.1 Alternative features for the study.

Alternative	Feature name	Size
Reconductor	New ampacity	628 A @ summer 924 A @ Winter
Load transfer		1.4 MW
ES	Sizes	2 MW @ 4 MWh 2 MW @ 8 MWh
PV	Max allocation per year	4 MW

4.1 Study results

After conducting the study, a total of 229 different valid technical combinations were found, and an extract of the results report is shown in Fig. 4.5. In Fig. 4.5, the alternatives colored in gray are the ones that failed to solve the proposed violation for the current year. The other colors highlight alternatives that successfully solved the problem during the year and allowed the analysis to move into the subsequent years.

The effectiveness of the alternatives used, given the feeder features, can also be obtained by navigating the DBT.

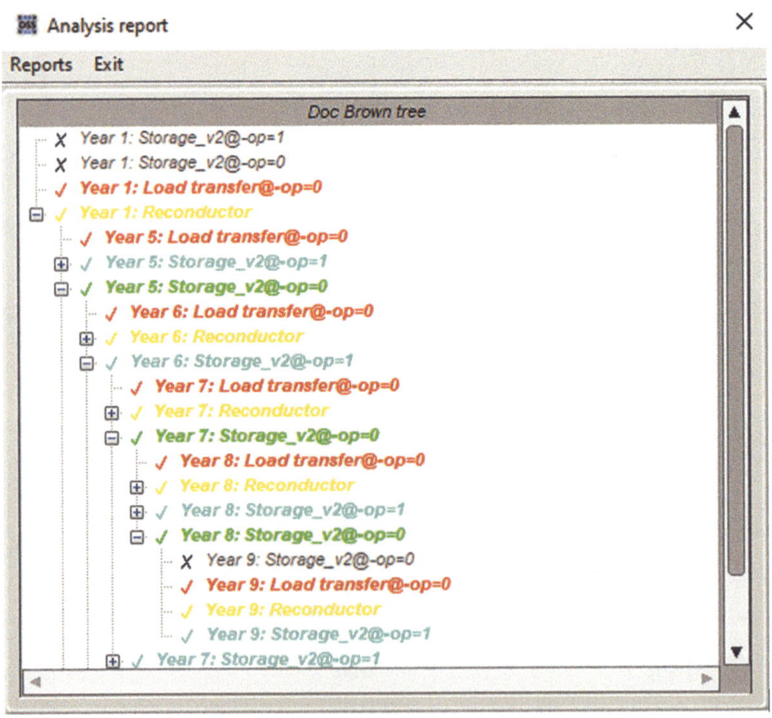

Fig. 4.5 Analysis results for the case study.
Credit: Authors.

For the case presented in this chapter, the load transfer solves all the violations for the year in which it is applied and the years to come. This is possible due to the load transfer size and the hosting capacity of an adjacent feeder.

The other successful alternative on which it was used every year was the reconductor. The cable selected for upgrading lines reporting thermal violations seems to be sufficient for relieving them during the year under analysis. The reconductoring module can also consider overload thresholds for updating lines that are close to their thermal limits before the overload takes place.

The logic for using ES to relieve thermal violations consists in placing the storage downstream at the end of the consecutive set of violations. The set of violations is formed by grouping overloaded line segments that are close to each other within a given range. The control settings for the storage device are current based setting the discharging target at the most overloaded line rating. Their features are user defined.

It is expected that the ES will peak shave the demand at critical times to relieve the violation. The charging cycle takes place when the current at the most overloaded element, which normally is upstream of the consecutive violations, is below a threshold set for such purpose. For this example, the charging cycle starts when the current at the monitored element is below 20% of its ampacity rating.

Similar guidelines are considered for PV placing; however, the control signal is given by the irradiance profile used as shown in Fig. 4.6. In this case, the daily demand peak and the irradiance profile are not coincident, making PV a nonviable alternative for this study.

Even when ES does not seem to be as effective as load transfer and/or reconductor every year it is applied, when combined with other alternatives, the study results reveal that its inclusion can contribute to successfully relieving the thermal violations.

For example, in one branch of the DBT obtained, by only using reconductor in years 1, 5, 6, 8, and 9, the total length of line segments upgraded is 2.44 mi. In another branch in which reconductor is used in years 1 and 5 plus ES used in years 6, 7, 8, and 9, only 0.83 mi of line segments are upgraded and 20 MWh of ES installed across the model.

The combination of alternatives is also an opportunity for accomplishing regulatory mandates for including DER in the distribution system.

For a better explanation of alternatives utilization during the study, consider Figs. 4.7 and 4.8. Fig. 4.7 shows the alternatives utilization per year, focusing on the years in which each alternative is used during the study. This graph reveals the years in which an alternative has at least one chance to relieve a violation successfully.

As can be seen in Fig. 4.7, load transfer and reconductor are the only alternatives capable of relieving the violations in year 1, which will keep with the circuit until year 5, where new violations appear.

From year 5 and onward, ES joins the set of alternatives capable of relieving the violations.

On the other hand, Fig. 4.8 displays the alternatives utilization rate across the DBT. This graph considers all the edges created across the DBT in which a corrective action took place and then evaluates the number of times each alternative was used for relieving the violation effectively.

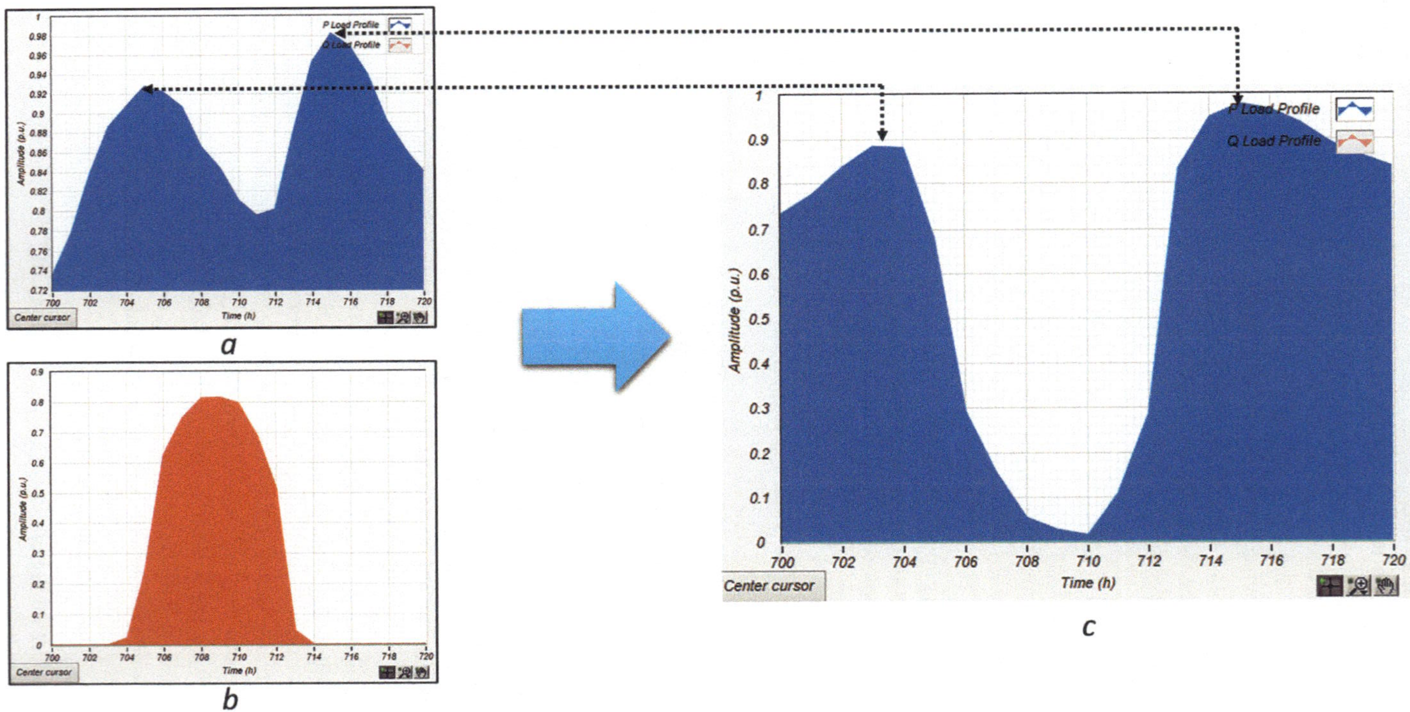

Fig. 4.6 (A) Daily demand peaks, (B) PV daily irradiance profile, and (C) PV generation cannot shave the peaks. Credit: Authors.

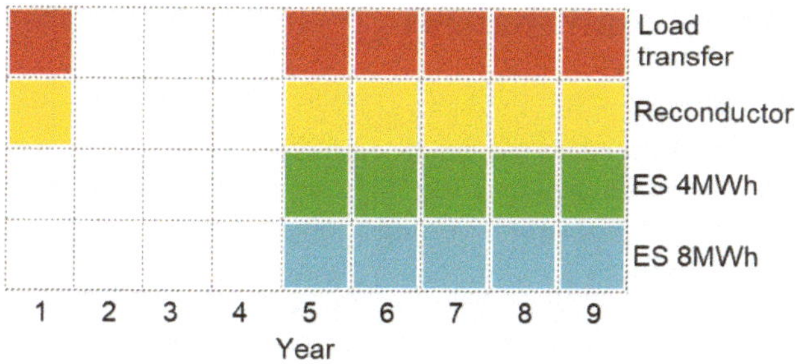

Fig. 4.7 Alternatives utilization per year.
Credit: Authors.

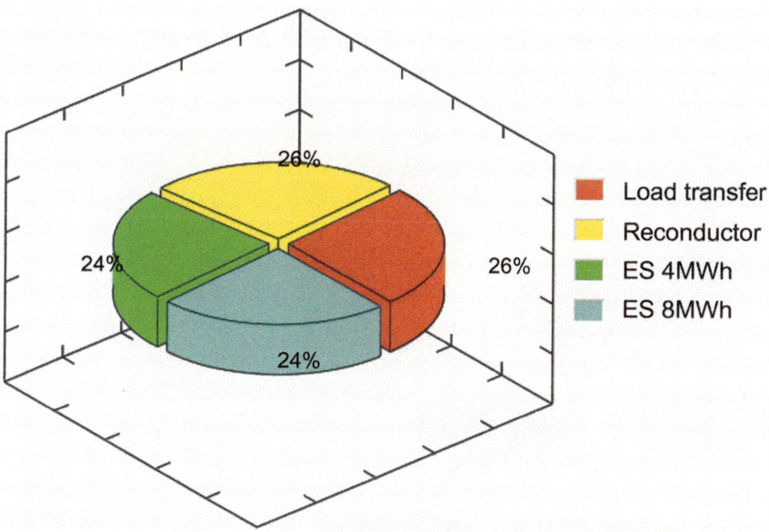

Fig. 4.8 Alternatives utilization rate across the DBT.
Credit: Authors.

As can be seen in Fig. 4.8, load transfer, reconductor, and the two sizes of ES selected for the study are very close in the utilization rate, suggesting that despite the case-year, some of the NWAs proposed (ES) have a good participation factor for reinforcing the circuit for the given planning horizon.

The graphs presented in Figs. 4.7 and 4.8, as well as the topology of the DBT, are tied to the circuit features and the set of alternatives selected for the study. At the end of the study, all the viable technical paths can be sorted by their economic cost, involving economic variables such as escalation, discounts, depreciation, life span, among others [15].

Using the sorted list of paths, the utility can evaluate which represents the best planning approach for not only getting the best investment but also covering the regulatory mandates/goals.

5 Analysis based on the DBT

The DBT is a decision tree represented as a graph for highlighting dependencies based on time and actions performed across the distribution system during the planning horizon. Each branch of the tree depicts a feasible technical path to go through the planning horizon relieving thermal and/or voltage violations. Branches can be represented as an occurrence matrix for describing the timely order in which a corrective action is required. For example, consider the DBT shown in Fig. 4.9. In this tree, the planning horizon is 9 years and there are only two alternatives considered for relieving thermal violations, reconductoring and load transfer. The labels at each node of the graph can be read as YA_AltB_R, where A is the year in which the alternative was applied, B is the alternative index as described at the legend, and R is a random number for differentiating each node.

In this case, the nodes are the years in which a mitigation action takes place and the links between years represent the actions. For illustrating this concept, consider the node Y9_Alt0_8 in Fig. 4.9; the occurrence matrix (Oc) for this branch will be as follows:

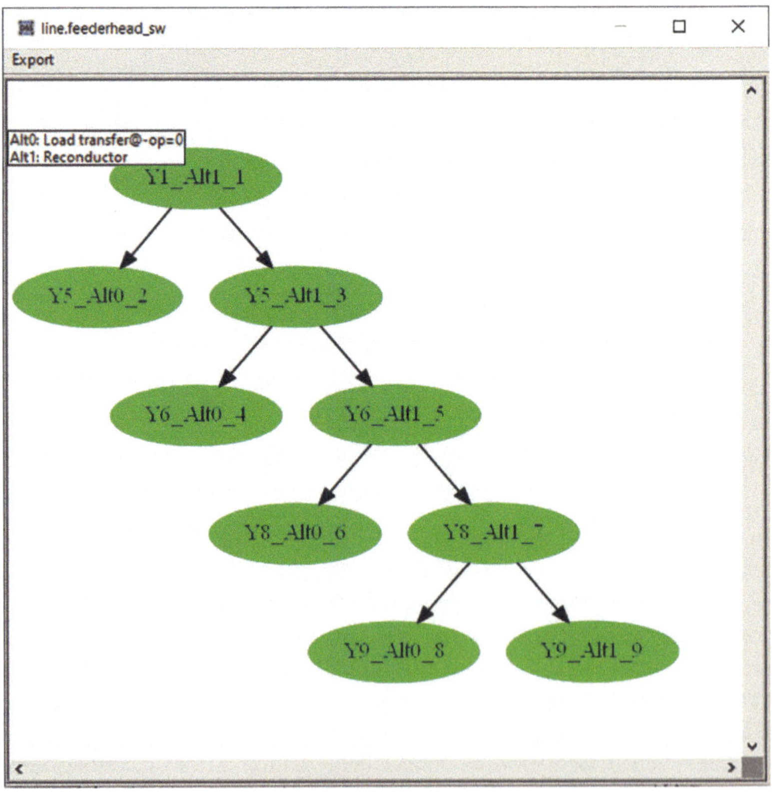

Fig. 4.9 Example of DBT represented as graph.
Credit: Authors.

$$Oc = \begin{matrix} \text{Load transfer} \\ \text{Reconductor} \end{matrix} \begin{bmatrix} 0 & 0 & 0 & 0 & 0 & 0 & 0 & 0 & 1 \\ 1 & 0 & 0 & 0 & 1 & 1 & 1 & 0 & 0 \end{bmatrix} \tag{4.4}$$
$$\qquad\qquad\qquad\qquad y1\ y2\ y3\ y4\ y5\ y6\ y7\ y8\ y9$$

In this matrix, the columns represent the years in which an action was required and the rows the alternative applied. When an alternative is applied in a particular year, the cell content will be different than 0. The data within Oc can be transported into two domains: the alternatives and time domains. The alternatives domain can be used to evaluate the utilization of an alternative across the branches. The time domain can be used to determine the probability of applying corrective actions on each year. These domains can be calculated as follows:

$$\text{Alternative domain} = Oc{*}Oc^T \tag{4.5}$$

$$\text{Time domain} = Oc^T{*}Oc \tag{4.6}$$

When transporting Oc into the alternative and/or time domain, it is possible to estimate the occurrence in time and type of alternative for the selected branch. However, these estimations are pointing to a particular branch required to perform the same calculation for each branch of the DBT independently, as shown in Fig. 4.10.

The total time and alternative occurrence matrices will then be obtained by the sum of all the different time and alternative matrices for all the successful branches in the DBT. This can be described as follows:

$$\text{Alternative occurrence} = aOc = \sum_{\#\text{branches}}^{n=1} Oc_n{*}Oc^T{}_n \tag{4.7}$$

$$\text{Time occurrence} = tOc = \sum_{\#\text{branches}}^{n=1} Oc^T{}_n{*}Oc_n \tag{4.8}$$

The probability of occurrence for an alternative within the aOc can be normalized using the number of branches as follows:

$$aOc(\%) = \left(\sum_{\#\text{branches}}^{n=1} Oc_n{*}Oc^T{}_n \right) / \#\text{branches} \tag{4.9}$$

Eq. (4.9) represents the probability of an alternative to be used within the obtained DBT. If the probability is greater than 1, it means that an alternative can be used more than once in the same path. If required, the matrix can be normalized using the largest number of occurrences if this overpasses the number of branches. The diagonal of aOc (%) will then indicate the probability of an alternative to be used for solving an issue within the case under study.

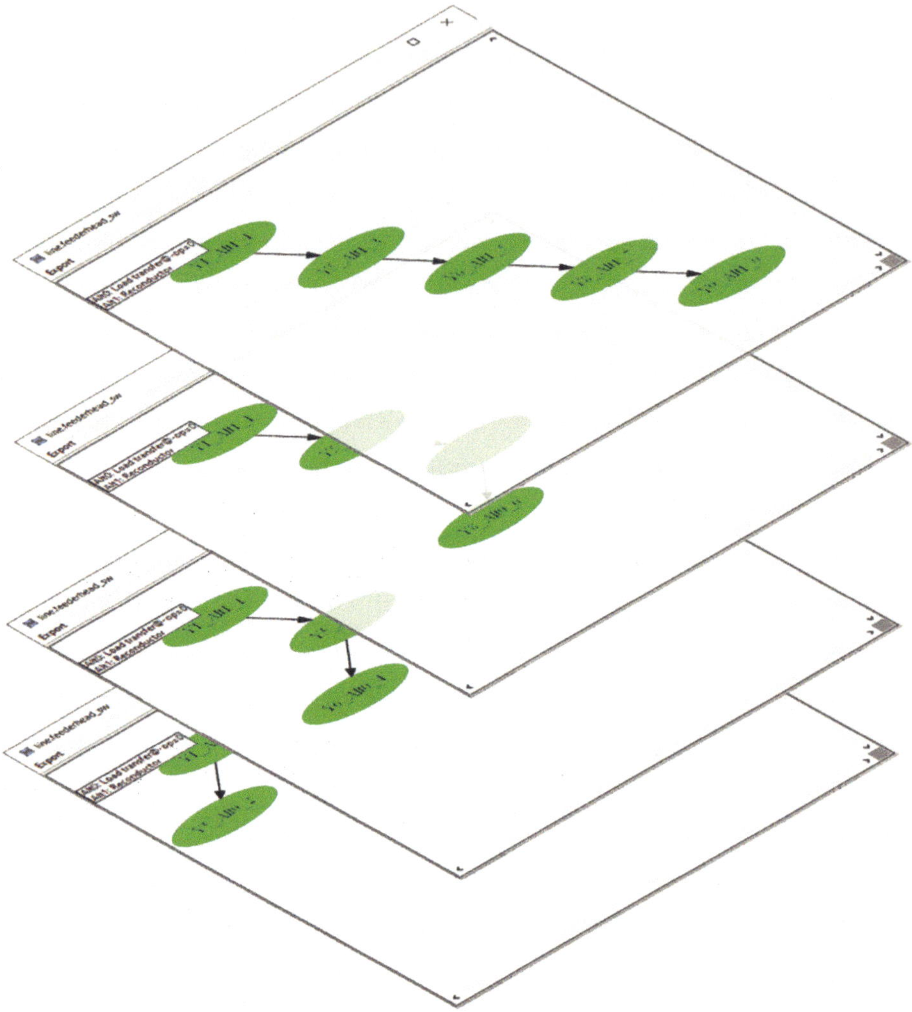

Fig. 4.10 Branch independency.
Credit: Authors.

Consider the alternative occurrence matrix shown in Fig. 4.11. At the top of Fig. 4.11, it is the *aOc* matrix represented as an intensity graph; the brightest the color indicates a more frequent occurrence across the planning study. The bar graph at the bottom is the normalized diagonal of *aOc*, and to the right of this graph is the legend for identifying each alternative in the bar chart. This graph shows that there is a high probability that for every path solving issues in the planning study there will be reconductoring, while other NWAs such as ES (different sizes) and PV have a lower probability of occurrence. Other mitigation alternatives can be evaluated in ADAPT such as demand side management programs, which were also considered in the example shown in Fig. 4.11.

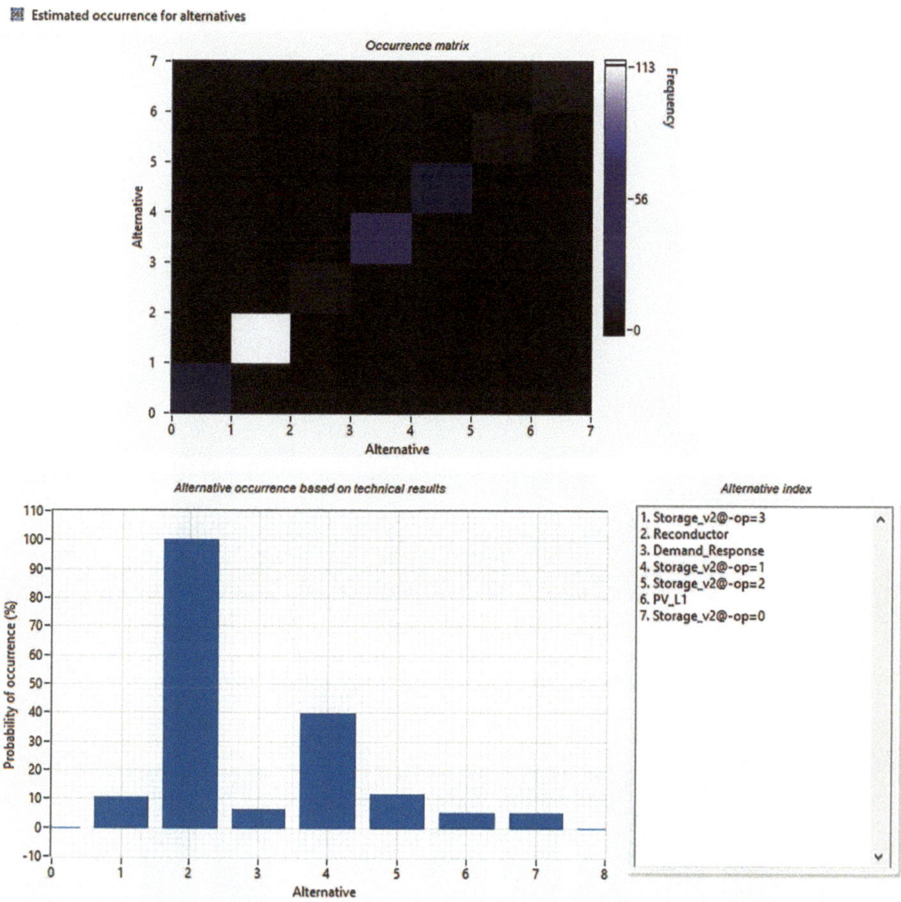

Fig. 4.11 Alternative occurrence.
Credit: Authors.

The same principle can be applied to the *tOc* as follows:

$$tOc(\%) = \left(\sum_{\#\text{branches}}^{n=1} {Oc^T}_n * Oc_n \right) / \#\text{branches} \tag{4.10}$$

In this case, the resulting matrix will reveal the probability of having to apply a mitigation action to each year, which in economic terms represents the years in which an investment will be required. This information can be gathered by extracting the diagonal of $tOc(\%)$.

Consider the time occurrence matrix shown in Fig. 4.12. At the top, it is the *tOc* matrix represented as an intensity graph. The brightest the color indicates a more

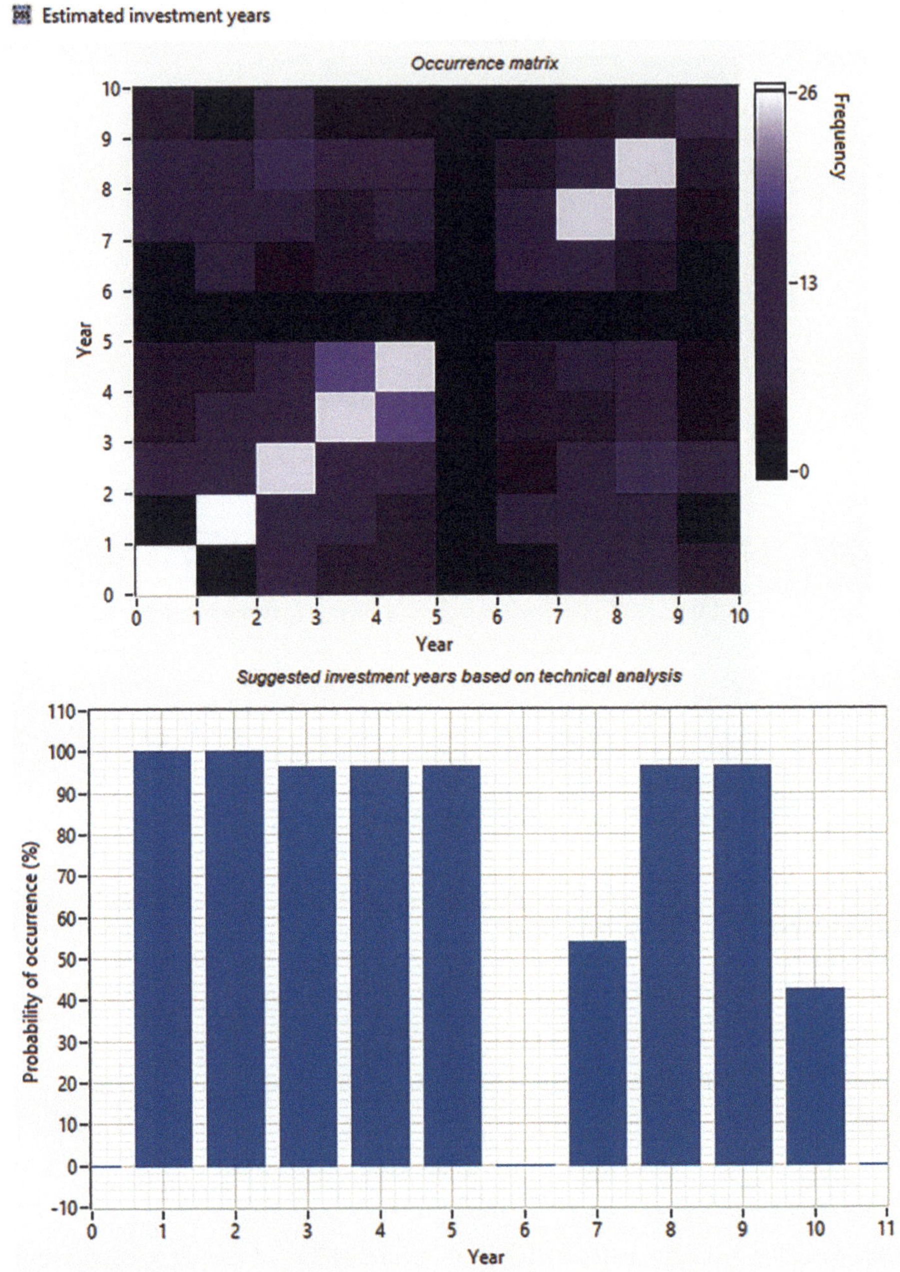

Fig. 4.12 Probability of investments required in time.
Credit: Authors.

frequent utilization of mitigation alternatives on that particular year during planning study. The bar graph at the bottom shows the probability of investment for the year. In this example, it is very likely to have investments in years 1–5 plus 8 and 9. Years 7 and 10 have a lower probability of requiring investments and year 6 does not require investments because during the analysis used for this example, year 6 never required a mitigation action to be applied. The lack of mitigation alternatives in year 6 can be the product of mitigation actions applied in previous years, the distribution model features, and the load growth, which for this example remains steady from year 5 to year 6. All these features were captured in the DBT used to derive the reports presented above.

6 Conclusions

This chapter has presented a methodology for identifying potential alternative deployment scenarios considering the dynamic characteristics associated with NWA. NWAs are progressively being considered by distribution and transmission planners for economically addressing acknowledged grid limitations. Gaining popularity due to technological advancements, falling costs, and supportive regulatory orders, regulators in several jurisdictions are now mandating that utilities evaluate the feasibility of DER-based alternatives prior to conducting any major grid reinforcements.

The alternative deployment is synthesized in a directed graph called the DBT, containing the different dimensions or technical viable paths that recreate time critical time instants dynamically, deriving new tangents/dimensions representing alternative timelines complementing an initial single investment plan.

The DBT also considers the boundaries set by the planning engineer, serving as the basis for detailed cost-benefit assessments and least-cost optimizations that will allow planners to identify the best investment plan according to the utility goals.

References

[1] M.O.W. Grond, J. Morren, H.J.G. Slootweg, Requirements for advanced decision support tools in future distribution network planning, in: 22nd International Conference and Exhibition on Electricity Distribution (CIRED 2013), 10–13 June 2013, 2013.

[2] Integrating Non-Wires Alternatives Into Utility Planning: 2020 EPRI Research Guide, EPRI, 2020 (3002018655).

[3] A. O'Connell, J. Taylor, J. Smith, L. Rogers, Distributed energy resources takes center stage: a renewed spotlight on the distribution planning process, IEEE Power Energy Mag. 16 (2018) 42–51, https://doi.org/10.1109/MPE.2018.2862439.

[4] M. Rey, S.M. de Oca, Á. Giusto, M. Vignolo, Distributed generation and demand response effects on the distribution network planning, in: 2018 IEEE PES Transmission & Distribution Conference and Exhibition – Latin America (T&D-LA), 18–21 September 2018, 2018.

[5] C.J. Dent, L.F. Ochoa, G.P. Harrison, J.W. Bialek, Efficient secure AC OPF for network generation capacity assessment, IEEE Trans. Power Syst. 25 (1) (2010) 575–583, https://doi.org/10.1109/TPWRS.2009.2036809.

[6] A. Keane, M. O'Malley, Optimal allocation of embedded generation on distribution networks, IEEE Trans. Power Syst. 20 (3) (2005) 1640–1646, https://doi.org/10.1109/TPWRS.2005.852115.
[7] W. El-Khattam, K. Bhattacharya, Y. Hegazy, M.M.A. Salama, Optimal investment planning for distributed generation in a competitive electricity market, IEEE Trans. Power Syst. 19 (3) (2004) 1674–1684, https://doi.org/10.1109/TPWRS.2004.831699.
[8] A. Keane, L.F. Ochoa, C.L.T. Borges, G.W. Ault, A.D. Alarcon-Rodriguez, R.A.F. Currie, et al., State-of-the-art techniques and challenges ahead for distributed generation planning and optimization, IEEE Trans. Power Syst. 28 (2) (2013) 1493–1502, https://doi.org/10.1109/TPWRS.2012.2214406.
[9] IEEE guide for conducting distribution impact studies for distributed resource interconnection, 2014, pp. 1–137, https://doi.org/10.1109/IEEESTD.2014.6748837. IEEE Std 1547.7-2013.
[10] R.A. Walling, R. Saint, R.C. Dugan, J. Burke, L.A. Kojovic, Summary of distributed resources impact on power delivery systems, IEEE Trans. Power Delivery 23 (2008) 1636–1644, https://doi.org/10.1109/TPWRD.2007.909115.
[11] R.C. Dugan, J.A. Taylor, D. Montenegro, Energy storage modeling for distribution planning, IEEE Trans. Ind. Appl. PP (99) (2016) 1, https://doi.org/10.1109/TIA.2016.2639455.
[12] J. Taylor, J.W. Smith, R. Dugan, Distribution modeling requirements for integration of PV, PEV, and storage in a smart grid environment, in: 2011 IEEE Power and Energy Society General Meeting, 24–28 July 2011, 2011.
[13] National Academies of Sciences, Engineering and Medicine, Models to Inform Planning for the Future of Electric Power in the United States: Proceedings of a Workshop, The National Academies Press, 2020, https://doi.org/10.17226/25880.
[14] J.L. Picard, I. Aguado, N.G. Cobos, V. Fuster-Roig, A. Quijano-López, Electric distribution system planning methodology considering distributed energy resources: a contribution towards real smart grid deployment, Energies 14 (7) (2021), https://doi.org/10.3390/en14071924.
[15] J. Smith, B. Rogers, J. Taylor, J. Roark, B. Neenan, T. Mimnagh, E. Takayesu, Time and location: what matters most when valuing distributed energy resources, IEEE Power Energy Mag. 15 (2017) 29–39, https://doi.org/10.1109/MPE.2016.2639178.
[16] J.M. Bloemink, T.C. Green, Effects of power electronic compensation on distribution network thermal and voltage violations, in: 2013 IEEE Power & Energy Society General Meeting, 21–25 July 2013, 2013.
[17] R. Balakrishnan, K. Ranganathan, A Textbook of Graph Theory, Springer, 2000.
[18] R. Diestel, Graph Theory, Springer, 2006.
[19] R.C. Dugan, T.E. McDermott, An open source platform for collaborating on smart grid research, in: 2011 IEEE Power and Energy Society General Meeting, 24–29 July 2011, 2011.

Green approaches in future power systems

5

Hamed Delkhosh and Mohsen Jorjani
Faculty of Electrical and Computer Engineering, Tarbiat Modares University, Tehran, Iran

1 Introduction

After the industrial revolution, fossil fuel usage has been explosively increasing to provide the required energy for expanding industrial activities and widening people's access to energy. However, in recent decades, human beings have profoundly realized the issues of the current energy regime. Firstly, these resources are finite and with the current consumption rate, they are estimated to run out soon. Secondly, a large part of these limited resources is located in specific countries, which leads to energy security concerns and intrigues the need for energy independence and sustainability. Thirdly, environmental issues such as climate change and global warming are strongly related to fossil fuel usage. Given these motivations, the utilization of green resources has become one of the main trends worldwide.

Green resources can be classified into three categories from the electricity energy perspective. The first category is focused on transforming the existing traditional power plants to clean resources using solutions such as carbon capture systems. The second category is energy efficiency, which can be considered as a cost-effective solution to meet sustainability goals. However, this solution is surprisingly rarely utilized now because of the stakeholders profit-seeking and the cultural background of energy usage originated from the political policies. The third category involves renewable resources usage, including hydro, wind, solar, bio, marine, and geothermal energy.

Moving toward utilizing these green resources is strongly related to the decentralization concept, which is the main focus of this book. Specifically, renewable energy resources and energy efficiency solutions are almost totally decentralized in nature. This can be better understood by looking at the concepts of moving from bulk generation and dumb load to distributed generation and responsive demand. In this regard, energy storage devices and hybrid AC/DC grids are facilitating components that provide the platform for various grid integration possibilities and the emergence of ideas such as microgrids. However, breaking the existing foundation of power grids and moving toward decentralization have many issues that should be explored precisely.

The topics mentioned above have been covered by a wide range of studies and researches in recent decades. Obviously, it is not possible to comprehensively address all of these topics in just one book chapter. Accordingly, this chapter aims at presenting a big conceptual picture of green approaches in future power grids. After

Decentralized Frameworks for Future Power Systems. https://doi.org/10.1016/B978-0-323-91698-1.00012-1

introducing the concept of green transformation, the discussion continues by presenting the main motivations, namely the energy issue and its related subfields, including finite resources, environmental concerns, and energy security. Then, the green resources, including clean power plants, renewable energies, and energy efficiency, are briefly discussed as the primary tools for moving toward a solution to the presented issues. Finally, the topic is discussed from the decentralization perspective (the main scope of this book) by investigating various aspects such as distributed generations, demand response, energy storages, hybrid AC/DC grids, grid integration, and other emerging issues. According to the primary purpose of this chapter, only the basic and required concepts about the mentioned topics are presented. However, appropriate complementary references are introduced for further reading where needed.

2 Green transformation

Given the title of this chapter, this section will try to provide a clear definition of "being green" and explain the general concepts of green transformation while discussing its relationship to what will be presented in this chapter. The word "green" can be viewed from two different perspectives. The first one is focused on the environment itself and pays attention to its conservation. This perspective does not show the direct connection between the importance of the environment and its relation to human life. The other perspective focuses on human beings and deals with the value of environment and ecosystem in their life for achieving their goals. Also, two different viewpoints on the current state of the environment can be differentiated from each other in terms of the severity of the present situation. The first viewpoint is optimistic about the situation and considers the environment as relatively robust. On the contrary, the situation is assumed to be critical and the environment is considered as being very vulnerable and endangered in the second viewpoint. The authors believe that the second view is more suitable for both defining and evaluating the situation of being green. This means that the condition of the environment is worrying and this has a very crucial impacts on human beings in the future. Based on these explanations, the need for green transformation is not only justifiable but also vital.

Green transformation can be defined as the movement of society toward sustainability in all dimensions in order to face various environmental issues, especially climate change. These dimensions can include aspects such as ecological, social, economic, and so on. Green transformation can be viewed from two different perspectives which are "type" and "direction" of change. From the perspective of the "type of change," two key political and technical aspects are involved. The political aspect of change has a significant impact on the overall process in terms of policy-making, legislation, financing, culture-making, and incentivizing. On the other hand, technical aspects of these changes are very challenging, time-consuming, and costly due to the nongreen nature of most of the existing infrastructures. In terms of the "direction of change," the bottom-up and top-down directions can be followed. The bottom-up direction requires more mobility in the individuals of a society, while the top-down direction is more dependent on governments and large businesses actions. The focus of this chapter is on the role of power grids in the overall green transformation plan, especially from the perspective of the utilization of green resources, as well as addressing various technical aspects that are needed to be changed in this regard, which requires the use of both bottom-up and top-down directions. Refs. [1, 2] are recommended for further studies on the green transformation concept.

3 Energy issues

Long before the industrial revolution, human beings mainly used organic energy, including man and animal power and wood-burning for various applications. Thereafter, the mineral fuels (including coal, oil, and natural gas) initiated the industrial revolution and empowered it through its continued road to development. The contribution of nuclear energy to the industrial revolution, especially in the electricity sector, cannot be ignored either. During that period, events such as (1) industrial activities dramatic development, (2) road and rail transportation development and pervasiveness, (3) electrical grids emergence and their uninterrupted growth, and (4) the wider access of people to various types of energy, caused the intensified use of mineral fuels. This, in turn, has led to three concerns: (1) the scarcity and depletion of these resources, (2) serious environmental concerns about fossil fuels usage, and (3) the security of sustainable energy supply worldwide. These concerns, which have led to a serious and growing move toward green approaches (especially renewable resources) in recent decades, will be briefly reviewed in this section. The historical turnaround in the energy regime from organic to mineral, and more recently to renewable, is discussed in detail in Ref. [3].

3.1 Finite resources

The total energy consumption of the world has been growing steadily in recent decades. Fig. 5.1 illustrates this trend over the last 50 years. The current global annual consumption of all types of primary energy resources is approximately 162,000 TWh. As seen in Fig. 5.1, the increase in total energy consumption cannot be attributed solely to more than 2.3 times growth in the world's population (from 3.3 to 7.7 billion) within these years; An almost 60% increase in the per capita energy consumption has a significant role in this growth. This increase is because of two main reasons: industrial activities development and the wider access of people to various types of energy sources.

Fig. 5.2 depicts the historical trend of various energy resources contribution to the world's energy supply. As it can be seen from this figure, fossil fuels (coal, oil, and gas resources) have always had a massive share in supplying the world's energy needs (this is currently about 84%). Despite concerns such as the 1970s energy crisis, oil continues to have the main role in supplying the world's energy. The declining share of coal has risen again since 2000 due to China's policies regarding its usage. Natural gas also has a steadily increasing trend due to fewer environmental issues. Meanwhile,

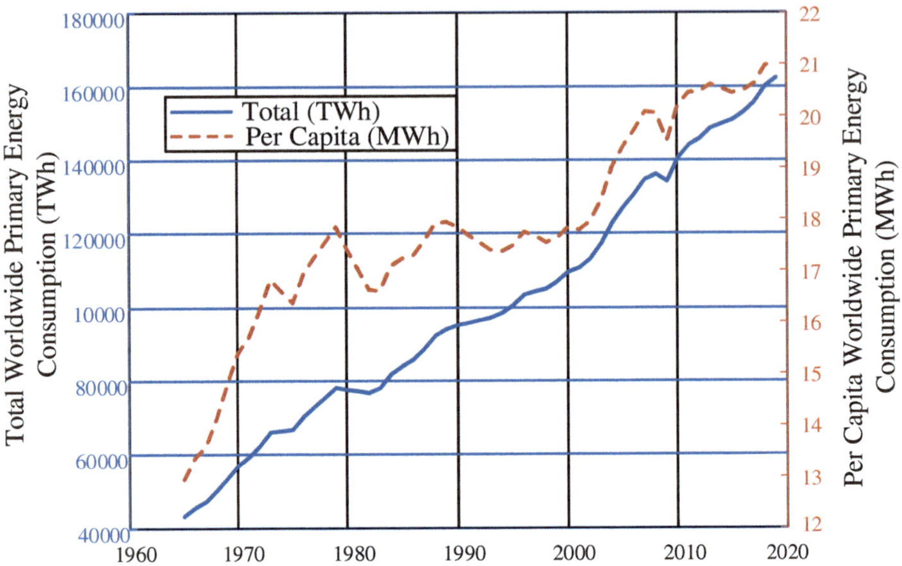

Fig. 5.1 The historical trend of primary energy carriers totals and per capita consumption.

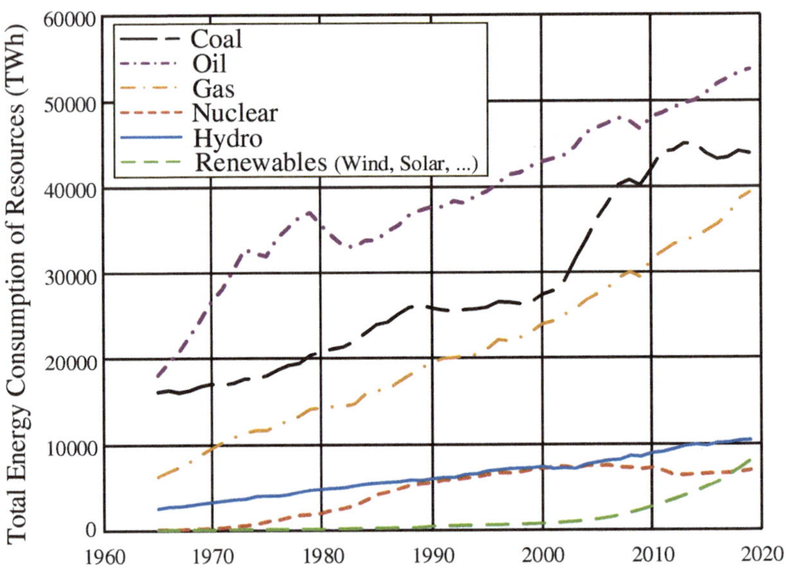

Fig. 5.2 The historical trend for various resources total energy consumption.

hydro resources usage seems to be saturated. Nuclear fuel usage is also declining due to concerns that will be discussed in the next paragraph. There is also a clear shift toward using renewable resources in the last two decades, but these resources still have a long way to go for providing a greater share of the world's energy.

Nuclear energy, which is mainly used to generate electricity, is highly controversial. Advocators argue that it should account for a larger share of the world's total energy (currently 4.5%) given the available uranium resources, very high energy density, and low carbon pollution. Critics do not advise the use of this resource for the following three issues: (1) the possibility of the process getting out of control for a variety of causes that may result in nuclear catastrophes such as Chernobyl (1986) and Fukushima (2011), (2) the possibility of nuclear weapons development due to the high similarities between this process and nuclear electricity generation technology, and (3) the complexity and dangers incorporated in disposing of radioactive waste. These conflicts are also evident in different countries policies. Although some countries (such as Germany) are currently closing their nuclear facilities, others (such as the United Kingdom) highly pursue their nuclear power plants development plans. According to Fig. 5.2, it can be seen that the share of nuclear energy in providing the total world energy need is a little declining. This trend is not expected to change significantly considering the current policies. Ref. [4] is suggested for more information in this regard.

The growing consumption of fossil fuels as the world's leading energy supplier, along with the finite and nonrenewable nature of these resources, has raised serious concerns in recent years. It is estimated that the existing oil and gas resources will only last for the next 30 and 60 years at current consumption rates. It should be noted that some nonconventional resources such as heavy oil, shale oil, and gas are also available (their available reserves are estimated to be three times the conventional resources) with much higher extraction costs [5]. Therefore, we should expect a decrease in the extraction rate of fossil fuel resources and an increase in their prices, causing even more issues in the near future. On the other hand, the available coal resource reserves are estimated to be several times greater than oil and natural gas resources. However, the high air pollution rate of coal resources due to their high emission of carbon dioxide brings many challenges that restrict their long-term usage.

Therefore, fossil fuel resources depletion is a serious challenge that will soon affect the entire world. This problem has been one of the main motivations for moving toward the utilization of renewable resources in recent decades due to their eternal nature. However, the movement toward using these resources should be much faster so that the mentioned issue does not become a global crisis in the near future. The statistical information used in the figures and their corresponding analysis is mainly extracted from Ref. [6].

3.2 Environmental concerns

The CO_2 emission of fossil fuels is the most critical environmental concern related to using these resources. As shown in Fig. 5.3, CO_2 emissions have grown exponentially since the early 19th century right after the industrial revolution. This growth has not stopped to this day, despite all the international efforts [such as the Kyoto Protocol (1997) and the Paris Agreement (2016)] made in recent decades. According to Fig. 5.3, it can be seen that almost all of the CO_2 emissions are due to the burning of fossil fuels such as coal, oil, and natural gas. The comparable share of coal and

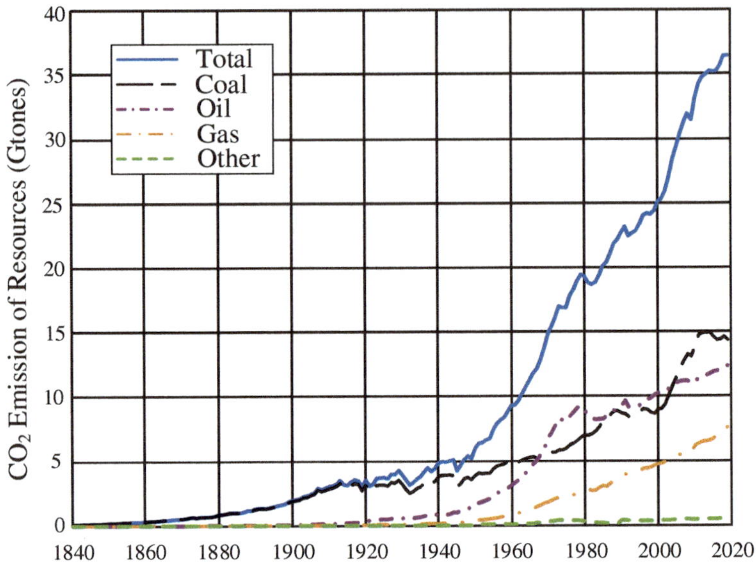

Fig. 5.3 Historical trend of various resources CO_2-emissions.

oil in CO_2 emissions, considering the fact that coal has a smaller share in the worldwide energy supply, indicates the very high air pollution rate of coal. The significantly lower share of natural gas in CO_2 emissions, considering its almost equal share in energy supplement with coal and oil, also indicates that this fuel is much cleaner than the other two.

The CO_2 emission of fossil fuels has caused global warming, which has a devastating effect on the entire planet's environment. The historical trend shown in Fig. 5.4 illustrates the approximately 1°C increase in the average global temperature over the last century. Many studies have shown that global warming cannot be caused solely by natural events, and the severe effect of human factors on this phenomenon has been proven.

In addition to the previously mentioned cases, other aspects of the environmental damage caused by fossil fuels are by no means negligible. Nitrogen and sulfur oxides (NO_x and SO_x) are the main causes of the destructive phenomenon of acid rain. Coal mining activities and natural gas leakages release methane gas emissions into the atmosphere which have a higher greenhouse effect than CO_2. In addition, the extracting process of these resources typically has significant environmental impacts on the land and habitats where they are located. Moreover, oil spills into the oceans that usually happens during their transportation have also attracted much attention in recent decades.

According to what was stated in the preceding paragraphs, the urgent need for finding a solution to the described issues is undeniable. Studies show that these issues have so far had some irreversible effects on the environment and delays in addressing them will only increase the cost of required remedial actions. Green approaches that can be

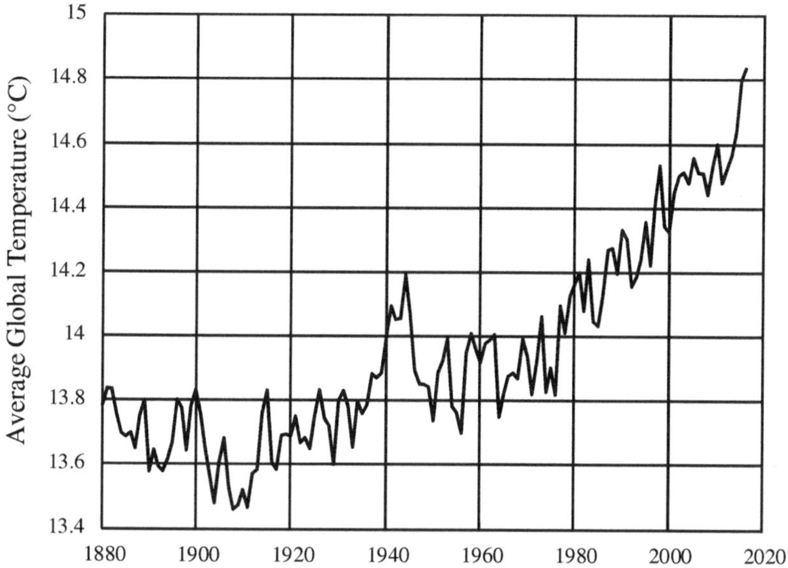

Fig. 5.4 Historical trend of average global temperature.

used to solve these issues can be divided into four categories: (1) conversion of fossil fuels into cleaner resources using technologies such as Carbon Capture Systems (CCS), (2) utilization of more renewable resources with much fewer emissions, (3) improving the efficiency of energy usage as a cost-effective and less used resource, and (4) reducing and stopping the deforestation process and enhancing the reforestation. Because of our focus on power grids, the first three issues are briefly discussed in the fourth section of this chapter. Refs. [7–9] are recommended for further studies on the environmental issues caused by fossil fuels usage.

3.3 Energy security

The energy security issue and its relationship with green approaches in future power systems can be discussed from two key aspects. The first is the need for different countries to have access to their required energy resources in order to achieve industrial development and economic growth. The second aspect is consumerism and its resulting class distinctions between rich and poor countries (and even rich and poor people within the same country) from the energy accessibility viewpoint.

Two key factors have raised serious concerns (especially for the industrialized countries without fossil fuel resources) regarding the first aspect. The first is the existing conflicts between Western and Islamic cultures. This contradiction has caused many issues due to the fact that a large part of underground world resources, especially oil and natural gas, are located in the Middle East with a predominantly Muslim population. In addition, disagreements between Russia, as one of the main suppliers of underground resources (especially natural gas), and the European Union over regional issues (such as Ukraine) have intensified these concerns. Therefore, one of the necessities of moving toward green approaches in future power systems is the need for each country to have access to sustainable and secure energy supply for advancing its industrial and economic goals.

Regarding the second aspect, it seems that political parties adopted policies aimed at satisfying people after the rapid industrial growth following the industrial revolution. These policies have led to consumerism in different societies resulting in an unnecessary increase in energy consumption. This, in turn, has caused significant differences between different countries per capita energy consumption. This difference cannot be attributed only to the fact that one country is more industrialized than the others since it also exists between countries at the same industrialization level. This difference is even reflected in different social classes (for example, urban and rural) of a nation. Moving toward green approaches, especially energy efficiency is essential for solving this issue. Unfortunately, this cost-effective solution has been neglected so far due to the economic and political benefits of various stakeholders. Refs. [9, 10] are recommended for further studies on energy security.

4 Green resources

In the previous section, energy issues were briefly discussed in three main categories: finite resources, environmental concerns, and energy security. These issues are the main motivations for using green resources. The purpose of this section is to provide an overview of various aspects of green resources in the following three categories: (1) clean power plants, (2) energy efficiency, and (3) renewables.

4.1 Clean power plants

Due to the high cost of some clean solutions such as renewable energies and, more importantly, the huge investments made by various stakeholders in fossil fuel exploration and production activities, it is not possible to abandon these resources easily in the short term. Therefore, the first green approach is to use these conventional resources in a clean way. This means making the most of existing capacities (remaining resources and developed infrastructure) using different advanced technologies to minimize the environmental damages. Such an approach is critical to the context in which the world currently finds itself. It facilitates the required transition period of moving from the initial point of complete reliance on fossil fuels to the final destination of extensive and comprehensive use of renewable resources. However, what future trends will look like given the current situation is still unclear as discussed in various references, including Ref. [11].

It was mentioned earlier that CO_2 is the primary concern among air pollutants. All industries (e.g., power plants) that use fossil fuels as the source of energy to produce their products (e.g., electricity) emit this pollutant. Accordingly, CCS have received particular attention in recent decades. These systems absorb CO_2 by compressing it. However, CO_2 management after its absorption (i.e., its transfer processes (in some

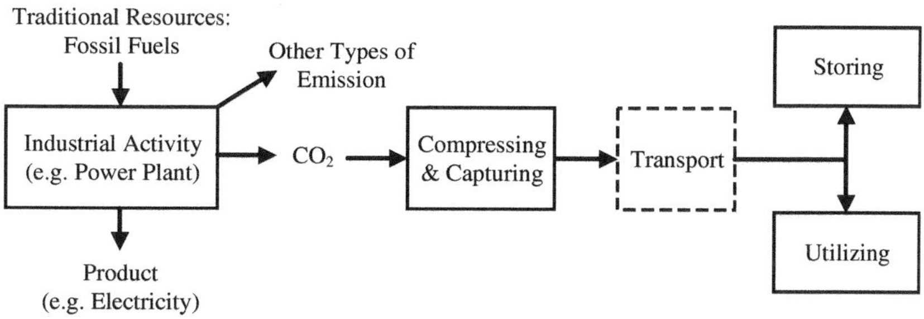

Fig. 5.5 Schematic representation of CO_2 production in industrial activities and its management.

cases) and storage or utilization) is the major problem. The storage is often done in abandoned mines or oil and gas fields. The product of this process can be utilized in combination with various chemicals, construction materials, algae nutrition for biofuel production, etc. The schematic representation of this process is illustrated in Fig. 5.5. It should be noted that such a process is an advanced and challenging technology that is as expensive as utilizing renewable energies. Researchers such as Wilberforce et al. [12] can be referred for further study in this regard. Here, the possibility of using different types of fossil fuels in a clean form is briefly described.

Among the fossil fuels, coal can be considered the dirtiest one. For the same level of energy production, coal emits approximately 1.5 and 2 times more CO_2 than oil and natural gas, respectively. Besides, its particles pollution, which is very harmful to the health, is much higher than oil and natural gas (approximately 100 and 10,000 times [7], respectively). Coal mining has also severe detrimental impacts on the land and the running waters. Therefore, it is almost impossible to use coal in a clean manner at a competitive cost with other options. However, the use of this fuel in some countries has increased in recent decades due to their accessibility to vast cheap coal sources, which is a huge cause for environmental concerns.

On the other hand, natural gas can act as a bridge in this transition period due to its characteristics. Meanwhile, the cost of using renewable resources will be reduced, and their widespread use will become economically possible during this period. In recent decades, the use of Compressed Natural Gas (CNG) and Liquefied Natural Gas (LNG) instead of liquid fuel derived from oil has become common for transportation purposes. A clean and cost-effective solution would also be the widespread use of this fuel in other industries (especially the electricity sector) combined with less complex carbon capture systems.

4.2 Energy efficiency

The general concept of energy efficiency is to produce more economically while using less energy. In fact, energy efficiency combines the two concepts of consumption saving and production efficiency improvement. Energy efficiency gained attention after the 1970s oil crisis presenting the idea of hidden fuel, and lots of researches in various fields such as industries [13], buildings [14], transportation [15], etc., have been focused on it ever since. Assessing that the potential for energy efficiency is not an easy task, it is estimated to be approximately 20% of total energy consumption, which is very impressive. This capacity becomes more attractive when combined with its economic viability. However, it seems that energy efficiency utilization is currently considered seriously just in acute and critical situations (i.e., when the demand cannot be met and the prices rise) and there are currently many obstacles to the full activation of its capacity. Some of these obstacles are the lack of policy, legislation, and investment to motivate the movement in this direction. Prioritizing rapid economic development and keeping the citizens satisfied are two key factors that prevent governments from pursuing energy efficiency policies. This can be clearly seen in various governmental subsidies for fossil fuels usage worldwide, to the extent that the amount of these subsidies is estimated to be more than the total investments made on energy efficiency [16]. That's mainly because of the need to return the enormous invested money on fossil fuels infrastructures that have been made in the past. The consumerism in society and the reluctance of consumers to change their consumption patterns are the other crucial obstacles to the use of energy efficiency.

Despite the obstacles described above, governments seem to pursue more restricting energy efficiency policies for various industries and the household sector, given the climate crisis and global warming. Also, the liberalized market and the supply and demand system, in which a shortage of supply is highly expected in the near future, make it necessary to move toward energy efficiency. The technologies used in the field of energy efficiency are also developing day by day, which creates more potentials in this field.

By all accounts, the potential and increasing capacity of energy efficiency is a key factor in the transition from fossil fuels to renewable energies. However, the more this capacity is enabled, the more effective are the discussed obstacles for its further activation.

Energy efficiency is highly developed in the electrical energy sector to increase the efficiency of generation and reduce energy consumption. The conversion of gas power plants into Combined Cycle (CCPP) units, Combined Heat and Power (CHP) units, combined AC/DC networks equipped with high-efficiency converters, etc., can be mentioned as some examples for efficiency improvement. The consumption (demand) management in the fields of production, household sectors, agriculture, etc., has also been the center of focus in recent years. Some of these cases are discussed in Section 5.

4.3 Renewables

Clean resources (as described in Section 4.1) are temporary solutions for improving the situation during the transition period. Energy efficiency (as described in Section 4.2) has not been well utilized despite being cost-effective due to political and economic incentives and its capacity saturation characteristic. Meanwhile, renewables are the prominent and widely accepted approach for solving the world's energy issues. The worldwide growing trend of using these resources in the last two decades is a sign of this claim. For instance, most European countries had renewables penetration targets ranging from 10% to 50% by 2020, which almost all of them were met.

The share of renewables in global energy production is currently approximately 11.5%. About 6.5% of this amount is the share of traditional hydro resources that have long been used alongside fossil fuels. Thus, renewables have accounted for an almost significant share of 5% of the world's total energy consumption in the last two decades. This means that governments have recognized renewables as a long-term solution for energy issues despite their different political and economic viewpoints. Of course, renewable energy still has a long way to go to overcome fossil fuels (currently supplying about 84% of the world's energy). The numbers given in this paragraph can be deduced from Fig. 5.2.

4.3.1 Sources

In order to examine the different types of renewables, their main energy sources must be first known. This is presented in Table 5.1. As can be seen, almost all renewables get their energy from the sun through different direct and indirect processes.

The renewability of these resources makes sense given the source of most of them, the sun, and its constant availability. It should be noted that fossil fuels also originate from the sun and can be considered as bioenergy produced by ancient plants and animals. However, the problem is that the time it takes for the process of absorbing energy from the sun and preparing these resources is so long that one cannot expect to reproduce it in a comparable period with human's lifetime. Hence, these resources are referred to nonrenewable and exhaustible.

4.3.2 Energy conversion

Fig. 5.6 illustrates the different pathways for converting various renewable and nonrenewable resources into other forms of energy. It can be seen that modern sources and methods of energy conversion have a significant share in these pathways. A key point about some of the most important renewable resources (hydro, wind, and solar

Table 5.1 Types of renewable resources with their main energy source and process.

Type			
General	**Detail**	**Source**	**Process**
Hydro		Sun	Water evaporation and condensation
Solar	Thermal	Sun	Direct heating
	Photovoltaic	Sun	Direct radiation
Wind		Sun	Unbalanced heating of the atmosphere
Marine	Wave	Sun	Unbalanced heating of the atmosphere
	Tidal	Gravitational	Different forces in various parts
	Ocean thermal	Sun	Heating water unevenly in different depth
Bioenergy		Sun	Photosynthesis
Geothermal		Earth core	Direct or indirect heating

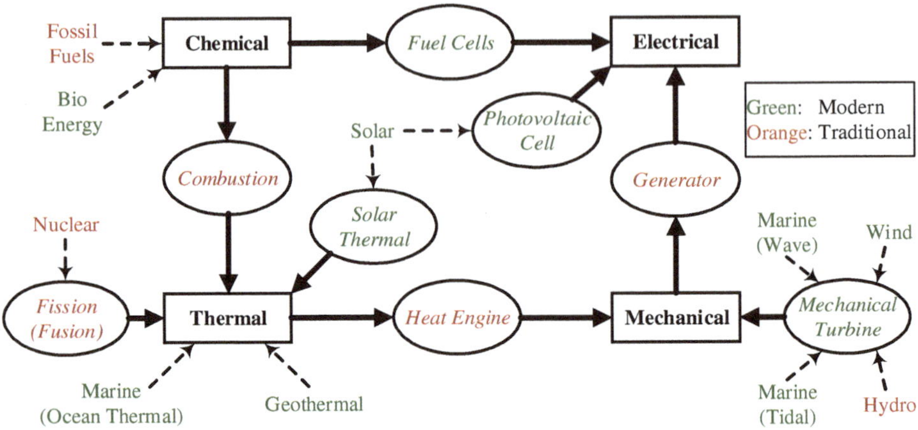

Fig. 5.6 Different pathways for converting resources into different forms of energy.

photovoltaic) is that they do not need to use heat engines. Due to the limitations imposed by Carnot's maximum efficiency theory, these engines have efficiencies of less than 50% (in the best circumstances, e.g., when using combined-cycle designs). Given the uncertainties related to the available times of most renewable resources (especially wind and solar), the role of storages in their sustainable energy supply is crucial. This is explained in Section 5.3. Fuel cells (which work by directly converting chemical energy to electrical energy) will also be further explained there.

4.3.3 Trends

In order to clarify the past and present status of renewable resources, the trend of changes in their installed capacity in the electrical energy sector over the last two decades is illustrated in Fig. 5.7. This figure can also show the possible future trend to some extent. It can be seen that hydropower has always had the most installed capacity among the renewables due to its long history of utilization. Besides, its usage has been increasing in recent years. However, there has been a saturation in its installed capacity in recent years. Wind and solar usage have the most growth among other renewables. These resources have become fully operational in the last two decades and currently have almost the same installed capacity. It should be noted that solar energy usage growth started with almost a decade delay compared to wind energy, but it is currently growing much faster. This rapid growth is also expected to continue in the near future. Bioenergy, which has received more attention in the transportation sector, is growing at a slow pace. Geothermal energy does not have a high installed capacity due to the limited number of well-suited sites with good potentials around the world. Marine energy is still mainly in the research phase and has not found a serious operational aspect yet. The data used to draw the figures in the current and the next section is extracted from Ref. [17].

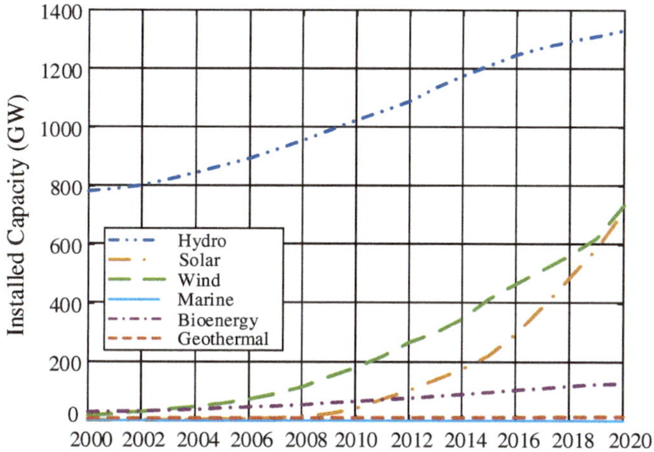

Fig. 5.7 Installed capacity of various renewable resources in the last two decades.

4.3.4 Hydro resources

Hydro resources get their energy from the sun, through the cycle of water evaporation from the earth's surface and its condensation and the resulting rainfall at higher altitudes. When the water is returning to sea level, its released potential energy can be utilized. Hydro resources enable the extraction of electrical energy in an extensive range, from very large power plants (e.g., China's 22,500 MW power plant) to tiny ones (e.g., 100 W). Large-scale hydroelectric power plants usually operate by constructing large dams with large reservoirs on rivers. The construction of such dams is very costly and, more importantly, has a significant impact on the environment and the management of downstream water resources. These power plants are sometimes used as large-scale pump storage designs for large-scale energy storage. Smaller hydro units often operate with run-of-the-river design (without dams). Impulse turbines (Pelton, Turgo, etc.) and reaction turbines (Francis, Kaplan, etc.) are the typical hydro turbine designs that are used considering the head and flow of water in the area.

Hydroelectric units have almost zero fuel costs compared to thermal power plants. The start-up time of these units is also much lower, and their flexibility (ramp rate) is much higher. Compared to other renewable units (especially wind and solar), their generated power is not intermittent. So, these units are dispatchable. However, precipitation patterns and many other constraints (such as the energy constraints due to the required downstream water capacity and some other specific constraints related to the cascaded power plants on the same river) must be considered in their planning and operation.

These resources have long been used alongside fossil fuels to generate electricity. The installed capacity of these resources is currently more than 1300 GW which is way higher than other renewable resources. Hydro resources have received so much attention that countries such as Norway and Paraguay are now generating almost all of their electricity from these resources. However, most of the attractive sites for these resources have been used so far, and we cannot expect very extensive developments in hydro resources in the future. Refs. [18, 19] are recommended for further studies on hydro resources.

4.3.5 Solar resources

Solar energy comes from the nuclear fusion that takes place in the sun. The generated energy raises the sun's temperature to about 6000 K, causing the star to emit light and heat to its surroundings including the earth. The maximum flux radiated to the Earth is about 1 kW/m^2. Therefore, the amount of energy that reaches the Earth in 1 h can provide the worldwide energy consumption for one whole year. However, the spatial and temporal dispersed nature of this energy makes its complete absorption very difficult. Thus, the spatial conditions on Earth (latitude and longitude) and solar conditions of the sun in the sky (azimuth and altitude angles) throughout the year must be taken into account for designing schemes to absorb this energy. Two main methods of extracting energy from this resource are solar thermal and photovoltaic systems, which will be briefly discussed below.

Solar thermal systems absorb radiation in the form of heat and transfer it to a fluid (usually air, water, or oil) for later use. These absorbers are divided into two groups in terms of the area of collision and absorption, namely, concentrating and nonconcentrating. Concentrating absorbers (often concave) are divided into different categories according to their tracking procedure of the sun. Fig. 5.8 shows some types of heat absorbers with various applications such as electricity generation, water heating, etc.

Photovoltaic systems convert solar energy into electrical energy directly and without any intermediate steps. PN-based solar cells operate based on doped semiconductors (similar to diodes). The photons emitted on these cells form electron–hole pairs that flow into the circuit due to the cell's built-in field and generate electric current if the radiation is continuous. These systems completely dominate solar thermal systems (99% of installed capacity) in the electricity sector.

Solar cells are usually electrically modeled as a current source parallel to a diode with series (for modeling the nonideal conductivity) and parallel (for modeling current leakage due to poor insulation) resistors that have an almost square-shaped current–voltage curve (see Fig. 5.9). As can be seen, there is a point in this curve where the power (voltage multiplied by the current) is maximum, which is called the Maximum Power Point (MPP). This point must be followed somehow to extract as much

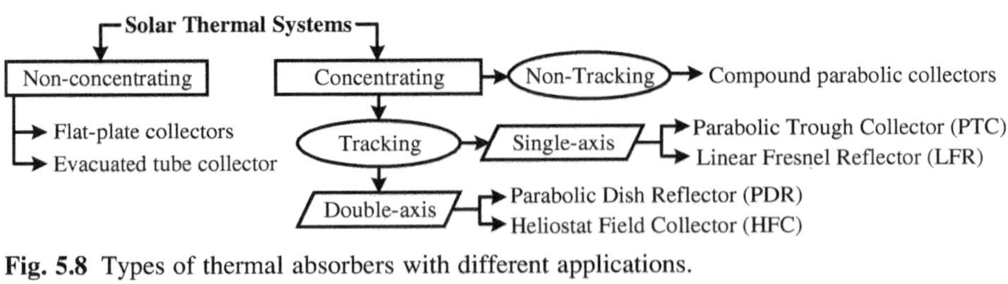

Fig. 5.8 Types of thermal absorbers with different applications.

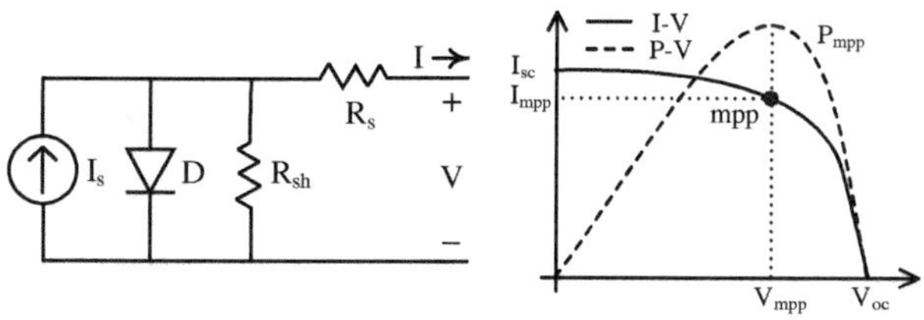

Fig. 5.9 Equivalent circuit and curves of a photovoltaic cell.

power as possible from these cells. Constant voltage method (the simplest method), Perturb & Observe technique (P&O), or hill climbing (the most popular method), and Incremental Conductance (IC) are among the most widely used MPP tracking procedures.

Solar modules can be constructed by connecting solar cells in series (higher voltage) or parallel (higher current) to achieve higher voltages and currents. These modules are typically manufactured in 36-cell (12 V) and 72-cell (24 V) clusters covering an area of about 0.5–1 m^2. They are capable of producing about 100 W/m^2 (almost 10% efficiency). Bypassing and blocking diodes are typically used with these modules to solve issues such as the diodes being forward biased at night (when there is no radiation) or complete and partial shadow of the module. These modules are also made series and parallel to each other at the installation site and form solar panels. Power electronic converters are typically used in order to connect to these panels and control them.

Solar cells are typically made of crystalline silicon (old and common) or thin film (cheap and new) materials. Advances in construction materials that have reduced cell prices have a great impact on solar resources wider usage. Refs. [20–22] are suggested for further study on solar energy.

4.3.6 *Wind resources*

Wind is a moving air that moves from high pressure to low-pressure areas due to the pressure difference. This difference in the air pressure stems from the nonuniform warming of the earth by the sun. Wind energy has long been used by windmills and sailboats. However, this energy has attracted attention as a means for generating electricity in just about the last 30 years. Like solar energy, the enormous potential to generate electricity from the wind is dispersed throughout the earth.

Unlike solar energy, which has an almost ubiquitous average or below-average potential for electricity generation worldwide, the potential of generating electric power from wind can be estimated as either good or very bad in one place. Furthermore, unlike solar energy, this resource is not quite applicable to very small and household scales. In recent years, offshore wind energy has received particular attention due to its advantages, such as faster and more stable winds at the sea and no visual or noise pollutions. Although offshore wind is currently about 5% of the wind energy installed capacity, a special count has been made on it for the future of this industry.

Describing the wind in an area is crucial for turbine designing purposes to extract as much energy as possible from it. Wind speed is usually described by distribution functions (usually Weibull and Riley distributions) based on historical data gathered from a specific area. The windrose graph is also used to indicate the prevailing wind speed and directions. The time scale of the wind in an area is usually described based on scattering indices and tools such as the Van Der Hoven diagram. Besides, different wind types classifications are used based on indicators related to its speed and power density.

Wind turbines can be classified into two main schemes: Horizontal Axis (HAWT) and Vertical Axis (VAWT). These turbines can be designed as upwind or downwind. There are various designs for HAWT (the axis of rotation is perpendicular to the ground and wind stream direction), the most important types of which are Savonius, Darius, and Gorlov. VAWTs (the axis of rotation is parallel to the ground and wind stream direction), are also available with specific designs and a different number of blades. VAWTs have the advantages of higher conversion efficiency and accessibility to higher wind speeds (due to their taller tower). In contrast, HAWTs have lower installation and maintenance costs (due to the generator and gearbox installation location) and are not sensitive to wind direction. Overall, the most common type of wind turbine is the three-blade upwind vertical axis wind turbine.

A wind turbine can only absorb a fraction (power factor) of the wind kinetic energy. It is proved that the maximum value of this power factor is theoretically 0.593 (Betz limit). Turbines must operate at an Optimal Tip Speed Ratio (OTSR) to get as close as possible to their maximum power. The lower rotation speed allows the wind to pass through the blades easily without transferring energy to them, and the higher rotation speed leads to blocking part of the wind and not extracting energy from it. The typical power curve of a wind turbine generator is shown in Fig. 5.10.

Due to insufficient torque and in order to prevent mechanical damage, no power is taken from a wind turbine before the cut-in speed and after the cut-out speed. Tracking methods such as Hill-Climb Search (HCS), Tip Speed Ratio (TSR), and Power Signal Feedback (PSF) are used for MPPT between cut-in and rated speeds. The turbine-generator power is limited to the nominal value between the rated and cut-out speeds by means of stall and pitch control methods (pitching the blade to reduce the amount of intersection) because of the generator limited power of the generator. The yaw control

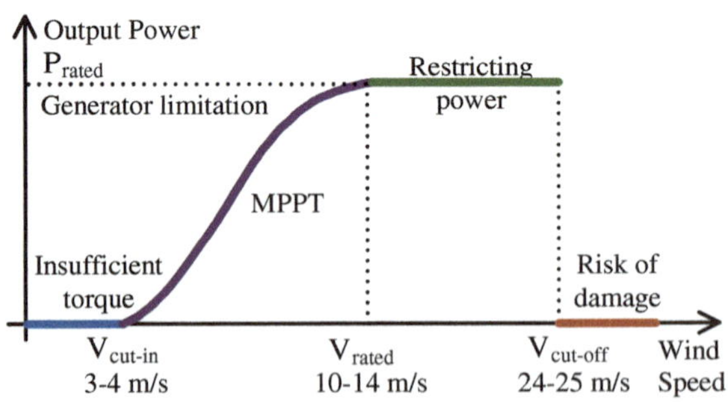

Fig. 5.10 A conventional power curve for a wind turbine generator.

Table 5.2 Different types of wind turbines-generators with their specifications and features.

Type	Generator	Speed variability		Reactive compensation and grid connection
		Device	Range	
A	Squirrel Cage Induction Generator (SCIG)	None	0%	Capacitor bank Soft starter
B	Wound Rotor Induction Generator (WRIG)	Variable resistance on rotor	10%–15%	Capacitor bank Soft starter
C	Doubly-Fed Induction Generator (DFIG)	Part Load Convertor (PLC) on rotor	30%–40%	No need for additional device
D	Permanent Magnet Synchronous Generator (PMSG)	Full Load Convertor (FLC) on stator	100%	No need for additional device

(changing the direction of the plane of entire blades) is also used to position the turbine in the wind direction.

Wind turbines generators can be classified into four categories from an electrical point of view. These categories and their specifications and features are listed in Table 5.2. In the last 30 years, technological progress (from types A to D) has led to an increase in the nominal capacity of individual wind units and the broader use of wind resources worldwide. Refs. [23, 24] are recommended for further study on wind energy.

4.3.7 *Marine resources*

As is already discussed in the previous sections, hydro resources extract the energy of rivers. On the contrary, marine resources use the energy of the seas and oceans. These resources can be classified into three categories, namely the wave, tidal, and ocean thermal. Energy extraction from these resources has not yet become operational on a large scale. Each of these resources is briefly mentioned below. Ref. [25] is suggested for further studies in this regard.

Waves are created by the wind interacting with the surface of water and are a combination of longitudinal and transverse mechanical waves. Thus, a particle on the water surface has an elliptical motion (up and down, front and back). The energy (potential and kinetic) that these waves transmit is proportional to the wavelength and the square of its height. So far, various equipment has been proposed to absorb this energy, which can often not withstand the damage caused by the waves in the long run. Three examples of such equipment are the following: point absorber buoy (going up and down with the waves and often using linear generators), Pelamis wave power (several floating pieces to drive hydraulic pumps), and oscillating water column (shore-mounted air chamber that is compressed by the waves and drives the air turbine).

The difference in the gravitational forces of the moon and the sun on different parts of the earth (due to different distances) leads to the deformation of the earth, especially the waters. This, along with the Earth's rotation, causes tidal currents in different parts of the earth at different times. One way to extract energy from these streams is to use turbines similar to HAWT and VAWT underwater. To extract more energy, dams (the barrage and lagoon types) are constructed to create a significant head on their sides. In addition, newer designs such as dynamic tidal power have also been proposed, which includes a large T-shaped dam perpendicular to the shore. The major concern regarding different tidal energy schemes is their impact on marine organisms and habitats.

The ocean thermal schemes use the temperature difference between the surface water and the ocean depth (about 20 K) to drive a heat engine (with the Rankin cycle). Special working fluids such as ammonia (NH_3) should be used in these schemes due to the low temperature at where they are installed. The low-temperature difference leads to very low efficiency (6%–7%) according to the Carnot limit. The practical efficiency is even lower than that, and this causes these projects to have little economic justification.

4.3.8 Biological resources

Bioenergy is a type of renewable energy that is obtained from materials of biological origin. These materials, called biomass, absorb the sun's energy, often through photosynthesis (plants, algae, and bacteria), and store it as chemical energy. The fuel that results from the application of various processes to biomass is called biofuel and can be in one of the three following forms: solid (such as wood and coal), liquid (such as bioethanol that is used in transportation, and biodiesel that can be used instead of traditional diesel fuel), and gas (such as biogas that is derived from the decomposition process).

Since CO_2 is absorbed and O_2 is released in photosynthesis, CO_2 from burning these resources can be ignored. In fact, these types of resources can be considered renewable and without pollution due to their much shorter period of CO_2 absorption and release compared to fossil fuels.

Fig. 5.11 illustrates various types of biomass resources, their conventional conversion technologies, and typical products. Refs. [26, 27] are suggested for further read on this topic.

4.3.9 Geothermal resources

Geothermal energy originates from the underground heat, which is affected by several different mechanisms, the most effective of which is the radioactive decay of different isotopes of Uranium and Thorium. Obviously, this resource is considered renewable due to its inexhaustibility. However, hardly accessible volcanic regions are the areas with the highest potential for extracting energy from this resource. Therefore, the use of these resources is not very convenient and economical. Although most of the

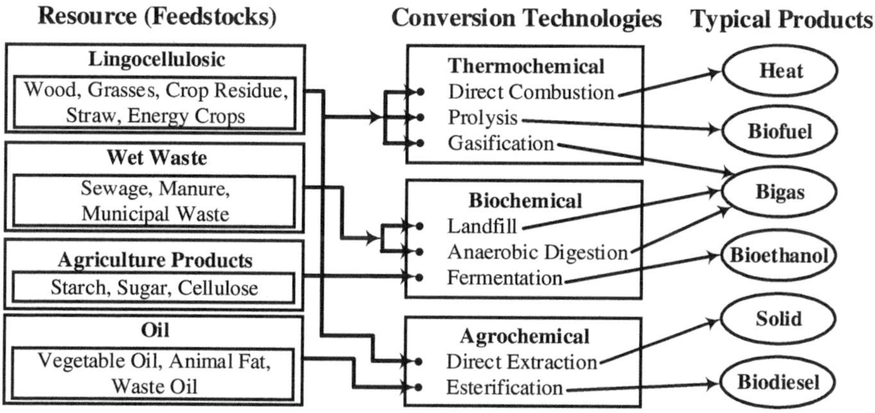

Fig. 5.11 Biomass resources and their conventional conversion technologies.

attractive sites of these resources have already been used, they still do not have a significant share in the worldwide electricity generated by renewable resources (they currently have an approximately 15GW of installed capacity). Besides, unlike most renewables, the energy derived from these resources is not intermittent.

These resources can be classified into two main categories: convective hydrothermal (heat transfer through boiling water in liquid, vapor, and mixed phases) and hot rock (heat transfer through contact of pumped water with solid rocks and magma). There are also three general schemes to extract electrical energy from these resources: dry steam (starting the turbine with pure erupted steam), flash steam (separating steam from mixed erupted water by means of one or two flash systems), and binary cycle (using separate heating and main Rankin cycles). Refs. [28, 29] are suggested for further read on geothermal energy.

5 Decentralization viewpoint

Some green approaches, such as CCS-equipped power plants, large wind and solar photovoltaic farms, and demand-side management of large-scale industrial plants, can be aggregated into the conventional power grid design with some considerations. However, the widespread use of renewable resources on a smaller scale and demand response in the smart grid context are redefining some concepts for future power grids. In fact, the need for more energy has led to the extensive development of power grids and their complexity. Decentralization seeks to reduce the complexities encountered in managing such a system by transforming it into independent subsystems without the need for central decision-making. The purpose of this section is to briefly outline the key concepts that facilitate the decentralization of the future green approaches-based power system. Therefore, by adopting a conceptual order, distributed generation, demand response, energy storage, role of power electronics, grid integration issues, and microgrids will be briefly discussed in this section.

5.1 Distributed generation

Traditionally, electrical power was generated in bulk and delivered to the final consumer through various networks at different voltage levels. Many factors, such as complexity and noncompliance with the varying needs of the grid, have called the suitability of this paradigm into question. Since power grids have been developed based on this concept for more than 100 years, fundamental changes will not be straightforward and easy. There is now a massive infrastructure for power generation and transmission that must be used for the rest of their useful life to return their capital investments. Fortunately, some of the green approaches are compatible with this conventional paradigm, which facilitates the transition period. However, decentralization has been accepted as one of the key principles of future power grids. One aspect of this decentralization is the shift from bulk generation to Distributed Generation (DG). This means that power generation is done on smaller scales and closer to the demand. This concept is consistent with the distributed (and sometimes dispersed nature) of most renewable resources (especially solar and wind) as the key aspect of green approaches.

DGs are also called small-scale generations, Distributed Energy Resources (DERs), embedded generations, etc. in the literature. However, there are differences between some of these definitions in different references. In general, DGs can be divided into two main categories: nonrenewable (such as reciprocating motors and microturbines) and renewable (such as solar photovoltaic, wind, bioenergy, small hydropower, etc.). Since different types of renewable resources are described in Section 4.3, nonrenewable DGs are briefly introduced below. The issues encountered when integrating these resources into the grid are also described in Section 5.5.

One of the main advantages of nonrenewable DGs over most renewable resources is their constant fuel availability and, hence, their dispatchability. This feature is critical if DG is to be used to power a remote area not connected to the grid or when it is intended to be used as the emergency power for sensitive facilities such as hospitals. It is also possible to use DGs for simultaneous generation of electricity and heat in CHP schemes due to their thermal basis of operation. This is especially attractive for small industries, commercial buildings, and residential building blocks due to the increased efficiency and reduced energy costs. Although these types of DGs allow for the paradigm shift and separation from centralized production, they have all the disadvantages of using fossil fuels, such as finite resources, energy security, and especially environmental issues.

Reciprocating engines are internal combustion engines that use multiple pistons to convert combustion pressure into rotational motion. These engines, which typically use natural gas or diesel fuels, can be classified into two categories: Spark-Ignition (SI) and Compression-Ignition (CI). Microturbines are small gas turbines that operate based on the Brighton cycle.

Microturbines can be classified into single-shaft and double-shaft types with a nominal capacity of 20–500 kW. Reciprocating engines have simpler technology, lower investment costs, and higher fuel efficiency. On the other hand, microturbines have higher reliability and stability, higher power-to-weight ratio, less pollution and noise, and lower maintenance costs. Refs. [30, 31] are suggested for further study on DGs.

5.2 Energy storages

According to the law of conservation of energy, the power consumption and generation in electrical grids must be equal at all times. In the event of an imbalance, the difference is compensated by the kinetic energy of the synchronous rotating devices which cause an undesirable change in frequency. It is advantageous to have energy storage devices to cover this difference by discharging and charging in case of power shortage or excess, respectively. However, energy storage is not a simple, cheap, and common task. Common types of storage devices, classified based on their mechanical, chemical, electrical and magnetic, and thermal nature, are illustrated in Fig. 5.12.

Storage devices can be classified into the following categories in terms of their energy capacity and power rating (charging and discharging): long-term response (used for energy management purposes, e.g., to charge at valley load and discharge at peak load), short-term response (used for providing bridge power to maintain continuity of supply), and very short-term response (used for enhancing the power quality). Obviously, longer-term applications require higher capacity, while shorter-term applications require higher power ratings.

The mechanical pumped-storage (closing the water cycle of traditional hydropower plants with the help of a pump that returns water from bottom to top) and compressed air (compression and storage of air in the depth of ground by a compressor and using it in the gas turbine cycle) storage devices are the only large-scale energy storages in terms of both the energy capacity and the rated power. Therefore, it is possible to use them economically for energy applications in the transmission grid.

The supercapacitor energy storage (increasing the capacitance by adding porous spongy conducting materials between the capacitor plates) and Superconducting Magnetic Energy Storage (SMES) (reducing energy storage losses in the magnetic field using a superconducting wire in a very low-temperature chamber) have very low energy capacity. They are primarily used in applications such as power quality

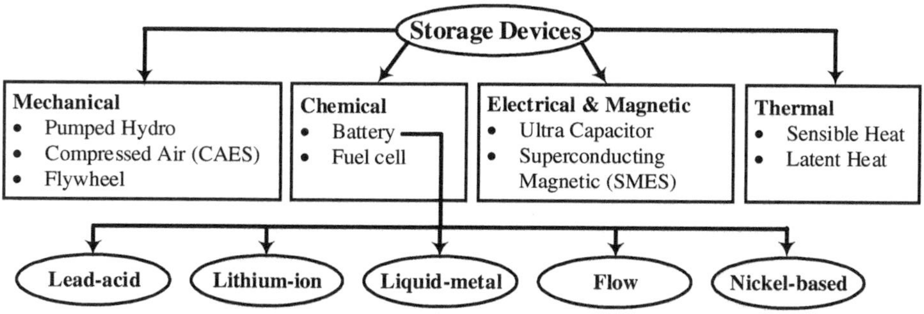

Fig. 5.12 Different types of energy storages.

improvement. The flywheel (storing energy in a heavy high-speed rotating body and preventing energy loss through creating a vacuum chamber and magnetically levitated suspension) has almost the same applications as supercapacitors and SMES, with relatively limited energy capacity.

Batteries (energy storage in an electrolyte interface between the two electrodes, the cathode and anode) are among the first sources of electrical energy used by humans. Different types of batteries the most important of which are lead-acid (commonly used as car batteries), lithium-ion (small rechargeable batteries), liquid-metal (with an easy construction process and long lifetime), flow (with electrolyte in a separate chamber requiring pump), and nickel-based (costly) usually have features (energy density, price, etc.) and advantages that allow using them for a variety of applications (often in small sizes).

Fuel cells convert the incoming chemical energy (often hydrogen gas) into electrical energy utilizing oxygen in a mechanism similar to batteries (by using electrolytes and electrodes) and produce water as the output. Unlike batteries, fuel cells require a constant fuel flow to power themselves and are more similar to renewable resources in this regard. However, because hydrogen does not exist in its natural form on the earth and its production requires the use of another energy resource such as natural gas, fuel cells are often classified as energy storages. Fuel cells can be designed in power ranges from a few watts to several megawatts and can be used even in vehicles.

Similar to storing electrical energy in the valley load and returning it at the peak load, thermal energy can also be stored at high temperatures (e.g., day time) and returned at low temperatures (e.g., night time). The sensible heat (energy storage based on temperature changes in the medium) and latent heat (energy storage based on phase change in the medium) schemes are the two common mechanisms in this regard. Latent heat schemes can store more energy with much fewer temperature changes.

Energy storage devices are key supplements to intermittent renewables and critical tools for implementing demand response schemes (Section 5.3). The widespread emergence of electric vehicles has also made the storages (especially chemical type) more economically justifiable and has accelerated their development. These electric vehicles and their parking lots can have lots of applications in future power grids. Therefore, energy storages play a critical role in moving toward green approaches in future power systems. Refs. [32, 33] are suggested for further study on energy storages.

5.3 Demand response

In the conventional power grid, the loads were often dumb and inactive. They had almost no interaction with the grid except to receive their required energy. The grid operators were only in contact with some loads (often large industrial loads at transmission level) during peak periods. In recent years, the concept of demand response, in which financial incentives encourage the consumers to adjust their demand and consuming pattern, has received particular attention due to the importance of energy efficiency as one of the key green approaches. The main goal of demand response is to encourage consumers to consume less energy during peak hours and/or to shift energy

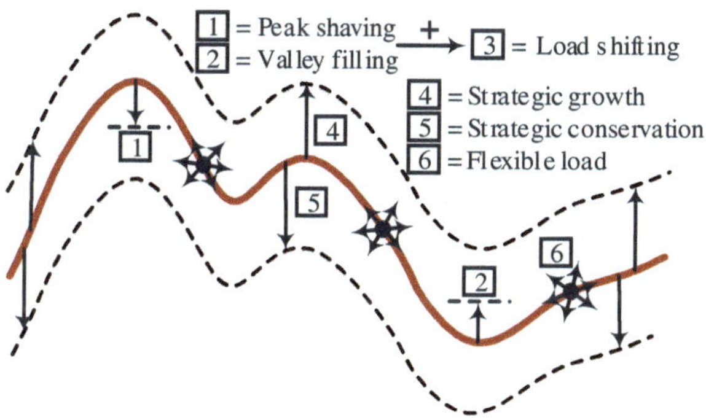

Fig. 5.13 Schematic representation of various demand-side management methods on the load profile.

consumption to off-peak hours such as night and weekend. It should be noted that demand response does not necessarily reduce the total energy consumption but can only smooth the power profile in a period of time (e.g., 1 day) in order to facilitate the supply of energy required.

Fig. 5.13 illustrates the schematic representation of some important types of load changes due to demand response. These include peak shaving (reducing the maximum amount of load, which reduces the planning needs), valley filling (increasing load amount during the off-peak period, which leads to flattening of the load profile and increasing the load factor of the grid), load shifting (transferring the load from the peak to the valley, which has the advantages of peak shaving and valley filling simultaneously without changing the total energy consumption), strategic growth and conservation of load (purposeful change of total grid energy consumption according to requirements), and flexible load (the consumer can change its consumption at any time as desired, which can have many technical applications).

The growing technology of smart sensors and meters has provided the loads with the ability to track and respond to the dynamic prices of the electricity market and has made it possible to take advantage of this potential resource. The proliferation of DGs and energy storage devices in distribution grids has also transformed the consumer (just consuming energy) into a prosumager (consumer, producer, and storage) and made it possible to use a variety of demand response schemes. These prosumagers enable decentralization in the form of microgrids which is described in Section 5.6. Refs. [34–37] are recommended for further study on demand response schemes.

5.4 Role of power electronics

As the most important green approach of the future, renewable energies are fundamentally different from conventional generations in terms of the energy conversion process and power control. Among the most important differences are the DC power generation by photovoltaic cells and the use of various generator types in different wind turbines. DC to DC converters, back-to-back converters, soft starters, active power filters, etc. are some of the various types of power electronic devices that enable different control mechanisms (power, current, voltage, etc.) to extract power from the renewables and improve the power quality. The role of these devices in providing various solutions for the renewable energies integration issues (see Section 5.5) is also very prominent.

On the other hand, the growth and development of power electronic devices utilization in controlling different types of loads (such as variable speed motor drives) as well as the increasing integration of DC loads (such as data centers) into the grid, has made the role of power electronic devices more substantial in future power systems decentralization. In fact, it can be said that the advancements in various aspects of power electronic devices have made the concept of AC/DC hybrid grids operational.

The future power grids load types will be completely different due to the expansion of DC loads and AC loads equipped with power electronic devices. For example, the frequency and voltage dependencies of loads controlled by power electronic devices are very different from the regular AC ones. For instance, conventional induction motors in a conventional grid are sensitive to the grid frequency and participate in frequency control through inertial response and damping, which is not the case with variable speed motors controlled by power electronic drives. Power electronic devices that allow loads to be controlled in various manners for different situations play a key role in meeting the potential challenges of such changes. Refs. [38, 39] are recommended for further studies.

5.5 Grid integration issues

Grid integration of decentralized and renewable generations is a complex process. That's because all of the current implemented technical solutions for various issues of the grid are designed based on the concept of centralized generations. As shown in Fig. 5.14, the renewables integration issues can be classified into two main categories: the issues related to the distribution grid (due to the emergence of DGs) and those related to the transmission grid (due to high penetration of DGs and large-scale renewable resources). These issues are briefly discussed below. Refs. [30, 40–42] are recommended for further studies.

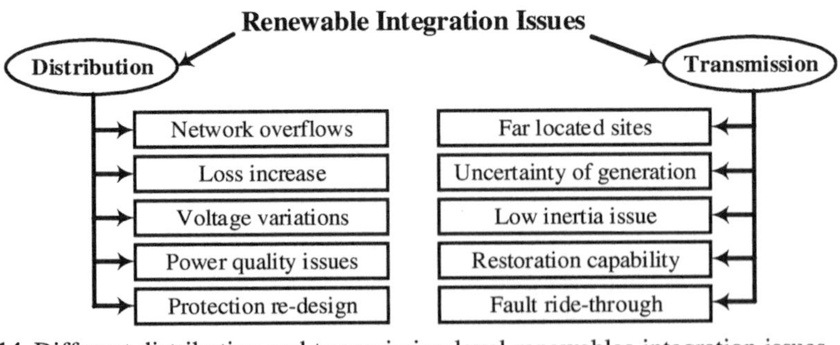

Fig. 5.14 Different distribution and transmission level renewables integration issues.

5.5.1 Distribution grid issues

The emergence of DGs in the distribution grid causes the generation to be closer to the final consumer and consequently reduces potential overloads and grid losses. However, increasing the penetration of DGs in the distribution grid without conducting sufficient technical studies may reverse this trend. According to studies, such an issue has a negligible impact on grid losses. However, line overflow must be considered as an important technical constraint in grid hosting capacity calculations.

The main voltage issue in conventional distribution grids is voltage drop with the increasing length of feeders. Various solutions such as tap changing transformers and capacitor banks have already been used to deal with this issue. However, overvoltage can be considered as one of the most severe concerns of DGs integration into the distribution grid (it reduces the hosting capacity to zero in some cases). This issue can be mainly solved using DGs voltage control capability (with and without power electronic devices).

DGs integration often leads to power quality issues such as increased harmonics and voltage fluctuations (flickers). However, these issues are not usually considered as severe. In fact, loads are often the primary source of these issues in the distribution grids. Power imbalance is a critical issue of using single-phase DGs (which are typically used in distribution grids). In addition, resonance problems that may arise due to capacitor banks associated with DGs should also be considered. Despite all these issues, DGs (especially the ones equipped with power electronic devices) can manage these issues and even can be used to improve the power quality in the distribution grid.

The distribution grid protection is based on overcurrent relays. DGs can interfere with these relays and cause their incorrect (when a fault occurs) or unwanted (under normal conditions) operation. That's because DGs can participate in the normal or fault currents flowing through protection relays according to the grid configuration. These issues are mainly related to DGs with synchronous generators and even the induction ones are directly connected to the grid. DGs that use power electronic devices for grid integration usually have a minimal contribution to the fault current.

5.5.2 Transmission grid issues

DGs integration into the transmission grid is typically desirable since it results in a reduction in the power flowing through transmission lines and the risk of transient instability. However, the increasing integration of DGs into the distribution grid and the emergence of large-scale renewable resources in the transmission grid come with new issues and challenges. For instance, large renewable resources sites (such as offshore wind farms and solar photovoltaics in deserts) are often far from the existing transmission infrastructure. Thus, the electrical connection of these remote sites to the grid is often costly and sometimes requires new technologies, such as underwater HVDC lines.

One of the main issues encountered by the increasing penetration of renewable resources into the grid is the addition of electricity generation uncertainty to the existing uncertainty of the load. The high penetration of these types of resources (mainly intermittent) leads to the emergence of different hourly, daily, weekly, monthly, seasonal, and annual power patterns, complicating the grid operation. Some studies have focused on increasing the accuracy of various forecasts such as wind speed, precipitation, cloud cover, etc. Other studies have tried to include these types of resources in different grid studies to enable the optimality of the grid considering their uncertainties.

The high penetration of renewable resources also affects different power grid stability types. That's because the integration of these resources usually causes new and less studied dynamic behavior of the grid by reducing the share of conventional synchronous generations. The main concern in this regard is reducing the grid inertia that causes severe issues in frequency stability. Unlike conventional synchronous generators, renewable resources connected to the grid through power electronic devices do not have a natural inertial response to grid power imbalances. This causes a significant increase in the Rate Of Change Of Frequency (ROCOF) and maximum frequency deviation after an incident in the grid. However, it is possible to make these resources imitate the behavior of conventional generators using the power electronic devices for control schemes such as virtual droop and inertia. Energy storages and frequency responsive loads can also be used to compensate for the mentioned issue. Ref. [43] is suggested for further readings on the topic of power system frequency control.

Increasing the penetration of nonrenewable DGs into the grid enhances the ability for fast and local restoration of electricity to essential loads after incidents. Renewable generations equipped with power electronic devices can also have a black-start feature using little energy. They can be used to recover the grid after blackouts with their relatively short start-up time. However, they cannot be counted on definitively for this purpose due to their generation dependence on the availability of energy sources.

Furthermore, the integrated renewable generations into the grid (especially large-scale) must have the necessary tolerance facing the disturbances and their consequences, i.e., under and over voltage and frequency conditions. This feature (fault ride-through) in renewables has fundamental differences from conventional generators due to the sensitivity of electric machines and power electronic converters used in renewable generations. This issue has become a research topic in recent years.

5.6 Microgrids

A microgrid (see Fig. 5.15) is a group of loads, DGs, and storage devices (a set of prosumagers) with defined electrical boundaries that act as a controllable entity in connection with the main grid. The microgrid can operate in both islanded and grid-connected modes through a breaker and often a converter. The differences between microgrids and active distribution networks are the existence of clearly defined electrical boundaries and the possibility of operating islanded from the main grid. For this purpose, the installed generation capacity of the microgrid must be more than its peak load. An energy management system is also vital in a microgrid. This system uses the information gathered from various sources (energy and ancillary

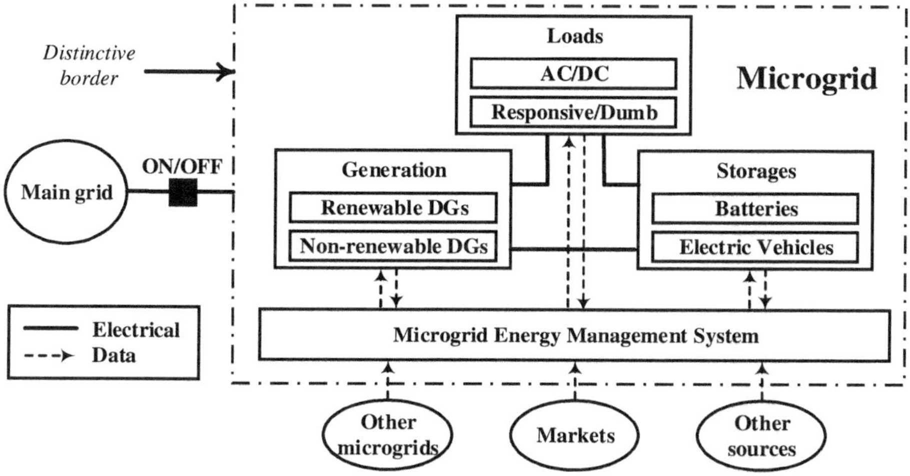

Fig. 5.15 Schematic representation of the microgrid components and its connections.

services markets, other neighboring microgrids, air forecasting systems, etc.) to send the desired control signals to microgrid equipment.

As it turns out, microgrids provide a platform for realizing the most important green approaches while solving their integration issues. They also give an operational meaning and applicability aspect to the concept of decentralization in the power grid due to their possibility of operating independently. In fact, microgrids are the independent blocks that make up the future smart grids. Refs. [44, 45] are recommended for further study on microgrids.

6 Conclusions

In this chapter, the need to move toward green approaches in future power systems was first discussed by explaining the major energy issues, namely, the limitation of fossil resources, environmental issues, and energy security concerns. Then, different types of green approaches (clean power plants, energy efficiency, and renewables) were described. Different types of renewable resources including hydro, solar, wind, marine, biological, and geothermal were also described. These green approaches were then addressed from a decentralization viewpoint. In this regard, DGs and demand response were described as the applied and decentralized versions of two green approaches (renewables and energy efficiency). Energy storages and power electronic devices were also described as the key devices that facilitate the implementation of green approaches. Furthermore, the renewable resources integration issues into the distribution and transmission grids were discussed. Finally, the concept of microgrids was presented as a key platform for the implementation of decentralized green approaches in future power systems. Overall, it can be concluded that moving toward green approaches is necessary and inevitable, and the growing movement toward its realization in the future power system in a decentralized manner can be done and is even happening now, despite the problems ahead. The main purpose of this chapter was to convey a general and macro view of the topic. Therefore, appropriate references have been introduced throughout the chapter for further studies.

References

[1] S. Gu, M. Xie, X. Zhang, Green Transformation and Development, Palgrave Macmillan, 2019.
[2] I. Scoones, P. Newell, M. Leach, The Politics of Green Transformations, Routledge, 2015.
[3] A.N. Penna, A History of Energy Flows: From Human Labor to Renewable Power, Routledge, 2019.
[4] R. Haas, L. Mez, A. Ajanovic, The Technological and Economic Future of Nuclear Power, Springer Nature, 2019, p. 385.
[5] D. Infield, L. Freris, Renewable Energy in Power Systems, John Wiley & Sons, 2020.
[6] Our World In Data (OWID), 2021. Available online at *https://github.com/owid*.
[7] R. Ehrlich, H.A. Geller, Renewable Energy: A First Course, CRC Press, 2017.
[8] N. Wood, K. Roelich, Tensions, capabilities, and justice in climate change mitigation of fossil fuels, Energy Res. Soc. Sci. 52 (2019) 114–122.
[9] M. Asif, Energy and Environmental Security in Developing Countries, Springer Nature, 2021.
[10] N. Mouraviev, Koulouri, A. (Eds.),, Energy Security: Policy Challenges and Solutions for Resource Efficiency, Springer, 2018.
[11] N. Abas, A. Kalair, N. Khan, Review of fossil fuels and future energy technologies, Futures 69 (2015) 31–49.
[12] T. Wilberforce, A. Baroutaji, B. Soudan, A.H. Al-Alami, A.G. Olabi, Outlook of carbon capture technology and challenges, Sci. Total Environ. 657 (2019) 56–72.
[13] K. Awuah-Offei, K. Awuah-Offei, Doyle, Energy Efficiency in the Minerals Industry, Springer, Rolla, 2018.
[14] U. Desideri, Asdrubali, F. (Eds.),, Handbook of Energy Efficiency in Buildings: A Life Cycle Approach, Butterworth-Heinemann, 2018.
[15] J.G. Gupta, S. De, A. Gautam, A. Dhar, A. Pandey, Introduction to sustainable energy, transportation technologies, and policy, in: Sustainable Energy and Transportation, Springer, Singapore, 2018, pp. 3–7.
[16] M. Yang, X. Yu, Energy Efficiency: Benefits for Environment and Society, Springer, 2015.
[17] International Renewable Energy Agency (IRENA), Available online at *https://www.irena.org/Statistics*.
[18] H.J. Wagner, J. Mathur, Introduction to Hydro Energy Systems: Basics, Technology and Operation, Springer Science & Business Media, 2011.
[19] F.R. Førsund, Hydropower Economics, Springer, 2015.
[20] R. Foster, M. Ghassemi, A. Cota, Solar Energy: Renewable Energy and the Environment, CRC Press, 2009.
[21] S.C. Bhatia, R.K. Gupta, Textbook of Renewable Energy, Woodhead Publishing India PVT. Limited, 2019.
[22] H. Tyagi, A.K. Agarwal, P.R. Chakraborty, S. Powar, Introduction to advances in solar energy research, in: Advances in Solar Energy Research, Springer, Singapore, 2019, pp. 3–11.
[23] T. Ackermann (Ed.), Wind Power in Power Systems, John Wiley & Sons, 2012.
[24] K. Starcher, Wind Energy: Renewable Energy and the Environment, CRC Press, 2018.

[25] A. Khaligh, O.C. Onar, Energy Harvesting: Solar, Wind, and Ocean Energy Conversion Systems, CRC Press, 2017.
[26] K.R. Hakeem, M. Jawaid, U. Rashid, Biomass and Bioenergy, Springer International Publishing, 2016.
[27] P.K. Sarangi, S. Nanda, P. Mohanty (Eds.), Recent Advancements in Biofuels and Bioenergy Utilization, vol. 232, Springer, 2018.
[28] A. Dincer, M. Ozturk, Geothermal Energy System, Elsevier, 2021.
[29] A. Manzella, A. Allansdottir, A. Pellizzone (Eds.), Geothermal Energy and Society, Springer International Publishing, 2019.
[30] M.H. Bollen, F. Hassan, Integration of Distributed Generation in the Power System, vol. 80, John Wiley & Sons, 2011.
[31] R. Bansal, Handbook of Distributed Generation, Springer International Publishing, Switzerland, Switzerland Valsan SP, 2017.
[32] M. Sterner, Stadler, I. (Eds.),, Handbook of Energy Storage: Demand, Technologies, Integration, Springer, 2019.
[33] Z. Hu, Energy Storage for Power System Planning and Operation, John Wiley & Sons, 2020.
[34] P. Du, N. Lu, H. Zhong, Demand Response in Smart Grids, vol. 262, Springer International Publishing, 2019.
[35] S. Nojavan, K. Zare (Eds.), Demand Response Application in Smart Grids: Operation Issues-Volume 2, Springer Nature, 2019.
[36] H.A. Aalami, M.P. Moghaddam, G.R. Yousefi, Demand response modeling considering interruptible/curtailable loads and capacity market programs, Appl. Energy 87 (1) (2010) 243–250.
[37] P.T. Baboli, M.P. Moghaddam, M. Eghbal, Present status and future trends in enabling demand response programs, in: 2011 IEEE Power and Energy Society General Meeting, IEEE, 2011, July, pp. 1–6.
[38] G. Rigatos, Intelligent Renewable Energy Systems: Modelling and Control, Springer, 2016.
[39] Q.C. Zhong, T. Hornik, Control of Power Inverters in Renewable Energy and Smart Grid Integration, vol. 97, John Wiley & Sons, 2012.
[40] J. Hossain, Mahmud, A. (Eds.),, Renewable Energy Integration: Challenges and Solutions, Springer Science & Business Media, 2014.
[41] D. Jayaweera (Ed.), Smart Power Systems and Renewable Energy System Integration, Springer International Publishing, 2016.
[42] M.I. Alizadeh, M.P. Moghaddam, N. Amjady, P. Siano, M.K. Sheikh-El-Eslami, Flexibility in future power systems with high renewable penetration: a review, Renew. Sustain. Energy Rev. 57 (2016) 1186–1193.
[43] H. Seifi, H. Delkhosh, Model Validation for Power System Frequency Analysis, Springer, 2019.
[44] N. Hatziargyriou (Ed.), Microgrids: Architectures and Control, John Wiley & Sons, 2014.
[45] M.S. Mahmoud (Ed.), Microgrid: Advanced Control Methods and Renewable Energy System Integration, Elsevier, 2016.

Blockchain for future renewable energy

6

Jianguo Ding[a,b] *and Vahid Naserinia*[b]
[a]Department of Computer Science, Blekinge Institute of Technology, Karlskrona, Sweden,
[b]School of Informatics, University of Skövde, Skövde, Sweden

1 Introduction

Renewable energy continues to grow in the energy market and will dominate the energy landscape in the future.

The EU's overall energy demands had been estimated to reach at least 20% by 2020, and the Nordic targets were higher: 30% for Denmark, 38% for Finland, 72% for Iceland, 67.5% for Norway, and 49% for Sweden. Nevertheless, all Nordic countries met or exceeded their targets 2 years earlier. These results are due to the flexibility of electricity markets in the region, the favorable starting position in the energy mix in sustainable power to others, and the increased demand for carbon-free energy, heating, and transportation. Nordic energy consumption is about 100 million tons of oil equivalent (Mtoe), with a low of 96.6 Mtoe in 2009 and a high of 104.2 Mtoe the following year after the global financial crisis [1,2].

Fig. 6.1 shows the trend of energy consumption in the Nordic countries. Governments around the world have set targets in official policy or legislation to increase installed renewable energy capacity by 2030. To meet these targets, about 721 GW of wind, solar, biomass, waste-to-energy, geothermal, and marine power plants would need to be built over the next decade [3].

Therefore, the growth trend of renewable energy is a sign that renewable energy will play an important role in the future energy landscape and will have a profound impact on the development of society as a whole. But, how to develop and use renewable energy efficiently is an urgent issue nowadays.

The traditional structure of the power grid, the framework for electricity trading, and the management system are largely centralized. The decentralized system is based on distributed production, electricity storage, and the desire of the market [4]. The centralized system relies on a third party (intermediary) to handle transactions between suppliers and consumers. In centralized energy trading systems, some vulnerabilities remain and affect consumer satisfaction because they can disrupt business [5]. Increasing renewable energy technology and the market will change the energy landscape and require new technical solutions to manage the noncentralized renewable energy system.

(1) Distributed generation: Known as embedded generation, on-site generation, scattered generation, and decentralization, distributed generation is the essential component of a

Decentralized Frameworks for Future Power Systems. https://doi.org/10.1016/B978-0-323-91698-1.00011-X

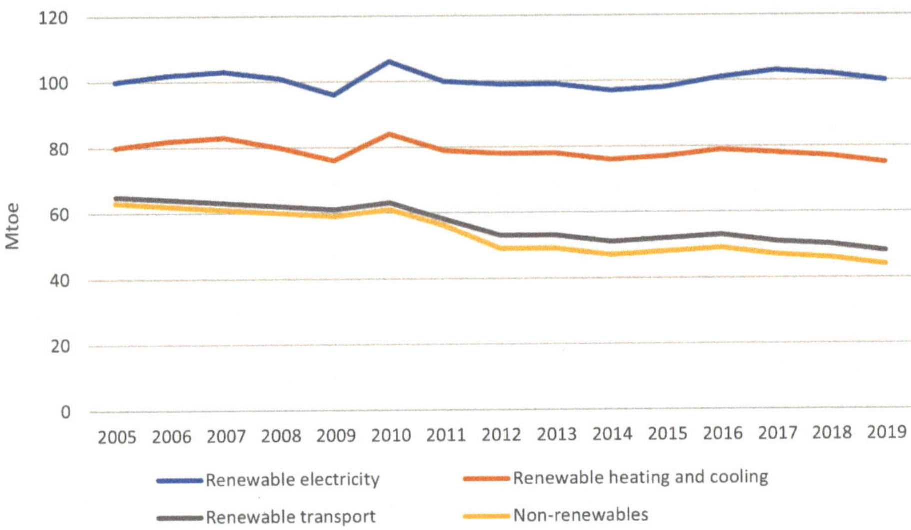

Fig. 6.1 Energy consumption in the Nordic countries.

decentralized energy system. The decentralized generation of heat and power is possible. However, heat cannot be transmitted over long distances and therefore has always been generated locally. The shift to decentralized power generation makes it possible to coordinate heat and power generation in cogeneration plants. In this way, electricity and heat production boost the system's efficiency because heat is a by-product of many electricity generation processes.

(2) Electricity storage: A fundamental constraint of electricity distribution is that electricity cannot be retained and generated on demand. Additional generation sources could create more problems in regulating supply to best meet demand in a decentralized system. Nevertheless, battery storage, compressed air, and pumped hydro-storage systems can help maintain the grid stability by storing energy when supply exceeds demand and feeding it back to the grid during peak periods. The storage of intermittent energy plants is especially useful, often generating at their highest capacity in the nonpeak hours. Storage can be decentralized to enhance efficiency and generation; it can be off-grid or grid-tied.

(3) Demand response: Demand-response systems provide another option for managing grid stability if decentralized generation is connected to the grid. Grid management has traditionally concentrated on supply management. But, emerging technologies, such as the smart grid and smart meters, enable power suppliers and users to monitor and communicate in real-time to improve system utilization. Many electricity users will sometimes also be energy producers with distributed generation and storage. In order to construct a really decentralized energy system, smart grid technology is needed to facilitate grid management.

(4) Electricity industry: A decentralized energy system is a relatively recent strategy in many countries. The power sector has traditionally concentrated on developing large, centralized power plants and transferring generation loads to consumers in the region through extensive transmission and distribution lines. Decentralized energy systems are aimed at bringing energy sources closer to end-users. End-users are scattered across an area so that energy generation can reduce inefficiencies in transmission and distribution and the corresponding economic and environmental impacts in a similar decentralized fashion.

Decentralized blockchain technology is worth speculating on renewable energy. It offers numerous benefits such as cost savings, efficiency, and an efficient procedure [6]. From an industrial perspective, blockchain offers various advantages, like flexibility, scalability, and security. Blockchain also offers energy suppliers an alternative for their neighbors to sell excess energy. Moreover, in a decentralized blockchain system, third-party services are no longer required.

2 Challenges in renewable energy with decentralized frameworks for operation, management, and business

Renewable energy not only shows a significant growth trend, but it will also bring important changes to the current power grid and the future power industry and commerce. The production, transmission, distribution, cooperation, coordination, and renewable energy consumption are very different from traditional power systems. These differences are reflected in their extensiveness, dynamics, distribution, interactivity, and integration.

However, these distributed and decentralized renewable energy systems present new challenges.

2.1 Systemic

These challenges are not caused by a single, individual factor, but by problems inherent in the entire system [7–9].

The state-controlled electricity market hinders the development of decentralized electricity systems because a distributed generation has encouraged many participants to become energy producers. Centralized management and control pose legal, administrative, and economic hurdles for developers. Therefore, technical, commercial, and trading mechanisms for off-grid and near-grid services need to be established to support IoE (Internet of Energy) development.

Management, control, and coordination among renewable energy system participants are complex and necessary. To identify violations and other regulatory issues, regulators increasingly require a large amount of information from E&R companies for evaluation. Gathering and cleaning up essential data is the brunt of modern technology and procedures. There is also a significant risk of data misuse, exposure of critical company data, and harm to the company. They can also be misused.

2.2 Quality

Power quality (PQ) is a very important aspect of distributed renewable generation systems because today's loads are more sensitive to PQ disturbances and nonlinear loads are increasing in power distribution networks.

Power quality challenges include: (1) Voltage and frequency oscillations in the integrated grid are caused by the unpredictable behavior of renewables as a result of often changing weather conditions. (2) Harmonics associated with the DG systems originating from the power electronic inverters used to integrate the renewables and inject active power into the grid, as well as from locally connected nonlinear loads at the point of common coupling (PCC).

Although voltage quality improvement by the VSM (virtual synchronous machine) method has great potential for stabilizing DG plants, if load balancing, harmonic, and reactive power compensation functions are performed in addition to improving voltage and frequency stability. There is also a need to develop some advanced control techniques for grid-connected inverters to improve the ability of DGs to handle both grid- and load-side disturbances, eliminating the need for additional custom power devices (CPDs) [10].

2.3 Technical

(1) Adequate planning: Without adequate planning, large-scale deployment of distributed power sources may result in unstable voltage distribution. The operational requirements of the entire power system will need to be redesigned to accommodate emerging technologies such as smart grids, renewable energy, energy storage, etc.

(2) Rapid demand response: Demand-response technologies require continuous, trusted access to the system via multiple communication technologies, which is not currently possible in many regions of the world.

(3) Decentralized operations: Decentralized operations are needed to provide adequate support for independent providers and consumers.

(4) Accelerating the energy supply chain: The efficient use of renewable energy depends on the ability to accelerate the process of energy production, storage, transmission, distribution, and consumption.

(5) Security: The security of the commercial chain from production to consumption is not in place. This covers the security of energy networks, financial transactions, and all online information.

(6) Resilience: An integrated and resilient renewable energy system will support cross-regional applications to adapt to dynamic energy markets. It is not possible to accurately calculate the total energy use in time. In turn, total consumption depends on hardware efficiency and climate disruption, depending on the availability of green or brown power in the region.

2.4 Economic

These are areas of dynamic economic conditions captured by income or other standard economic indicators [7,9]. Compared to centralized plants, the capital cost per kilowatt is generally higher for distributed electricity, mainly because of connection costs. Due to the high capital costs and longer life cycles, upgrading infrastructure to a more efficient transmission and distribution system is complex.

In addition to power distribution and transmission systems, integration with interconnected distributed generation resources is also considered for system stability reasons. Power transmission and transaction costs between distributed and small-scale generation and consumption are difficult to control.

2.5 Stability

The rotational inertia of variable renewable generators is significantly lower compared to synchronous generators. In the event of an imbalance between market forces, this leads to faster frequency fluctuations. Variable renewable generation is diverted or throttled when frequency fluctuations are too rapid [11,12].

Due to the short-term fluctuations in renewable generation, there is a decrease in frequency regulation reserves. Variable renewable generation, on the other hand, does not provide any frequency reserves at all. Small variable renewable energy generators do not have remote control interfaces. As a result of the uncontrolled injection of fluctuating renewable generation, unpredictable power flows occur, resulting in shortened equipment life, trips, or equipment damage [2,12].

Because variable renewable generation is site dependent, transmission distances between generation and consumption sites become longer, resulting in increased transmission losses. Variable renewables must be shut down when they operate outside a certain voltage band. For this reason, variable renewable generators shut down more frequently as voltage fluctuations increase. A larger grid area also results in shorter equipment life and the possibility of damage [2,13,14].

Interactions between renewable energy providers and the power grid are becoming more frequent, leading to unnoticed power fluctuations [10]. A lack of control can lead to shortened asset life, outages, or faults. As a result, there is a lack of coordination in setting voltage trip limits. This results in variable renewable generators tripping more frequently as voltage fluctuations increase. These cascading trips result in violations of dynamic stability standards or an increase in stability events.

2.6 Imbalance

The stability of renewable energy sources is not sufficient. This has led to higher reserve requirements and a more erratic imbalance between generation and load. Activation, dispatch, or curtailment of power must be balanced to maintain output.

Variable renewable generators have limited dispatch capability as their main resource supply shifts [15,16]. To compensate for unexpected outages of other power plants, variable renewable generators are limited. These unplanned mismatches between generation and load lead to unplanned activation of balancers and control reserve systems.

As renewable generation increases, the power requirements for conventional generation change, such as the need for a faster ramp-up. Therefore, generation-load imbalances, dispatch, or curtailment may occur in the short term [2,12,16].

Because variable renewable generation is unpredictable, forecast accuracy decreases. Power activation, dispatch, or curtailment are the result of these unanticipated imbalances between generation and load [10,15,17].

Insufficient long-term generation capacity indicates that as renewable generation increases, the performance criteria for conventional generation also change, such as nighttime or seasonal balancing of generation. These performance requirements must be met; otherwise, long-term generation-load imbalances are foreseeable [12,15].

3 Blockchain technology

Blockchain is a digital ledger of transactions recorded on the blockchain and transmitted throughout the computer network. Each block on the chain contains many transactions. Whenever a new transaction takes place on the blockchain, each participant's ledger is a record of the transaction. It is a comprehensive technology that integrates various research results such as consensus methods, encryption, and distributed data storage methods [18–20].

The key features of blockchain:

(1) Decentralization: Since the blockchain system uses the model of a P2P network, no control center is required. The blockchain is managed by decentralized nodes through consensus protocols and runs peer to peer, without trust authorization and without central trusted nodes.

(2) Immutability: As long as most nodes are not malicious, the content of the block cannot be changed. Each networked node in the blockchain has the same status in the system. All nodes in the system will save the created data block. Therefore, all nodes capture and store the transaction data, and the database is more robust.

(3) Untrusted but secure: In a blockchain, it is not necessary to blindly trust certain entities. The exchange of information between nodes in the blockchain system follows a predetermined procedure to prevent the identity of either party from being revealed. On the contrary, procedural standards are used in the blockchain to establish mutual trust. Nodes do not rely on a trusted intermediary for communication, and all records/transactions are protected using the principle of asymmetric encryption. Strong protocol algorithms protect the enforceability of blockchain data, nondestructive changes, and outside attacks. For example, the Bitcoin blockchain can only be achieved if 51% of the computing power is controlled. Therefore, the cost of data manipulation is much higher than the potential benefits. Therefore, it is unlikely that participants will attempt to manipulate data to improve the security of blockchain data.

(4) Openness, transparency, and auditability: The blockchain system uses trusted algorithms with open and transparent operating rules to govern transaction behavior. Data exchange between system nodes does not require trust. All data is open to all participants, and nodes are encrypted in the system, except for confidential information. Using the hash value of the block header, anyone can query the blockchain information. To update information, each node must authenticate each other, so the information of the whole system is very transparent and a node cannot mislead other nodes. As more and more nodes can join the network, the blockchain network can spread freely. Network nodes can verify the authenticity of records to ensure that the block is not altered. By opening all records to all users, transparency allows these blocks to be verified by any node in the network. Therefore, the entire transmission process of the transaction object can be fully tracked and recorded, making it easier to monitor the transaction.

(5) Resilience: Any disruption or malicious activity can be easily detected and recovered. The robustness and resilience come from the decentralization of the architecture; there is no single point of failure, and all nodes store their entire chain on their premises. Blockchain technology is particularly promising in industries where peer-to-peer networks (such as grid-connected power generators and consumer networks) rely on shared data sets.

(6) Automated contract execution: Smart contracts can be created on the blockchain, specifying the obligations of each participant and the terms and conditions for execution.

Blockchain technology automatically evaluates the contract terms. When it is determined that all contract requirements are met, the blockchain technology automatically fulfills the contract terms. Smart contracts improve the efficiency of contract execution and are critically executed without robust third-party monitoring.

4 Potential application of blockchain for future renewable energy

4.1 Electric vehicle

The widespread use of electric vehicles (EVs) will be a valuable response to the challenges of green travel and energy conservation. However, EV users have problems with charging. The number of charging points for electric cars remains modest compared to gas stations, which is the main bottleneck limiting the widespread use of electricity. Many EVs currently charge at stack operators and payment systems, but charging standards vary, leading to significant disadvantages for users. There is greater public acceptance of a unified, low-cost payment platform based on blockchain technology. On-time leasing of individual smart batteries for charging and common routing technology could also reduce the situation of a limited number of batteries for charging [21].

In the relationship between EVs and the grid, there are no "vehicle to grid" incentives; battery quality cannot be guaranteed when using electric battery cascades. Energy Blockchain Labs, for example, has developed a blockchain for car reactions, battery blockchain data storage, and virtual currency stimulus certification [22].

The JuiceNet application in California, a market-driven IoT platform, was provided by eMotorWorks. With this blockchain technology, charging station owners can use JuiceNet to rent out charging station time to drivers [23]. This means that the number of charging stations will increase. As a result, charging station owners can be compensated for their economic benefits at the same time. JuiceNet uses blockchain technology to store and process data in real time, making it safer and more accessible for individuals to connect to and share charging services. Any motorist can download the JuiceNet application, monitor the chargers on the map, and select a charging station nearby.

Both renewable energy and EVs urgently need similar solutions to expand their respective applications.

4.2 Decentralized (peer-to-peer) energy transaction

For a centralized, traditional energy trade, the energy Internet has become increasingly complex with the access to a large number of customers. Difficulties arise when a centralized organization is established, such as excessive operating costs and insufficient information protection. In addition, the commercial enterprise lacks confidence if there is no centralized management organization. These concerns can be addressed by introducing blockchain technology into energy sector. The P2P energy trading model based on the blockchain system provides a low-cost, affordable, open, and trustworthy trading platform for energy networks. Zhao et al. [24] proposed a double-layer framework for blockchain-based energy transactions in multimicrogrids to ensure decentralized trading, information transparency, and mutual trust between nodes in the trading market.

We view energy as a commodity and do not consider energy commodities as property. In fact, energy exchange cannot be isolated from specialized monitoring processes, so such studies are not sufficient to address the operational use of energy trading. Kuznetsova et al. [25] evaluated the security transactions in the grid and advised insufficient centralized power trading management. In addition, a management body with limited capacity needs to perform security checks and monitor the stopping of transactions. Zia et al. [26] studied the barriers to the adoption of energy trading over the Internet. They identified a three-tier structure based on blockchain technology (transaction, extension, and blockchain). The weak central authority is a specialized node within the blockchain system and monitors the parties involved in energy trading to ensure that the transactions run smoothly.

4.3 Certification and trading of carbon emissions

The rapid increase and concentration of greenhouse gases have led to a worldwide climate catastrophe. The United Nations has developed a carbon dioxide emission rights certification and trading mechanism under the Kyoto Protocol to encourage energy conservation and emission reduction, using the money for the purpose of emission control [27]. Based on the carbon emissions of different industries, China allocates a certain amount of carbon emission rights to emitting enterprises. The emission rights of firms with emission balancing rights can be purchased from firms that issue more than their allowances, or they can be sanctioned. Conversely, companies with additional emission rights can make a profit by selling them.

There are some challenges in the carbon market, such as the huge workload of emissions certification and the difficulty of tracking transaction data. Blockchain technology can provide a framework for smart carbon emissions trading and certification. This technology can track all tonnage of carbon and trading information, preventing information manipulation and asymmetry. For example, China Certified Emission Reductions (CCERs) are sold as "carbon tickets" for digital assets. Each carbon ticket has a unique ID, which is time-stamped and recorded as a blockchain. Automatically, smart contracts are used for carbon trading [24].

This example shows the potential contribution to carbon emissions using blockchain. Each contributing node sends many error-prone files, while the normal development length exceeds 1 year. Meanwhile, it takes a long time to build traditional carbon assets, including companies, government regulators, trading of carbon assets, third-party verification and certification, etc. This framework will significantly reduce the cycle time for generating CO_2 emissions and reduce the development cost of CO_2 assets by 20%–30% [28].

4.4 Physical information security

To make fair decisions, the system is guided by correct information. However, incorrect information can have disastrous effects on the system if it disrupts or attacks the information system. The most important security strategy for power grid companies is to create special communication links to isolate internal and external networks. The use of data acquisition systems in a transmission network increases system costs during construction. At the same time, disruptions and attacks can occur when power system operators or public networks are used to communicate or exchange information. The origin of Stuxnet, an industrial computer malware, was discovered in 2010. This can penetrate a network of external computers via infected USB devices and attack the company's Internet. In 2015, the electricity information system in Ukraine was hacked, causing a massive power outage in the country [29]. Thus, electricity systems have improved their ability to withstand cyberattacks.

Decentralization, high redundancy, security, and privacy protection help to solve some of the security problems of the information and physical systems. However, there is currently limited research on the use of blockchain technology to solve security challenges in the energy Internet. Yao et al. [30] consider blockchains on the Internet from the perspective of private key loss, privacy violation, and protocol attacks on security problems. It also presents a strategy that constructs energy consisting of three components: architectural security, ontological security, and access control. The security measures control the entire lifecycle of the power blockchain communication system to efficiently meet the requirements for the future development of the energy blockchain.

4.5 Energy transmission

In the face of increasingly critical environmental problems, the development method resulting from conventional fossil fuel energy is unsustainable. Therefore, there is a tendency to replace fossil fuels with clean and sustainable energy. While research in sustainable energy is progressing, relatively stable techniques such as wind or solar energy have problems with decentralized geographical coverage, poor management and consumption rates, and high energy costs, leading to major limitations in the use and commercialization of new energy sources [31]. In 2008, a future renewable energy transmission and management system was called the Internet of Energy model [32]. The Internet of Energy consists of the new structure of power generation networks, networked storage systems, and the Internet.

Conversely, the quality of this platform is challenged by corporate and user service. The energy Internet encompasses a broader range of power and participants than the existing power grids. It changes the approach of information exchange to create a new energy supply with highly integrated and complementary information across multiple energy sources [33]. Blockchain-based energy systems will offer promising solutions for managing energy transmission among heterogeneous renewable energy systems.

4.6 Power-to-X

Power-to-X uses electricity to produce other products to store the energy. Power-to-X can be power-to-ammonia, power-to-chemicals, power-to-fuel, power-to-gas, power-to-hydrogen, power-to-liquid, power-to-methane, power-to-food, power-to-power, and power-to-syngas. Power-to-X offers the unique possibility of storing excess renewable energy generated when demand is relatively low, which can increase the economic value of renewable energy generation by facilitating classical storage arbitrage [34]. With power-to-X, enormous amounts of energy may be kept for lengthy periods of time, making it suitable for seasonal storage. Because many places have significant seasonal variability of solar, wind, and hydroelectric output, this could be especially advantageous for systems with high renewable penetration. In addition, as renewable energy deployment continues to expand, power-to-X also has the potential to help decarbonize other sectors of the economy, such as industry and transportation, that have long plagued carbon reduction experts.

Applications of blockchain smart contract technology include tracking and tracing of energy types, conversion history, and data support for transparency, credibility, and automated enforcement of future carbon transactions.

4.7 Internet of Energy

It is also argued that the Internet of Energy (IoE) is the basic research guideline to address the current problems in the energy sector. Faizan et al. [35] say that society will enter an IoE system that combines new energy and communication technologies with conventional fuel depletion and ongoing global environmental degradation. Japanese academic institutions have focused on the creation of the digital grid system and proposed the energy Internet "Power Router" by the Japan Digital Grid Alliance. Hu et al. [36] provide a preliminary description of the energy Internet: the energy Internet is a closed loop in which renewable energy is widely used as the main energy source, and other gas grids and transportation systems are closely connected. It is built on the Internet and other advanced IT systems, and the power grid is at the center. More specifically, the energy Internet described in Ref. [37] includes an Internet-connected smart power grid platform, big data, cloud computing, and other leading ICT platforms. Advanced electromechanical devices and smart management technologies in the future power system will be used to integrate horizontal, cross-source, complementary, comprehensive, hierarchical synchronization and data, source-grid-neutron-to-storage.

As shown below, the IoE offers features such as precise metering, wide-scale multisource cooperation, intelligent control, and open trading [27,30,35,38].

(1) Wide-area multisource cooperation: IoE provides wide coverage, high reliability, and many participants and coordinates the planning operations of large-scale power generation bases, power transfer, and end-use power utilization, including multisource cooperation. Through cooperation, participants can reach an agreement and maximize profits.

(2) Accurate measurement: This is the prerequisite for regulating the circumstances of the operation of various energy and computation systems. Therefore, the IoE must also address participants' consensual reliance on metering data and accurate measurement.

(3) Open trade: The technology takes advantage of modern society with huge centralized grids and local distribution networks. Without discrimination, any energy source can access the Internet. Energy producers can also be electricity consumers, and consumer involvement is greatly enhanced. The IoE provides various energy services anytime and anywhere to promote demand, support services, and buy and sell electricity. By reducing the peaks and troughs of the energy IoE, its operational performance will be improved.
(4) Intelligent control: Smart approaches such as large-scale data processing and web learning are essential and enable a wide range of dispersed energy sources. From generation to use, the entire process depends on intelligent control technology for flexible and efficient conversion of energy and optimization of transmission performance.

Blockchain can support the construction of a distributed IoE with autonomy, scalability, and intelligent control to realize the planning, management, and application of future renewable energy.

5 Implementation of blockchain for renewable energy

Smart grids and conventional grids differ primarily in the two-way flow of information between power grids and users [39]. Blockchains have contributed to the growth of distributed accounting as an alternative technological innovation after the Internet that is tamper-proof, traceable, highly trusted, and decentralized. Blockchains can enhance data security in the power grid and contribute to the implementation of a trustworthy, effective, and reliable smart grid system. Several studies have already been conducted on blockchains in the energy sector.

First, the creation of power grids across the energy industry is at the heart of the process. Smart grids are also being developed in the context of blockchains. Some researchers have studied the growth of smart grids under blockchains, but this research is not systematized and is still in the early stages. Most blockchain studies and smart grid combinations focus on using blockchain technology in a single part of the grid value chain, such as building secure smart grid data access and exchange systems based on alliance blockchains and secure distributed keyless blockchain signature methods. The evolving smart grid systems under the blockchain are not difficult to identify, but there is a scattering of existing research and a lack of integration.

Second, the organization, personnel, information, and resource system form the supply chain of the smart grid. Supply chains are inherently complicated, and inefficient supply chains can lead to major trust crises and need to be better shared and verified. In the application process, the blockchain works simultaneously with all the agents of smart networks, and different agents cooperate and influence each other. However, researchers have conducted limited research on the coordinated development of multiagent systems from a stakeholder perspective. From the perspective of longevity, there are few studies on the integration of blockchain and smart grids, which need to be considered from three aspects: economic, social, and environmental.

Third, the scope for private chains is limited because the authenticity of the information can be easily doubted. Unlike private chains, public chains have much higher trust but cannot protect the privacy and security of participants. The above challenges have hindered the application and diffusion of public chains in the productive economy. Therefore, alliance chains exhibit "partisan decentralization" characteristics. While an alliance chain may cover only a limited number of subjects, security can be enhanced, costs reduced, and reliability increased.

5.1 Blockchain system architecture

The renewable energy blockchain system is an integrated technical system that includes renewable energy systems, communication network systems, blockchain systems, and blockchain-based application systems. The systems are interconnected and communicate through various data streams (see Fig. 6.2).

In a blockchain-based renewable energy system, electricity generation and consumption data are transmitted through smart meters. Each stakeholder is a node in the blockchain. All relevant transactions are transmitted to the grid. The transmitted data, including electricity statistics and transaction data, is then confirmed in the blockchain ledger according to the consensus protocol. The consensus protocol can be switched between PoW and other consensus algorithms in the simulation. The blockchain ledger manages transactions, balances, and historical information.

The community committee can set up the blockchain system as an alliance blockchain. All participants register in the system and receive authorization and authentication information.

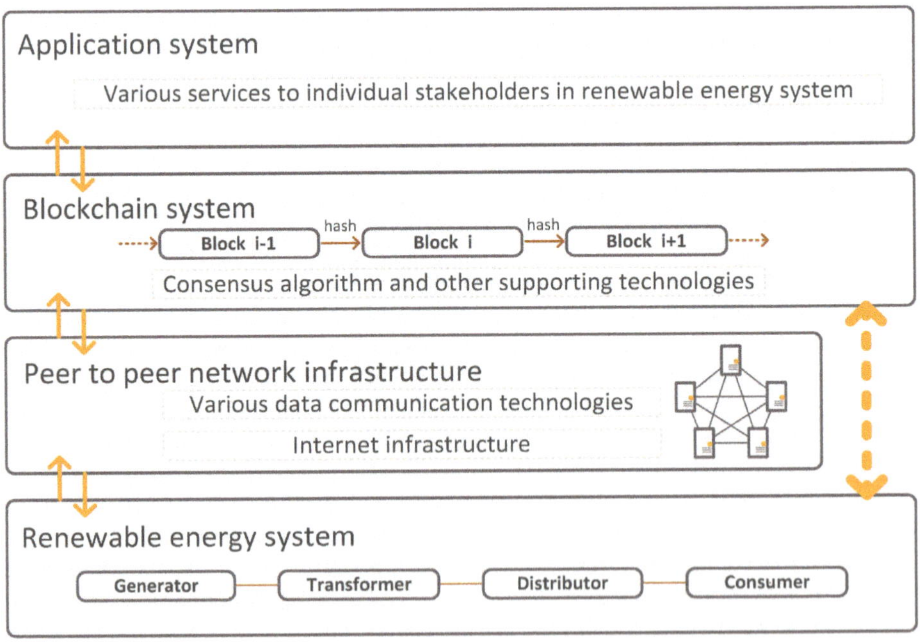

Fig. 6.2 Blockchain system architecture.

5.2 Data feed to the blockchain from the power grid

The blockchain system is a system for storing and managing data. Once the data enters the blockchain system, it is recorded in history. The history of the blockchain cannot be changed. Therefore, the data feed is directly related to the data quality of the blockchain and the quality of the system services. Data injection involves several important aspects:

(1) What data needs to be injected into the blockchain system? This involves the actual business requirements of the application, i.e., what data will be used in the future.
(2) How will the data be entered? Manual input and automated input are the common methods. To ensure data accuracy and avoid human error, automated input is the ideal way. This places requirements on existing renewable energy systems; these systems can provide automated interfaces for data output that feeds directly into the blockchain system. Automated data input from smart meters and VPP (virtual power plan) to blockchain will speed up data processing and its accuracy [40,41].
(3) How can the correctness of the data be ensured? This requires appropriate operations such as screening, inspection, cleaning, and validation of the data entered. In the early days of data entry into the blockchain, blocks may be revoked for technical reasons, so it is also important to ensure security before and after data entry. Relevant data protection technologies will help protect the security of the data entered.

5.3 Consensus selection

For a blockchain system, the consensus protocol is crucial. It is the rule that governs the creation of new valid blocks and a consistent blockchain leader. Different consensus mechanisms, including Proof of Work (PoW), Proof of Stake (PoS), Delegated Proof of Stake (DPoS), and Practical Byzantine Fault Tolerance (PBFT), are deployed in various blockchain applications. The main and most used protocols include PoW and PoS. Several other consensus protocols are offered, such as Proof of Elapsed Time (PoET), Proof of Burn (PoB), Proof of Authority (PoA), Proof of Capacity (PoC), etc. Essentially, the consensus protocol defines a voting system to select the next accountant. In PoW, nodes (e.g., miners) compete with each other to solve hash puzzles and the first to have a solution can create a new block and propagate it across the network. PoS requires nodes to lock specific pills into the network rather than having a puzzle-solving competition to select the bookkeeper [32]. A node's share affects its probability of being selected to verify transactions and create a new block. The verifier can only collect its coins and a payout, which precludes data forgery if the network recognizes the block.

The consensus protocol defines the key performance characteristics of a blockchain. The use of PoW in Bitcoin's consensus protocol has been widely criticized (a 51% attack means that the miner controlling over 51% of network computations can create blocks with fake transactions and invalidate blocks from honest miners). In terms of speeding up transactions and conserving resources, on the other hand, PoS is more powerful than PoW.

Although a significant number of transactions are performed with Bitcoin, the data processing rate in a blockchain is estimated to be about seven per second. In contrast,

Ethereum averages up to 15 transactions per second. These low processing rates are mainly related to the consensus mechanism (PoW) used in Bitcoin technology. Nodes need to compute the PoW algorithm to add a block to the blockchain, which requires a lot of computing power and time. According to Mollah et al. [42], 30 billion kWh of electricity was consumed to process 30 million transactions, which is about 0.13% of the world's energy consumption. In the energy sector, the volume of transactions per second is particularly high for large-scale operations, as thousands of customers participate simultaneously in the process of energy procurement and distribution. Thus, the nodes in the consensus and validation process represent a large overhead. The PoW algorithm could be replaced by the PoS or PoA algorithm to solve this problem. These algorithms require far less computing power and support higher transaction rates. The energy sector, with transaction speeds of several thousand per second, is exclusively served by another blockchain platform called EnergyWeb blockchain. It uses the consensus PoA mechanism, which makes processing so fast.

The impact of a consensus protocol on blockchain performance and the requirements for the consensus protocol are summarized as follows: (1) It should be energy efficient, as the primary goal is to promote sustainability. (2) PoW consensus protocols should be avoided. The system can provide a transaction platform to process a certain number of requests within an acceptable time period. (3) The consensus protocol establishes criteria for performance, latency, and scalability. (4) Security is an important consideration because many people and personal data are involved.

5.4 Blockchain security and maintenance

The blockchain is a distributed ledger database system, and also poses some threats to data security.

(1) Cyberattacks on the blockchain. Large-scale cyberattacks cannot disrupt the online services of the blockchain and even cause data loss.

(2) Verification of legal blocks. New blocks are constantly being added to the blockchain. The big issue is to make sure the legal blocks are reasonably selected and verified. This task is mainly performed by consensus protocols. Potential vulnerability or incomplete design of the consensus algorithms can jeopardize this.

(3) Forks of the blockchain. The protocol or version update of the blockchain can lead to a hard fork of the blockchain. The creation of new blocks in the blockchain can lead to a natural soft fork. Whether it is a hard fork or a soft fork, the structure of the blockchain may change, thereby affecting the performance of the blockchain system.

(4) Protection of smart contracts. Smart contracts contain important application logic and services. Any improper modification of smart contracts can lead to the disruption of blockchain services. Therefore, it is important to ensure the security and integrity of smart contracts. Of course, hash technology can adequately ensure the security of smart contract data. But for contracts with large granularity outside the chain, maintaining the security of the contract is still a challenge.

(5) Data privacy and ID management in blockchain are challenging. The data entering the blockchain is transparent to all users. If the ID management is not in place, a large amount of private data can leak out. Generally, the user ID or data ID and privacy-related data have been translated and stored off-chain before entering the blockchain. If this process fails,

privacy cannot be protected. While blockchain technology cannot guarantee privacy directly, advanced cryptographic measures can be incorporated to protect data. The following are some of the approaches to safeguard the privacy of the devices concerned: Zero-Knowledge Proof (ZKP), the Elliptic Curve Digital Signature Algorithm (ECDSA), and the linkable signature ring.

5.5 *Legal and regulatory*

The regulatory agencies promote active user involvement in the energy market and the development of community energy infrastructure. When it comes to drastic modifications in the main grid framework, the grid system does not facilitate energy exchange between producers and consumers, nor does it support the adoption in that framework of the distributed leader [42]. In particular, for the P2P trading system, new types of contracts must be devised, and energy pricing modifications must be made to promote such services. In the existing grid system, such issues are highly regulated. In these circumstances, although blockchain technology has shown value in informing a microgrid, it is very difficult to incorporate the technology into the power system framework without modifying the power systems. The ideal future for a solid renewable energy business is based on a good understanding of technology, operations, strategy, policy, and regulation.

6 Conclusions

Although the concept of blockchain is still most often associated with cryptocurrencies such as Bitcoin, it has been highlighted in theoretical and practical experiments in many contexts and sectors. One of the most important aspects of implementing a blockchain system is in the energy sector. Blockchain and related technologies are needed to handle the demand for decentralized applications and services in global renewable energy innovation. Blockchain can effectively deal with some of the key issues facing the renewable energy landscape, such as the contradiction between a large number of decentralized energy producers and drastically changing energy demand, coping with complex cross-system energy flows, balancing and transaction security, minimizing the environmental risks caused by energy use, and dealing with the demand for sustainable structures and the demand for greater flexibility.

Due to data encryption, accountability, and adaptability, there are also some system drawbacks and technical disadvantages. The relative novelty of the concept manifests itself in technical challenges, many of which have been solved, at least in principle. In general, the concept of blockchain for energy sectors is very attractive for the future. Still, the solution of the system challenges largely depends on key social and technological trends, the current and future power driving factors, and overall trends and constraints in the operation of the system. To influence these journeys, the academic and practical supporters of blockchain energy must combine further research and development with practical and real-life applications to demonstrate the real benefits and performance of the concept to relevant stakeholders and policymakers. Especially consider the concept of "design security" from a technology and IT perspective. In applying blockchain for energy sectors, cybersecurity cannot be ignored.

Acknowledgment

The work was supported by Energiforsk (https://energiforsk.se/en) in Sweden.

References

[1] C.K. Leander, M., Renewable Energy in the Nordics 2021, Nordic Energy Research, 2021.
[2] Å.G. Tveten, Renewable Energy in Northern European Power Markets: Effects, Challenges and Integration Options, 2020 (Ph.D. thesis).
[3] T. Ajadi, V. Cuming, R. Boyle, D. Strahan, M. Kimmel, M. Logan, A. McCrone, Global Trends in Renewable Energy Investment 2020, Frankfurt School – UNEP Collaborating Centre for Climate & Sustainable Energy Finance, 2020.
[4] R. Talat, M. Muzammal, Q. Qu, W. Zhou, M. Najam-ul-Islam, S.H. Bamakan, J. Qiu, A decentralized system for green energy distribution in a smart grid, J. Energy Eng. 146 (1) (2020) 04019036.
[5] R. Khalid, N. Javaid, A. Almogren, M.U. Javed, S. Javaid, M. Zuair, A blockchain-based load balancing in decentralized hybrid P2P energy trading market in smart grid, IEEE Access 8 (2020) 47047–47062.
[6] S. Rahmadika, D.R. Ramdania, M. Harika, A blockchain approach for the future renewable energy transaction, J. Phys. Conf. Ser. 1175 (1) (2019) 012122. IOP Publishing.
[7] M. Andoni, V. Robu, D. Flynn, S. Abram, D. Geach, D. Jenkins, P. McCallum, A. Peacock, Blockchain technology in the energy sector: a systematic review of challenges and opportunities, Renew. Sustain. Energy Rev. 100 (2019) 143–174.
[8] S.R. Sinsel, R.L. Riemke, V.H. Hoffmann, Challenges and solution technologies for the integration of variable renewable energy sources—a review, Renew. Energy 145 (2020) 2271–2285.
[9] H. Zimmermann, J. Hoppe, Blockchain–Opportunities and Threats for the Energy Transition, Germanwatch e.V, 2018.
[10] M. Bajaj, A.K. Singh, Grid integrated renewable DG systems: a review of power quality challenges and state-of-the-art mitigation techniques, Int. J. Energy Res. 44 (1) (2020) 26–69.
[11] B.B. Adetokun, C.M. Muriithi, J.O. Ojo, Voltage stability assessment and enhancement of power grid with increasing wind energy penetration, Int. J. Electr. Power Energy Syst. 120 (2020), 105988.
[12] B.M.S. Hodge, H. Jain, C. Brancucci, G.S. Seo, M. Korpås, J. Kiviluoma, et al., Addressing technical challenges in 100% variable inverter-based renewable energy power systems, Wiley Interdiscip. Rev. Energy Environ. 9 (5) (2020), e376.
[13] M.S. Alam, F.S. Al-Ismail, A. Salem, M.A. Abido, High-level penetration of renewable energy sources into grid utility: challenges and solutions, IEEE Access 8 (2020) 190277–190299.
[14] S. Choudhury, A comprehensive review on issues, investigations, control and protection trends, technical challenges and future directions for microgrid technology, Int. Trans. Electr. Energy Syst. 30 (9) (2020), e12446.

[15] D. Asija, R. Viral, Renewable energy integration in modern deregulated power system: challenges, driving forces, and lessons for future road map, in: Advances in Smart Grid Power System, Academic Press, 2021, pp. 365–384.
[16] W. Cole, A.W. Frazier, Impacts of increasing penetration of renewable energy on the operation of the power sector, Electr. J. 31 (10) (2018) 24–31.
[17] S. Impram, S.V. Nese, B. Oral, Challenges of renewable energy penetration on power system flexibility: a survey, Energy Strategy Rev. 31 (2020), 100539.
[18] R. Alt, E. Wende, Blockchain technology in energy markets—an interview with the European Energy Exchange, Electron. Mark. 30 (2) (2020) 325–330.
[19] M. Shahidehpour, M. Yan, P. Shikhar, S. Bahramirad, A. Paaso, Blockchain for peer-to-peer transactive energy trading in networked microgrids: providing an effective and decentralized strategy, IEEE Electrif. Mag. 8 (4) (2020) 80–90.
[20] B. Teufel, A. Sentic, M. Barmet, Blockchain energy: blockchain in future energy systems, J. Electrochem. Sci. Technol. 17 (4) (2019), 100011.
[21] E.A. Soto, L.B. Bosman, E. Wollega, W.D. Leon-Salas, Peer-to-peer energy trading: a review of the literature, Appl. Energy 283 (2020) 116268.
[22] V. Peter, J. Paredes, M.R. Rivial, E.S. Sepúlveda, D.A.H. Astorga, Blockchain Meets Energy: Digital Solutions for a Decentralized and Decarbonized Sector, German-Mexican Energy Partnership (EP) and Florence School of Regulation (FSR), 2019, pp. 1–45.
[23] H. Patil, S. Sharma, L. Raja, Study of blockchain based smart grid for energy optimization, Mater. Today: Proc. 44 (2021) 4666–4670.
[24] Z. Zhao, J. Guo, X. Luo, J. Xue, C.S. Lai, Z. Xu, L.L. Lai, Energy transaction for multi-microgrids and internal microgrid based on blockchain, IEEE Access 8 (2020) 144362–144372.
[25] E. Kuznetsova, M.A. Cardin, M. Diao, S. Zhang, Integrated decision-support methodology for combined centralized-decentralized waste-to-energy management systems design, Renew. Sustain. Energy Rev. 103 (2019) 477–500.
[26] M.F. Zia, M. Benbouzid, E. Elbouchikhi, S.M. Muyeen, K. Techato, J.M. Guerrero, Microgrid transactive energy: review, architectures, distributed ledger technologies, and market analysis, IEEE Access 8 (2020) 19410–19432.
[27] H. Chen, X. Wang, Z. Li, W. Chen, Y. Cai, Distributed sensing and cooperative estimation/detection of ubiquitous power internet of things, Prot. Control Mod. Power Syst. 4 (1) (2019) 1–8.
[28] X. Kong, J. Zhang, H. Wang, J. Shu, Decentralized multi-chain data management framework for power systems, CSEE J. Power Energy Syst. 1 (2019) 1–11.
[29] A. Kumari, R. Gupta, S. Tanwar, S. Tyagi, N. Kumar, When blockchain meets smart grid: secure energy trading in demand response management, IEEE Netw. 34 (5) (2020) 299–305.
[30] Y. Yao, C. Gao, T. Chen, J. Yang, S. Chen, Distributed electric energy trading model and strategy analysis based on prospect theory, Int. J. Electr. Power Energy Syst. 131 (2021), 106865.
[31] M.U. Hassan, M.H. Rehmani, J. Chen, DEAL: differentially private auction for blockchain-based microgrids energy trading, IEEE Trans. Serv. Comput. 13 (2) (2019) 263–275.
[32] F. Knirsch, C. Brunner, A. Unterweger, D. Engel, Decentralized and permission-less green energy certificates with GECKO, Energy Inform. 3 (1) (2020) 1–17.
[33] Y. Li, W. Yang, P. He, C. Chen, X. Wang, Design and management of a distributed hybrid energy system through smart contract and blockchain, Appl. Energy 248 (2019) 390–405.

[34] S. Byfield, D. Vetter, Flexibility Concepts for the German Power Supply in 2050: Ensuring Stability in the Age of Renewable Energies, Acatech–Deutsche Akademie der Technikwissenschaften, Munich, Germany, 2016.
[35] M. Faizan, T. Brenner, F. Foerster, C. Wittwer, B. Koch, Decentralized bottom-up energy trading using Ethereum as a platform, J. Energy Mark. 12 (2) (2019).
[36] B. Hu, C. Zhou, Y.C. Tian, X. Hu, X. Junping, Decentralized consensus decision-making for cybersecurity protection in multimicrogrid systems, IEEE Trans. Syst. Man Cybern.: Syst. 51 (4) (2020) 2187–2198.
[37] J. Li, Z. Zhou, J. Wu, J. Li, S. Mumtaz, X. Lin, et al., Decentralized on-demand energy supply for blockchain in Internet of Things: a microgrids approach, IEEE Trans. Comput. Soc. Syst. 6 (6) (2019) 1395–1406.
[38] Y. Wu, Y. Wu, J.M. Guerrero, J.C. Vasquez, Digitalization and decentralization driving transactive energy Internet: key technologies and infrastructures, Int. J. Electr. Power Energy Syst. 126 (2021), 106593.
[39] K. Heck, E. Mengelkamp, C. Weinhardt, Blockchain-based local energy markets: decentralized trading on single-board computers, Energy Syst. 12 (3) (2021) 603–618.
[40] T. Cioara, M. Antal, V.T. Mihailescu, C.D. Antal, I.M. Anghel, D. Mitrea, Blockchain-based decentralized virtual power plants of small prosumers, IEEE Access 9 (2021) 29490–29504.
[41] Q. Yang, H. Wang, T. Wang, S. Zhang, X. Wu, H. Wang, Blockchain-based decentralized energy management platform for residential distributed energy resources in a virtual power plant, Appl. Energy 294 (2021), 117026.
[42] M.B. Mollah, J. Zhao, D. Niyato, K.Y. Lam, X. Zhang, A.M. Ghias, et al., Blockchain for future smart grid: a comprehensive survey, IEEE Internet Things J. 8 (1) (2020) 18–43.

Electricity market issues in future power systems

7

Ali Karimi, Nader Tarashandeh, and Yousef Noorizadeh
Electrical and Computer Engineering, University of Kashan, Kashan, Iran

1 Introduction

Independent system operators (ISOs) and regional transmission organizations (RTOs) mainly use the centralized optimization method to run the electricity market in power systems. All the required information is collected in this method, and then the central operator performs the operational decisions. However, such a centralized framework in emerging power systems faces the fundamental challenges listed below:

(1) *Communication and computation challenges as the power system size increases:* As the power system dimensions increase, communication requirements for data collection and computational complexity for solving a large-scale problem such as security-constraint unit commitment (SCUC) and economic dispatch (ED) will increase dramatically. For example, from January 16, 2012 to February 15, 2012, CAISO had 7 UC disruptions and 35 ED disruptions. The mentioned incidents were mainly due to defects in software infrastructure and the high volume of information.

(2) *Political and technical challenges related to the coordination of multiarea systems:* In recent years, electricity infrastructures in different regions and countries have been connected to improve reliability and economic efficiency. In multiarea markets, each region is first operated separately by its respective ISO/RTO. From a political perspective, it is not rational to use a centralized approach to this practice, because for an area, all information must be shared with other areas.

(3) *Changing the perspective of energy production by distributed energy resources (DERs):* In recent years, the growth rate of DERs such as photovoltaic panels, smart appliances, electric vehicles, and storage systems has increased significantly. These resources provide low-voltage energy services and can be remotely controlled via the Internet of things. If used wisely, these resources will be able to reduce costs, increase reliability, and increase the penetration rate of renewable energy. Therefore, encouraging investors to invest more in this area will bring many benefits to the power system. However, if the aggregator or utility makes the payment for the DER services, the transparency and accuracy of this payment will always be ambiguous from the prosumer viewpoint; this is because it is possible to move the market away from the cost-minimizing equilibrium toward a profit-maximizing one by the monopoly aggregator. In fact, with the current structure of electricity markets, incentives for active prosumer participation have not been sufficient.

Given these challenges, decentralized optimization can be considered as an alternative to a centralized approach to the operation of electricity markets at the levels of wholesale, multiarea, and local distribution markets. In the decentralized approach, a large-scale power system is divided into smaller subregions. Then, the coordination of

Decentralized Frameworks for Future Power Systems. https://doi.org/10.1016/B978-0-323-91698-1.00005-4

limited informative results of each subproblem will lead to the optimized solution. The exchange of limited information between subregions significantly reduces communication needs. The method of decomposing power systems and forming subregions can vary depending on the type of market. In multiple or multiarea markets, one aspect of decomposition is created based on a geography-based method; thus, a large power system is geographically divided into several smaller areas. Tie-lines connect these areas. Fully decentralized approaches that apply optimization decomposition techniques are divided into primary-based and dual-based algorithms. Primary-based algorithms include LR (Lagrangian relaxation) and OCD (optimality condition decomposition) techniques. On the other hand, one of the most important dual-based algorithms is ALD (augmented Lagrangian decomposition). Moreover, ADM (alternative direction multiplier) and APP (auxiliary problem principle) methods are utilized to solve the ALD problems. In the decentralized approaches, some limited boundary information, including boundary voltage (magnitude and angle), tie-line flows, and power exchange prices, can be exchanged among the areas to coordinate tie-line scheduling.

Moreover, the system is usually decomposed based on network node/nodes in the local distribution markets and microgrids. In this way, each prosumer can act as an independent agent and directly buy and sell energy with other DERs and existing markets. In fact, in this case, a peer-to-peer (P2P) market is formed. In P2P markets, it is possible to form a community by merging several nodes, and the community manager is responsible for exchanging energy. In recent years, the advent of technologies such as blockchain has increased the security and transparency of exchanges. Blockchain or distributed ledger technology is an emerging technology for decentralized computing and data storage secured by cryptographic signatures and distributed consensus mechanisms. The use of this technology in energy markets will reduce the risks of cyber-attacks and tampering by participants. On the other hand, blockchains enable the execution of smart contracts on P2P networks in applications such as energy exchanges and settlement, eliminating the need for a third party for these applications.

Given the above explanations, this chapter describes the decentralized frameworks for the future of electricity markets, both in the economic exchanges of multiple wholesale markets with aiming tie-line scheduling and in local electricity markets for smart grids.

2 Multiarea market

Maintaining the security and the reliability of power systems to provide continuous electricity supply to consumers is one of the essential tasks and concerns of the transmission system operators [1]. Between 2012 and 2040, the demand for energy consumption will increase approximately by 30% [2]. Therefore, the annual expansion of electricity generation capacity is significant. Another way to meet the increasing demand of countries is to import electricity through multiarea electricity markets. In these markets, participants in each area send their production and consumption offers to the market manager. Then, the settlement process of the markets is executed, and the amount of exchanged energy and settlement prices among the areas will be determined. This idea has been well-received due to the relatively large benefits of multiarea markets around the world. Technically, the benefits of energy market integration include increasing stability and reliability, improving generated electricity, and reducing energy losses. On the other hand, it increases social welfare and expands exports from an economic point of view. Also, from a political perspective, it will develop collaboration and relationships among areas. All activities in each area of a multiarea system affect prices, stability, and degrees of independence in other connected areas. Thus, tie-line scheduling (TS) can play a significant role in determining the economic exchanges of energy. TS is basically an economic dispatch (ED) problem that must be coordinated between the operators [3]. In multiple markets, the coordination and the way of exchanging information between control areas are different. If the connected control areas are all under the supervision of one system operator, the extent and coordination of these control areas are easier. Suppose each connected control area has an independent market, and these areas are within a country. In that case, the amount of information exchange is more limited and must be more private than before. The most difficult situation arises when each control area has an independent market in a separate country. In this case, the amount of information exchange among areas is more limited, and each area usually wants its market information to remain confidential.

In general, the coordination of multiarea systems can be either centralized or decentralized [4]. In the centralized approach, the system operator can access all network information and schedules the whole interconnected network. This approach is usually hard to implement in situations where the multiarea system involves several countries due to the confidentiality of information. On the other hand, the computational burden of centralized market execution becomes very heavy. In the decentralized approach, coordination among regional operators is essential. The coordination of this approach could be with the presence or absence of a coordinator entity.

The coordinator entity performs the TS by managing the information of the control areas' operators. Depending on the assigned tasks, the coordinator has different access to network and participants' information for each connected power system. Explicit and implicit auctions are two concepts discussed nowadays in the case of decentralized exchanges of power between countries with the coordinator [5–7]. In the explicit auction of transmission lines, according to the available transfer capability (ATC) announced by transmission network operators, the purchase and sale of transmission rights are made separately from the energy market, while in the implicit auction, the allocation of transmission lines is settled simultaneously with the energy. This method may be done with different purposes such as congestion management, coordination of electricity markets in the regions to increase the efficiency of power exchanges, etc. For example, in Karimi Varkani et al. [8], the possibility of forming several electricity markets in each control area is discussed. In this way, producers and consumers in each area can participate in several electricity markets simultaneously. This way of participation in different markets within an area exists in European electricity markets. Finally, suppose the market results cause congestion on the tie-lines. In that case, the coordinator entity with the market mechanism will allocate the capacity of the congested tie-line between the electricity markets. According to scientific documents and reports, this method is one of the main options for future interregional transactions in Europe. With the implicit auction in two or more regions, the market coupling is created. In other words, first, each of the independent markets, regardless of the transmission network, is implemented. Then, the coordinator entity performs the coupling based on the net export curves of each of the markets and ATCs between the areas. One of the main challenges in the multiple-market structure is modeling the external players in each market and keeping the multiarea system secure. For this purpose, in Karimi et al. [9], the coordinator provides the external equivalent network data for each area. With this information, each market operator can model the external players in its market-clearing problem.

Due to data management and privacy issues, the TS problem sometimes needs to be solved in a decentralized approach without any coordinator entity. In this case, the independent system operators of connected areas exchange power and schedule the tie-lines by interacting with each other. To implement a multiarea market with a decentralized method and without the presence of a coordinator entity, methods based on optimization decomposition can be used. As mentioned in the introduction, LR, ALD, and OCD algorithms are often used as decomposition-based optimization methods in the fully decentralized approach. These methods are obtained based on the primary and the dual decomposition techniques. In fact, local problems are first implemented, and dual variables are used to coordinate local optimizations to find the globally optimal result. Moreover, purely market-based approaches are used in some decentralized methods instead of optimization decomposition methods [10]. The efficiency of these methods will be seriously affected if there are more than two adjacent areas. The focus of this section is on decentralized methods without the coordinator. Therefore, more details in this regard are provided in the following.

2.1 Multiarea market without coordinator entity

Due to the more limited exchange of information among areas and protecting network information in each control area, decentralized coordination without a central coordinator entity in power exchanges among different countries has received more attention in recent years. As mentioned, decomposition-based methods (such as LR, ALD, and OCD) along with the methods based on market mechanisms are used to implement the above schemes in interregional markets. In Conejo and Aguado [11], the TS is modeled using the LR method in a completely decentralized manner without the presence of a coordinating entity. In fact, each region has an independent market operator and models its line losses using a cosine approximation as an additional load for each bus. The Lagrangian coefficients of each market have also been updated with the subgradient method. The same authors continue their work in Aguado et al. [12], implementing the LR problem on a multiarea network with AC constraints. Lagrange coefficients have also been updated using the cutting plane method.

The OCD method is obtained based on the modification of the LR. This method was first discussed in Bakirtzis and Biskas [13] and further detailed in Biskas et al. [14] and Zhou et al. [15]. The most important advantages of this method are the lack of need for coordinators and tuning parameters. To improve the LR method's convergence, the quadratic penalty criterion is added to the objective function, and the problem is defined as an ALD. The alternative direction multiplier (ADM) [16,17] and auxiliary problem principle (APP) methods [18,19] can solve the ALD problems. The important question is which of the mentioned methods has the best utility and efficiency in different network topologies and different operating conditions for TS. To evaluate the efficiency of the mentioned methods, it is necessary to compare the results of each method in multiarea markets with the optimal result. The optimal result is extracted from the implementation of multiarea markets in a centralized manner. Therefore, it is first assumed that the entire multiarea system is centrally dispatched under the supervision of a central operator, considering all system constraints. Then, the central market is decomposed according to each of the mentioned methods. Hence, the central model of the interconnected system is first formulated; then, decentralized methods are implemented. Also, given that the ultimate objective is to reach the central optimal results by coordinating subproblems, some system constraints such as voltage and reactive power can be ignored. Therefore, in all the methods described in this chapter, DC load flow is used in the market model.

2.2 Central model for multiarea market

For this purpose, to achieve the optimal power flow on the tie-lines, it is assumed that in a multiarea interconnected power system, there are no system and market operators in each area, and one system operator dispatches the whole system. In this case, the tie-lines are seen as the internal transmission lines of the system. In Fig. 7.1, the aggregation of all interconnected networks is plotted by a central operator.

The formulation of this section aims to minimize the cost of power generations throughout the system if demands are assumed to be inelastic. Therefore, the problem is implemented as a DC-OPF. The objective function of the central dispatch problem subject to system constraints is considered as (7.1)–(7.5) [13]

$$\text{Min} \sum_{n=1} C_n(P_n) \tag{7.1}$$

$$\mathbf{B} \cdot \boldsymbol{\theta} = \mathbf{P} - \mathbf{D} \tag{7.2}$$

$$\theta_{ref} = 0 \tag{7.3}$$

$$\left| \frac{1}{x_{ij}} (\theta_i - \theta_j) \right| \leq F_{ij}^{\max} \tag{7.4}$$

$$P_n^{\min} \leq P_n \leq P_n^{\max} \tag{7.5}$$

where C_n is the production cost of unit n, P_n is the power output of unit i, θ_i is the angle of bus i, θ_{ref} is the reference bus angle, **B** is network susceptance matrix, **D** is network load vector, $F_{ij}^{\max}$ is maximum power flow capacity for transmission line ij, and $P_n^{\max}/P_n^{\min}$ is maximum/minimum power output of unit i. The load balance constraint

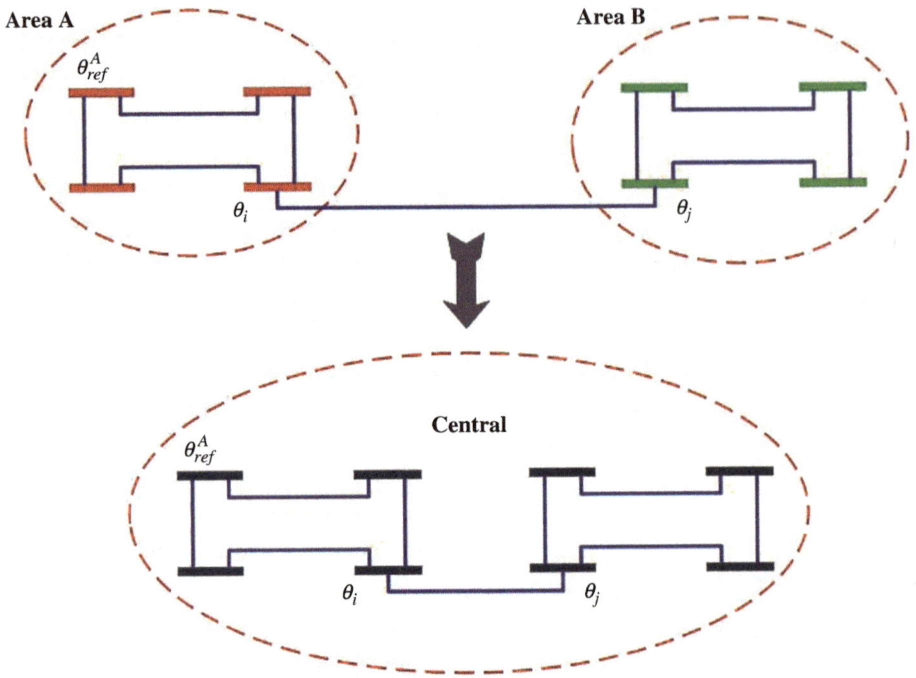

Fig. 7.1 Integration of a multiarea system by the central operator.

in each bus is represented in (7.2). Eq. (7.3) also indicates the reference bus voltage angle, and (7.4) indicates the permissible range of power flows on transmission lines. Finally, (7.5) represents the generation range of power units.

2.3 Decentralized market by the OCD method

As the first step in separating the OPF of the main problem by the OCD method, the central problem becomes an equivalent central problem with new additional variables. For this purpose, two new variables are defined for each tie-line. Therefore, for each tie-line ij, where $i \in A$ and $j \in B$ (A and B are control regions), two new variables T_{ij}^{A} and T_{ji}^{B} are defined. T_{ij}^{A} indicates the power flow of the tie-lines between the two busbars i and j, from region A to B, and T_{ji}^{B} represents the power flow of the tie-line from region B to A. For simplicity and clarity of concept, it is assumed that there are two connected areas. As shown in Fig. 7.2, the new connection variables for the existing tie-line are defined [13].

The central problem is rewritten so that local and independent variables are distinguished from the coupling variables. Therefore, (7.6)–(7.12) in the OCD method represent the equivalent central market problem:

$$\text{Min} \sum_{A \in Areas} \sum_{n \in A} C_n(P_n) \tag{7.6}$$

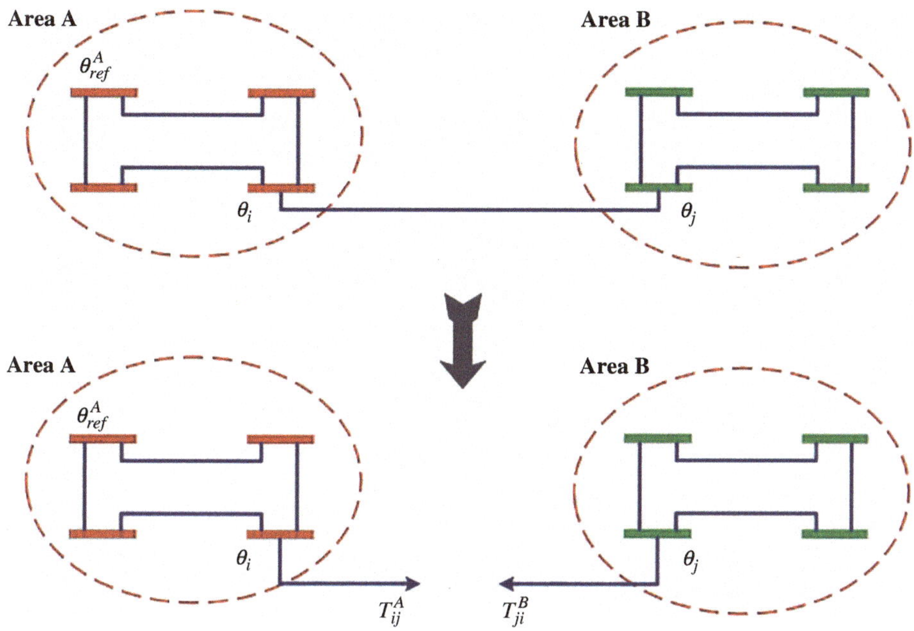

Fig. 7.2 Creating new coupling variables for each tie-line in the OCD method.

$$\mathbf{B^A} \cdot \boldsymbol{\theta}^{\mathbf{A}} + \mathbf{R^A} \cdot \mathbf{T^A} = \mathbf{P^A} - \mathbf{D^A} \tag{7.7}$$

$$\theta_{ref}^{A} = 0 \tag{7.8}$$

$$\left| \frac{1}{x_{ij}} \left(\theta_i^A - \theta_j^A \right) \right| \leq F_{ij}^{\max} \tag{7.9}$$

$$P_n^{\min} \leq P_n \leq P_n^{\max} \tag{7.10}$$

$$|T_{ij}^A| \leq T_{ij}^{\max} \tag{7.11}$$

$$\frac{1}{x_{ij}} \left(\theta_i^A - \theta_j^B \right) - T_{ij}^A = 0 \tag{7.12}$$

where $\mathbf{T^A}$ shows the power flow vector on the tie-lines in area A, $\mathbf{R^A}$ is a coefficients matrix for mapping the power of tie-lines to nodes' power, $T_{ij}^{\max}$ indicates the maximum capacity for tie-line ij, and $\mathbf{B^A}$ represents the admittance matrix corresponding to area A except for tie-lines.

For each area, by adding an expression for the amount of import or export of the tie-lines, the load balance constraint in (7.2) is maintained. The parameter F_{ij} is also used for transmission lines, and the variable T_{ij} is used for tie-lines. According to the equivalent central market problem (7.6)–(7.12), it is observed that (7.7)–(7.11) are separable for each region. Nevertheless, the constraint (7.12), which shows the relation of the tie-lines with the boundary angles, is inseparable. In optimization decomposition problems, constraints that are not parsable are called coupling constraints. By adding the Lagrangian coefficients of the constraint (7.12) to the objective function, the Lagrangian function can be formed, and Karush-Kuhn-Tucker (KKT) conditions are created. The Lagrangian coefficients of (7.12) represent the price of power exchange on each tie-line.

Assuming the existence of two interconnected regions to form the desired Lagrangian function and the KKT conditions to show how the formulas of each region can be separated, the equivalent problem for the two regions can be rewritten as follows.

$$\text{Min}\left[f_A(x_A) + f_B(x_B)\right] \tag{7.13}$$

$$g_A(\mathbf{x_A}) \leq 0 \tag{7.14}$$

$$g_B(\mathbf{x_B}) \leq 0 \tag{7.15}$$

$$h_A(\mathbf{x_A}, \mathbf{x_B}) = 0 \tag{7.16}$$

$$h_B(\mathbf{x_A}, \mathbf{x_B}) = 0 \tag{7.17}$$

where $\mathbf{x_A}$ and $\mathbf{x_B}$ are decision vector variables for area A and area B, respectively ($\mathbf{x_A} = [(P_A)^T, (\theta^A)^T, (T^A)^T]^T$, $\mathbf{x_B} = [(P_B)^T, (\theta^B)^T, (T^B)^T]^T$), and $f_A(\mathbf{x_A})/f_B(\mathbf{x_B})$ is operation cost function of area A/area B. In (7.14) and (7.15), $g_A(\mathbf{x_A})$ and $g_B(\mathbf{x_B})$ represent independent constraints (7.7)–(7.11) in the equivalent central market problem. Also, (7.16) and (7.17) indicate coupling constraints (instead of (7.12) in the equivalent central market problem). The Lagrangian coefficients of the coupling constraints are α_{ij}^A and α_{ji}^B for regions A and B, respectively; therefore:

$$h_A(\mathbf{x_A}, \mathbf{x_B}) = \frac{1}{x_{ij}}\left(\theta_i^A - \theta_j^B\right) - T_{ij}^A = 0 \tag{7.18}$$

$$h_B(\mathbf{x_A}, \mathbf{x_B}) = \frac{1}{x_{ij}}\left(\theta_j^B - \theta_i^A\right) - T_{ji}^B = 0 \tag{7.19}$$

Now, for (7.13)–(7.17) to be broken down into two independent and regional problems, first, the independent constraints and the operation costs in each area must be separated. Then, to assign the coupling constraints to each area, Lagrangian coefficients must be added as constant values in the objective function of each adjacent region problem. In this case, the KKT conditions are fully established. Hence, the local problem for area A is as follows:

$$\text{Min}\left[f_A(\mathbf{x_A}) + \hat{\alpha}_{ji}^B \cdot h_B(\mathbf{x_A}, \hat{\mathbf{x}}_\mathbf{B})\right] \tag{7.20}$$

$$g_A(\mathbf{x_A}) \leq 0 \tag{7.21}$$

$$h_A(\mathbf{x_A}, \hat{\mathbf{x}}_\mathbf{B}) = 0 \tag{7.22}$$

For area B:

$$\text{Min}\left[f_B(\mathbf{x_B}) + \hat{\alpha}_{ij}^A \cdot h_A(\hat{\mathbf{x}}_\mathbf{A}, \mathbf{x_B})\right] \tag{7.23}$$

$$g_B(\mathbf{x_B}) \leq 0 \tag{7.24}$$

$$h_B(\hat{\mathbf{x}}_\mathbf{A}, \mathbf{x_B}) = 0 \tag{7.25}$$

where $\hat{\alpha}_{ij}^A$ is the power exchange price from area A to area B from the previous iteration, $\hat{\alpha}_{ji}^B$ is the power exchange price from area B to area A from the previous iteration, and $\hat{\mathbf{x}}_\mathbf{A}$ and $\hat{\mathbf{x}}_\mathbf{B}$ are vectors of the variables obtained in area A and area B from the previous iteration, respectively.

Each area runs its market in the first iteration, assuming that the flow of tie-lines and exchange prices are zero. Thus, the market of each area runs completely separately without any exchange of information. This assumption is for the first iteration only. Then, from the next iterations, each area must share three categories of information with its adjacent areas after running its market:

- The angle of its boundary buses for each tie-line with the adjacent area: θ_i^A (θ_j^B)
- Power flow of the tie-line with the adjacent area: T_{ij}^A (T_{ji}^B)
- Power exchange price with the adjacent area: α_{ij}^A (α_{ji}^B).

The Lagrangian coefficients, boundary angles, and power exchange price for a typical area A in the OCD method are updated using (7.26)–(7.28) in iteration k.

$$\hat{\alpha}_{ij}^A = \left(\alpha_{ij}^A\right)^{k-1} \tag{7.26}$$

$$\hat{\alpha}_{ji}^B = \left(\alpha_{ji}^B\right)^{k-1} \tag{7.27}$$

$$\hat{\theta}_i^A = \left(\theta_i^A\right)^{k-1} \tag{7.28}$$

At the end of each iteration, each region compares the power flow values of its tie-lines, and convergence is achieved if the condition of (7.29) is met.

$$\left|T_{ij}^A + T_{ji}^B\right| \leq \varepsilon \tag{7.29}$$

2.4 Decentralized market by the LR method

The LR method has been one of the oldest methods available for decomposing optimization problems. In this method, the amount of shared information is limited. The implementation of this method is to separate the central problem and give independence to the market of each region.

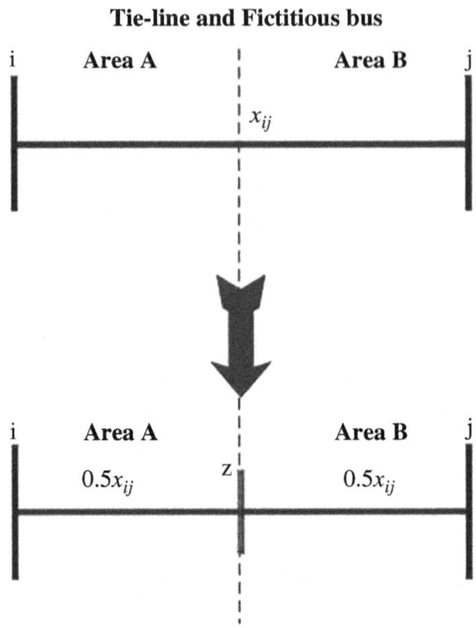

Fig. 7.3 division of each tie-line into two equal parts in the LR method.

In the first step of the LR method to separate the central problem into smaller problems, a fictitious bus is considered between each tie-line; so each fictitious bus divides its tie-line into two equal parts. In fact, this fictitious bus is the boundary between the two interconnected regions. No load is considered on this fictitious bus. Each region in its market considers only a fictitious bus of adjacent regions, which preserves the integrity of the independence of each region. Fig. 7.3 shows the division of each tie-line into two equal parts [11].

With the addition of each fictitious bus between each tie-line, (7.30) related to the load balance and generation must be considered in each boundary bus:

$$h_{ij} = \frac{1}{0.5x_{ij}}(\delta_i - \delta_z) + \frac{1}{0.5x_{ij}}(\delta_j - \delta_z) = 0 \tag{7.30}$$

where δ_i is the boundary bus angle i, δ_z is the fictitious bus angle, and x_{ij} is the impedance value between buses i and j. By placing (7.30) in (7.1)–(7.5), the general formulation of the main central problem is rewritten as

$$\text{Min} f(\mathbf{x}) \tag{7.31}$$

$$h(\mathbf{x}) = 0 \tag{7.32}$$

$$g_1(\mathbf{x}) = 0 \tag{7.33}$$

$$g_2(\mathbf{x}) \leq 0 \tag{7.34}$$

where $f(\mathbf{x})$ indicates (7.1), $h(\mathbf{x}) = 0$ represents (7.30) (the coupling constraint), $g_1(\mathbf{x}) = 0$ shows constraints (7.2) and (7.3), and $g_2(\mathbf{x}) \leq 0$ represents (7.4) and (7.5). First, the Lagrange function must be organized for solving the problem. To create the Lagrangian function, (7.32) with the Lagrangian coefficients ($\boldsymbol{\alpha}^T$) is added to the objective function according to (7.35):

$$L(\boldsymbol{\alpha}, \mathbf{x}) = f(\mathbf{x}) + \boldsymbol{\alpha}^T h(\mathbf{x}) \tag{7.35}$$

According to the LR decomposition method, to solve the Lagrangian function and obtain the appropriate values of Lagrangian coefficients, a dual function must be formed. The adaptive dual function is equal to:

$$\phi(\boldsymbol{\alpha}) = \mathrm{Min}_{\mathbf{x}} L(\boldsymbol{\alpha}, \mathbf{x}) \tag{7.36}$$

$$g_1(\mathbf{x}) = 0 \tag{7.37}$$

$$g_2(\mathbf{x}) \leq 0 \tag{7.38}$$

The Lagrangian coefficients in the dual function must be increased in such a way as to minimize the initial problem. As a result, the dual problem becomes a maximization problem. Suppose the main problem is convex. It can be shown that the optimal result of the main problem is equal to the optimal result of the dual problem. In general, the process of the LR algorithm is implemented as follows:

- Set the initial values for the vector of Lagrange coefficients;
- Solve the problem (7.36)–(7.38) for constant vector values α (this is equivalent to calculating or solving a dual function);
- Update the Lagrangian coefficient vector α;
- Check the convergence so that if the dual function and the coefficient vector α do not change much in two consecutive iterations, the algorithm stops; otherwise, step 2 is executed.

Therefore, initially, the market in the first area, which is considered the reference area, is implemented. Then, after determining the value of the fictitious bus angle on the tie-lines, the next adjacent area receives this angle and considers it as an input for its problem. It should be noted that the markets run sequentially. That is, each region must wait for the results of the implementation of neighboring markets. Then, the convergence of market results is examined. Due to the shorter time to achieve convergence in the subgradient method, this method is usually used to update the Lagrangian coefficients [11]. According to this method, the desired Lagrangian coefficients in the next iteration ($k + 1$) are updated as follows:

$$\alpha^{(k+1)} = \alpha^{(k)} + v^{(k)} \frac{s^{(k)}}{|s^{(k)}|} \tag{7.39}$$

where $v^{(k)}$ must be chosen that meets the following two conditions:

$$\lim_{k \to \infty} v^{(k)} \to 0 \text{ and } \sum_{k=1}^{\infty} v^{(k)} \to \infty \tag{7.40}$$

Considering s is a subgradient, a common option to select $\nu^{(k)}$ will be as follow:

$$\nu^{(k)} = \frac{1}{A + B \cdot k} \tag{7.41}$$

where A and B are constant parameters that are usually set in $A = 0.5$ and $B = 0.25$ to meet the condition in (7.40) [11]. The value of subgradients per iteration for the problem according to (7.42) is equal to the value of the power flow difference in the fictitious bus.

$$s^{(k)} = h_{ij}^{(k)} \tag{7.42}$$

The local independent problem for region A, with a fictitious bus for each tie-line, is as follows:

$$\text{Min}\left(\sum_{n \in A} f_n(p_n) + \sum_{ij \in A} \alpha_{ij} \frac{1}{0.5x_{ij}}(\delta_i - \delta_z)\right) \tag{7.43}$$

$$\mathbf{B^A} \cdot \boldsymbol{\theta}^A + \mathbf{R}^A \cdot \mathbf{Tie^A} = \mathbf{P}^A - \mathbf{D^A} \tag{7.44}$$

$$\theta_{ref}^A = 0 \tag{7.45}$$

$$\left|\frac{1}{x_{ij}}\left(\theta_i^A - \theta_j^A\right)\right| \le F_{ij}^{\max} \tag{7.46}$$

$$P_i^{\min} \le P_i^A \le P_i^{\max} \tag{7.47}$$

$$\left|\frac{1}{0.5x_{ij}}(\delta_i - \delta_z)\right| \le T_{ij}^{\max} \tag{7.48}$$

In (7.44), the balancing constraint for imports and exports is considered for area A. In this equation, the matrix $\mathbf{Tie^A}$ is used instead of the matrix $\mathbf{T^A}$. Matrix $\mathbf{T^A}$ itself is a decision variable; whereas, matrix $\mathbf{Tie^A}$ is obtained from the boundary decision variables according to (7.49):

$$\mathbf{Tie^A} = \frac{1}{0.5x_{ij}}(\delta_i - \delta_z) \tag{7.49}$$

2.5 Decentralized market by ALD (with APP and ADM methods)

In the LR method, the inseparable variables of the regions were separable for each region using the formation of a dual function. To improve the convergence of LR, the quadratic penalty criterion is added to the objective function, and the problem is defined as an ALD. In general, ADM and APP methods are used to solve the ALD optimizations.

In the APP method, a fictitious bus is considered for each tie-line. Therefore, with the addition of each fictitious bus between each tie-line, the load and generation balance constraint must be observed on each boundary bus. As a result, the formulation of the main problem is established in the form of (7.31)–(7.34). Now, using the balancing constraint in (7.32), the dual function of the problem is created according to (7.50). Eq. (7.50) is equivalent to an amplified Lagrangian function [19].

$$\phi(\alpha) = \text{Min}\left[f(\mathbf{x}) + \alpha \cdot h(\mathbf{x}) + \frac{\eta}{2}h^2(\mathbf{x})\right] \tag{7.50}$$

Since quadratic expressions exist in the objective function, they cannot be broken down into different regions. In order to separate the above problem into the number of available areas, the APP method is used. In this method, a limited number of problem variables are fixed to the optimal values of the previous iteration. It causes the central problem to be separable to the number of areas. Moreover, by updating the Lagrangian coefficients in each iteration, (7.50) is maximized. Now using fixed values (parameters $\hat{\delta}$), the separate objective function for area A is given below.

$$\text{Min}\begin{bmatrix} \sum\limits_{i\in A} f_i(p_i) + \sum\limits_{ij\in} \alpha_{ij}\frac{1}{0.5x_{ij}}\left(\delta_i^A - \delta_z\right) \\ +\sum\limits_{ij\in}\frac{2\gamma}{x_{ij}^2}\left[\left(\delta_i^A - \delta_z\right) - \left(\left(\hat{\delta}_i^A\right)^{k-1} - \left(\hat{\delta}_z\right)^{k-1}\right)\right]^2 \\ +\sum\limits_{ij\in}\frac{4\eta}{x_{ij}^2}\left[\left(\delta_i^A - \delta_z\right)\times\left(\left(\left(\hat{\delta}_i^A\right)^{k-1} - \left(\hat{\delta}_z\right)^{k-1}\right) + \left(\left(\hat{\delta}_i^B\right)^{k-1} - \left(\hat{\delta}_z\right)^{k-1}\right)\right)\right] \end{bmatrix} \tag{7.51}$$

where γ and η are constant coefficients that are selected experimentally. The separable problems of each region converge to the central problem with an iterative algorithm and limited exchanged information. The APP algorithm is run sequentially, which means that it is required to receive the results from the adjacent areas to execute a local problem. Each region executes its market problem, then sends the results of the boundary busbars (δ) and the fictitious buses (δ_z) between the tie-lines to the interconnected areas. After running the last area market, each area uses its received information and market results to update its Lagrangian coefficients. How to update Lagrange coefficients is based on the subgradient method. Eq. (7.52) shows the update of the Lagrangian coefficients of each region.

$$\alpha_{ij}^{A,B(k+1)} = \alpha_{ij}^{A,B(k)} + \rho\left[\left(\left(\hat{\delta}_i^A\right)^k - \left(\hat{\delta}_z\right)^k\right) + \left(\left(\hat{\delta}_j^B\right)^k - \left(\hat{\delta}_z\right)^k\right)\right] \tag{7.52}$$

To start the algorithm, in the first iteration, the values of Lagrange coefficients and constant values of the angles of the boundary and fictitious busbars are considered equal to zero. The algorithm also continues until the number of iterations exceeds the desired limit or the convergence constraint in (7.53) is realized. The purpose of the constraint in (7.53) is to equalize the angles of the two ends of each tie-line with the fictitious bus angle located in it for all existing tie-lines.

$$\sum_{A_n} \sum_{(i,j)\in Ties} \sqrt{\frac{2}{x_{ij}}\left(\hat{\delta}_i^{A_n} - \hat{\delta}_z\right)^2 + \frac{2}{x_{ij}}\left(\hat{\delta}_j^{A_k} - \hat{\delta}_z\right)^2} \leq \varepsilon; A_n \in \{A_1, A_2, \ldots, A_N\} \quad (7.53)$$

The ADM is another algorithm for solving ALD problems. In this method, as in the APP, the central problem is decomposed into the number of available areas using the quadratic expressions of the coupling constraints. Also, unlike the LR and APP, the fictitious bus between the tie-lines is not considered in the ADM. Each area in its local problem, in addition to the boundary bus in its area, considers a fictitious bus connected to the tie-line in the adjacent area. Fig. 7.4 shows a tie-line between a two-area network. As shown in Fig. 7.4, boundary buses are considered separately by each of the areas.

To illustrate the concept of the ADM method, it is first assumed that there exists a two-area network with a tie-line, as shown in Fig. 7.4. In this case, by considering the angles of boundary buses by each area, two sets of variables are created for each

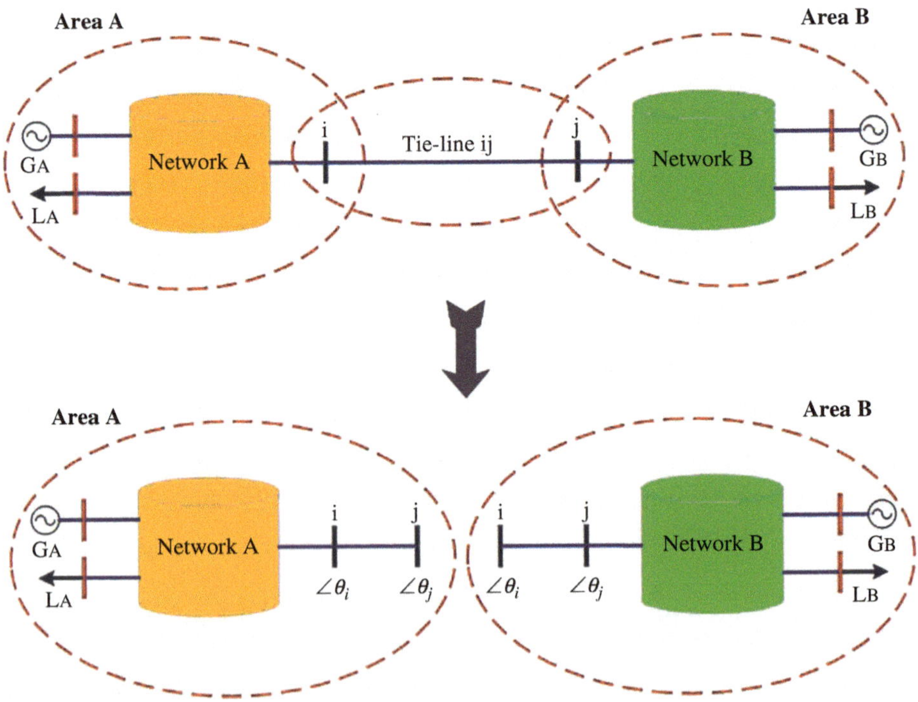

Fig. 7.4 Divide boundary buses in the ADM.

tie-line. Thus, set (θ_{iA}, θ_{jA}) for region A and set (θ_{iB}, θ_{jB}) for region B are local variables representing the angles of the a and b buses on the tie-line ij. Therefore, the coupling constraints of the above two regions are formulated as (7.54) and (7.55) [3,16].

$$\theta_{i_A} - \theta_{i_B} = 0 \tag{7.54}$$

$$\theta_{j_A} - \theta_{j_B} = 0 \tag{7.55}$$

The constraints in (7.54) and (7.55) express the concept that to converge the problem with a coordinated algorithm, the angles of the boundary buses obtained by each of the regions must be equal. As a result, these coupling constraints are each considered with Lagrangian coefficients and quadratic penalty expressions in the objective function of each local problem. The following equation represents the objective function of the local problem in area A

$$\text{Min}\left[\sum_{n\in A} f_n(p_n) + \sum_{i,j\in A}\left[\alpha_A\left(\theta_{Ai} - \overline{\theta_i}\right) + \frac{\rho}{2}\left(\theta_{Ai} - \overline{\theta_i}\right)^2\right]\right] \tag{7.56}$$

In (7.56), $\overline{\theta_i}$ represents the average angles obtained for boundary bus i in all areas connected to this bus by a tie-line. Therefore, it is obtained from (7.57).

$$\overline{\theta_i} = \sum_{i\in\phi(m)} \frac{\theta_{i,m}}{|\phi(m)|} \tag{7.57}$$

where $|\phi(m)|$ shows the number of the area connected to the boundary buses. If (7.56) is rewritten based on Fig. 7.4, objective functions (7.58) and (7.59) are obtained for area A and area B, respectively.

$$\text{Min}\left[\sum_{n\in A} f_n(p_n) + \alpha_{A_i}\left(\theta_{Ai} - \overline{\theta_i}\right) + \frac{\rho_A}{2}\left(\theta_{Ai} - \overline{\theta_i}\right)^2 + \alpha_{A_j}\left(\theta_{Aj} - \overline{\theta_j}\right) + \frac{\rho_B}{2}\left(\theta_{Aj} - \overline{\theta_j}\right)^2\right] \tag{7.58}$$

$$\text{Min}\left[\sum_{n\in B} f_n(p_n) + \alpha_{Bi}\left(\theta_{Bi} - \overline{\theta_i}\right) + \frac{\rho_B}{2}\left(\theta_{Bi} - \overline{\theta_i}\right)^2 + \alpha_{B_j}\left(\theta_{Bj} - \overline{\theta_j}\right) + \frac{\rho_B}{2}\left(\theta_{Bj} - \overline{\theta_j}\right)^2\right] \tag{7.59}$$

where ρ_A and ρ_B are the penalty coefficients. The penalty coefficients can be assumed to be fixed values, but they are usually considered different for various networks. These coefficients also can significantly affect the convergence of the problem. For this purpose, in some references [3,16], a dynamic method is used to assign penalty coefficients. In this way, the Lagrangian coefficients are updated according to (7.64)–(7.66) by calculating the primal residues ($P_{Ri}^{(k)}$ and $P_{Rj}^{(k)}$) and the dual residues ($D_{Ri}^{(k)}$ and $D_{Rj}^{(k)}$) by (7.60)–(7.63). In (7.66), τ and μ are constant terms that are selected empirically. Also, ε^P and ε^D show the maximum amount of acceptable initial residual and the maximum amount of acceptable dual residual, respectively.

$$P_{R_i}^{(k)} = \left|\theta_{i_A}^{(k)} - \overline{\theta}_i^{(k)}\right| \leq \varepsilon^P \tag{7.60}$$

$$P_{Rj}^{(k)} = \left|\theta_{j_A}^{(k)} - \overline{\theta}_j^{(k)}\right| \leq \varepsilon^P \tag{7.61}$$

$$D_{Ri}^{(k)} = \left|\rho_A^{(k)}\left(\overline{\theta}_i^{(k)} - \overline{\theta}_i^{(k-1)}\right)\right| \leq \varepsilon^D \tag{7.62}$$

$$D_{Rj}^{(k)} = \left|\rho_A^{(k)}\left(\overline{\theta}_j^{(k)} - \overline{\theta}_j^{(k-1)}\right)\right| \leq \varepsilon^D \tag{7.63}$$

$$\alpha_i^{(k+1)} = \alpha_i^{(k)} + \rho_A^{(k)}\left(\theta_{i_A}^{(k)} - \overline{\theta}_i^{(k+1)}\right) \tag{7.64}$$

$$\alpha_j^{(k+1)} = \alpha_j^{(k)} + \rho_A^{(k)}\left(\theta_{j_A}^{(k)} - \overline{\theta}_j^{(k+1)}\right) \tag{7.65}$$

$$\rho_A^{(k+1)} = \left\{\begin{matrix} \tau \cdot \rho_A^{(k)} & \text{if}\left(P_{R_i}^k > \mu \cdot D_{R_i}^{(k)}\right) \& \left(P_{R_j}^k > \mu \cdot D_{R_j}^{(k)}\right) \\ \dfrac{\rho_A^{(k)}}{\tau} & \text{if}\left(D_{R_i}^{(k)} > \mu \cdot P_{R_i}^k\right) \& \left(D_{R_j}^{(k)} > \mu \cdot P_{R_j}^k\right) \\ \rho_A^{(k)} & \text{otherwise} \end{matrix}\right\} \tag{7.66}$$

In general, the ADM method tries to bring the angles of the boundary buses closer to the mean angles by performing the problem of each region. The remarkable advantage of the ADM method is that local markets run simultaneously in all areas. As a result, unlike the LR and APP methods, each area does not wait for other areas to respond. Moreover, the status of information confidentiality is more appropriate in this method because it only shares the angles of boundary buses among areas.

2.6 Experience of US markets by implementing TO and CTS methods

For its technical and economic benefits, energy exchanges between American markets have been going on for years. For example, the markets of PJM, MISO, NYISO, and ISO-NE use both tie-optimization (TO) and coordinated-transaction-scheduling (TSO) methods to optimize interarea exchanges in real-time markets. These methods are implemented between two adjacent neighboring markets. Here, first, TO is described, and then the principles of implementing the CTS method on the TO are stated [20].

The TO method is basically designed for the tie-line scheduling in a two-area system. Due to the structure of the American markets, there is no coordinator entity. As a result, this method is implemented in a completely decentralized manner. The running time of this method is before the real-time markets of each area. In this way, the output of this method is considered as the input of real-time markets.

In this method, the total power exchange capacity between the two areas is considered. Therefore, if there is more than one tie-line between two areas, only one equivalent tie-line is modeled with the sum of all tie-line capacities. The two buses connected to this equivalent tie-line are called proxy buses in each area. In this section, for simplicity, only one tie-line between the two areas is considered.

Initially, the operator of each area separately tries to obtain energy export and import prices by using the estimated bidding of power plants. To do this, the operator of each area calculates the LMP of the proxy bus by placing different loads on this representative bus. The load range includes positive and negative values. Positive values are for modeling the power export mode, and negative values are for modeling the power import mode. At this stage, the operator of each area only monitors its network, and even the tie-line is not considered.

By obtaining the LMP of proxy buses for export and import quantities, the operator of each area forms a price curve based on the amount of export and import power. Then, one of the two areas provides the resulting price curve to the adjacent area. The operator who receives the price curve determines the optimal amount and price of the exchanged power flow by using the price curve of its area and the intersection of two curves.

For example, there are nine tie-lines between the NYISO and ISO-NE regional markets. Depending on the power price difference of the two areas, the flow of tie-lines can flow from east to west or vice versa. An example of the determination of the tie-line flow between the two regions, when the estimated power prices in the ISO-NE are higher than NYISO, is shown in Fig. 7.5. In this case, the two price curves intersect on the right side of the chart, indicating the desired power flow from left to right. Also, the horizontal axis shows the amount of exchange power between the two regions up to the total transmission capacity (TTC). Furthermore, it is possible to determine the desired power flow direction due to the higher or lower curves on the graph and at the zero point of the horizontal axis. The intersection points of the vertical axis with the curves would show the LMP for the boundary bus (proxy bus) if each area were to run separately without considering the tie-lines. Fig. 7.6 also shows the situation where the estimated power price is higher in the NYISO, and the power flow is from right to left.

After determining the amount of power for optimal interarea exchange, each area executes its problem in the real-time market to know the amount of import or export power in total.

In the CTS method, the same logic as in TO is considered. In the CTS, market participants' bids for external exchanges play a decisive role, while in the TO method, there are no offers for an external exchange. External offers in the CTS method consists of three parts:

- The amount of power required for the exchange.
- The desired exchange direction.
- The amount of the price difference that the participant is willing to exchange.

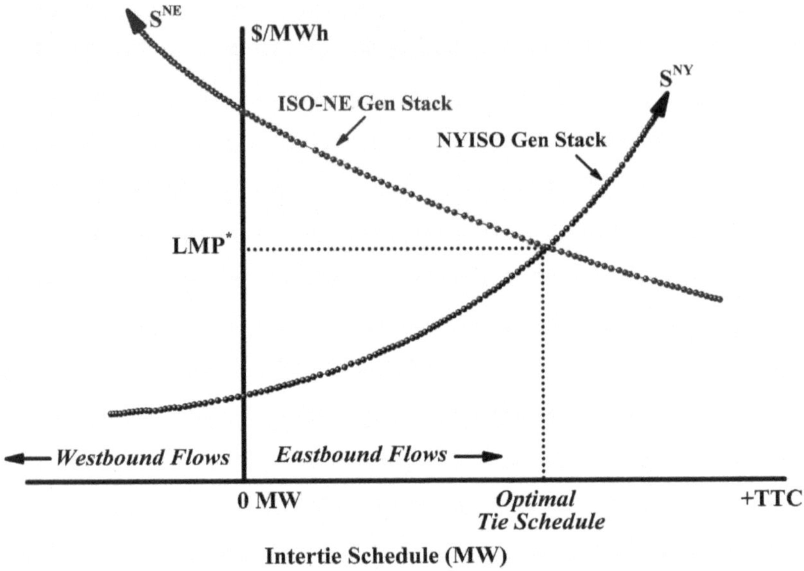

Fig. 7.5 Power flow when power prices in the ISO-NE are higher than NYISO (determined by the TO method).

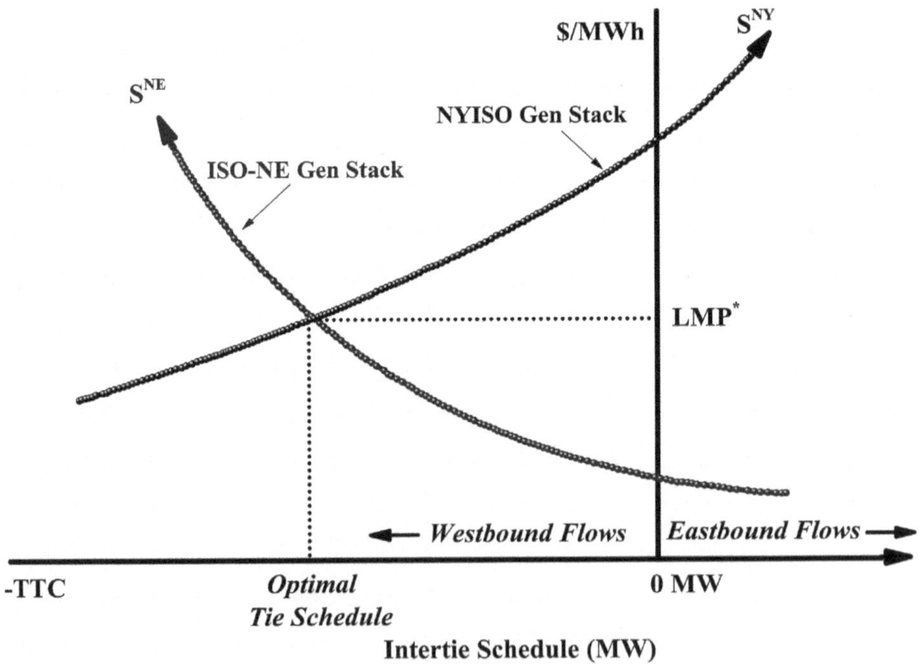

Fig. 7.6 Power flow when power prices in the ISO-NE are lower than NYISO (determined by the TO method).

CTS offers are virtual, and producers, consumers, and market traders can also make these offers. The market operator who receives the price curve of the adjacent area is responsible for collecting CTS bids from the two areas. The proxy operator sorts the CTS-aligned offers from low to high. Then, considering that one of the conditions of these offers is to have a price difference in the two areas, the operator selects the winners of the CTS offers that are in the desired flow direction (the optimal flow direction is determined from the vertical chart position).

2.7 Comparison of the decomposition-based methods

The formulation and important features of some decentralized methods for implementing multiarea markets were described in the previous sections. This section considers the implementation of OCD, ADM, and APP methods on a 15-bus three-area test system to examine their performance in more detail. This system has 15 buses, 15 internal lines, three tie-lines, and 12 generation units. The single diagram of the system is shown in Fig. 7.7. More detail about this system can be found in Karimi et al. [9] and Karimi Varkani et al. [8].

As seen from Fig. 7.7, the connection of the three areas is in the form of a loop. Therefore, this system can be an appropriate case for simulation due to the loop among the areas and the high capacity of tie-lines compared to the demand of each area. The flow of lines is highly dependent on the price offer of production units. In fact, offer values can determine the direction of power flow in tie-lines. In Table 7.1, three different price categories are considered for each area's units [21]. The prices assumed in this table can increase interarea exchange. On the other hand, in addition to the main case, in another case, the capacity of the tie-lines doubles. The higher capacity of these lines makes the problem more challenging to achieve optimal power exchange.

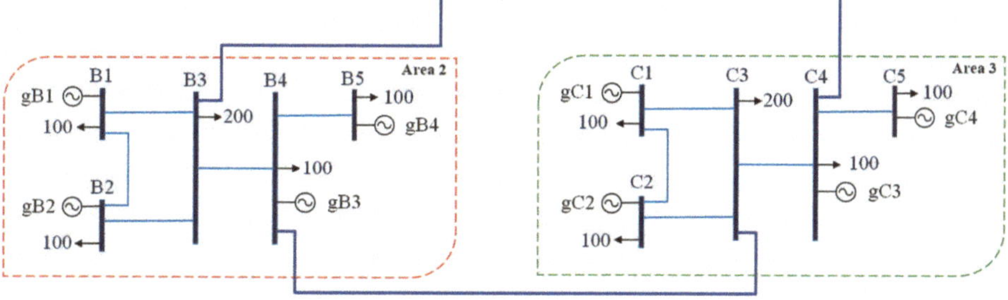

Fig. 7.7 Fifteen-bus three-area test system.

Table 7.1 Price offer of power units ($/MWh).

Unit	Area A	Area B	Area C
gX1	5	11	30
gX2	4	10	30
gX4	15	20	40
gX5	8	18	35

To examine the efficiency of each method, the results of centralized dispatch, as the optimal and reference results, considering two different capacities of the tie-lines, are shown in Table 7.2.

The results of OCD, ADM, and APP methods for the test system under different reference bus selections and different capacities of tie-lines are presented in Figs. 7.8–7.11. In Figs. 7.8 and 7.9, the efficiency and the number of iteration in each method are compared, respectively, considering that the capacity of tie-lines is 200 MW. In Figs. 7.10 and 7.11, the mentioned comparison has been made by considering the capacity of tie-lines in the amount of 400 MW. Moreover, the trend of convergence of tie-line power flows between areas in the case that the capacity of tie-lines is considered 400 MW ($S_{base} = 100$ MVA) for OCD, ADM, and APP methods are specified, respectively in Figs. 7.12–7.14. According to Figs. 7.8–7.11, the results of OCD have a high dependence on the choice of reference bus and the capacity of tie-lines. Also, a limited convergence rate was achieved in this method (shown with the Not Conv. phrase). Although this method can be suitable in two-area or multiarea networks with no loops, it can be challenging to solve in the looped multiarea system. However, according to Fig. 7.12, the power flow of tie-lines after 33 iterations has reached the optimum values obtained from running the centralized market (Table 7.2). Thus, Fig. 7.12a shows the power flow of the tie-line A-B having reached 133.3. Also, considering Fig. 7.12b and c, the flow of the tie-lines between areas A to C and B to C has reached 400 and 200 MW, respectively. It should be noted that the reference bus, in this case, is B4.

Table 7.2 Central market results.

Tie-lines capacity (MW)	$A \rightarrow C$ (MW)	$B \rightarrow C$ (MW)	$A \rightarrow B$ (MW)	Total operation cost ($)
200	200	200	0	21,300
400	400	200	133.3	16,850

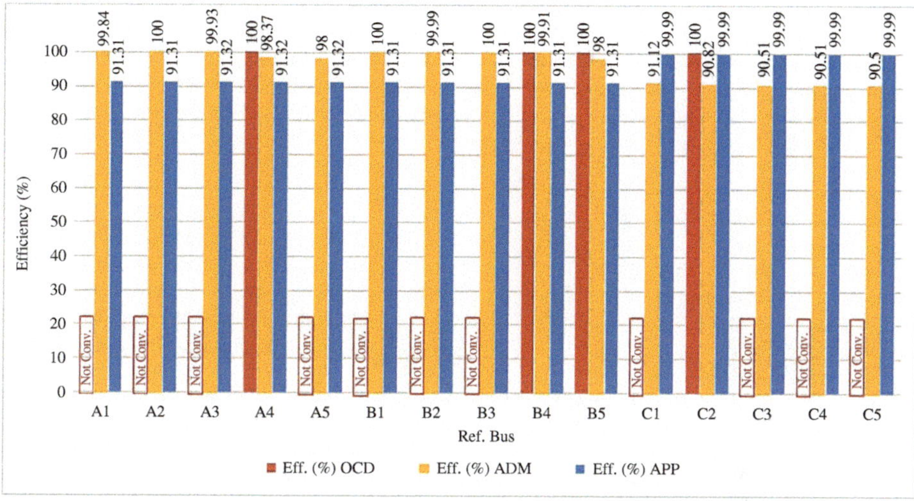

Fig. 7.8 Comparing the efficiency of OCD, ADM, and APP methods (tie-lines capacity = 200 MW).

Another conclusion that can be drawn is that the ADM has a higher efficiency than the APP, but its iteration number to achieve convergence is very high. In order to allow a better comparison, Table 7.3 also shows the average and the standard deviation of efficiencies and the number of iterations of ADM and APP methods. From the results, it can be concluded that although the average efficiency of both

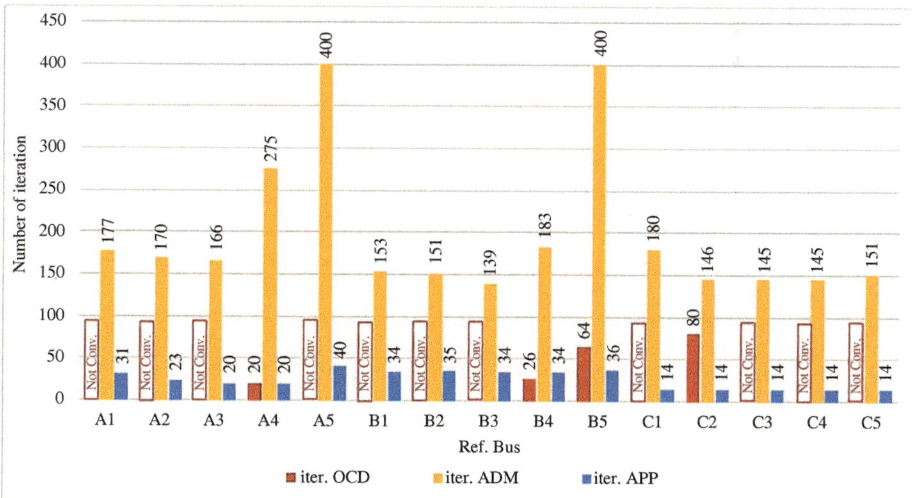

Fig. 7.9 Comparing the number of iteration of OCD, ADM, and APP methods (tie-lines capacity = 200 MW).

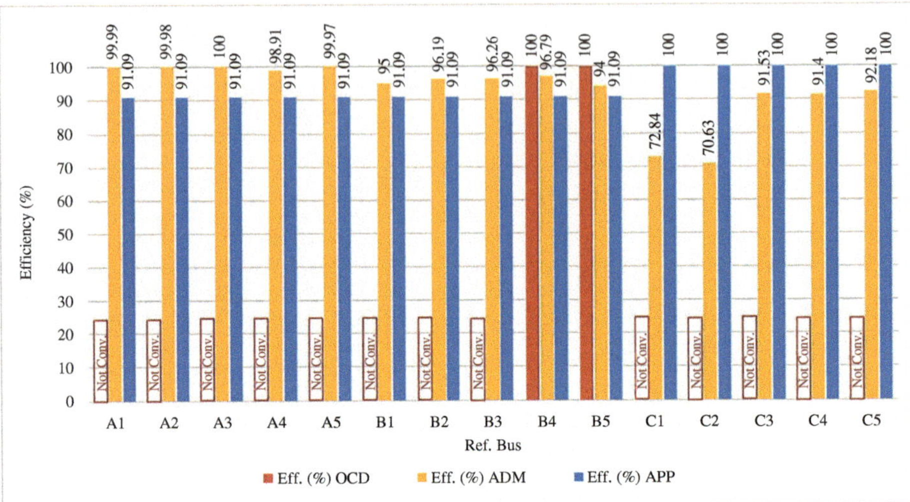

Fig. 7.10 Comparing the efficiency of OCD, ADM, and APP methods (tie-lines capacity = 400 MW).

ADM and APP methods is nearly the same, the number of iterations of the ADM is much higher. Also, the standard deviations of the mentioned parameters in the ADM are much higher than the APP. Therefore, the APP can be an acceptable method among the decomposition-based methods for solving the decentralized approach. More importantly, other factors, such as network topology, may also influence

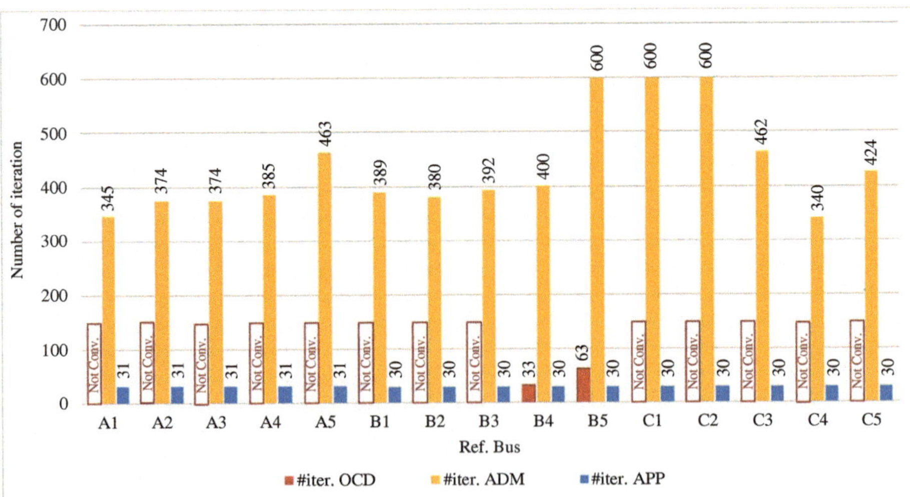

Fig. 7.11 Comparing the number of iteration of OCD, ADM, and APP methods (tie-lines capacity = 400 MW).

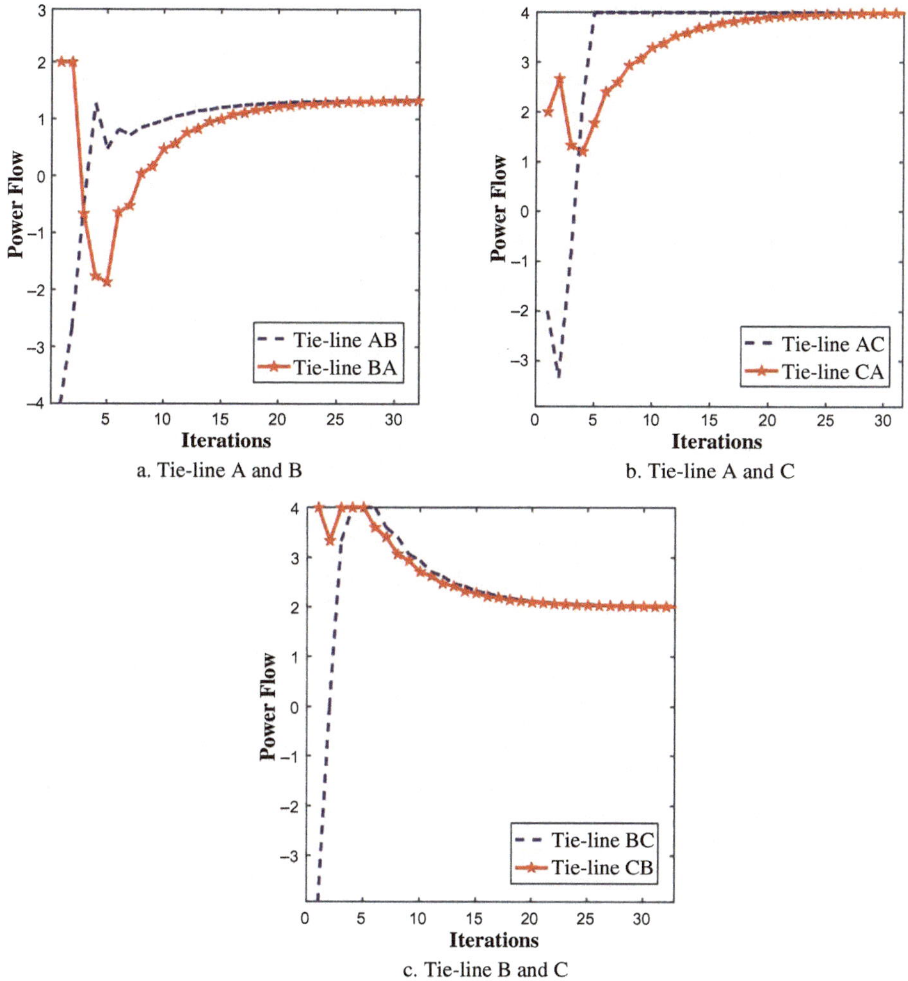

Fig. 7.12 Convergence of tie-lines power flow in the OCD method.

the results of decomposition-based methods. Therefore, one method may not work best in all situations.

As in the case of Fig. 7.12, it is possible to review and compare the results of Figs. 7.13 and 7.14 with the results of implementing the three-area market in a centralized manner. Fig. 7.13 illustrates how the power flow of the tie-lines converges on considering B3 as the reference bus, whereas, in Fig. 7.14, C1 is considered as a reference bus.

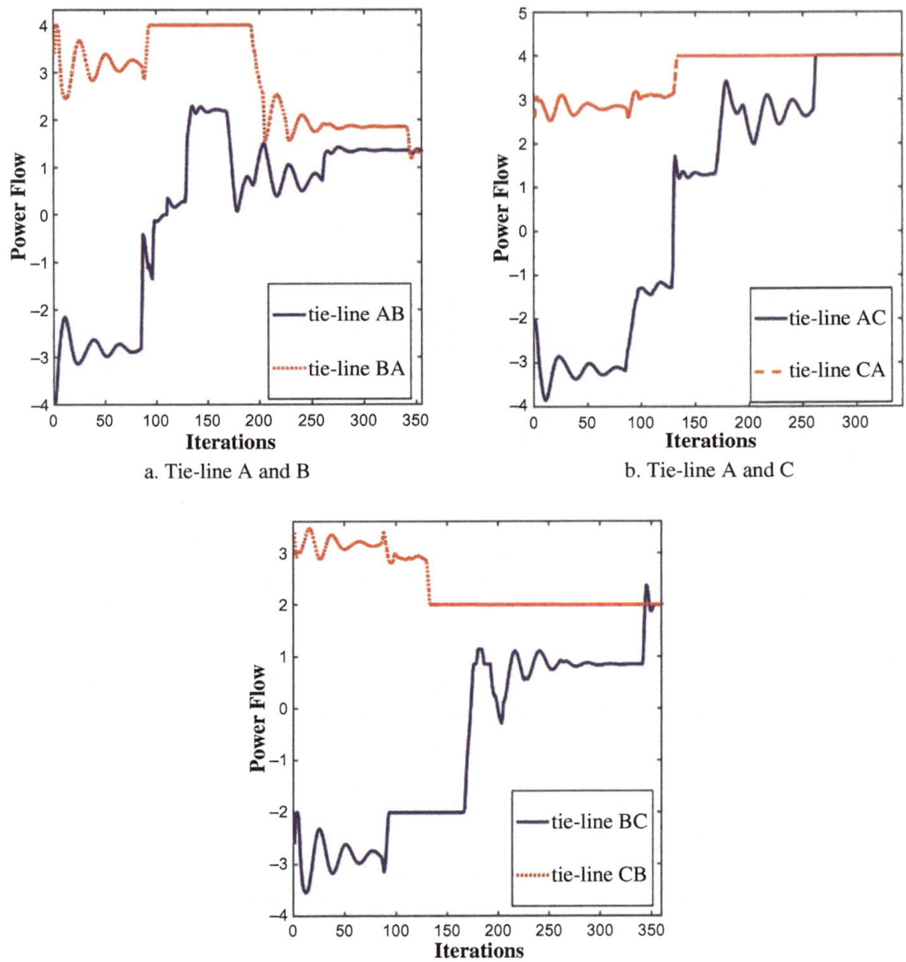

a. Tie-line A and B

b. Tie-line A and C

c. Tie-line B and C

Fig. 7.13 Convergence of tie-lines power flow in the ADM method.

3 Local electricity markets for smart grids

In recent years, the energy production landscape has been reshaped by increasing the penetration rate of DERs such as solar panels, smart appliances, electric vehicles (EV), and energy storage systems (ESSs). For example, in California, more than 21% of the renewable energy growth is provided by distributed photovoltaic generation. There are several reasons that justified greater use of DERs in the power system, such as economic motivations, improved reliability, and reduction of system losses. On the other hand, with the growth of the energy demand, generation capacity did not always increase appropriately. On these occasions, market operators consider possibilities for deeper integration of DER. Increasing the penetration rate of DERs, and

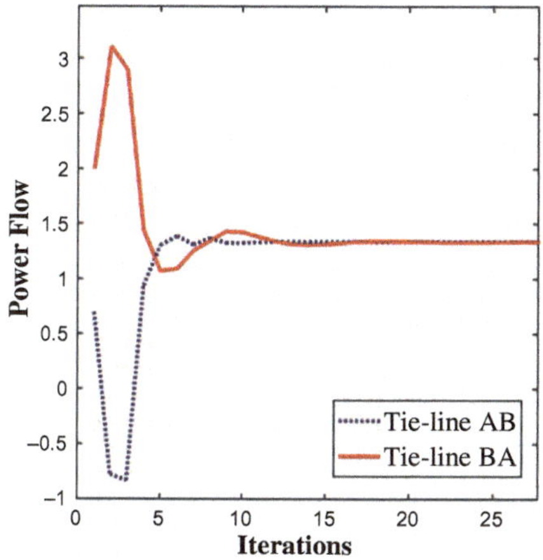

a. Tie-line A and B

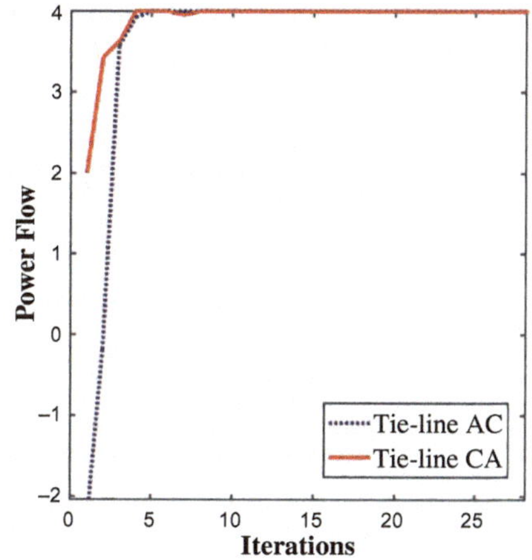

b. Tie-line A and C

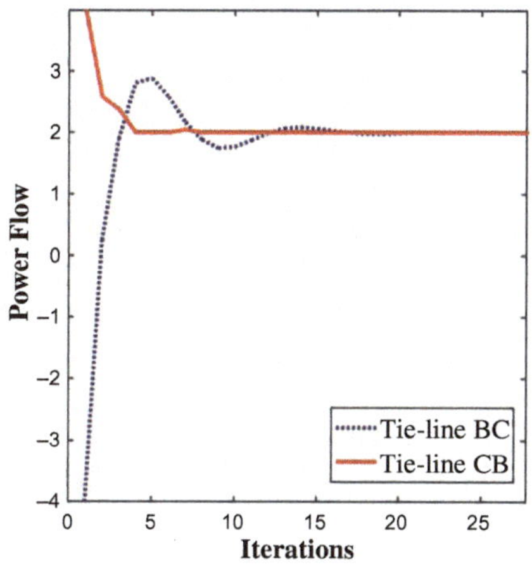

c. Tie-line B and C

Fig. 7.14 Convergence of tie-lines power flow in the APP method.

Table 7.3 Comparison of average results between APP and ADM.

Capacity (MW)	Method	Efficiency (%)	#Iterations	Standard deviation of efficiency	Standard deviation of #iterations
200	APP	94.20	25	4.09	9
	ADM	96.5	198	4.16	102
400	APP	94.06	30	4.20	0.5
	ADM	93.04	435	8.87	89

advances in information and communication technology (ICT) have led to a rapid increase in the number of prosumers. The prosumers are customers with proactive behavior. In fact, in addition to consumption, they also produce electricity [22–26].

With the penetration of DERs and increasing power data, building a new, reliable, and effective energy trading model to uncover the potential of distributed generation is essential. If resource allocation and pricing are still performed based on the centralized and top-down approach in the electricity market, the prosumer behaves as a passive receiver. On the other hand, if the payment for DER services is made by aggregators or utilities, the transparency, and accuracy of this payment will always be ambiguous from the prosumer viewpoint. In fact, with the current structure of electricity markets, incentives for active prosumer participation have not been sufficient. According to the above, the new structure should offer more independence and freedom of action to market participants. Also, it guarantees fair payments to all parties. Using electricity markets within the decentralized structure and integrating it with the collaborative principles allow for a bottom-up approach that would give more independence and transparency to the market and empower prosumers [25,27,28].

Unlike multiarea markets, where decomposition is based on distinct areas and geography, the system is usually decomposed based on network nodes in the decentralized local market. In this way, each prosumer can act as an independent agent and directly buy and sell energy with other DERs and existing markets. In fact, in this case, a P2P market is formed. In decentralized local markets, it is possible to consider a set of nodes that are either geographically close to each other or have the same stakeholders and formed a community-based P2P market. It is still possible to use the decomposition methods described in the previous section in the decentralized local market, such as LR, OCD, and ADM. In recent years, the advent and spread of distributed ledger technologies, particularly blockchain technology, has increased the security and transparency of exchanges. Blockchains are an emerging platform for decentralized structures, where data is stored in a secured state. Cryptographic signatures and distributed consensus mechanisms ensure information security in the blockchain platform. The use of this technology in energy markets will reduce the risks of cyber-attacks and tampering by participants. On the other hand, blockchains enable the execution of smart contracts on P2P networks in applications such as energy exchanges and settlement and eliminate the need for third-party applications. In the following, more details about the P2P structure and blockchains will be provided.

3.1 P2P markets

As mentioned in the previous section, to enhance the operation of DERs, the use of a decentralized alternative structure instead of the conventional centralized structure was proposed. This on the P2P consumer-centric market viewpoint depends on the P2P structure. P2P electricity trading defines a decentralized structure based on an interconnected platform (market), where all exchanges take place directly, without the presence of an intermediary. The first reference that offered the P2P concept for power systems dates back to 2007. The P2P electricity trading model was created

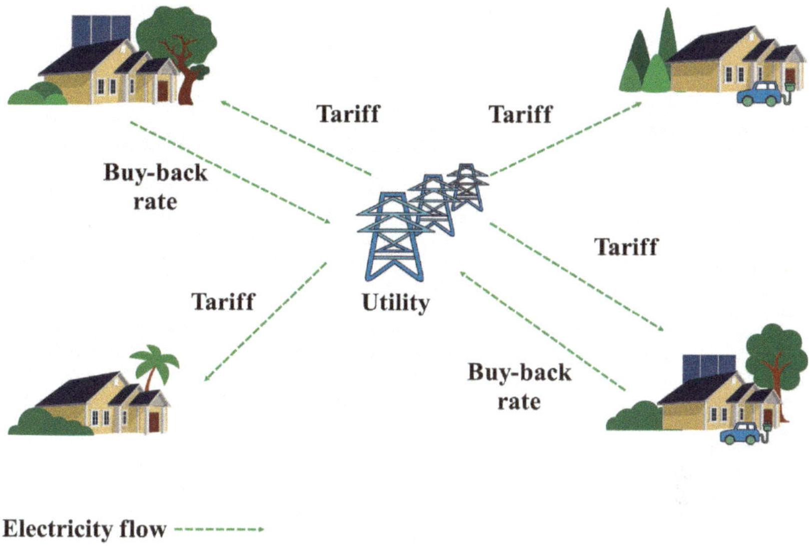

Fig. 7.15 Traditional trading model of consumer and prosumer with utilities.

due to the increasing penetration of DERs connected to distribution networks [29]. P2P electricity trading empowers prosumers and provides more incentives to promote the further deployment of DERs. If the P2P structure is not used and the conventional structure is used, as shown in Fig. 7.15, consumers would procure electricity from utilities via fixed tariffs or time-of-use tariffs. In contrast, prosumers sell surplus electricity back to the grid at a "buy-back rate." Despite consumer tariffs being higher than the buy-back rates, these consumer tariffs do not reflect other benefits these DERs bring to the power system. In the P2P market, prosumers can directly trade electricity with other consumers or prosumers to attain a better and more transparent outcome [30–32].

So far, three structures have been presented for the P2P markets: (1) full P2P market, (2) community-based market, and (3) hybrid P2P market. In these structures, the degree of decentralization and also the topology are different.

3.1.1 *Full P2P market*

In this market structure, two peers negotiate directly with each other to buy and sell electricity and other services, as shown in Fig. 7.16. Also, they can agree on the amount and price of energy exchanged without centralized supervision. A general mathematical model of a full P2P market design is described below [28,34].

$$\min_{D} \sum_{n \in \Omega} C_n \left(\sum_{m \in \omega_n} P_{nm} \right) \tag{7.67}$$

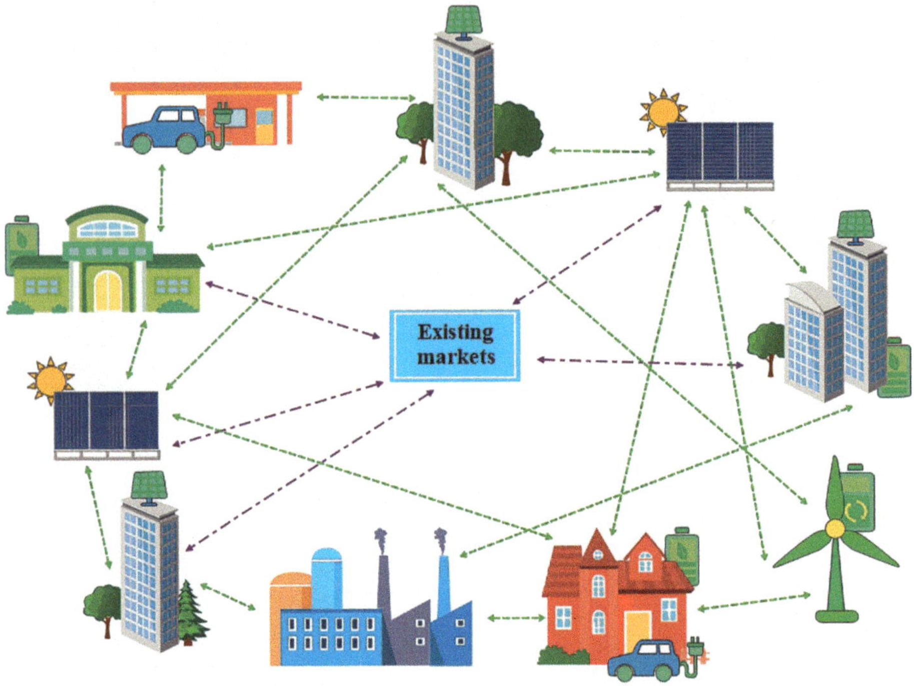

Fig. 7.16 Full P2P market [26,28,33].

Subject to:

$$\underline{P_n} \leq \sum_{m \in \omega_n} P_{nm} \leq \overline{P_n} \quad \forall n \in \Omega \tag{7.68}$$

$$P_{nm} + P_{mn} = 0 \quad \forall (n, m) \in (\Omega, \omega_n) \tag{7.69}$$

$$P_{nm} \geq 0 \quad \forall (n, m) \in (\Omega_p, \omega_n) \tag{7.70}$$

$$P_{nm} \leq 0 \quad \forall (n, m) \in (\Omega_c, \omega_n) \tag{7.71}$$

where $D = (P_{nm} \in \mathbb{R})_{n \in \Omega,\, m \in wn}$ with P_{nm} correspond to the transactions between agent n and m. The correlated dual variable λ_{nm} exhibits the price for each bilateral trade. In general, the outcome of each negotiation process can lead to different prices. The function C_n is mostly related to the production cost (or willingness to pay), which is commonly a quadratic function in terms of bilateral trades P_{nm} to demonstrate production/consumption costs. The positive value of the P_{nm} corresponds to the sale or production of energy (7.70), while a negative value is equal to purchase or consumption. Ω refers to all peers, while, Ω_p and Ω_c are for producers and consumers respectively. Also, ω_n contains the trading partners of a certain peer n. In Fig. 7.17, a simple example of the P2P trading between four peers is shown, in which peers 1 and 2 have

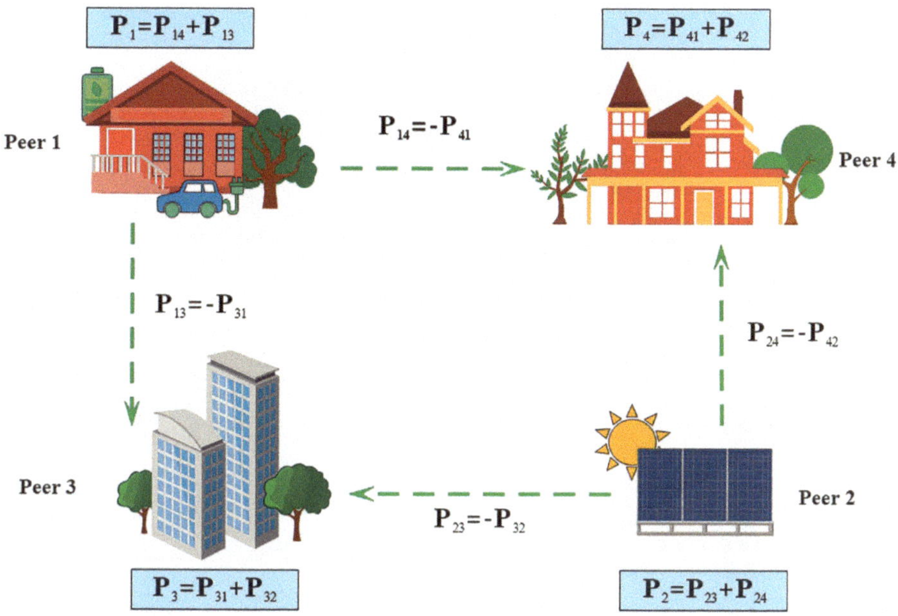

Fig. 7.17 Illustrative sample of a Full P2P model [28].

appeared in the role of producer, and peers 3 and 4 are consumers. In order to solve the given optimization problem, the decomposition methods expressed in the previous sections, such as LR, OCD, and ADM, can be used, which are consistent with the structure of the P2P markets. In this state, each node only shares the power and price that is wanting to trade.

3.1.2 Community-based market

In this scheme, a set of agents with common locational characteristics (i.e., geographically close) can form a community [35]. Also, the formation of a community can be based on the same goals and interests. For instance, a group of nodes willing to share clean energy can make a community though they are not in a similar geographical area. As another example, the various levels in the multimicrogrids (MMGs) network containing multiple stakeholders, such as the main power grid, the distribution company, and diverse types of MGs, can create a separate community [36]. In every community, there is a community manager. According to Fig. 7.18, the management of trading inside the community, as well as between a community and the rest of the system, is the responsibility of the community manager.

A general mathematical model of a community-based market design is described below [38].

$$\min_{D} \sum_{n\in\Omega} C_n(p_n, q_n, \alpha_n, \beta_n) + G(q_{imp}, q_{exp}) \tag{7.72}$$

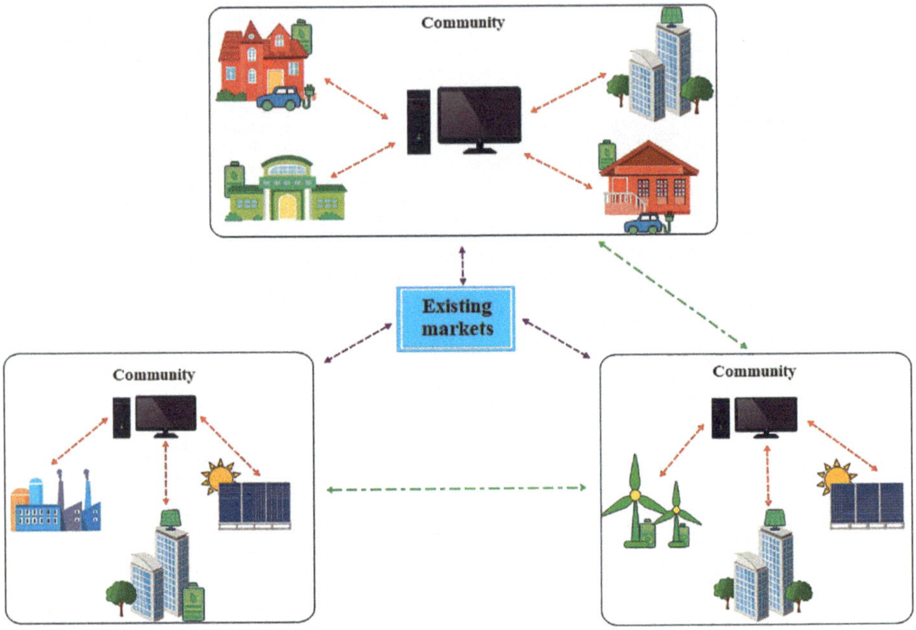

Fig. 7.18 Community-based market design [28,33,37].

Subject to:

$$p_n + q_n + \alpha_n - \beta_n = 0, \quad \forall n \in \Omega \tag{7.73}$$

$$\sum_{n \in \Omega} q_n = 0 \tag{7.74}$$

$$\sum_{n \in \Omega} \alpha_n = q_{imp} \tag{7.75}$$

$$\sum_{n \in \Omega} \beta_n = q_{exp} \tag{7.76}$$

$$\underline{p_n} \leq p_n \leq \overline{p_n} \quad \forall n \in \Omega \tag{7.77}$$

where $D = (p_n, q_n, \alpha_n, \beta_n \in \mathbb{R})_{n \in \Omega}$ and p_n corresponds to the generation or consumption of peer n. In order to clarify the concept, in Fig. 7.19, a small community is given as a conceptual explanation of the community-based market design. In this design, each agent can be traded within the community through q_n. Also, it is possible for any peer to trade with the outside through α_n and β_n, which are respectively the power export and import. The objective function (7.72) accounts for the cost associated with all decision variables. The components of this cost include a quadratic cost function of

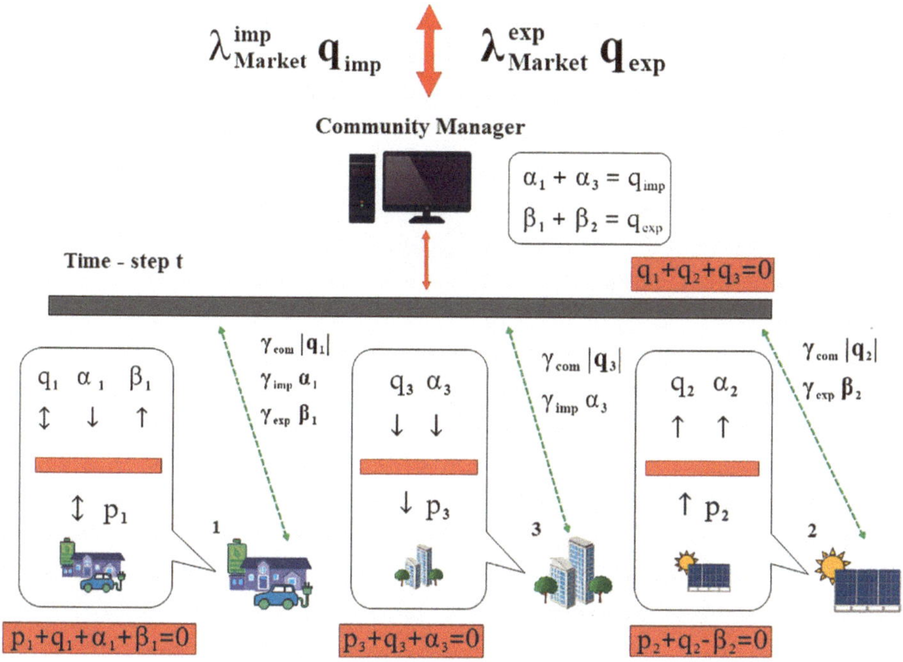

Fig. 7.19 Conceptual explanation of a community-based market design [28].

p_n and a transaction cost γ_{com} associated with q_n. Also, for transactions outside the community (α_n, β_n), one can use weighting coefficients γ_{imp} and γ_{exp} to translate the member's preference toward the outside world. In addition to the above, the community manager also has a function associated with the energy exchanged with the outside world $G(q_{imp}, q_{exp})$.

As mentioned, one of the applications of the community-based market is its use in MMGs with different stakeholders. Fig. 7.20 shows a typical MMGs which contains multistakeholders, and each stakeholder is responsible for the operation and coordination of its own equipment. Therefore, the system can be divided into network, supply, and utility level communities. There are multiple AC and DC MGs at the utility level, and each MG has a renewable energy resource, ESSs, and electric load. MGs exchange power with each other through bidirectional lines (BL). On the other hand, each MG exchanges power through interaction lines (IL) with the supply level. After coordination of neighboring MGs through BL, it is possible that there is an excess or shortage of energy. Excess or shortage of energy at the utility level will be sold to or purchased from the supply level through IL. To dispatch all IL identically, the supply-level conversion lines link AC and DC MGs, and a diesel engine maintains the adequate power supply to MGs at the supply level. In addition, the network-level links AC buses through grid-connected lines (GL), thus achieving electric power trading.

Due to security issues, each stakeholder only shares the power of the tie-lines, and therefore, distributed optimization is often used to solve these problems. Using SD

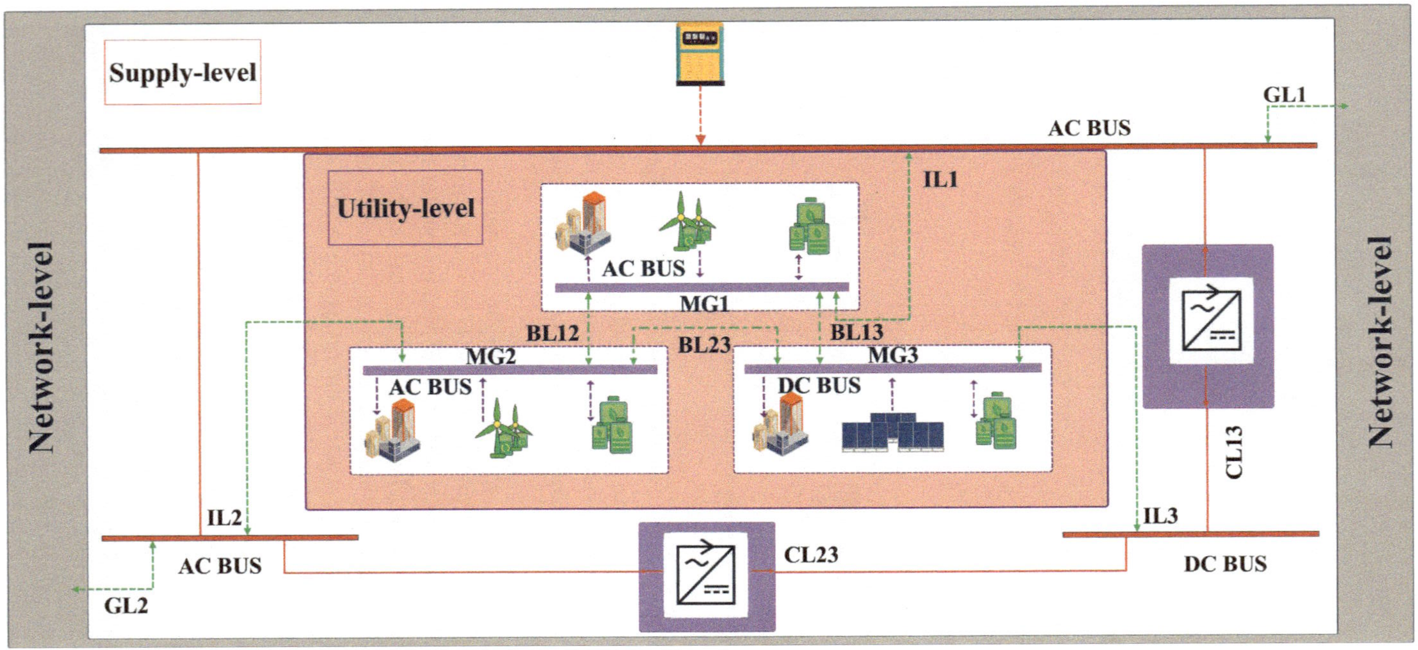

Fig. 7.20 General structure of AC/DC MMGs.

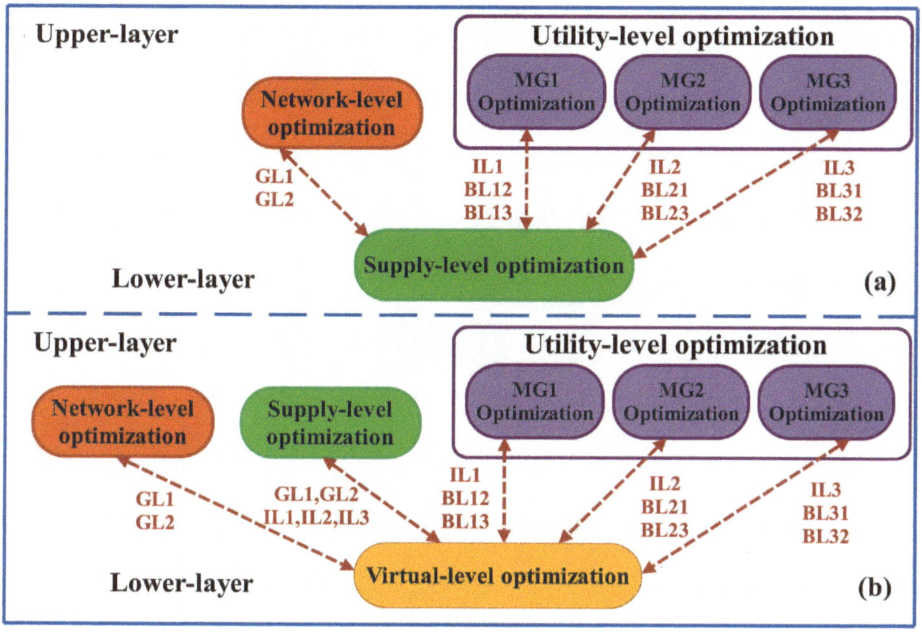

Fig. 7.21 Framework of SD and PD optimization.

(sequential distributed) and PD (parallelizing distributed) optimizations are two common approaches for solving such a multistakeholder scheduling problem in a distributed way. For SD optimization, since the supply level has a direct physical relationship with the network level and utility level, it can be prioritized as a coordination center in the lower level; while, the network level and utility level are assumed to be located in the upper layers. The SD optimization framework is shown in Fig. 7.21a [36].

Unlike the SD, in the PD optimization, a virtual coordinator has the task of creating coordination between the supply, utility, and network levels, and its framework is shown in Fig. 7.21b. In this method, the virtual coordination is installed in the lower level and helps to minimize power deviation in all shared tie-lines.

3.1.3 Hybrid P2P market

This design combines two previous approaches, wherein each layer, community, and single peers may interact directly with each other, as shown in Fig. 7.22. The scheme can be considered as a two-level approach. At the upper level, individual peers or energy collectives engage in P2P transactions between themselves and interact with existing markets. At the bottom level, the energy collectives behave like the community-based approach previously introduced, where a community manager is responsible for the trading inside its community. Although there is no generic formulation for this scheme, a simple form of hybrid design can be

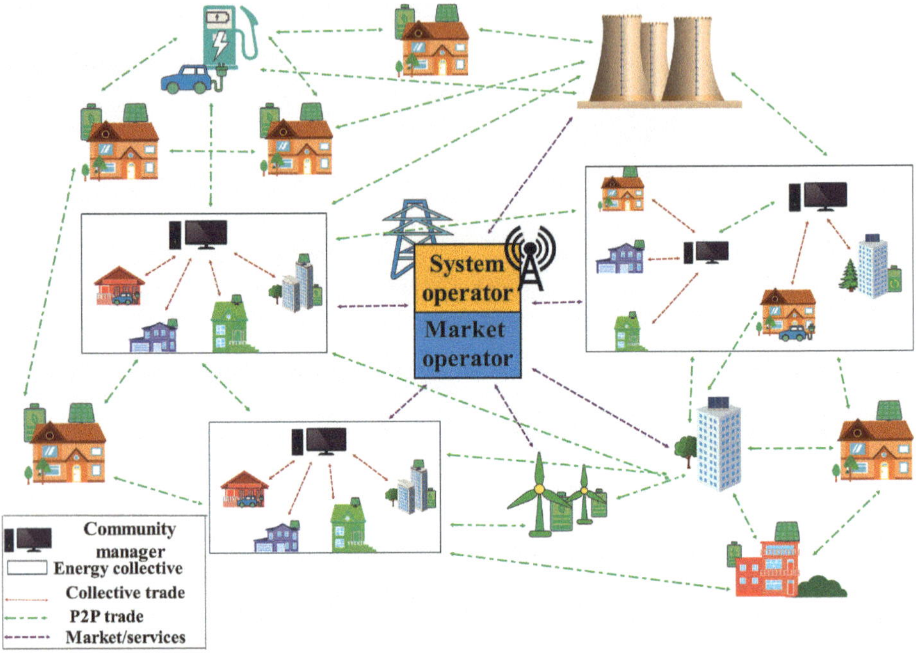

Fig. 7.22 Hybrid P2P market design [28].

achieved by combining the two previous formulations. For this purpose, two levels are considered. Only communities are considered at the bottom level, whereas the upper level assumes a P2P negotiation between individual peers and community managers.

$$\min_{D} \sum_{n\in\Omega^u} C_n^u\left(\sum_{m\in\omega_n} p_{nm}\right) + \sum_{n\in\Omega^b} c_n^b(p_n, q_n, \alpha_n, \beta_n) \tag{7.78}$$

Subject to upper level-full P2P design:

$$\text{Constraints in } (7.67)-(7.71) \quad \forall n\in\Omega^u \tag{7.79}$$

$$\sum_{m\in\omega_n} p_{nm} = q_{exp}^n - q_{imp}^n \quad \forall(n,m)\in(\Omega_{co},\omega_n) \tag{7.80}$$

Subject to bottom level-community based design:

$$\text{Constraints in } (7.72)-(7.77) \quad \forall n\in\Omega^b \tag{7.81}$$

where Ω^u and Ω^b are set for all peers in the upper and bottom levels, respectively.

3.2 Use of the P2P concept in multiarea markets

In the previous sections, explanations were provided on how to improve the performance of local markets using the P2P structure. It is also possible to use this concept in multiarea markets. So far, different types of P2P markets have been discussed. In general, if a market participant could offer a bilateral contract to another participant in a decentralized multiarea market, then P2P contracts could also appear in the range of wholesale multiarea markets. Currently, bilateral contracts exist in European markets, but these contracts are within each local TSO boundary. Therefore, with the decentralization trend of multiarea markets, these bilateral transactions could occur beyond the TSO boundaries.

The implementation of P2P algorithms could eventually replace the form of bilateral transactions within wholesale electricity markets. In this case, P2P transactions could be increased (as bilateral transactions) due to the higher speed and transparency for real-time markets. This could lead to a new challenge created by managing a decentralized system with a condition of many P2P transactions. The system operator would face the challenge of adjusting the electrical grid with all the power flow constraints by implementing many P2P transactions. Currently, the level of bilateral transactions in electricity markets has a low rate compared to whole markets. Therefore, the system operator can handle and adjust those transactions without a heavy burden. But increasing P2P transactions in wholesale markets can create new challenges for system operators because these transactions must comply with network constraints. Therefore, this issue increases the responsibility of system operators. Of course, it is expected that by defining a new platform of P2P in the decentralized structure of wholesale multiarea markets, the challenge of the responsibility of system operators can be solved, and network constraints, under the supervision of the operators, can be considered in the interaction among participants.

3.3 Use of blockchains and edge computing in P2P market design

According to the above, the decentralized implementation of energy markets makes better operation of DERs and increases market implementation efficiency and transparency. Under the decentralized structure, common computing centers are no longer enforceable to improve decision-making efficiency. Thus computing resources need to be sunked from centralized cloud computing servers to market entities. There is also a need for a credible trading platform to increase transaction security. Given the capabilities of blockchain technology and edge computing, combining these with the P2P market design can address the mentioned challenges. Blockchains are a combination of distributed ledger technology and ICT, which is a suitable platform for a decentralized electricity market structure. Edge computing is a computation paradigm that pushes the cloud resources as close to an Internet of things (IoT) device as possible. While the main focus and responsibility of blockchains are on the security, privacy, and consensus for validating transactions, the main benefits of using edge computing in running the P2P market are implementing computing services in distributed entities and conducting efficient control in a decentralized manner. Blockchains and edge computing cooperation can establish an intelligent market that is secure and increases independence. These features will be appropriate for a decentralized electricity market framework. Fig. 7.23 illustrates the implementation of the P2P market scheme by considering blockchain and edge computing technologies. As shown in this figure, the task of IoT smart meters is to exchange information between market participants. In order to understand the use of blockchains in energy exchanges, first, the general instruction of blockchains is described. Then, its practical applications on the implementation of P2P markets are discussed [39,40].

3.4 General structure of blockchain technology

Blockchains are an emerging technology for decentralized computing, where the data are stored in an encrypted chain of blocks and distributed into a P2P network. Cryptographic signatures and distributed consensus mechanisms secure this technology. Fig. 7.24 shows the general structure of the blockchain. As can be seen, in the first place, one of the users of the P2P network submits a request for an exchange. This request is encrypted based on the recipient's address, the amount of the transaction and the private key of the transaction requester and is distributed throughout the P2P network.

The main parts of the blockchain include validation of the transaction, saving the transaction in a new block or adding it to the existing blocks, and updating the hash of each block. Consensus algorithms involve a process for the generation and the validation of a block from and by the network nodes. There are a number of different consensus algorithms, each one having its own advantages, disadvantages, and particular characteristics. The consensus algorithm plays a substantial role in the performance characteristics of a blockchain. These characteristics include scalability, transaction speed, transaction finality, and security. Many approaches for the consensus problem have been proposed, which can be classified as lottery-based and voting-based. In the voting-based approach, a consensus is reached through the vote and participation of all members. Consensus with voting from all members is difficult and sometimes impossible. In this case, the member who wants to add the block must prove that it has better conditions than other network members to do so. Therefore, these types of consensus algorithms are called lottery-based algorithms.

The most famous lottery-based approaches include proof of work (PoW) public blockchains (used by most cryptocurrency systems, such as Bitcoin or Ethereum). The PoW algorithm builds on top of the concept of the miner. Miner is a node that creates new blocks. The PoW algorithm requires that the miner finds an acceptable nonce as the result of the hash output, and the miner is rewarded for solving cryptographic puzzles in order to validate transactions and create new blocks. In a blockchain, the information in each block is converted into a set of numbers and letters or "hashes" by mathematical functions. By adding a transaction to the previous transaction set, the hash of each block must be updated, which is done by the miner. After the mining process is complete, a new block is added to the previous block set. Another feature of blockchains is the use of smart contracts. Smart contracts are stored on a blockchain that specifies the rules of transactions [41,42].

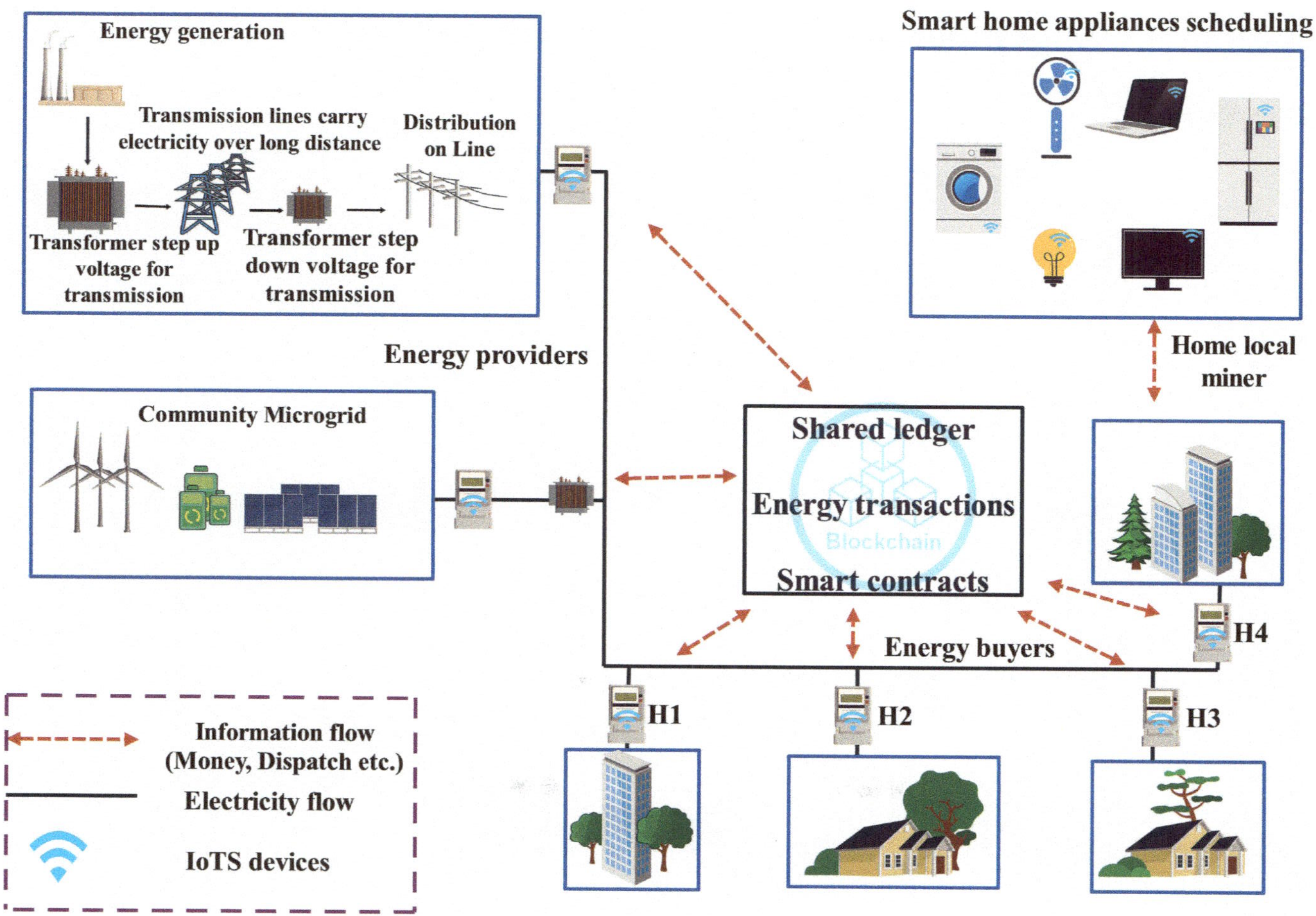

Fig. 7.23 Implementation of P2P using blockchains and smart meters.

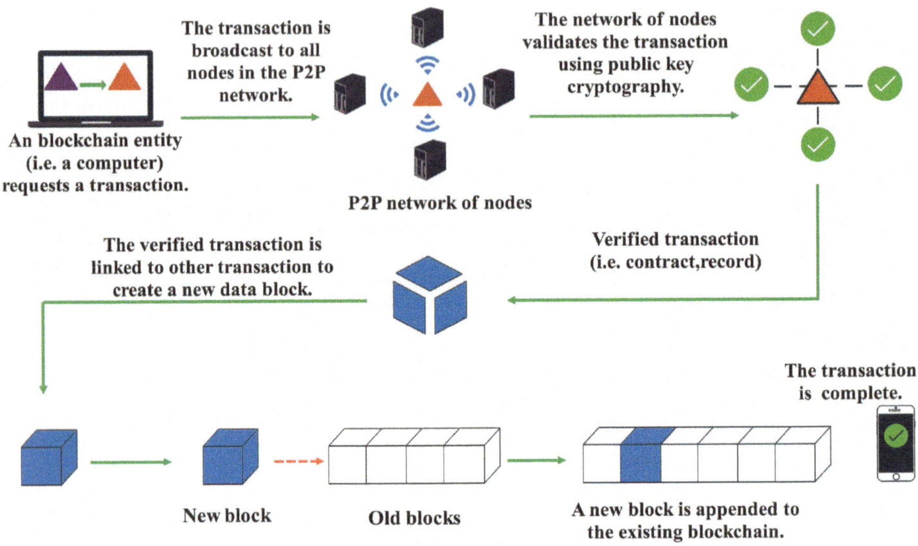

Fig. 7.24 General instruction of blockchains.

3.5 Application of blockchains in the implementation of energy markets

In order to use blockchains in the implementation of energy markets, two general schemes can be presented. In the first scheme, which is more suitable for implementing the P2P market in smart grids, the need to produce or consume energy in each node is determined using a combination of smart meters and smart contracts and after approval is recorded in the blockchain. In the second scheme, which is more suitable for implementing multiarea markets, after the decomposition of the whole network based on existing nodes or areas, only the data exchange and aggregation step between nodes or areas take place in the blockchain.

3.5.1 First scheme

The market mechanisms are implemented through a smart contract. At any given time (e.g., every 15 min), power balance information in any peer of the P2P market is measured by smart meters and analyzed through a smart contract. Based on the analysis of production and supply quantities in each node, there will be four cases. In the first case, the user's resources can mitigate its own demand. In this case, there is no exchange between the mentioned node and other nodes as well as the existing market. In the second case, if the smart meter detects energy shortage, the node enters into the energy market intending to buy energy. In the third case, upon detecting surplus energy generation by prosumer, the corresponding node enters into the energy market intending to sell energy and deliver power to the grid. Also, there is a possible scenario in which no exchange of power is made between node and grid in the fourth case.

As soon as the participants reach a consensus on accepting the new block, it is appended to the chain of blocks shared by all participants. In the same manner, it will also be possible to clear the market by sending recorded amounts of production and consumption by smart meters to smart contracts. This process provides an ineffaceable and transparent record of all the transactions [43].

3.5.2 Second scheme

In the previous sections, explanations were provided regarding the breakdown of centralized markets into decentralized markets. This decomposition can be done both based on the control areas in the multiarea markets and can be done in more detail based on each node in the smart grid. For example, in the multiarea markets, each area operates its own market separately and then exchanges limited information with other areas based on various available methods. This exchange of information will continue until convergence and coordination between regions is achieved. Blockchain capabilities can now be used to increase the security of the data exchange against cyberattacks or tampering by participants; in such a way, the aggregation stage and achieving convergence occur in the blockchain. Also, in this case, the capabilities of the smart contract can be used for settlement, so there is no need for any centralized utility for settlement. ADM is suitable for implementation on a blockchain platform, as it guarantees convergence, although this method has a slow convergence speed. According to the mentioned points, the use of blockchains in the mentioned application causes all participants to have sufficient control over the algorithm's progress, and as a result, the transparency of the market implementation will increase [25].

References

[1] S. Gope, A.K. Goswami, P.K. Tiwari, S. Deb, Rescheduling of real power for congestion management with integration of pumped storage hydro unit using firefly algorithm, Int. J. Electr. Power Energy Syst. 83 (2016) 434–442.

[2] D.S. Kirschen, G. Strbac, Fundamentals of Power System Economics, John Wiley & Sons, 2018.

[3] Z. Li, W. Wu, B. Zeng, M. Shahidehpour, B. Zhang, Decentralized contingency-constrained tie-line scheduling for multi-area power grids, IEEE Trans. Power Syst. 32 (1) (2016) 354–367.

[4] X. Lai, L. Xie, Q. Xia, H. Zhong, C. Kang, Decentralized multi-area economic dispatch via dynamic multiplier-based Lagrangian relaxation, IEEE Trans. Power Syst. 30 (6) (2014) 3225–3233.

[5] M. Kurzidem, Analysis of Flow-Based Market Coupling in Oligopolistic Power Markets, ETH Zurich, 2010.

[6] G. Oggioni, Y. Smeers, Market coupling and the organization of counter-trading: separating energy and transmission again, in: 17th Power Systems Computation Conf, 2010.

[7] G. Oggioni, Y. Smeers, Degrees of coordination in market coupling and counter-trading, Energy J. 33 (3) (2012).

[8] A. Karimi Varkani, H. Seifi, M.K. Sheikh-El-Eslami, Locational marginal pricing-based allocation of transmission capacity in multiple electricity markets, IET Gener. Transm. Distrib. 8 (5) (2014) 983–994.
[9] A. Karimi, H. Seifi, M.K. Sheikh-El-Eslami, Market-based mechanism for multi-area power exchange management in a multiple electricity market, IET Gener. Transm. Distrib. 9 (13) (2015) 1662–1671.
[10] Y. Ji, L. Tong, Multi-area interchange scheduling under uncertainty, IEEE Trans. Power Syst. 33 (2) (2017) 1659–1669.
[11] A.J. Conejo, J.A. Aguado, Multi-area coordinated decentralized DC optimal power flow, IEEE Trans. Power Syst. 13 (4) (1998) 1272–1278.
[12] J.A. Aguado, V.H. Quintana, A.J. Conejo, Optimal power flows of interconnected power systems, in: 1999 IEEE Power Engineering Society Summer Meeting. Conference Proceedings (Cat. No. 99CH36364), 1999.
[13] A.G. Bakirtzis, P.N. Biskas, A decentralized solution to the DC-OPF of interconnected power systems, IEEE Trans. Power Syst. 18 (3) (2003) 1007–1013.
[14] P.N. Biskas, A.G. Bakirtzis, N.I. Macheras, N.K. Pasialis, A decentralized implementation of DC optimal power flow on a network of computers, IEEE Trans. Power Syst. 20 (1) (2005) 25–33.
[15] M. Zhou, M. Wang, J. Li, G. Li, Multi-area generation-reserve joint dispatch approach considering wind power cross-regional accommodation, CSEE J. Power Energy Syst. 3 (1) (2017) 74–83.
[16] M. Doostizadeh, F. Aminifar, H. Lesani, H. Ghasemi, Multi-area market clearing in wind-integrated interconnected power systems: a fast parallel decentralized method, Energy Convers. Manag. 113 (2016) 131–142.
[17] Z. Li, M. Shahidehpour, W. Wu, B. Zeng, B. Zhang, W. Zheng, Decentralized multiarea robust generation unit and tie-line scheduling under wind power uncertainty, IEEE Trans. Sustainable Energy 6 (4) (2015) 1377–1388.
[18] A. Ahmadi-Khatir, A.J. Conejo, R. Cherkaoui, Multi-area energy and reserve dispatch under wind uncertainty and equipment failures, IEEE Trans. Power Syst. 28 (4) (2013) 4373–4383.
[19] A. Ahmadi-Khatir, A.J. Conejo, R. Cherkaoui, Multi-area unit scheduling and reserve allocation under wind power uncertainty, IEEE Trans. Power Syst. 29 (4) (2013) 1701–1710.
[20] M. White, R. Pike, Inter-regional interchange scheduling analysis and options, in: ISO New England and New York ISO, ISO White Paper, 2011.
[21] Y. Noorizadeh, A. Karimi, Decentralized coordinated tie-line scheduling based on decomposition methods in case of multiple markets, in: 2020 28th Iranian Conference on Electrical Engineering (ICEE), 2020.
[22] N.Y.S.E.P. Board, The Energy to Lead: New York State Energy Plan, 2015.
[23] T. Chernova, E. Gryazina, Peer-to-peer market with network constraints, user preferences and network charges, Int. J. Electr. Power Energy Syst. 131 (2021), 106981.
[24] J. Coignard, E. Munsing, J. MacDonald, J. Mather, Co-simulation framework for blockchain based market designs and grid simulations, in: 2018 IEEE Power & Energy Society General Meeting (PESGM), 2018.
[25] E. Münsing, J. Mather, S. Moura, Blockchains for decentralized optimization of energy resources in microgrid networks, in: 2017 IEEE Conference on Control Technology and Applications (CCTA), 2017.
[26] Y. Parag, B.K. Sovacool, Electricity market design for the prosumer era, Nat. Energy 1 (4) (2016) 1–6.

[27] S. Cheng, B. Zeng, Y. Huang, Research on application model of blockchain technology in distributed electricity market, in: IOP Conference Series: Earth and Environmental Science, 2017.
[28] T. Sousa, T. Soares, P. Pinson, F. Moret, T. Baroche, E. Sorin, Peer-to-peer and community-based markets: a comprehensive review, Renew. Sust. Energ. Rev. 104 (2019) 367–378.
[29] H. Beitollahi, G. Deconinck, Peer-to-peer networks applied to power grid, in: Proceedings of the International conference on Risks and Security of Internet and Systems (CRiSIS), 2007.
[30] C. Giotitsas, A. Pazaitis, V. Kostakis, A peer-to-peer approach to energy production, Technol. Soc. 42 (2015) 28–38.
[31] IRENA (International Renewable Energy Agency, A. D.), Innovation Landscape Brief: Peer-to-Peer Electricity Trading, 2020.
[32] T. Morstyn, N. Farrell, S.J. Darby, M.D. McCulloch, Using peer-to-peer energy-trading platforms to incentivize prosumers to form federated power plants, Nat. Energy 3 (2) (2018) 94–101.
[33] Z. Guo, P. Pinson, S. Chen, Q. Yang, Z. Yang, Chance-constrained peer-to-peer joint energy and reserve market considering renewable generation uncertainty, IEEE Trans. Smart Grid 12 (1) (2020) 798–809.
[34] E. Sorin, L. Bobo, P. Pinson, Consensus-based approach to peer-to-peer electricity markets with product differentiation, IEEE Trans. Power Syst. 34 (2) (2018) 994–1004.
[35] R. Verschae, T. Kato, T. Matsuyama, Energy management in prosumer communities: a coordinated approach, Energies 9 (7) (2016) 562.
[36] H. Qiu, F. You, Decentralized-distributed robust electric power scheduling for multi-microgrid systems, Appl. Energy 269 (2020), 115146.
[37] M.F. Zia, E. Elbouchikhi, M. Benbouzid, J.M. Guerrero, Microgrid transactive energy systems: a perspective on design, technologies, and energy markets, in: IECON 2019-45th Annual Conference of the IEEE Industrial Electronics Society, 2019.
[38] F. Moret, P. Pinson, Energy collectives: a community and fairness based approach to future electricity markets, IEEE Trans. Power Syst. 34 (5) (2018) 3994–4004.
[39] M. Afzal, Q. Huang, W. Amin, K. Umer, A. Raza, M. Naeem, Blockchain enabled distributed demand side management in community energy system with smart homes, IEEE Access 8 (2020) 37428–37439.
[40] S. Chen, Y. Bai, Y. Zhang, X. Liu, J. Zhang, T. Gao, Z. Fang, Y. Dai, A framework of decentralized electricity market based on the collaborative mechanism of blockchain and edge computing, in: 2019 IEEE International Conference on Service Operations and Logistics, and Informatics (SOLI), 2019.
[41] M. Andoni, V. Robu, D. Flynn, S. Abram, D. Geach, D. Jenkins, P. McCallum, A. Peacock, Blockchain technology in the energy sector: a systematic review of challenges and opportunities, Renew. Sust. Energ. Rev. 100 (2019) 143–174.
[42] M. Foti, C. Mavromatis, M. Vavalis, Decentralized blockchain-based consensus for optimal power flow solutions, Appl. Energy 283 (2021), 116100.
[43] A. Alam, M.T. Islam, A. Ferdous, Towards blockchain-based electricity trading system and cyber resilient microgrids, in: 2019 International Conference on Electrical, Computer and Communication Engineering (ECCE), 2019.

Role of game theory in future decentralized energy frameworks

8

Waqas Amin[a], Muhammad Afzal[b], Li Jain[b], Qi Huang[b], Hoay Beng Gooi[c], Yi Shyh Foo Eddy[c], and Khalid Umer[b]

[a]Department of Electronics & Power Engineering, PN Engineering College, National University of Sciences & Technology, Karachi, Pakistan, [b]Sichuan Provincial Key Lab of Power System Wide-Area Measurement and Control, School of Mechanical and Electrical Engineering, University of Electronic Science and Technology of China, Chengdu, China, [c]School of Electrical and Electronic Engineering, Nanyang Technological University, Singapore, Singapore

1 Introduction

The underlying success of future decentralized energy framework is the increased involvement of participants. The role of these participants in the future decentralized energy framework can be categorized as energy buyers (consumers), energy sellers (suppliers), or both. With the involvement of participants, the future decentralized energy framework can be treated as a market where different entities strategically interact with each other to promote market activities (i.e., selling or purchasing energy). Understanding and analyzing the participants' behavior strategies can be an effective tool to estimate the participants' expectations. Game theory is an effective analytical tool that helps us to investigate and analyze these activities and participants' behavior.

This chapter is focused on the role of game theory in future decentralized energy frameworks. An overview of the game theory model, along with their different types, are provided. Also, a different way of representing the game is given. The role of game theory in the decentralized energy frameworks is further explored, which helps the reader to understand game theory's role in a decentralized energy framework.

2 What is the game theory model?

Game theory is a branch of mathematics that helps to identify and analyze the strategic behavior of the participants (players) who interact with each other to trade some product (energy) and decide its quantity and price after offering them different offers (strategies) based on their expected benefits (utility/ payoff) [1]. Generally, a game model is presented as

$$\Xi = \langle N, A, U \rangle \tag{8.1}$$

where N represents the number of players consisting of consumers, sellers [2, 3], and electrical appliances [4, 5]. A represents the action profile or strategy set of the players.

Decentralized Frameworks for Future Power Systems. https://doi.org/10.1016/B978-0-323-91698-1.00014-5

It can be offered and proposed prices, quantity or preference of devices. *U* is the expected utility of the players, which can be considered as saving in energy bills or profitability of sellers.

Based on the players' strategies, a game can be classified either as (i) a pure strategy game or (ii) a mixed strategy game. In a pure strategy game, complete information about how a player he/she will play a game is provided. It can be considered as a plan a player will follow through the game. In a mixed strategy game, a player may have different strategies that he/she opt to play according to the other players' strategy. Further, in a mixed strategy game, a probability distribution function can be used to find out the most likely strategy played by a player [6] (Fig. 8.1).

Other than the pure and mixed strategies, different types of strategies in a game can be classified as (i) dominant and dominated strategy, (ii) minimax strategy, and (iii) maximin strategy. The dominant strategy presents a type of strategy in the game that ensures a better payoff if the player opts for this strategy over the other available strategies. In contrast, the dominated strategy presents the strategy that leads a player to get the worse outcome [7, 8]. Maximin is a player's strategy that leads him/her to increase the probability of getting the maximum profit and reduce the risk of loss [9]. Minimax strategy presents a strategy that leads the player to get the maximum profit with minimization of loss.

In a game, a player's strategy heavily depends upon the availability of the information. This information may include the opponent player's strategy. It may also have some other information about the system for which a game is played (e.g., availability of grid supply, energy prices, supply and demand ratio, etc.). Based on the availability

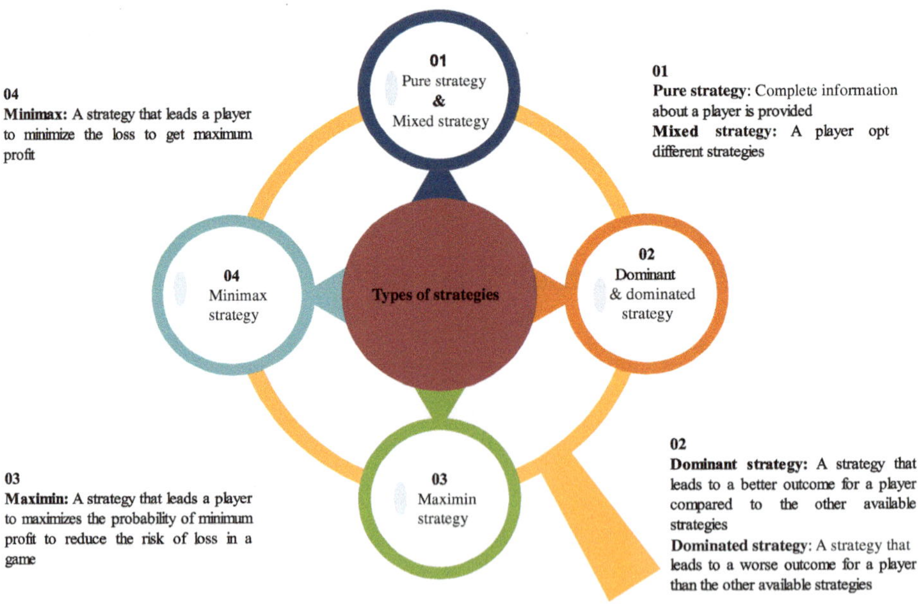

Fig. 8.1 Types of strategies.

of information, a game can be classified either as (i) perfect information game and (ii) imperfect information game [10, 11]. In a perfect information game, a player has all the necessary information about the other players' possible moves or other information described earlier. Whereas, in an imperfect game, a player has inaccurate or no information about the other players' possible strategies.

2.1 Nash equilibrium

John Forbes Nash Jr., a mathematician, proposed a concept that determines the optimal game solution known as Nash equilibrium. This concept states that a player does not gain anything if he/she deviates from his/her strategy while keeping the other players' strategy unchanged [7]. Let A_i be the set of all possible strategies of a player i, where $i = 1, 2, 3, \ldots, N$. Let u_i be the expected maximum payoff of a player i if he choose, the strategy a^*. Then, the strategy a^* will be the Nash equilibrium if

$$u_i(a_i^*, a_{-i}^*) \geq u_i(a_i, a_{-i}^*) \;\; \forall \; a_i \epsilon A_i \tag{8.2}$$

A game may have either one or more Nash equilibrium.

3 Types of games

A game can be classified into different types, which is based on (i) presentation, (ii) the involvement of the participants, and (iii) its mode of operation. Based on the presentation, a game can be classified into (i) extensive form game and (ii) normal form game. As shown in Fig. 8.2, an extensive form game presents a tree structure in a game theory where different choices are made at different points in a tree (corresponding to each node). The expected payoff or utility is presented at the end of each branch. Generally, in an extensive form game, perfect information is required as each player can see the decision taken by the previous player. In game theory, the normal form

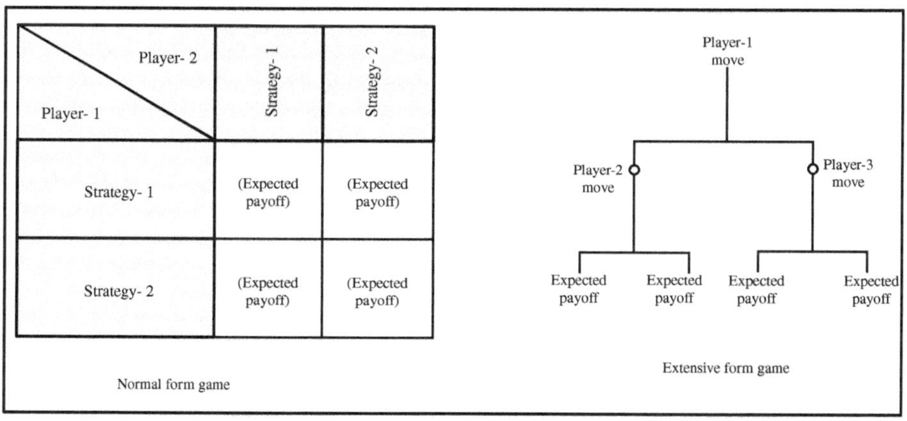

Fig. 8.2 Normal form versus extensive form.

representation of a game uses matrix structure to present a game. In the given matrix, the players are placed on the row and column. The matrix cell presents the players' strategies and their expected payoff following their played strategies.

An extensive form game among energy producers and consumers has been proposed in [2] as shown in Fig. 8.3. In the proposed game, each energy producer offers a profit-sharing ratio to the energy consumer. The consumer has only two options either to accept the offer or to reject the offer. If the consumer accepts the offer, he/she will get some revenues from the earning of energy producer. If he/she rejects the offer, the expected payoff of the consumer will be zero, which is beneficial for both the consumer and the producer. Hence, it gives the "strict Nash equilibrium" of the proposed game.

A normal form of noncooperative game formulation among electric vehicles, storage units, and array of batteries has been proposed in [12]. In the proposed game, the owner of storage units can decide the maximum amount of energy he/she intended to trade in energy market to maximize his/her expected payoff. This payoff reflects the tradeoff between the expected revenue and its associated cost. The proposed game shows its significant improvements as the expected payoff reaches up to 130.2% more as compared to convention greedy approach.

4 Types of games based on participants' involvement

In future decentralized energy framework, the participants can interact with each other in several ways, for example, one to one interaction, or in the form of coalition to trade energy among themselves. Based on their interaction, a game can be classified as (i) cooperative game, (ii) noncooperative game, or (iii) evolutionary game.

4.1 Cooperative game

Nuemann and Morgenstern being the founders proposed the first concept of cooperative games [13]. In cooperative games, there is a contract among the participants who binds them in cooperation to achieve certain goals and objectives of the game. In the future decentralized energy framework, a cooperative game model is deployed where several participants interact with each other in the form of coalitions to perform activate trading activities and other energy-related objectives, as shown in Fig. 8.4.

As in a cooperative game, participants interact with each other in coalition; therefore, the game's outcome is also used for a coalition. However, several methods such as Shapley value, Shapley Shublik power index, and marginal contribution can be used to find out the individual outcome. A cooperative game can be formulated among the number of players called multiplayer cooperative game. For two players' cooperative game formulation, generally, the Nash bargaining equilibrium method is used to search for the solution of the game [14–17]. For the multiplayer game, the solution can be obtained by the dominant strategy-based solution.

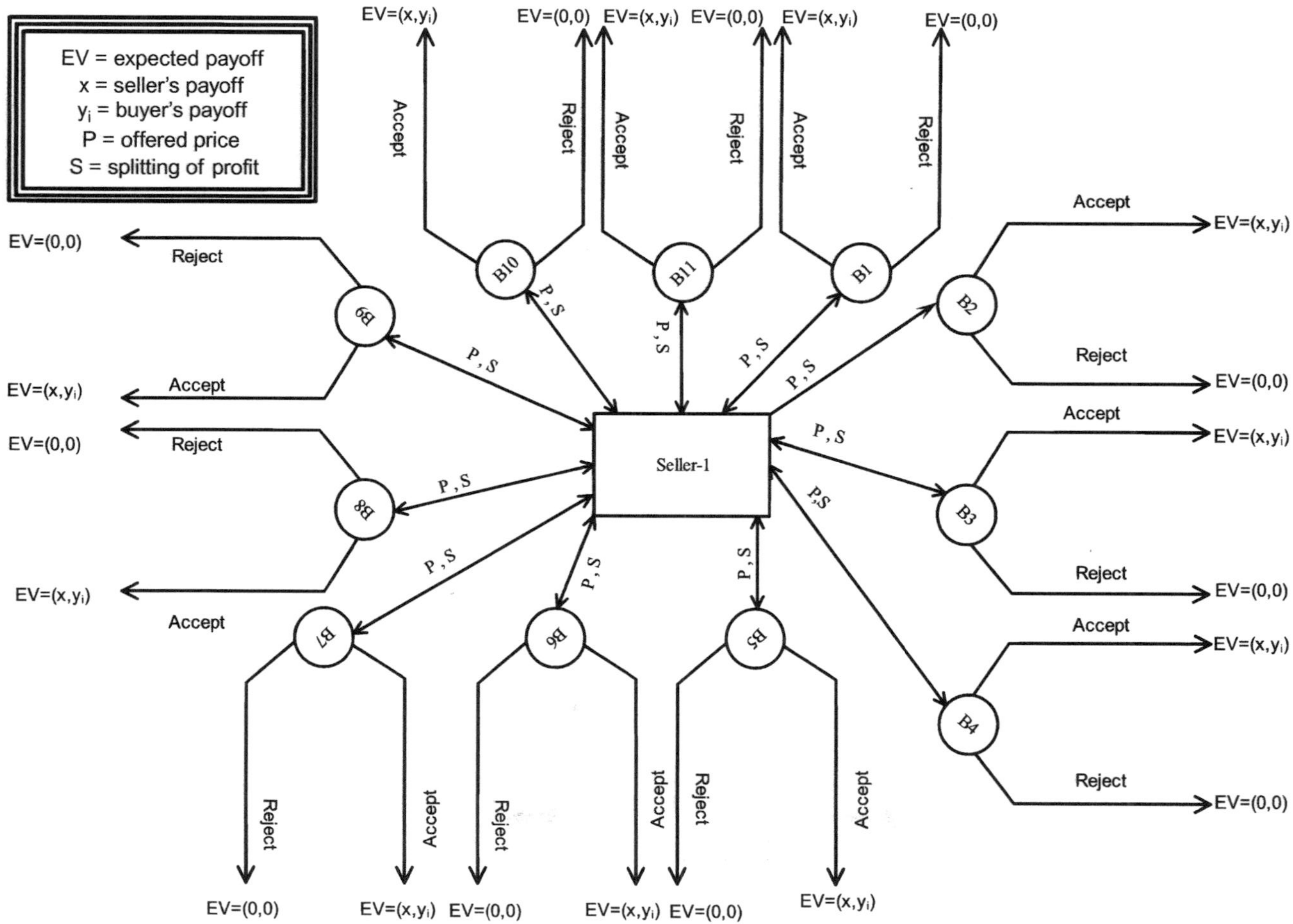

Fig. 8.3 Extensive form.

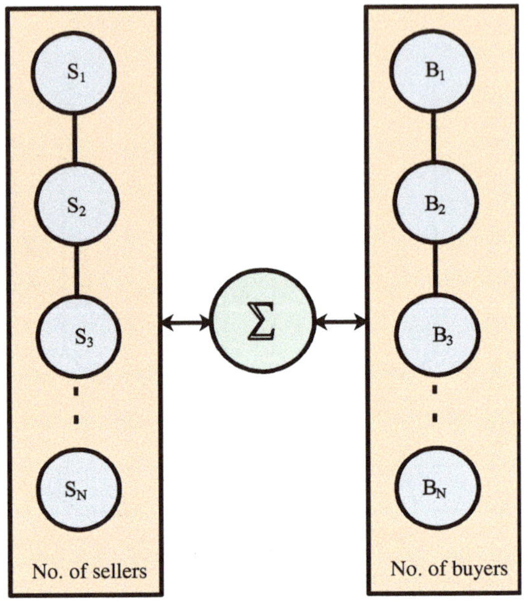

Fig. 8.4 Cooperative game.

4.1.1 Uses of cooperative games in decentralized energy framework

In a decentralized energy framework, a cooperative game can be used in several ways, for example, to find the energy trading prices, allocation of energy, appliances scheduling, etc. Table 8.1 presents some examples of the uses of a cooperative game in the decentralized energy framework proposed by several researchers.

4.2 Noncooperative game

A noncooperative game has different attributes from the cooperative one in which players do not form a group and interact with each other. In a noncooperative game, each player, that is, seller or buyer, operate independently regardless of the other peer players' benefits. A player can interact with several other players at a same time to meet his/her objectives, for example, selling or purchasing energy, etc. Fig. 8.5 presents a general overview of a noncooperative game.

Based on available information, and action played by the participant's, noncooperative games can be classified into two types [34]. (i) A static game where players select the action simultaneously. Players are not aware of the actions of the other players' strategy. (ii) Dynamic games, where players can use the historical information of the game and play their actions to get the optimized results.

Nash equilibrium [14], a very useful concept, provides a solution to noncooperative games. It provides an equilibrium point where none of the players can increase his/her benefits by unilateral deviation. Due to the complexity, variability, and diversification of practical problems, it is difficult to solve the Nash equilibrium. Therefore, many scholars prove Nash equilibrium by considering the game as a function and propose

Table 8.1 Uses of cooperative games in decentralized energy framework.

Ref.	Proposed model	Uses of cooperative game
Peng and Tao [18]	Proposed an interregional power transaction model based on bottom-up modeling	Cooperative game is used to quantify the cost of energy prices in the spot market with some economic assumptions
Kattuman et al. [19]	Proposed a method of assigning power flows in an electricity network to particular generators and loads, assuming perfect mixing at each node	Cooperative game with Shapley value is used to demonstrate that the proposed model has all the desirable characteristics required in cost allocation schemes
Faria et al. [20]	Proposed a model that integrates the interests of hydro agents with the needs of the regulatory agencies to find the solution that gives the right incentives to the optimal system development	Cooperative game theory is employed to determine the firms energy rights in hydropower generation systems
Nash [21]	A model is proposed to simulate the cooperative behaviors of multiple grid-connected microgrids to achieve a higher energy efficiency	A cooperative game model of grid-connected microgrids is constructed to minimize the total operation cost
Kristiansen et al. [22]	A framework is proposed to examine the benefits of substitute flexibility providers, such as fast-ramping gas turbines, hydropower, and demand-side management, by using a generation and transmission capacity expansion planning model	Shapley value from cooperative game theory is deployed to examine the impact of incorporating all possible deployment sequences
Kim et al. [23]	A two-stage energy operation model in a local microgrid system is proposed to optimize the energy usage in a particular area	Cooperative game theory with the Shapley value algorithm is deployed to distribute the revenue and payment in a real-time period based on the actual contribution of the individual prosumer
Bao and Cui [24]	Proposed a model to allocate the transmission losses in the energy market	Cooperative game with Shapley value distribution is used to distribute the losses to the participants in the energy market
Cui et al. [25]	Direct load control (DLC) is proposed in day-ahead real-time electricity	A cooperative game among the users and retailer is presented to minimize the cost of energy

Continued

Table 8.1 Continued

Ref.	Proposed model	Uses of cooperative game
Han et al. [26]	Propose an energy coalition model in which energy storage system owners operate collaboratively to reduce the total coalitional energy cost	Cooperative game and Shapley value distribution are used to reward the players equitably
Prete et al. [27]	Introduced a model in the regulated energy market that provides energy, ancillary services, heat, and enhanced reliability to the consumers	Used cooperative game to assess the interaction of the participants in the energy market and quantify how microgrid development affects prices and cost under different sets of assumptions
Wang et al. [28]	Proposed a Shapley value-based model that determines incentives for end user in the energy market when wholesale prices for energy increases	Cooperative game theory is used to achieve the maximum profit equilibrium for each participant in the grand coalition game
Peng and Tao [29]	Proposed a model for China's spot energy market that improves the energy retailers' competition	Used cooperative game theory and Shapley value to allocate the benefits in the deregulated energy market
Wang et al. [30]	Proposed a microgrid operation strategy model that implements the time of use pricing (TOU) when demand-side users transfer the load, which is used to maximize the revenue and optimize the reliability of microgrid	Cooperative game model is used to optimize the configuration of the users' load under TOU tariffs
Zhang et al. [31]	Proposed an energy purchasing strategy model for energy sellers and discusses the changes in the interest of various energy retailers under a cooperation alliance	Cooperative game is used to make an alliance, and the Shapley value method is used to distribute the profit among the cooperative alliance
Taheri et al. 32	Proposed a model to investigate the power generation companies' bidding strategies in a pool-based energy market	Used cooperative game to make an alliance among the power generation companies
Neimane et al. [33]	Proposed a model for energy supply system planning in the market environment	Used cooperative game to form a coalition between energy supply companies

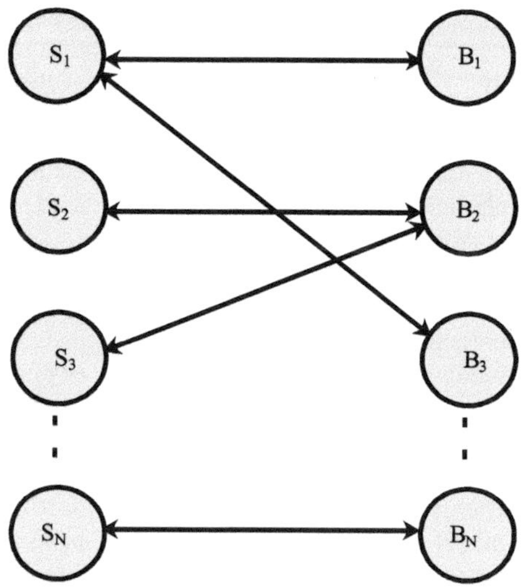

Fig. 8.5 Noncooperative game.

a fixed-point solution to find the game's equilibrium [35]. However, this method seems inappropriate for practical application. For this purpose, different researchers proposed a series of methods to compute the Nash equilibrium of a noncooperative game, for example, fictitious strategies and the players' best response as a Nash equilibrium have been proposed in Hopkins [36] and Frihauf et al. [37], respectively.

4.2.1 Uses of noncooperative games in decentralized energy framework

Several participants interact in a decentralized energy framework as one-to-one interaction to perform the trading activities in the energy market. These activities mainly include price determination for energy trading, energy allocation, deciding the electrical appliances' operating time, etc. Table 8.2 presents different uses of noncooperative game in the decentralized energy framework.

4.3 Evolutionary game

Maynard Smith and Price, as founders, proposed the first theory of the evolutionary game [54]. The proposed evolutionary game theory can be considered as an organic combination of dynamic evolution process and general game theory [55]. In cooperative and noncooperative game theory, the problem is formed within the framework of bounded rationality rather than complete rationality. In the evolutionary game, the game formulation is based on biological evolution. It means that players in this game continuously adjust their strategies accordingly to the game's environmental change and follow the change in strategies of the other players with limited availability of knowledge about the information of other players and his/her reasoning ability [56, 57]. The evolutionary game consists of four stages, as shown in Fig. 8.6. The

Table 8.2 Uses of noncooperative game in decentralized energy framework.

Ref.	Proposed model	Uses of noncooperative game
Abapour and Mohammadi-Ivatloo [38]	A competitive demand response model (DR) based on price elasticity and customer benefit is proposed in the network operator energy market	Static noncooperative game with incomplete information is used to model the competition among the distributed energy resources
Motalleb and Ghorbani [39]	Proposed a model for determining the optimal bidding strategies for selling stored energy	Noncooperative game with incomplete information to engage sellers in the competition is used
Shafie-Khah and Catalão [40]	To get the optimized strategies of the participants in the energy market, a stochastic model is proposed that investigates the behavior of wind power producers	An incomplete information noncooperative game is used to study the interaction of participants in the day-ahead energy market
Li and Shahidehpour [41]	Proposed a model that analyzes the competition among the electricity generating companies in a transmission constrained network with incomplete information	A noncooperative game with incomplete information is used to promote competition among participants
Neuhoff et al. [42]	Proposed a model to analyze the competitive behavior of sellers	Noncooperative game is used to introduce competition among sellers
Lise et al. [43]	Proposed a static game-theoretic model for liberalizing the European electricity market and presented the reduction in smog emissions and lower acidifying with lowered prices for consumers as the effects of the proposed model	Noncooperative game is used to derive the competition among the firms in the electricity market
Nanduri and Das [44]	A reinforcement-based model is proposed to assess the role of participants in the day-ahead energy market	A stochastic noncooperative game is used for trading energy
Adika and Wang [45]	Proposed a model in the smart grid environment with high integration of renewable energy resources to reduce the price of electricity	A noncooperative game based on rationality concept among the households is used to reduce their electricity bills
Motalleb et al. [46]	Proposed a competition model among the demand response to sell stored aggregated energy directly to the other aggregators in the energy market	Noncooperative game is used to find the optimal bidding strategies among the energy aggregators

Table 8.2 Continued

Ref.	Proposed model	Uses of noncooperative game
Marzband et al. [47]	Proposed a game-theoretic model for optimal operation of home microgrids and their interoperability in the active distribution grid	A noncooperative game is used to achieve the optimal solution
Zhang et al. [48]	Proposed a game-theoretic model for economic operations of residential distribution system with high participation of distributed energy prosumers and identified the new roles of utilities and distributed electricity prosumers in the future retail electricity market	Noncooperative game is used to clear the retail electricity trading price
Marzband et al. [49]	A framework is proposed for retail energy market with large penetration of DERs from sellers side and demand-side management from consumers side	Noncooperative game is used to maximize the payoff of the participants
Su and Huang [50]	Proposed a model for the retail energy market, for residential distribution systems and distributed energy suppliers	Noncooperative game is proposed to promote competition among the participants
Soliman and Leon-Garcia [51]	A demand-side management model is proposed for consumers having energy storage equipment	Noncooperative game is used to obtain the optimal schedule strategy to minimize the cost of energy bills
Atzeni et al. [52]	A model is proposed for the smart grid that focuses on traditional users having some sources of distributed energy resources with storage equipment	A noncooperative game is used to analyze the optimal strategies of traditional users
Wu et al. [53]	Proposed a new game-theoretic model to analyze the interactions among electrical vehicles (EVs) and the aggregator in a vehicle-to-grid (V2G) system	A noncooperative game is used to develop the interaction of EVs and aggregators

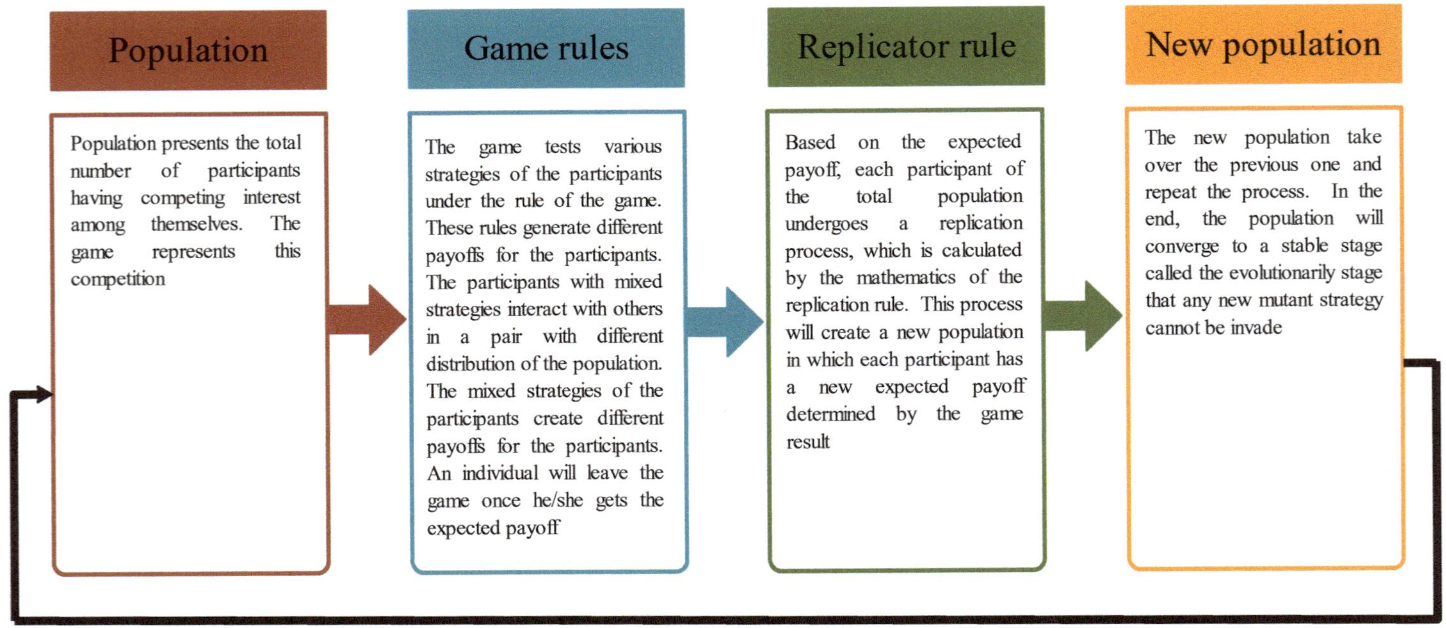

Fig. 8.6 Evolutionary game.

equilibrium point of the evolutionary game is known as stable evolutionary equilibrium, which is a refined form of Nash equilibrium.

4.3.1 Uses of evolutionary games in decentralized energy framework

Evolutionary games have been widely used in the decentralized energy framework to analyze the strategic behaviors of the participants. Table 8.3 presents some uses of evolutionary games in the decentralized energy framework.

Table 8.3 Uses of evolutionary game in decentralized energy framework.

Ref.	Proposed model	Uses of evolutionary game
Menniti et al. [58]	A genetic algorithm-based model is proposed to analyze the behavior of two or more producers operating in the same energy market	An evolutionary game is used to create competition among the energy producers
Chai et al. [59]	Proposed a two-level game model to create interaction between the grid and home users	Evolutionary game is used to formulate the interaction among the home users
Dai et al. [60]	Proposed a real-time pricing scheme for energy trading in smart grid with multiple retailers and multiple residential users	An evolutionary game with private information is formulated among power sellers and residential users as buyers
Wang et al. [61]	Proposed a model to evaluate the impact of social networks on the diffusion of real-time energy pricing and presented that a higher degree of consumer social network will result in slower diffusion of real-time price	Evolutionary game is used to form a social network among consumers
Cheng and Yu [62]	Proposed a model that considers three parties, that is, energy consumers, power grid, and new power supply entity (NPSE) and shows the factors affecting payoff	Evolutionary game is used to solve the incomplete information and find the optimal price for energy trading
Wang et al. [63]	Proposed a model to analyze the bidding strategies in the energy market with price-elastic demand	An evolutionary game with incomplete information is used to analyze the bidding strategies
Zaman et al. [64]	Proposed a model that determines the optimal bidding strategies in the competitive energy market to maximize the profit of each bidder	Evolutionary game is used to analyze the strategic behaviors of the bidders
Xin-Gang et al. [65]	Proposed a model that evaluates the behavior strategies of power producers in an energy market	Evolutionary game is used to study the symbiotic evolution among power producers

5 Conclusions

Determining and analyzing the participants' strategic behavior has a significant role in the future decentralized energy framework. The analysis of these behaviors not only helps the stakeholders to design and implement the guidelines and policies in the future decentralized energy framework but also ensures the participants' positive utilities, which increases their motivational level and satisfaction toward the participation in the decentralized energy framework. Game theory proves itself as an effective tool to analyze strategic behaviors. This chapter presents and discusses the basic game theory model. Then, different types of strategies have been discussed. Based on the representation of a game, a game can be classified into two forms, which have been discussed in this chapter. A game can be classified as a cooperative and noncooperative game based on the involvement of the participants, which have been discussed in detail. Further, different games and their uses in the future decentralized energy framework have been explored deeply in this chapter.

References

[1] W. Amin, Q. Huang, M. Afzal, A.A. Khan, K. Umer, S.A. Ahmed, A converging non-cooperative & cooperative game theory approach for stabilizing peer-to-peer electricity trading, Electr. Power Syst. Res. 183 (2020) 106278, https://doi.org/10.1016/j.epsr.2020.106278.

[2] W. Amin, Q. Huang, K. Umer, Z. Zhang, M. Afzal, A.A. Khan, S.A. Ahmed, A motivational game-theoretic approach for peer-to-peer energy trading in islanded and grid-connected microgrid, Int. J. Electr. Power Energy Syst. 123 (2020) 106307, https://doi.org/10.1016/j.ijepes.2020.106307.

[3] W. Amin, Q. Huang, M. Afzal, A.A. Khan, Z. Zhang, K. Umer, S.A. Ahmed, Consumers' preference based optimal price determination model for P2P energy trading, Electr. Power Syst. Res. 187 (2020) 106488, https://doi.org/10.1016/j.epsr.2020.106488.

[4] M. Afzal, K. Umer, W. Amin, M. Naeem, D. Cai, Z. Zhenyuan, Q. Huang, Blockchain based domestic appliances scheduling in community microgrids, in: 2019 IEEE Innovative Smart Grid Technologies—Asia (ISGT Asia), 2019, pp. 2842–2847, https://doi.org/10.1109/ISGT-Asia.2019.8881074.

[5] M. Afzal, Q. Huang, W. Amin, K. Umer, A. Raza, M. Naeem, Blockchain enabled distributed demand side management in community energy system with smart homes, IEEE Access 8 (2020) 37428–37439, https://doi.org/10.1109/ACCESS.2020.2975233.

[6] C. Zhang, J. Wu, Y. Zhou, M. Cheng, C. Long, Peer-to-peer energy trading in a microgrid, Appl. Energy 220 (2018) 1–12, https://doi.org/10.1016/j.apenergy.2018.03.010.

[7] E. Maskin, Nash equilibrium and welfare optimality, Rev. Econ. Stud. 66 (1) (1999) 23–38.

[8] J. Guerrero, A. Chapman, G. Verbic, Peer-to-peer energy trading: a case study considering network constraints, in: Proceedings of the Asia-Pacific Solar Research Conference, Sydney, Australia, 2018, pp. 4–6.

[9] B.R. Sutherland, Energy trading as a multiplayer game, Joule 3 (8) (2019) 1817–1818.

[10] M. Yan, M. Shahidehpour, A. Paaso, L. Zhang, A. Alabdulwahab, A. Abusorrah, Distribution network-constrained optimization of peer-to-peer transactive energy trading among multi-microgrids, IEEE Trans. Smart Grid 12 (2) (2021) 1033–1047, https://doi.org/10.1109/TSG.2020.3032889.

[11] Y. Yu, G. Li, Z. Li, A game theoretical pricing mechanism for multi-microgrid energy trading considering electric vehicles uncertainty, IEEE Access 8 (2020) 156519–156529, https://doi.org/10.1109/ACCESS.2020.3019815.
[12] Y. Wang, W. Saad, Z. Han, H.V. Poor, T. Başar, A game-theoretic approach to energy trading in the smart grid, IEEE Trans. Smart Grid 5 (3) (2014) 1439–1450, https://doi.org/10.1109/TSG.2013.2284664.
[13] J. von Neumann, O. Morgenstern, Theory of Games and Economic Behavior, Princeton University Press, Princeton, NJ, 1944.
[14] J.F. Nash Jr., The bargaining problem, Econometrica 18 (2) (1950) 155–162.
[15] J.F. Nash Jr., Two-person cooperative games, Econometrica 21 (1) (1953) 128–140, https://doi.org/10.2307/1906951.
[16] N.P. Yu, L. Tesfatsion, C.C. Liu, Financial bilateral contract negotiation in wholesale electricity markets using Nash bargaining theory, IEEE Trans. Power Syst. 27 (1) (2012) 251–267, https://doi.org/10.1109/TPWRS.2011.2162637.
[17] Z. Zhang, J. Shi, H.H. Chen, M. Guizani, P. Qiu, A cooperation strategy based on Nash bargaining solution in cooperative relay networks, IEEE Trans. Veh. Technol. 57 (4) (2008) 2570–2577.
[18] X. Peng, X. Tao, Cooperative game of electricity retailers in China's spot electricity market, Energy 145 (2018) 152–170, https://doi.org/10.1016/j.energy.2017.12.122.
[19] P.A. Kattuman, R.J. Green, J.W. Bialek, Allocating electricity transmission costs through tracing: a game-theoretic rationale, Oper. Res. Lett. 32 (2) (2004) 114–120, https://doi.org/10.1016/S0167-6377(03)00095-6.
[20] V.A.D. Faria, A.R. de Queiroz, L.M.M. Lima, J.W.M. Lima, Cooperative game theory and last addition method in the allocation of firm energy rights, Appl. Energy 226 (2018) 905–915, https://doi.org/10.1016/j.apenergy.2018.06.065.
[21] Y. Du, Z. Wang, G. Liu, X. Chen, H. Yuan, Y. Wei, F. Li, A cooperative game approach for coordinating multi-microgrid operation within distribution systems, Appl. Energy 222 (2018) 383–395, https://doi.org/10.1016/j.apenergy.2018.03.086.
[22] M. Kristiansen, M. Korpås, H.G. Svendsen, A generic framework for power system flexibility analysis using cooperative game theory, Appl. Energy 212 (2018) 223–232, https://doi.org/10.1016/j.apenergy.2017.12.062.
[23] J. Kim, Y. Heo, G.H. Lee, J. Lee, H. Oh, J. Choi, Two stage market model in microgrid using cooperative game theory, in: 2018 IEEE 7th Global Conference on Consumer Electronics (GCCE), 2018, pp. 138–140, https://doi.org/10.1109/GCCE.2018.8574874.
[24] X. Bao, W. Cui, Allocation of transmission losses in electricity system based on cooperative game theory, in: 2010 Asia-Pacific Power and Energy Engineering Conference, 2010, pp. 1–4, https://doi.org/10.1109/APPEEC.2010.5448705.
[25] Q. Cui, X. Wang, X. Wang, Y. Zhang, Residential appliances direct load control in real-time using cooperative game, IEEE Trans. Power Syst. 31 (1) (2016) 226–233, https://doi.org/10.1109/TPWRS.2015.2391774.
[26] L. Han, T. Morstyn, M. McCulloch, Constructing prosumer coalitions for energy cost savings using cooperative game theory, in: 2018 Power Systems Computation Conference (PSCC), 2018, pp. 1–7, https://doi.org/10.23919/PSCC.2018.8443054.
[27] C.L. Prete, B.F. Hobbs, A cooperative game theoretic analysis of incentives for microgrids in regulated electricity markets, Appl. Energy 169 (2016) 524–541, https://doi.org/10.1016/j.apenergy.2016.01.099.
[28] J. Wang, et al., Ensuring profitability of retailers via Shapley value based demand response, 2019, https://doi.org/10.1016/j.ijepes.2018.12.031.
[29] X. Peng, X. Tao, Cooperative game of electricity retailers in China's spot electricity market, 2018, https://doi.org/10.1016/j.energy.2017.12.122.

[30] S.D. Wang, W. Du, L. Lin, J.H. Li, W.Q. Chen, X. Gao, Optimal allocation of photovoltaic energy storage microgrid under the demand side response based on cooperative game, 2018, https://doi.org/10.7667/PSPC162051.
[31] C. Zhang, X.Y. Han, C.F. Bai, B. Hu, Y.F. Chai, J.Q. Zhou, Purchasing strategy model for power supplier based on the cooperative game theory in an open electricity market, 2017, https://doi.org/10.11930/j.issn.1004-9649.2017.06.177.08.
[32] I. Taheri, M. Rashidi Nejad, A. Badri, A. Abdollahi, The study of GENCOs' bidding strategies in a pool-based electricity market using cooperative and non-cooperative game theory, J. Electr. Syst. Signals 1 (1) (2013).
[33] V. Neimane, A. Sauhats, G. Vempers, I. Tereskina, G. Bockarjova, Approach for energy supply system planning based on cooperative game theory, 2008, p. 6, https://doi.org/10.1109/EEM.2008.4579032.
[34] L. Cheng, T. Yu, Game-theoretic approaches applied to transactions in the open and ever-growing electricity markets from the perspective of power demand response: an overview, IEEE Access 7 (2019) 25727–25762, https://doi.org/10.1109/ACCESS.2019.2900356.
[35] K.C. Border, Fixed Point Theorems With Applications to Economics and Game Theory, Cambridge University Press, Cambridge, UK, 1985, pp. 22–50.
[36] E. Hopkins, A note on best response dynamics, Games Econom. Behav. 29 (1999) 138–150, https://doi.org/10.1006/game.1997.0636.
[37] P. Frihauf, M. Krstić, T. Başar, Nash equilibrium seeking in noncooperative games, IEEE Trans. Autom. Control 57 (5) (2012) 1192–1207.
[38] S. Abapour, B. Mohammadi-Ivatloo, A Bayesian game theoretic based bidding strategy for demand response aggregators in electricity markets, Sustain. Cities Soc. (2019) 101787, https://doi.org/10.1016/j.scs.2019.101787.
[39] M. Motalleb, R. Ghorbani, Non-cooperative game-theoretic model of demand response aggregator competition for selling stored energy in storage devices, Appl. Energy 202 (2017) 581–596, https://doi.org/10.1016/j.apenergy.2017.05.186.
[40] M. Shafie-Khah, J.P.S. Catalão, A stochastic multi-layer agent-based model to study electricity market participants behavior, IEEE Trans. Power Syst. 30 (2) (2015) 867–881, https://doi.org/10.1109/TPWRS.2014.2335992.
[41] T. Li, M. Shahidehpour, Strategic bidding of transmission-constrained GENCOs with incomplete information, IEEE Trans. Power Syst. 20 (1) (2005) 437–447, https://doi.org/10.1109/TPWRS.2004.840378.
[42] K. Neuhoff, J. Barquin, M.G. Boots, A. Ehrenmann, B.F. Hobbs, F.A.M. Rijkers, M. Vázquez, Network-constrained Cournot models of liberalized electricity markets: the devil is in the details, Energy Econ. 27 (3) (2005) 495–525, https://doi.org/10.1016/j.eneco.2004.12.001.
[43] W. Lise, V. Linderhof, O. Kuik, C. Kemfert, R. Östling, T. Heinzow, A game theoretic model of the Northwestern European electricity market–market power and the environment, Energy Policy 34 (15) (2006) 2123–2136, https://doi.org/10.1016/j.enpol.2005.03.003.
[44] V. Nanduri, T.K. Das, A reinforcement learning model to assess market power under auction-based energy pricing, IEEE Trans. Power Syst. 22 (1) (2007) 85–95, https://doi.org/10.1109/TPWRS.2006.888977.
[45] C.O. Adika, L. Wang, Non-cooperative decentralized charging of homogeneous households' batteries in a smart grid, IEEE Trans. Smart Grid 5 (4) (2014) 1855–1863, https://doi.org/10.1109/TSG.2014.2302449.
[46] M. Motalleb, A. Annaswamy, R. Ghorbani, A real-time demand response market through a repeated incomplete-information game, Energy 143 (2018) 424–438, https://doi.org/10.1016/j.energy.2017.10.129.

[47] M. Marzband, M. Javadi, S.A. Pourmousavi, G. Lightbody, An advanced retail electricity market for active distribution systems and home microgrid interoperability based on game theory, Electr. Power Syst. Res. 157 (2018) 187–199, https://doi.org/10.1016/j.epsr.2017.12.024.
[48] N. Zhang, Y. Yan, W. Su, A game-theoretic economic operation of residential distribution system with high participation of distributed electricity prosumers, Appl. Energy 154 (2015) 471–479, https://doi.org/10.1016/j.apenergy.2015.05.011.
[49] M. Marzband, M. Javadi, J.L. Dominguez-Garcia, M.M. Moghaddam, Non-cooperative game theory based energy management systems for energy district in the retail market considering DER uncertainties, IET Gener. Transmiss. Distrib. 10 (12) (2016) 2999–3009, https://doi.org/10.1049/iet-gtd.2016.0024.
[50] W. Su, A.Q. Huang, A game theoretic framework for a next-generation retail electricity market with high penetration of distributed residential electricity suppliers, Appl. Energy 119 (2014) 341–350, https://doi.org/10.1016/j.apenergy.2014.01.003.
[51] H.M. Soliman, A. Leon-Garcia, Game-theoretic demand-side management with storage devices for the future smart grid, IEEE Trans. Smart Grid 5 (3) (2014) 1475–1485, https://doi.org/10.1109/TSG.2014.2302245.
[52] I. Atzeni, L.G. Ordóñez, G. Scutari, D.P. Palomar, J.R. Fonollosa, Demand-side management via distributed energy generation and storage optimization, IEEE Trans. Smart Grid 4 (2) (2013) 866–876, https://doi.org/10.1109/TSG.2012.2206060.
[53] C. Wu, H. Mohsenian-Rad, J. Huang, Vehicle-to-aggregator interaction game, IEEE Trans. Smart Grid 3 (1) (2012) 434–442, https://doi.org/10.1109/TSG.2011.2166414.
[54] J.M. Smith, G.R. Price, The logic of animal conflict, Nature 246 (5427) (1973) 15–18, https://doi.org/10.1038/246015a0.
[55] P.D. Taylor, L.B. Jonker, Evolutionary stable strategies and game dynamics, Math. Biosci. 40 (1–2) (1978) 145–156, https://doi.org/10.1016/0025-5564(78)90077-9.
[56] L.F. Cheng, T. Yu, Exploration and exploitation of new knowledge emergence to improve the collective intelligent decision-making level of web-of-cells with cyber-physical-social systems based on complex network modeling, IEEE Access 6 (2018) 74204–74239, https://doi.org/10.1109/ACCESS.2018.2879025.
[57] L.F. Cheng, T. Yu, Nash equilibrium-based asymptotic stability analysis of multi-group asymmetric evolutionary games in typical scenario of electricity market, IEEE Access 6 (2018) 32064–32086, https://doi.org/10.1109/ACCESS.2018.2842469.
[58] D. Menniti, A. Pinnarelli, N. Sorrentino, Simulation of producers behaviour in the electricity market by evolutionary games, Electr. Power Syst. Res. 78 (3) (2008) 475–483, https://doi.org/10.1016/j.epsr.2007.04.005.
[59] B. Chai, J. Chen, Z. Yang, Y. Zhang, Demand response management with multiple utility companies: a two-level game approach, IEEE Trans. Smart Grid 5 (2) (2014) 722–731, https://doi.org/10.1109/TSG.2013.2295024.
[60] Y. Dai, Y. Gao, H. Gao, H. Zhu, Real-time pricing scheme based on Stackelberg game in smart grid with multiple power retailers, Neurocomputing 260 (2017) 149–156, https://doi.org/10.1016/j.neucom.2017.04.027.
[61] G. Wang, Q. Zhang, H. Li, Y. Li, S. Chen, The impact of social network on the adoption of real-time electricity pricing mechanism, Energy Procedia 142 (2017) 3154–3159, https://doi.org/10.1016/j.egypro.2017.12.383.
[62] L. Cheng, T. Yu, Nash equilibrium-based asymptotic stability analysis of multi-group asymmetric evolutionary games in typical scenario of electricity market, IEEE Access 6 (2018) 32064–32086, https://doi.org/10.1109/ACCESS.2018.2842469.

[63] J.H. Wang, Z. Zhou, A. Botterud, An evolutionary game approach to analyzing bidding strategies in electricity markets with elastic demand, Energy 36 (5) (2011) 3459–3467, https://doi.org/10.1016/j.energy.2011.03.050.
[64] F. Zaman, S.M. Elsayed, T. Ray, R.A. Sarker, Co-evolutionary approach for strategic bidding in competitive electricity markets, Appl. Soft Comput. 51 (2017) 1–22, https://doi.org/10.1016/j.asoc.2016.11.049.
[65] Z. Xin-Gang, R. Ling-Zhi, Z. Yu-Zhuo, W. Guan, Evolutionary game analysis on the behavior strategies of power producers in renewable portfolio standard, Energy 162 (2018) 505–516, https://doi.org/10.1016/j.energy.2018.07.209.

Toward customer-centric power grid: Residential EV charging simulator for smart homes

9

Maarit Jäntti, Anssi Jäntti, and Miadreza Shafie-khah
School of Technology and Innovations, University of Vaasa, Vaasa, Finland

1 Introduction

Renewable energy sources, the Internet of things, and electrification are acknowledged as the development trends that are affecting energy markets globally. These drivers will most probably increase the need for demand and storage flexibility. The role of the end users in the energy markets is changing from passive consumers to active prosumers and therefore the new solutions in the energy markets need to be designed in a user-centric way. The motives, incentives, requirements, and criteria of consumers need to be taken into considerations and furthermore, how consumers could support the business goals of the different energy domain actors [1].

In this chapter, firstly, a literature review is presented about demand response, the role of governments and regulators including incentives, benefits to the electricity consumers to participate in the electricity market, and also the regulatory barriers. Afterward, a smart home demand response simulation including electric vehicles is presented. An EV charging simulator program has been developed for the economics of charging an electric car under different conditions. The program is used to simulate the depleting and charging of an electric car based on the habits of its user. The program contains several different profiles related to power economics, and the implications of these can be evaluated under the specified conditions. The program tries to automatically load the Nord Pool spot price database and it takes as inputs several parameters regarding the car and its usage extracted from real data of daily driving. The battery size and charger capacity are taken from a standard Nissan Leaf equipped with the basic onboard AC charger intended for home use from a normal 16A wall outlet. The chapter focuses on the transition toward decentralized power grids where the role of consumers in making a decision is significant.

Decentralized Frameworks for Future Power Systems. https://doi.org/10.1016/B978-0-323-91698-1.00017-0

2 Literature review

2.1 Demand response

One solution to make the power grid more secure, efficient, and cleaner is to make electricity consumption more flexible. Balancing electricity consumption to reduce peaks in demand or to take advantage of renewable energy sources is described as demand-side flexibility [2].

From consumer's viewpoint, demand flexibility or demand response (DR) means that the electricity consumers shift the timing of their electricity consumption behavior when required, for example, washing laundry or heating water at different times than usual [1] or smart charging strategies that shift the time of day when EVs are charging from the grid [3]. Fig. 9.1 shows different layers of demand response implementations.

2.1.1 Flexibility management

The balancing of the power grid is currently achieved by adjusting production to match consumption. Increasing renewable energy sources such as solar and wind are difficult to forecast accurately and cannot be controlled in the same way as fossil-fuel power plants. This means that the flexibility needs to come from the storage solutions and demand-side solutions. According to Immonen et al. [1], the one way to

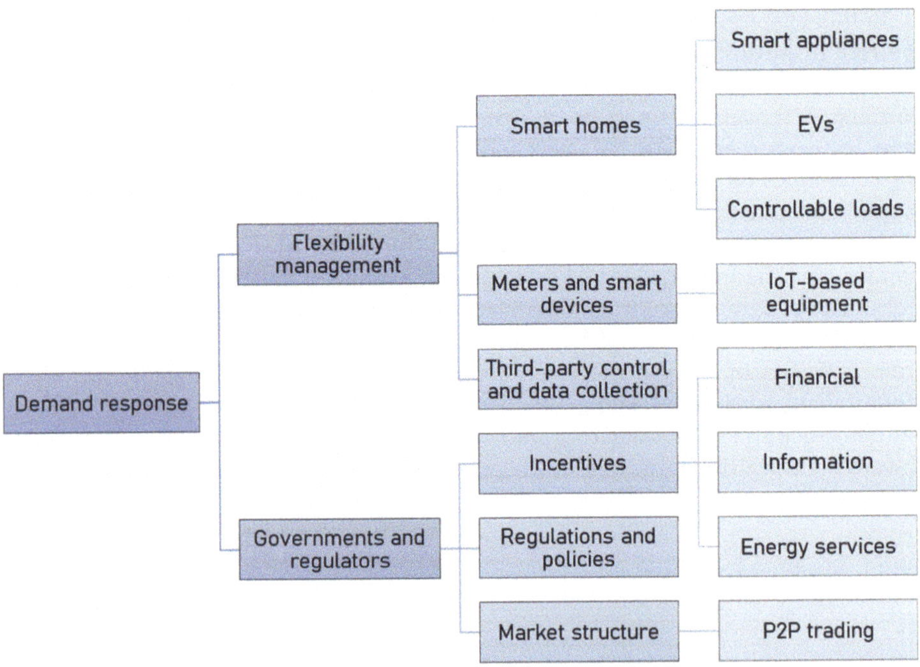

Fig. 9.1 Demand response implementation.

conduct this flexibility management are batteries, but with current technologies, this solution is not feasible cost-wise and batteries waste more natural resources in comparison to managing the demand side using existing flexibility resources. Therefore, it is important to exploit all existing flexibility resources from the demand side such as electric heaters, HVAC (heating, ventilation, and air conditioning system), water boilers, and different smart appliances. Immonen et al. [1] have presented that consumers could provide their home devices and energy storage directly as flexibility potential, for being an active participant in the energy market. This is resulting in the need for real-time mechanisms for flexibility management and new service providers that take care of functionality between households and the market. For example, aggregators, virtual power plant operators, and technical service providers have the potential to play an important role in future energy markets [1]. Also, the International Energy Agency has stated that new business models such as aggregation, virtual power plants (VPPs), and other distributed energy resources are a great promise for enabling demand-side flexibility.

One enticing way of implementing the demand response is the utilization of the batteries of the increasingly popular electric vehicles (EV). The batteries of the cars can be charged when the prices of the electricity are low, and the batteries can be discharged back to the network when prices surge. Simulations show that under suitable conditions, this kind of system may provide basically free electricity for the operation of the EV or even generate considerable profit from the selling of the electricity even when considering the investment on the required equipment and wear on the batteries [4].

2.1.2 Smart homes

In order to properly implement the demand response management (DRM), we first need to have controllable devices at the residential buildings, and the buildings need to be connected to the external world [5]. Residential buildings have been slowly filling with various smart and connected gadgets and appliances since the first futuristic American concept homes of the 30s [6]. This trend has been picking up speed in the last decades due to technological advancements [6] and a recent study released by the electronics manufacturer Xiaomi [7] showed that more than half of the surveyed US customers had purchased at least one smart device during the COVID-19 pandemic.

In addition to making the life of the residents easier and more comfortable, these so-called smart homes (SH) may also have the potential participant in the operation of the electrical grid more than ever. Smart homes may generate power and offer the ability to externally adjust the level of power consumed by the building [6]. Smart homes, with their various connected devices and continuous internet connections, provide an ideal bridge between the electrical network and demand response implementation, enabling the smart homes to function as part of the smart grid [5]. To further increase the deployment of smart homes, Lai et al. [5] also advocated for a large-scale deployment of the fiber power cable to the home (FPCTTH) which combines residential power delivery cabling to the fiber optic cabling used, for example, for internet and TV.

Several sources can be found in the literature regarding the simulation of various demand response systems in smart homes. For example, Gudi, Wang, Devabhaktuni, and Depuru [8] have used Java language to develop a simulator which is a particle swarm optimization (PSO) in order to create a schedule for various home appliances so that the simultaneous power consumption and the price of the electricity would be as small as possible. Further expansion to this work was provided by using Matlab from Tsui and Chan [9] who utilized convex programming (CP) while also integrating parameters regarding how the users accept the various appliances being off, i.e., for example, the resulting increase in the temperature while the air conditioning is not operating. Gayathri et al. [10] presented an idea of a centralized controller which utilized Matlab programming and game theory's algorithms for scheduling loads of various appliances at home.

Hoosain and Paul [11] implemented a prototype of a tier-based smart plug system, in which the users can connect different types of appliances to the grid based on their importance, and then the utilities can remotely shut down various devices based on the criticality of the power shortages. The system also included an option for the users to definable timetables for indicating the times at which they wanted certain appliances to be excluded from the demand response system.

Fernandesa, Moraisa, Valea, and Ramos [12] described a system called Supervisor Control and Data Acquisition House Intelligent Management (SHIM), developed at the Institute of Engineering—Polytechnic of Porto (ISEP/IPP). This system is able to simulate various functionalities of a smart home, connect to the real world, and control actual hardware. The authors implemented a demand response functionality to the simulator and introduced the concept of a dynamic load priority (DLP). In DPL, the priority of the various loads in the system can change during a DR event instead of remaining fixed. For example, a refrigerator that has just executed its cycle gets a low priority and the priority keeps increasing when the temperature inside the refrigerator increases close to a critical limit.

Wang et al. [13] presented an interesting idea for taking advantage of the thermal inertia of a smart home for shifting the scheduling of an air conditioner. In this work, thermal energy inside a house is considered a type of virtual energy storage system. They used a mixed-integer programming language (MILP) for optimizing the schedule. The same methodology could also be likely used for heating.

2.1.3 Meters and smart devices

A smart metering system is an electronic system that is capable of measuring electricity feed into the grid and consuming electricity from the grid. The system is able to transmit and receive information, monitor, and control. The system brings benefits for the energy system and the users, for example, with smart meters, consumers are able to get regular measurements of their energy usage. Smart meters can also provide almost real-time feedback on energy consumption and enable consumers to manage better their energy usage, save energy, and lower their electricity bills. For those consumers that are interested to be actively involved in the electricity market, smart meters can offer even more, for example, allow to adapt the energy usage to different energy prices during a day and enable to consume when the prices are at the lowest. Smart meters are also relevant to those consumers who generate electricity, for example, by solar panels. A smart meter can measure the electricity that household supplies to the grid and inform the grid manager. For network operators, one benefit is that operators receive more precise information about activities in different parts of the network. Thus, operators can better allocate the investments and manage their infrastructure to fulfill the requirements of the customers [14].

According to Immonen et al. [1], consumers incentives to invest in meters and smart devices that allow the third party to control their devices are financial savings; saving energy; environmental benefits and ecology; residential comfort and conditions; smart home, making everyday life easier; monitoring and reporting electricity consumption.

Immonen et al. [1] have identified different obstacles to investing in meters and smart devices. Financial obstacles are high investment costs, too small incentives, too long payback time, and contract issues. Other obstacles are identified as safety and reliability of devices, security issues, difficulties in installing devices, compatibility issues, and lack of experience [1]. Tounquet and Alaton [15] have also stated in their report for the European Commission that data privacy is considerable concern among consumers. They have also indicated that consumers are concerned about electromagnetic radiation and the broader impact of smart meters on health [15].

2.1.4 Third-party control and data collection

According to Immonen et al. [1], consumers would allow third party controlling their home devices if it occurred without disruption and is beneficial to them, and the most preferred is a discount in electricity bill or some other financial compensation. In addition, concerns about the environment and the reduction of emissions are increasingly motivating consumers. Also, consumers are interested in some services such as automatic heat control and network reliability. In the research are also identified concerns about third party controlling and these are: the desire to maintain control; insecurity of the Internet; lack of experience/knowledge; costs vs benefits; problems and uncertainty; data collection and utilization; objectives and motives; concern about the property; contracts; stuck with the same provider [1].

Immonen et al. [1] have identified issues that should be taken into account in data collection: security, data privacy, and data protection regulation (GDPR); the trustworthiness of the data collector; disclosure of information forbidden without the consent of the consumer; consumers' access to their own data; consumers' want to decide what data is collected about them and what purposes it can be used and only collecting relevant data.

2.2 The role of operators, governments, and regulators

To help consumers respond to price signals or signals from system operators, governments and regulators should study the feasibility of using ICT platforms and smart contracts. Governments and regulators should also develop flexibility offerings and implement time-based rate programs as well as regulatory structures that monetize flexibility at the point of use. Governments should also promote consumer and third-party access to smart-metering data and dynamic pricing and other signals [3].

According to the Council of European Energy Regulators [16], the key principle in tariff design is cost reflectivity that leads to economic efficiency. Other principles are nondistortion, cost recovery, nondiscrimination, transparency, predictability, and simplicity, and regulators need to find a balance between these principles [16].

Consumers' attitudes toward flexibility management and related services highly rely on incentives and benefits. Financial incentives seem to be the most effective way to change electricity consumers' behavior. In addition, consumers are increasingly interested in awareness about energy markets, environmental issues, and energy consumption. New business models are required to enable the consumers to become independent actors in the new energy markets [1].

2.2.1 Incentives

Financial

Many researchers have indicated that financial incentives seem to be the most effective way to change electricity consumers' behavior and to engage them to participate in demand flexibility [17].

Immonen et al. [1] presented in their research that the electricity consumers rationalized their refusal on the basis of the costs caused by the device and system investments should be paid by the party that gains the most benefit, for example, electricity company, the DSO, the controlling party, or the service provider. They think that consumer should somehow receive their share of the forthcoming benefits and that government support was also required, for example, tax reliefs.

Information

Kendel, Lazaric, and Maréchal [18] showed that detailed feedback on the energy consumption helped the electricity consumers to understand their behavior and the structure of their consumption and helped them identify the inefficiencies in their consumption. The results of their research show that all participants reduced their electricity consumption and the main reason for that seemed to be direct learning from feedback about the consumption and/or indirect learning through self-monitoring. However, learning can occur also by intrinsic motivation without any external feedback. The study highlighted that the full engagement of participants was an important aspect to reductions in electricity consumption. The experiment revealed some sources of overconsumption and participants become more aware of their potential wasting of energy [18]. Verbong, Beemsterboer and Sengers [17] also noted that providing consumers detailed feedback on their electricity consumption will lead to a reduction in energy use.

Immonen et al. [1] have identified in their research how electricity consumers follow price developments and events in the electricity market. According to research, the most used data source are newsletters of the electricity company. Other well-used information sources according to the research are the Internet and other general media.

Furthermore, information is obtained by using the application of TSO in hourly prices, inspecting the electricity bills, and information that comes from offer requests from electricity companies. Electricity consumers also need information about demand flexibility and encouragement to become active participants in the energy market [1].

Gangale et al. [19] have studied consumer engagement in smart grid projects in Europe. The aim of the research was to collect information about key aspects of customer engagement. According to their research, to engage customers, it is important to raise awareness and provide information about new technologies. In order to change consumer behavior, consumers need to be aware of their energy consumption and how it affects the environment, energy security, and potential money savings. Although, information and feedback are not necessarily enough to change consumer behavior. Energy providers need to build trust and put consumer motivations in the center when making decisions [19].

Energy services

Different energy services can be considered as incentives and according to Immonen et al. [1], the most interesting energy services among electricity consumers are monitoring electricity prices and optimizing own consumption; automatic advance detection of electrical equipment malfunction; automatic heat adjustment; guidance services and electric car charging services [1].

2.2.2 Regulations and policies

The use of digital technologies in the energy sector causes new risks that need to be addressed through safety, security, and liability policies. It requires collaboration from regulators to address the new complexities that are caused by flexible electricity services [2].

Electricity consumers would be interested to sell their surplus electricity to their neighbors and self-producing communities but Immonen et al. [1] wrote the article it was not possible (at least in Finland) due to legal constraints. In addition, moving the time of consumption is not profitable with current taxes and transmission costs because these costs stay at the same level, only the energy itself may be cheaper [1].

According to Zepter et al. [20], European Commission aims to actively reduce the regulatory barriers to promoting consumers to participate in the electricity markets [20].

2.3 Peer-to-peer (P2P) trading

Prosumers can actively participate in the energy market by a new energy management technique for smart grids called peer-to-peer (P2P) trading. Prosumers can gain benefits by selling their excess energy to other consumers or by reducing the energy demand. P2P trading can be also beneficial to generators, retailers, and distribution network system providers because of decreasing peak demand, lower investment and operational costs, reducing reserve requirements, and improving power system reliability. However, P2P platform is an untrustworthy system because it is expected that prosumers will trade their surplus energy with one another with a very low influence from a central controller. Also, the large number of users causes challenges to model the decision-making process parameters that are affected by rational choices of prosumers that might conflict with the interest of other prosumers in the network [21].

Zepter et al. [20] have modeled a market design that allows the participation of the community of prosumers in the day-ahead and intraday markets. In the model, prosumers use P2P trading and battery storage to balance generation and load within the community and to maximize the use of wind and solar power generation. Their model aims to minimize expected costs for electricity consumption within the community. The model takes into consideration supply-demand balancing decisions, trade restrictions and rules on local P2P trading, and restrictions for the physical boundaries of storage systems. The electricity supply, in this case, comes from four possible sources: direct wind/PV consumption, electricity grid consumption, storage discharge, and P2P purchases. According to the simulation, when combining P2P trading and storage solutions, most of the demand can be met within the community and grid consumption seems to be necessary only during evening peak hours [20].

3 Smart home demand response simulation

The works in the reviewed smart home demand response-related literature seem to assume that the price of the electricity is either known beforehand or that the cycling of the power of the devices at the residential location is directly controlled as an ancillary service by an external party. It is true that in practice the prices might be known in advance within some timeframe, for example, in the case of the Nord Pool's day-ahead prices or other similar systems, but this might not be the case everywhere. We wanted to consider the case where either the price is not known in advance or the information is not available for some reason, for example, due to being in a remote location without an internet connection. What can the end user do in this case if they want to optimize their power consumption? Forecasting the price of electricity has certainly been researched a lot and complex algorithms likely exist, but is it possible to develop some naïve and computationally light approach that would yield adequate results?

We wanted to consider the case of charging an EV at home and optimizing the schedule in order to save on the cost of the electricity. We chose this firstly because the authors of this work are using a 2015 Nissan Leaf EV for their daily driving and thus this subject is interesting on a personal level. We also chose this as being a fairly complex case, if the developed system is able to optimize the electricity cost for the EV, the same system should be fairly easily applicable for other more simplistic devices such as laundry machines or water heaters. The core components of the proposed intelligent charging controller are shown in Fig. 9.2.

In order for the intelligent charging system to be useful and convenient to the users, the system probably should have some indication to its current status and the timing of the charging. The user interface would be naturally implemented through to an

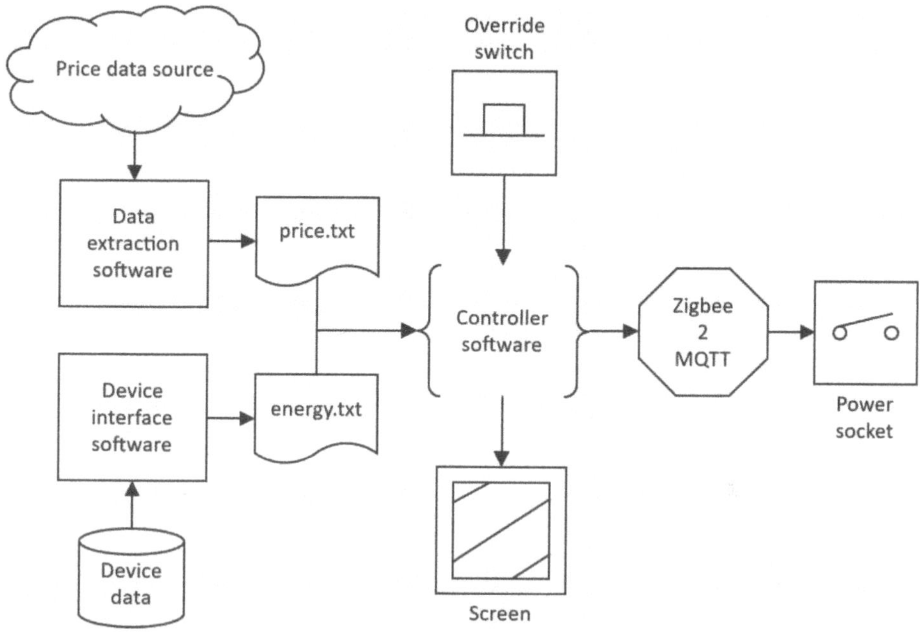

Fig. 9.2 Core components of the proposed intelligent EV charging.

existing smart home system, in case, the intelligent charging controller is integrated into such a system. The output of the controller software is a control signal, which is used to enable and disable the charger. In case this system is integrated directly into a charging unit, this could likely directly control an output relay of the charger. However, in order to remain as generic as possible, the output from the software could also be used to control a remote power socket.

The controller software itself resides at the heart of the implementation. This software needs to be able to read the input files, produce graphics to the screen, send control signals as outputs. One integral part of the software is the maintaining of the price information. A similar structure based on the spot prices was used as in the previously described sections and the day-ahead spot-price information needs to be loaded from a file provided daily to the software, but in addition to the day-ahead prices, the software also needs to be able to maintain the price information for the ongoing day. The software needs to detect when new data is available in the input file, it needs to take into account the changing of the day, and it's very useful if the software is able to maintain its state over host system restarts.

In a stand-alone solution (in case the intelligent charging controller is not integrated into some third-party home automation system), the controller software can be implemented to, for example, some embedded platform.

The simulations in the literature aim to be (or at least they should be) generic and provide universally applicable results. We wanted to focus specifically on the electricity contracts locally available residential customers. Thus, we included in our simulator the pricing schemes from the local supplier (Vaasan Sähkö Oy) and the day ahead Nord Pool spot pricing for the region of Finland. The results of this work might be also more broadly applicable, but the specifics of this work were selected purely for personal reasons.

Li et al. [4] argue that most of the vehicles are parked and available for a plug-in for most of the day. However, we would highly question this statement; even though vehicles are in general not in use for most of the time, they are not necessarily available for plug-in other than when specifically parked at a charging station or at home where the charger is available. Charging stations at workplaces or public locations are not very common yet and for example, the sockets present in the parking spaces in Finland are only intended to be used with motor heaters, thus usually not either capable or allowed to be used for charging BEVs. This is something we wanted to take into account in our simulations.

3.1 Nord Pool spot prices

For the purposes of implementing the simulator, the Nord Pool spot prices for Finland were extracted and analyzed. The following section describes this process in more detail.

Nord Pool is the power exchange for Nordic and Baltic countries. At the time of writing this, the Nord Pool website offers a public record of the day-ahead prices spanning back to the beginning of 2019. The prices for the following day are typically published at 12:42 CET. The available data can be manually viewed and downloaded 1 week at a time from the Nord Pool website. Automatic extraction of the data (and/or use for commercial purposes) is prohibited under Nord Pool. The site claims that the data is available in Excel format, but closer inspection of the downloaded files shows that the content is actually in HTML format.

The data available on the Nord Pool website contains the following for each day; 24 hourly spot prices, minimum price, maximum price, average price, peak, off-peak 1, and off-peak 2. An object structure was created reflecting this data. The class definition in C# is shown in Fig. 9.3.

The day-ahead price history for Finland was manually downloaded from the Nord Pool between the duration of 18.11.2019 and 4.1.2021. The authors of this work wanted to include at least one full calendar year worth of data for this project. Further analysis of a larger dataset might yield more accurate results or additional insights regarding the data, but these were deemed to be outside the scope of this work.

A tool was then developed for extracting the data from the raw Nord Pool files into the object structures mentioned previously. The tool assumes that the files are either directly downloaded from the Nord Pool (HTML content with .xls file extension in which case it tries to rename the files to .html) or that they have already been renamed to .html. The tool then proceeds to read all the values from the provided files and constructs an object database based on the found data. After data extraction, the tool performs a simple error correction for missing data and sorts the objects based on the date in descending order. Finally, the database can be saved in an XML format. The data parser tool is shown in Fig. 9.4.

```
public class DayAhead : IComparable
{
    public DateTime Date = DateTime.MinValue;

    public float[] Prices = { float.NaN, float.NaN, float.NaN, ...

    public float Min = float.NaN;
    public float Max = float.NaN;
    public float Average = float.NaN;
    public float Peak = float.NaN;
    public float OffPeak1 = float.NaN;
    public float OffPeak2 = float.NaN;

    public DayAhead() {}

    public int CompareTo(object obj)
    {
        if (obj == null || !(obj is DayAhead))
            return 1;

        return DateTime.Compare(Date, (obj as DayAhead).Date);
    }
}
```

Fig. 9.3 Object model representing the Nord Pool day-ahead price data.

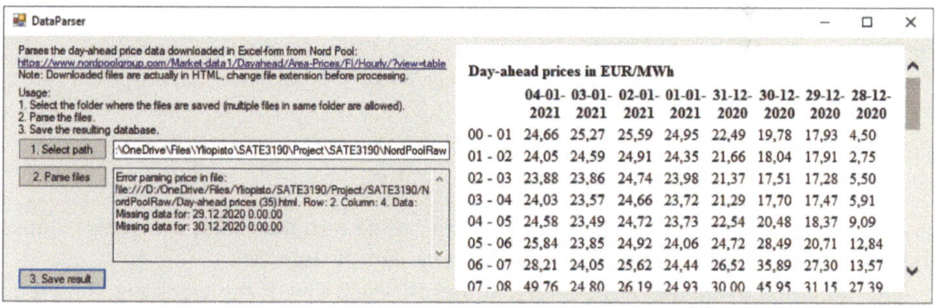

Fig. 9.4 Program developed for parsing the spot price data downloaded from Nord Pool.

After extracting the data, another tool was developed for analyzing the gathered data. This tool starts by taking as an input the database created by the previously mentioned tool and then goes through all the available data. The analyzer tool calculates averages of the minimum, maximum, peak, etc. prices for each day of the week. The tool also calculates hourly weighted averages for each day. The daily averages of the different values are shown in Fig. 9.5.

From these, we can determine, that on average, the electricity seems to be cheapest on Saturdays and Sundays as the maximum, average, and peak values are highest

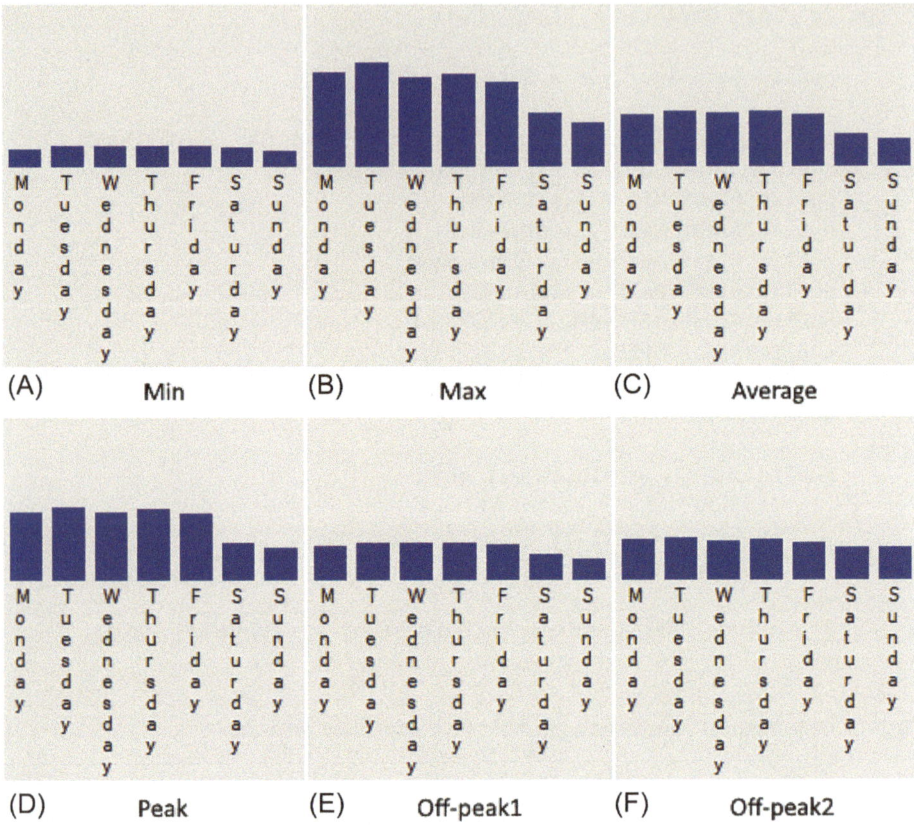

Fig. 9.5 Averaged daily electricity price data from Nord Pool for the Finnish region: (A) minimum price, (B) maximum price, (C) average price, (D) peak price, (E) off-peak 1 price, and (F) off-peak 2 price.

between Monday and Friday. However, based on the minimum and off-peak values, cheap electricity might be available during any day of the week.

The weighted averages of the hourly prices for each day of the week are shown in Fig. 9.6. The diagram for Monday–Friday look all fairly similar with only minor variations. Diagrams for Saturday and Sunday however seem quite different from the rest. From these diagrams, we can determine, that on average, the cheapest electricity is available on weekdays approximately 6 h between 23:00 and 04:59. During Saturdays this window is slightly shifted to approximately between 01:00 and 05:59 and 1 h even later during Sundays. Fig. 9.6 shows a strange jump in the price between 23:00 h on Friday and 00:00 h on Saturday, this is however just because daily weighted averages were used for the calculation and the prices on Saturdays are in general lower than the prices on Fridays. If weekly weighted averages were used instead, this jump would not be present in the diagram.

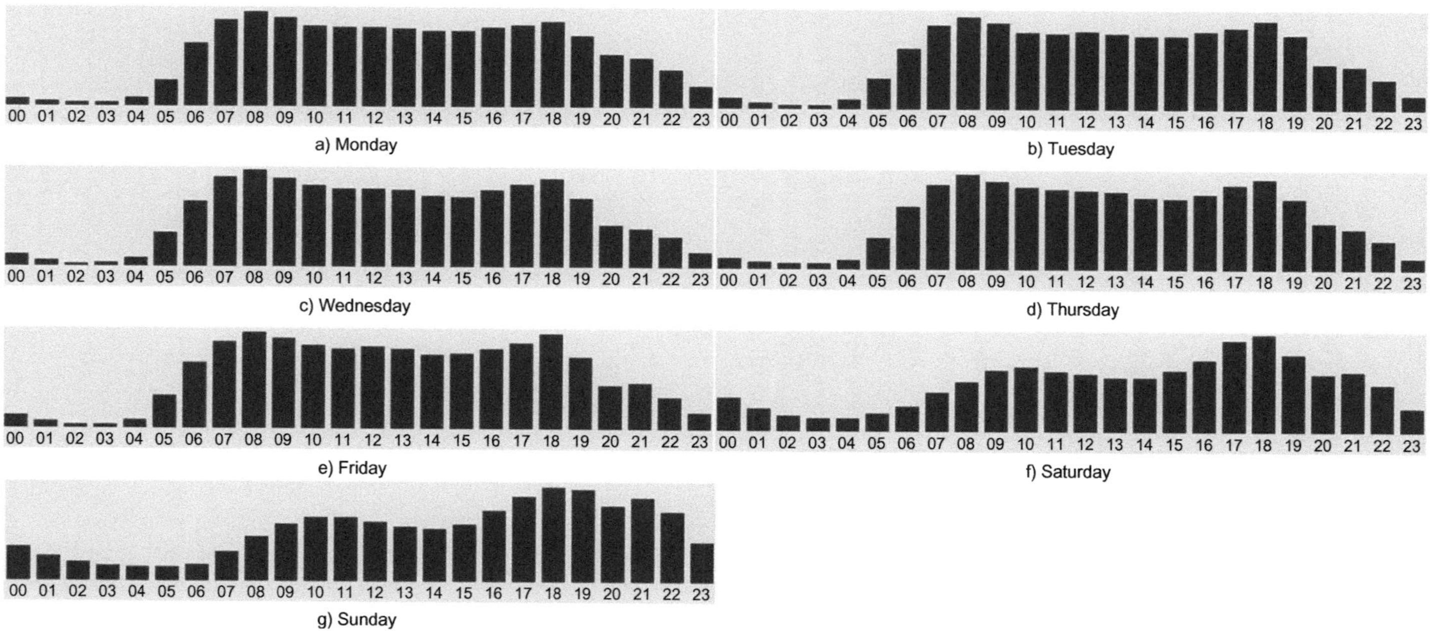

Fig. 9.6 Normalized average hourly electricity prices from Nord Pool for the Finnish region: (A) Monday, (B) Tuesday, (C) Wednesday, (D) Thursday, (E) Friday, (F) Saturday, and (G) Sunday.

3.2 EV charging simulator

An EV charging simulator program was developed for the economics of charging an electric car under different conditions. The program can be used to simulate the depleting and charging of an electric car based on the habits of its user. The program contains several different profiles related to power economics, and the implications of these can be evaluated under the specified conditions. An image of the simulator is shown in Fig. 9.7.

The program tries to automatically load the Nord Pool spot price database mentioned in the previous chapter and it takes as inputs several parameters regarding the car and its usage. The battery size and charger capacity are taken from a standard 2015 Nissan Leaf equipped with the basic onboard AC charger (no quick charge or DC charging) intended for home use from a normal 16A wall outlet.

The simulation software also takes as inputs the energy consumption (kWh per km), the average driving distance for each day of the week, and the times at which the car may be disconnected/plugged into the grid (**Grid connection Off** specifies the time at which the car is disconnected from the charger, while the **Grid connection On** specifies the time at which the car may be connected back to the charger, **dash or other nontime value** can be used to indicate that the car remains plugger for the entire day). The latter two of these depend purely on the driving habits of each individual user and the consumption varies heavily depending on various factors such as the climate in which the car is being used/stored, amount of the cabin heating/cooling, and

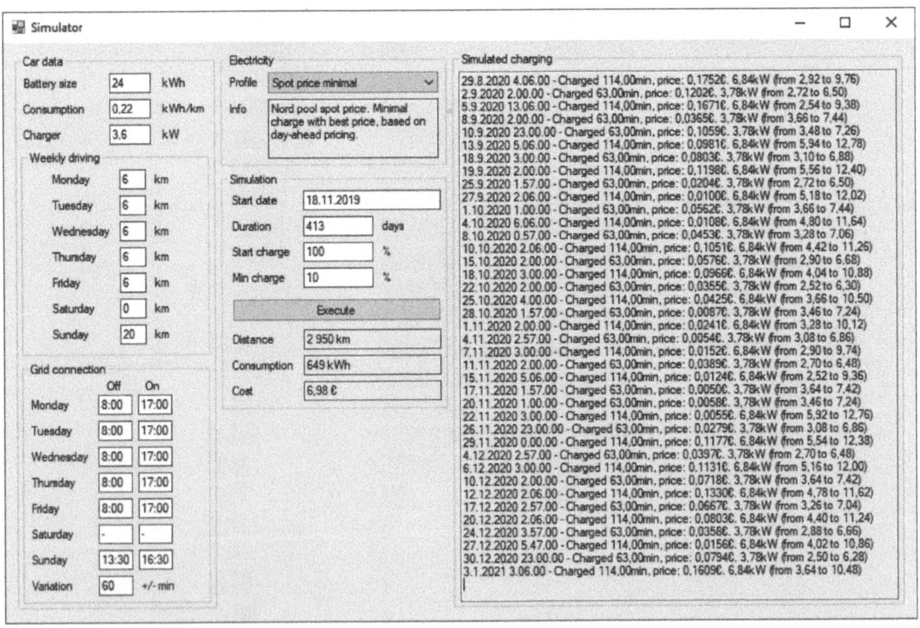

Fig. 9.7 EV charging simulator, default settings, and output after running the simulation.

the type of driving (heavy/light on the gas/breaks, city/highway, etc.). The default values of the simulator are based on the actual usage of the author of this work.

The user of the simulator can also specify the starting date of the simulator, duration of the simulation in days, the starting charge of the battery when beginning the simulation, and the minimum charge under which the simulator tries to maintain the level of the battery considering the daily driving figures. The start date and duration are automatically taken from the automatically loaded Nord Pool spot price database. Some simulations can be run outside this range, but some require that the specified dates can be found from the price database.

The simulator functions by taking the start date with the provided starting battery charge. The simulator then steps through each day within the selected timeframe, decrementing the battery charge level based on the daily driving habits described by the user. At the end of each simulated day, the simulator checks whether the following days driving will drain the battery under the specified minimum charge. In case this condition is detected, the program tries to simulate battery charging within the grid connection times provided by the user. By this behavior, the program does not assume that the car would be plugged into the grid and charged fully every day (as this probably is not how EV users generally behave), but instead, the program attempts to simulate the fact that the user likely knows how much they are going to use the car in the near future and when the battery should be charged in relation to that.

The most interesting configurable simulation parameter is the electricity profile, which affects the cost of charging and the more precise time window within which the actual simulated charging takes place. This is further explained in the following chapter.

The program produces an output of the distance of the simulated driving, consumed electricity, and electricity cost under the selected profile. The program also outputs the time, price, and amount of power for each simulated charging event.

3.2.1 Electricity profiles

The simulator contains a total of seven individual electricity economical profiles used for charging the EV. The first of these is the fixed price single tariff profile from the local electricity reseller; 5.19 c/kWh "Yleissähkö-Perussäkö" from Vaasan Sähkö Oy. This is the most simplistic profile and under this, the charging is simulated always starting from the moment of the EV being plugged into the grid.

The following two profiles are slightly more complicated; they simulate the double tariff electricity from the local reseller; "Yösähkö-Perussähkö" from Vaasan Sähkö Oy, 5.53 c/kWh between 7 and 22 and 4.67 c/kWh between 22 and 7. The first of these two profiles simulate the charging starting always at the grid connection, but the latter delays the charging to start at 22:00 h with the cheaper nighttime tariff.

The fourth available electricity profile assumes that the user is able to buy electricity-based directly on the Nord Pool spot prices. Spot price electricity supply contracts are available in Finland, such as the "Aktiivi" from Oomi Oy. In this profile, the simulated charging happens immediately after grid connection with the spot price of the hour, and the price changes at every hour when the charging proceeds.

The last three profiles leverage also the Nord Pool spot prices, but these try to optimize the actual time of charging instead of starting to charge at the time of grid connection. One of these profiles takes a simplistic approach by timing the charging of the EV based on the price trend analysis described in the previous chapter. This is the naïve optimization algorithm mentioned previously. Another profile looks at the minimum amount of charge that needs to be applied to the battery during each charging event in order not to fall under the specified minimum battery charge, the profile then looks for the window within the grid connection which will yield the cheapest electricity for achieving the calculated charge. The last profile does the same as the previously described, with the exception that it tries to always charge to full instead of just charging the bare minimum.

3.2.2 *Simulation results*

Several different conditions were simulated with the developed tool during its development. The most interesting of these are listed in Table 9.1. The columns indicate the selected power profile in a shortened form. These stand for Local **single** tariff, Local **dual** tariff, Local **dual** tariff **opt**imized, **Spot** price, **Spot** price **trend**, **Spot** price **min**imal, **Spot** price optimal **full**. The rows correspond with different amounts of driving. The **low driving** amount represents the driving profile as shown in Fig. 9.7 (6 km on Monday–Friday, 0 km on Saturday, 20 km on Sunday), total 2950 km and 649 kWh. The **medium driving** amount is quadrupled the low drive (24 km on Monday–Friday, 0 km on Saturday, 80 km on Sunday), total 11,800 km and 2596 kWh for the simulated timeframe. Corresponding with roughly 11,800 km yearly driving. The **high driving** amount is probably pushing what is reasonable with the EV in question (45 km on Monday–Saturday, 90 km on Sunday), total 21,240 km and 4673 kWh simulated.

Table 9.2 lists the values on the previous table in relation to the base price, which is the single tariff electricity from the local reseller. From these, we can determine that the dual tariff electricity is a bad idea if one does not also intend to time their EV charging purely to the cheaper night tariff, and even in that case, the savings are fairly marginal. Simply by switching to the spot pricing one may save up to 40% on their electricity, and with slight optimization, the total cost can be reduced to ⅓ or ¼ of the original. The highest savings can be achieved with the minimal charging scheme, but this will probably also place the most stress on the battery. Spot pricing with a fully optimized charging window is probably the most reasonable, and the naïve trend-based algorithm

Table 9.1 Results of the simulations.

	Single	Dual	Dual opt	Spot	Spot trend	Spot min	Spot full
Low drive	33.39 €	34.87 €	30.04 €	21.10 €	9.41 €	6.99 €	8.41 €
Med drive	134.73 €	141.94 €	121.23 €	81.86 €	40.33 €	32.89 €	35.70 €
High drive	242.52 €	255.00 €	218.22 €	155.08 €	73.27 €	61.31 €	65.30 €

Electricity cost of different driving amounts simulated with different electricity profiles.

Table 9.2 Relative changes in the cost of electricity from the previously described simulated cases.

	Single (%)	Dual (%)	Dual opt (%)	Spot (%)	Spot trend (%)	Spot min (%)	Spot full (%)
Low drive	100	104	90	63	28	21	25
Med drive	100	105	90	61	30	24	26
High drive	100	105	90	64	30	25	27

provides also considerable savings while being easy to implement in the software and not requiring connection to any external system while in operation.

3.2.3 *Further improvements*

The simulated charger is currently assumed to be 100% efficient, which is not the case in real life. Taking this into account in the simulator would be fairly trivial and thus was left out from this implementation.

Only the price of the electricity itself is considered in the calculation. Various taxes, commissions, monthly fees, and the price of the electric power transmission should also be calculated in order to give the user a better picture of the cost of operating the vehicle. These were also considered to be outside the scope of this work.

The simulator also has a parameter for varying the grid connection times. Using this adds randomness to the grid connection times and thus to the resulting total cost under certain electricity profiles. After further consideration it was determined that this parameter is rather useless as is; the varied values should trend toward the fixed (variation set to 0) after an infinite amount of iterations. Instead, the program could attempt to calculate the variation of the resulting total price based on the variations on the grid connection. However, the search space would likely increase exponentially and efficiently implementing this is probably quite difficult without employing more sophisticated optimization methods.

The simulator aims to always charge the battery back to the full charge specified by the battery size parameter, but depending on the various input conditions, the simulated battery may also be left in a partially charged state. This does not happen with the default values present in the simulator but may result, for example, from the case where daily driving consumes the battery more than what can be charged under the specified grid connection times. These conditions were not tested during the development and the simulator will likely start to behave badly if such conditions are being met excessively.

The simulator is not really a "production ready" software. Error cases are not very clearly reported to the user, the software might even crash, and several improvements could be made to the code in order to optimize the execution times. However, it was not the goal here to develop a commercial grade software, but instead more a proof-of-concept type testing tool.

4 Conclusions

The consumer's role in the future energy market will increase but energy market actors need to decide how to encourage and motivate consumers to become active players in the market. Consumers need guiding and support as being active members of the energy market.

Consumers' attitudes toward flexibility management and related services highly rely on incentives and benefits. Financial incentives seem to be the most effective way not only to change electricity consumers' behavior but also to increase awareness about energy markets, environmental issues, and energy consumption. New business models are required to enable the consumers to become independent actors in the new energy markets. For example, aggregators, virtual power plant operators, and technical service providers have the potential to play an important role in future energy markets.

There are still multiple challenges that need to be addressed before consumers can actively participate in energy markets. For example, although both the smart grids and homes have already been a hot research topic for a while and are also becoming more and more common in practice, there seems to be a total lack of standardization regarding the communication protocols and interfaces which would be required in order to connect the two of these for enabling the implementation of demand response. Several technologies and implementations have been proposed for solving this, but there does not seem to be one preferred over the others.

An EV charging simulator program was developed for evaluating the economics of charging an electric car under various conditions. One of the goals of this work was to explore and evaluate whether a naïve scheduling algorithm could be developed in such a way that savings in the electricity cost could be achieved even without having the utilities directly controlling the scheduling. In the simulated cases, we observed considerable savings even in the case when foresight regarding the development of the electricity price was lacking. The methodology could likely quite easily be applied to any schedulable electricity needs, in addition to just the EV charging.

References

[1] A. Immonen, J. Kiljander, M. Aro, Consumer Viewpoint on a New Kind of Energy Market, Elsevier B.V., 2020. Retrieved 4 January 2021.

[2] The European Consumer Organisation, The consumer voice in Europe, in: Electricity Aggregators: Starting off on the Right Foot with Consumers, 2018. Retrieved December 29, 2020, from https://www.beuc.eu/publications/beuc-x-2018-010_electricity_aggregators_starting_off_on_the_right_foot_with_consumers.pdf.

[3] International Energy Agency, IEA: Demand response, 2020, Retrieved Joulukuu 29, 2020, from https://www.iea.org/reports/demand-response.

[4] Z. Li, M. Chowdhury, P. Bhavsar, Y. He, Optimizing the performance of vehicle-to-grid (V2G) enabled battery electric vehicles through a smart charge scheduling model, Int. J. Automot. Technol. 16 (2014) 827–837.

[5] J. Lai, H. Zhou, W. Hu, D. Zhou, L. Zhong, Smart Demand Response Based on Smart Homes, Hindawi Publishing Corporation, 2015.

[6] S.J. Darby, Smart technology in the home: time for more clarity, Build. Res. Inf. 46 (1) (2018) 140–147, https://doi.org/10.1080/09613218.2017.1301707.
[7] Wakefield Research, New Survey Finds 70% of Consumers Improved Home during COVID-19, More than Half Used Smart Devices, Xiaomi, New York, 2021. Retrieved January 2021, from https://blog.mi.com/en/2021/01/05/new-survey-finds-70-of-consumers-improved-home-during-covid-19-more-than-half-used-smart-devices/.
[8] N. Gudi, L. Wang, V. Devabhaktuni, S.S. Depuru, Demand response simulation implementing heuristic optimization for home energy management, in: North American Power Symposium, IEEE, 2010.
[9] K. Tsui, S. Chan, Demand response optimization for smart home scheduling under real-time pricing, in: IEEE Transactions on Smart Grid, Volume: 3, Issue: 4, IEEE, 2012, pp. 1812–1821.
[10] N. Gayathri, V.V. Vineeth, N. Radhika, A novel approach in demand side management for smart home, Procedia Technol. 21 (2015) 526–532.
[11] M. Hoosain, B.S. Paul, Smart homes: a domestic demand response and demand side energy management system for future smart grids, in: International Conference on the Domestic Use of Energy (DUE), IEEE, 2017.
[12] F. Fernandesa, H. Moraisa, Z. Valea, C. Ramos, Dynamic load management in a smart home to participate in demandresponse events, Energy Build. 82 (2014) 592–606.
[13] H. Wang, K. Meng, F. Luo, Z. Xu, Z.Y. Dong, G. Verbic, K. Wong, Demand response through smart home energy management using thermal inertia, in: Australasian Universities Power Engineering Conference (AUPEC), IEEE, 2013.
[14] European Commission, European Commission Website, 2020, Retrieved from Energy. Smart grids and meters: https://ec.europa.eu/energy/topics/markets-and-consumers/smart-grids-and-meters/overview_en.
[15] F. Tounquet, C. Alaton, Benchmarking Smart Metering Deployments in the EU-28, Publications Office of the European Union, Luxembourg, 2020.
[16] Council of European Energy Regulators (CEER), CEER Paper on Electricity Distribution Tariffs Supporting the Energy Transition, 2020, Retrieved from: https://www.ceer.eu/documents/104400/-/-/fd5890e1-894e-0a7a-21d9-fa22b6ec9da0.
[17] G.P. Verbong, S. Beemsterboer, F. Sengers, Smart Grids or Smart Users? Involving Users in Developing a Low Carbon Electricity Economy, Elsevier, 2012. Retrieved January 4, 2020.
[18] A. Kendel, N. Lazaric, K. Maréchal, What Do People Learn by Looking at Direct Feedback on Their Energy Consumption? Results of a Field Study in Southern France, Elsevier, 2017, https://doi.org/10.1016/j.enpol.2017.06.020. Retrieved 12 31, 2020, from.
[19] F. Gangale, A. Mengolini, I. Onyeji, Consumer Engagement: An Insight From Smart Grid Projects in Europe, Elsevier, 2013.
[20] J.M. Zepter, A. Lüth, P. Crespo del Granado, R. Egging, Prosumer Integration in Wholesale Electricity Markets: Synergies of Peer-to-Peer Trade and Residential Storage, Elsevier, 2019.
[21] W. Tushar, T.K. Saha, C. Yuen, D. Smith, V.H. Poor, Peer-to-Peer Traiding in Electricity Networks: An Overview, 2020. Retrieved December 29, 2020.

Glossary

BEV Battery electric vehicle—Commonly shortened just to EV.
CP Convex programming—Mathematical optimization technique which can be utilized if the problem can be presented as certain types of convex functions.

DLP Dynamic load priority—A concept by which the priority of various loads is being dynamically adjusted during a demand response event.
DR Demand response—Adjustment of customers' electricity demand as a response to the changing conditions of the electrical network.
DRM Demand response management—A process for managing devices in order to enable the demand response functionality.
EV Electric vehicle—A machine used for transportation, uses one or more engines and electricity for propulsion.
FPCTTH Fiber power cable to the home—A type of cabling which combines the delivery of electrical power and fiber optic connection for residential buildings.
PSO Particle swarm optimization—An iterative optimization technique where candidate solutions ("particles") move around the search-space like certain birds or fish.
SG Smart grid—Electrical grid which includes for example various smart components and renewable resources.
SH Smart home—A residential building with varying level of home automation and connected appliances.
SHIM Supervisor Control and Data Acquisition House Intelligent Management—A smart home simulation system developed at the Polytechnic of Porto.
V2G Vehicle to grid—Transferring power from EVs to the electric grid, using the EVs as battery storage for the grid.

Equivalent dynamic modeling of active distribution networks for TSO-DSO interactions

10

F. Conte, F. D'Agostino, and F. Silvestro
Department of Electrical, Electronic, Telecommunication Engineering and Naval Architecture, University of Genova, Genova, Italy

Nomenclature

δ angle between E' and V [rad]
E' voltage behind X' [p.u.]
E_f excitation voltage [p.u.]
ω grid electric angular velocity [p.u.]
H inertia time constant [s]
T'_d direct-axis transient time constant [s]
T_m mechanical torque [p.u.]
V bus voltage [p.u.]
X reactance [p.u.]
X' transient reactance [p.u.]

1 Introduction

Distribution networks are today active components of the electrical power system, due to the presence of distributed energy resources (DERs), expected to be comparable or even exceeding the demand in the near future [1]. In this scenario, an active distribution network (ADN) dynamically responds to system disturbances, and affects, in this way, the dynamical behavior of the entire power system. As a consequence, the result of classical representation of distribution networks, as passive aggregated loads, may not be adequate for predicting and controlling the modern power system.

From the perspective of transmission system operators (TSOs), characterizing the behavior of ADNs has been identified as one of the major challenges, as well as the management of DERs for the distribution system operators (DSOs) [2]. Moreover, distributed generation has reduced the share of production from conventional forms of generation, involved in providing system ancillary services at the transmission level. This fact has led to the definition of guidelines [3] and control strategies [4] devoted to unlock the potential of distribution-level resources to cover the ancillary services

Decentralized Frameworks for Future Power Systems. https://doi.org/10.1016/B978-0-323-91698-1.00002-9

deficit. In this context, while the DSO is responsible for continuity of service in distribution networks, the TSO continues to have the responsibility for system security, coordination of balancing, frequency control, and system recovery. To ensure system reliability and to improve the coordination at the TSO-DSO interface, there is a strong need for greater observability [5] and knowledge of distribution networks. In particular, TSOs need detailed models that able to provide a sufficient description of the dynamical response of the distribution system to disturbances and/or regulation signals.

To model the distribution system is particularly challenging, since it is composed of a significant number of different devices, and is organized into time-varying configurations. Moreover, TSOs do not have a knowledge of each ADN as sufficient as to realize a highly detailed model, which would be, in any case, extremely large and not suitable for real-time analysis. Therefore, one of the most promising solutions is the development of equivalent dynamic models (EDMs).

An EDM of an ADN should be able to dynamically determine the profiles of active and reactive powers exchanged at the connection point, based on the measurements of voltage magnitude and frequency collected at the same point. The main goal is to allow replacing the full (unknown) highly detailed model of an ADN, or a portion of an ADN, with a simple EDM. The ability to reduce ADN complexity and dynamic response through the application of EDM represents an essential part to properly tackle the problems of dimensionality. In fact, the use of EDM allows the development of distributed controllers (e.g., model predictive-based approach) for each ADN that can play a key role in a future decentralized power system.

The development of EDMs for ADNs has attracted the attention of the scientific community, which has proposed several different solutions. The report [6] and the relevant paper [7] provide an extended survey of different techniques developed for representing ADNs. Generally, *measurements-based methods* are preferred to other approaches since they do not require a detailed knowledge of the architecture and the composition of the ADN. A great amount of measurements provided by information communication technologies are used to identify the EDM using a suitable system identification technique [7]. In [8–10], EDMs are identified through artificial neural networks (ANNs), which are considered particularly suited for modeling nonlinear dynamics. Indeed, measurements of the aggregated response of battery energy storage systems (BESSs) are used to develop two ANNs in [11], for active and reactive power, respectively, while in [12, 13] neural networks are used to identify EDMs for microgrids, which can be considered as a special case of ADNs [6]. *Prony analysis* is applied in [14, 15] to design an EDM for distribution network cells and microgrids, respectively. *Modal analysis* based on the Henkel norm approximation is used in [16]. A *coherency-based* method is applied in [17]. *Variable-order transfer functions* models are used in [18], while an enhanced *exponential recovery model* is introduced in [19] and further applied in [20].

All mentioned techniques are based on the *black-box* approach, which does not assume any preestablished form of physical-related models. Differently, in the *gray-box* approach, a model structure is selected using available basic physical knowledge of the ADN. This class of methodologies is defined and recommended in [7, 21, 22]. In particular, in [7], it is remarked that black-box ANN-based approaches have been initially the key players in representing ADNs, since, in general, they can efficiently learn nonlinear dynamics from experience, without the need of any physical knowledge of the system to be represented. However, in order to obtain accurate results, large datasets are required to be informative enough to allow the model to correctly reproduce different settings of the same ADN [7]. Therefore, when the available physical knowledge allows to select a physically parameterized model structure, gray-box modeling approaches can be beneficial for purposes of deriving dynamic equivalents for ADNs [7, 23, 24]. One of the key features is that, with the physical interpretation, gray-box models are practical to be integrated in dynamic simulation tools. An example of gray-box generic ADN model is proposed in [25, 26] and improved in [27], as will be detailed in Section 2. A further gray-box model is used in [28] for obtaining microgrids EDMs. In [29], a nonlinear gray-box model including discrete-time events triggered by inverter control is proposed for representing an ADN with dispersed photovoltaic (PV) units.

In this chapter, the problem of investigating the equivalent dynamic modeling for ADNs starts with the introduction of an *unconstrained gray-box linear modeling method*, where a nonlinear model is initially defined, then it is linearized, and the parameters of the resulting linear model are identified.

Subsequently, an *operational constrained gray-box nonlinear modeling method* is described. This method introduces the system operating constraints and avoids any linearization of the model. Differently from the previous one, this approach allows an effective relation with the actual configuration of the ADN to be kept. In this way, the prior knowledge of the structure of the ADN can be effectively included in the identification procedure. The result is a model that can be adapted to different configurations and operating conditions of the ADN, and then to be embedded into the simulation of the entire power system. Moreover, by using a physical-based nonlinear structure, all model parameters maintain a physical meaning. This makes possible to simulate generation and load variations arising within the ADN, possibly caused by control and automation actions.

The two methodologies are tested on two simulation environments, where detailed models of the target distribution networks are implemented. Modeling and simulation results will be described in detail. Finally, the nonlinear modeling method is adapted and validated with experimental data on a real microgrid facility, which comprises a low-voltage distribution network, a synchronous generator, two batteries, a photovoltaic (PV) plant, and static loads.

The chapter is organized as follows. Section 2 describes the unconstrained gray-box linear modeling method; Section 3 introduces the operational constrained gray-box nonlinear modeling method and its extension to microgrids; and Section 4 reports simulation and experimental results.

2 Unconstrained gray-box linear modeling method

In this section, the method proposed in [27] is presented. Some definitions here adopted will be used also in the following section, where the approaches of [30, 31] will be introduced. In particular, all models will be defined assuming the availability of the grid electric angular velocity ω and of the voltage V at the point of common coupling (PCC), measured by proper high-resolution meters (e.g., PMU). These p.u. quantities are computed with respect to the bus nominal voltage V^{nom} [V] and the nominal electrical angular velocity Ω^{nom} [rad/s].

2.1 Linear equivalent dynamic model

The EDM adopted in this approach is the one proposed in [25, 26]. It consists of a converted-connected synchronous generator (SG), in parallel with a composite dynamic load, as shown in Fig. 10.1. This last is made up of a ZIP load in parallel with an asynchronous motor (AM).

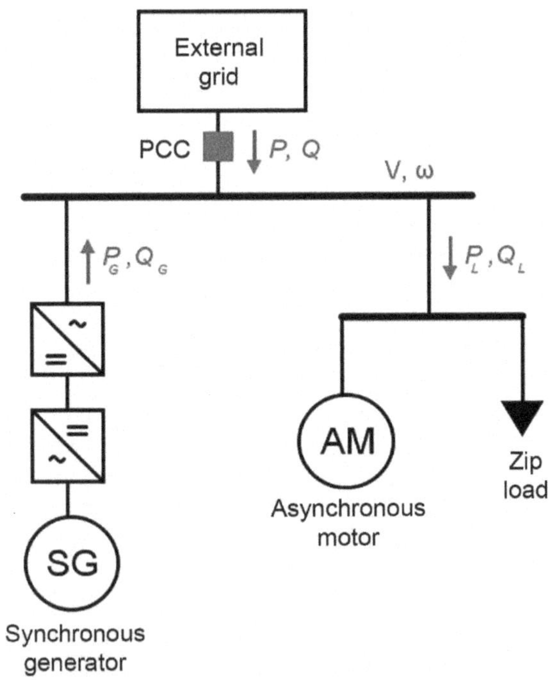

Fig. 10.1 EDM scheme for the unconstrained gray-box linear modeling method.

The behavior of the composite load model is described as follows:

$$\dot{E}'_a = \frac{1}{T'_{da}}\left(-\frac{X_a}{X'_a}E'_a + \left(\frac{X_a - X'_a}{X'_a}\right)V\cos(\delta_a)\right) \tag{10.1}$$

$$\dot{\delta}_a = \Omega^{\text{nom}}(\omega_a - \omega) - \left(\frac{X_a - X'_a}{X'_a}\right)\frac{V}{T'_{da}E'_a}\sin(\delta_a) \tag{10.2}$$

$$\dot{\omega}_a = -\frac{1}{H_a}\left(\frac{E'_a V}{X'_a}\sin(\delta_a) + T_{ma}\right) \tag{10.3}$$

where subscript a stays for asynchronous and ω_a [p.u.] is the AM angular velocity. The active and reactive powers P_L [W] and Q_L [var] absorbed by the composite load model are given by

$$P_L = P_Z V^2 + P_I V + P_P - S_a^{\text{nom}}\left(\frac{V}{X'_a}E'_a\sin(\delta_a)\right) \tag{10.4}$$

$$Q_L = Q_Z V^2 + Q_I V + Q_P + S_a^{\text{nom}}\left(\frac{V^2}{X'_a} - \frac{V}{X'_a}E'_a\cos(\delta_a)\right) \tag{10.5}$$

where $P_{Z/I/P}$ and $Q_{Z/I/P}$ are the ZIP model load component with constant impedance, constant current, and constant power, respectively, and S_a^{nom} [VA] is the nominal apparent power of the AM.

The model of converted-connected generator is composed by a third-order SG model and a back-to-back full converter model:

$$\dot{E}'_s = \frac{1}{T'_{ds}}\left(E_f - E'_s\left(\frac{X_s}{X'_s}\right) + \frac{(X_s - X'_s)}{X'_s}V\cos(\delta_s)\right) \tag{10.6}$$

$$\dot{\omega}_s = \frac{1}{H_s}\left(T_{ms} - \frac{VE'_s}{X'_s}\sin(\delta_s) - D\omega_s\right) \tag{10.7}$$

$$\dot{\delta}_s = \Omega^{\text{nom}}\omega_s \tag{10.8}$$

$$V_{dc} = \frac{1}{CV_{dc}}\left(V_{dg}I_{dg} + V_{qg}I_{qg} - V_{dG}I_{dG} - V_{qG}I_{qG}\right) \tag{10.9}$$

where subscript s stays for synchronous, D is the damping factor [p.u.], C is the converter capacitance at the DC link, V_{dc} and I_{dc} are the capacitor DC voltage and current, V_{dG}, I_{dG} are the d-axis voltage and current at the grid side of the converter, V_{qG}, I_{qG} are the q-axis voltage and current at the grid side of the converter, V_{dg}, I_{dg} are the d-axis voltage and current at the generator side of the converter, and V_{qg}, I_{qg} are the q-axis voltage and current at the generator side of the converter.

The active and reactive power P_G [W] and Q_G [var] delivered by the converted-connected generator are given by

$$P_G = S_s^{\text{nom}}\frac{V}{X'_s}E'_s\sin(\delta_s) + V_{dc}I_{dc} \tag{10.10}$$

$$Q_G = S_s^{\text{nom}} \left(\frac{V}{X'_s} E'_s \cos(\delta_s) - \frac{V^2}{X'_s} \right) + K_q V_{dc} I_{dc} \tag{10.11}$$

where K_q is a scaling factor and S_s^{nom} [VA] is the nominal apparent power of the SG.

The EDM of the ADN can be finally represented by a nonlinear seventh-order state-space model [25, 26]. The system state, input, and output vectors x, u, and y are, respectively, defined as follows:

$$\begin{aligned} x &= \left[E'_a \;\; \delta_a \;\; \omega_a \;\; E'_s \;\; \omega_s \;\; \delta_s \;\; V_{dc}\right]^\top, \\ u &= [V \;\; \omega]^\top, \;\; y = [P \;\; Q]^\top \end{aligned} \tag{10.12}$$

where the active and reactive power absorbed by the whole ADN P [W] and Q [var] are computed as $P = P_L - P_G$ and $Q = Q_L - Q_G$. In [26], the model is linearized around a steady-state point and the constant unknown quantities are grouped into 20 parameters $\theta_1, \theta_2, \ldots, \theta_{20}$ to be identified. The resulting linear system has the following structure:

$$\dot{x} = Ax + Bu + V \tag{10.13}$$

$$y = Cx + Du + W \tag{10.14}$$

where A, B, C, D, V, and W are composed by zeros and the mentioned parameters θ_i, $i = 1, 2, \ldots, 20$.

The role of the EDM is to return the profiles of active and reactive powers P and Q absorbed by the ADN, given the values of V and ω, measured at the PCC. In other words, we are interested in the input-output relation of system (10.13), (10.14). Based on this observation, the set of parameters to be identified θ_i, $i = 1, 2, \ldots, 20$ can be reduced by extracting the observable and reachable part of system (10.13), (10.14) [32].

Indeed, Eqs. (10.13), (10.14) are not fully observable and reachable. More precisely, the observability and reachability matrices

$$\mathcal{O} = \left[C^T A^T C^T \;\; A^{T^2} C^T \;\; \cdots \;\; A^{T^6} C^T\right]^T \tag{10.15}$$

$$\mathcal{R} = \left[B \;\; AB \;\; A^2 B \;\; \cdots \;\; A^6 B\right] \tag{10.16}$$

have ranks 3 and 4, respectively. By applying a standard state-space decomposition procedure [32], it is possible to extract the observable and reachable part of the system, which results to have dimension 2 and the following form:

$$\begin{aligned} \dot{\tilde{x}} &= \begin{bmatrix} \theta_1 & 0 \\ 0 & \theta_3 \end{bmatrix} \tilde{x} + \begin{bmatrix} \theta_7 & 0 \\ \theta_9 & 0 \end{bmatrix} u + \begin{bmatrix} 0 \\ \theta_{10} \end{bmatrix}, \\ y &= \begin{bmatrix} \theta_{11} & \theta_{13} \\ \theta_{12} & \theta_{14} \end{bmatrix} \tilde{x} + \begin{bmatrix} \theta_{17} & 0 \\ \theta_{18} & 0 \end{bmatrix} u + \begin{bmatrix} \theta_{19} \\ \theta_{20} \end{bmatrix} \end{aligned} \tag{10.17}$$

where $\tilde{x}$ is the reduced-order system state vector. Therefore, the set of parameters to be identified is reduced to 13 elements, collected into the vector θ. Being a reduced model, parameters in θ have not physical meaning. Notice that such a reduction will reduce the computational time of any parameters identification algorithm.

2.2 Identification procedure

To identify parameters in θ, an iterative procedure is adopted. First of all, an initial estimate is computed using the gray-box identification procedure (*grayest*) provided by the system identification toolbox of MATLAB. Then, such an initial estimate is refined using an iterative nonlinear optimization procedure based on the Levenberg-Marquardt algorithm [33]. Supposing to have a training set composed by a set of N measurement of the input V and ω and of the outputs P and Q, the objective function to be minimized is

$$J_\theta = \sum_{k=0}^{N-1} (P_k - \hat{P}_{k,\theta})^2 + (Q_k - \hat{Q}_{k,\theta})^2 \tag{10.18}$$

where P_k and Q_k are the active and reactive powers measured at time k, and $\hat{P}_{k,\theta}$ and $\hat{Q}_{k,\theta}$ are the active and reactive powers obtained at time k by simulating the EDM given the parameter set θ.

3 Operational constrained gray-box nonlinear modeling method

In this section, the approach introduced in [30] and the adaptation to microgrids in [31] are presented.

3.1 Nonlinear equivalent dynamic model

The EDM proposed in [30] is an extension of the one introduced in [25, 26]. As shown in Fig. 10.2, it is composed by an asynchronous machine, a ZIP load, and an SG. The asynchronous machine can optionally work as motor (AM) or as generator (AG); in this second case, the generator is converter connected to the bus. The SG can be optionally directly or converter connected to the bus, in order to cover different types of DERs.

In Section 3.1.1, the mathematical models adopted for the three components of the EDM are introduced. Modeling and operating constraints on variables and parameters, required to obtain the typical dynamics expected from each model, are then reported in Section 3.1.2. Such constraints have a key role since the final objective is the identification of the model parameters. The idea is indeed to use constraints to restrict the parameters feasible space, and to obtain an equivalent model which cannot move to no meaning or potentially unstable working points (e.g., with $\delta = 90$ degrees). The result

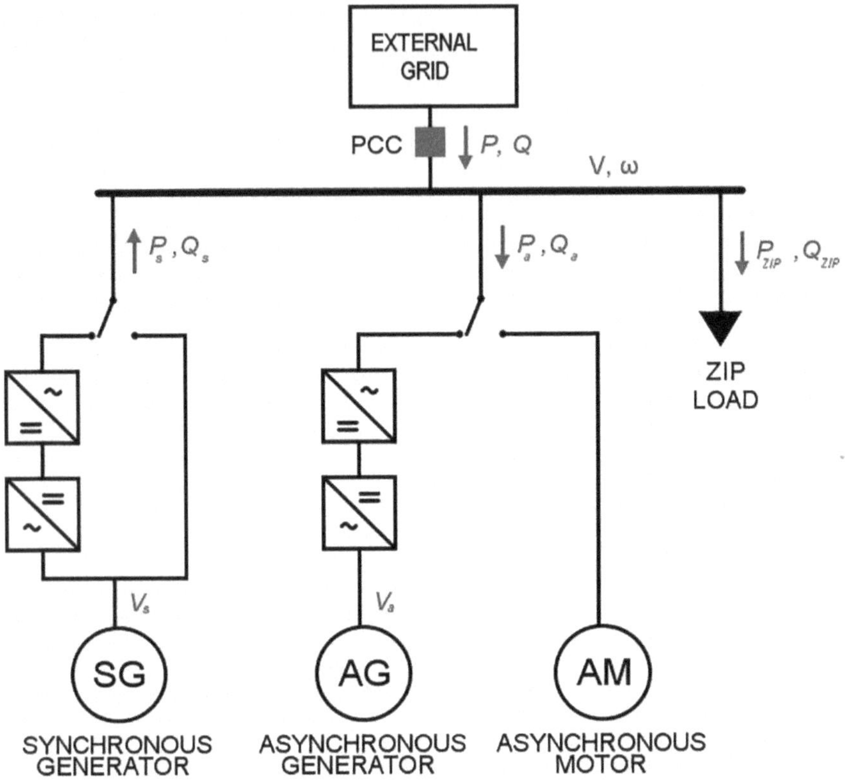

Fig. 10.2 EDM scheme for the operational constrained gray-box nonlinear modeling method.

will be effective representation of the typical and stable dynamics of a system composed by synchronous and asynchronous generators, and dynamic loads. Constraints are defined at steady-state conditions, and subscript 0 indicates steady-state values.

3.1.1 Model equations

The EDM equations are listed in the following, where subscripts a and s stay for asynchronous and synchronous, respectively.

Dynamic equations:

$$\textbf{AM/AG:} \qquad \dot{E}'_a = \frac{1}{T'_{da}}\left(-\frac{X_a}{X'_a}E'_a + \frac{X_a - X'_a}{X'_a}V_a\cos(\delta_a)\right) \tag{10.19}$$

$$\dot{\omega}_a = -\frac{1}{H_a}\left(\frac{E'_a V_a}{X'_a}\sin(\delta_a) + T_{ma}\right) \tag{10.20}$$

$$\dot{\delta}_a = \Omega^{\text{nom}}(\omega_a - \omega) - \frac{X_a - X'_a}{X'_a}\frac{V_a}{T'_{da}E'_a}\sin(\delta_a) \tag{10.21}$$

$$\begin{cases} \dot{V}_a = -\frac{1}{\tau_a}(V_a - V) & \text{if converter} - \text{connected (AG)} \\ V_a = V & \text{if directly connected (AM)} \end{cases} \tag{10.22}$$

$$\textbf{SG:} \qquad \dot{E}'_s = \frac{1}{T'_{ds}}\left(E_f - \frac{X_s}{X'_s}E'_s + \frac{X_s - X'_s}{X'_s}V_s\cos(\delta_s)\right) \tag{10.23}$$

$$\dot{\omega}_s = \frac{1}{H_s}\left(T_{ms} - \frac{V_s E'_s}{X'_s}\sin(\delta_s) - D\omega_s\right) \tag{10.24}$$

$$\begin{cases} \dot{\delta}_s = \Omega^{\text{nom}}(\omega_s - 1) & \text{if converter} - \text{connected} \\ \dot{\delta}_s = \Omega^{\text{nom}}(\omega_s - \omega) & \text{if directly connected} \end{cases} \tag{10.25}$$

$$\begin{cases} \dot{V}_s = -\frac{1}{\tau_s}(V_s - V) & \text{if converter} - \text{connected} \\ V_s = V & \text{if directly connected} \end{cases} \tag{10.26}$$

Output equations (power exchanges):

$$\textbf{AM/AG:} \qquad P_a = -S_a^{\text{nom}}\frac{V_a}{X'_a}E'_a\sin(\delta_a) \tag{10.27}$$

$$Q_a = S_a^{\text{nom}}\gamma_a\left(\frac{V_a^2}{X'_a} - \frac{V_a}{X'_a}E'_a\cos(\delta_a)\right) \tag{10.28}$$

$$\textbf{SG:} \qquad P_s = S_s^{\text{nom}}\frac{V_s E'_s}{X'_s}\sin(\delta_s) \tag{10.29}$$

$$Q_s = S_s^{\text{nom}}\left(\frac{V_s E'_s}{X'_s}\cos(\delta_s) - \frac{V_s^2}{X'_s}\right) \tag{10.30}$$

$$\textbf{ZIP:} \qquad P_{ZIP} = P_Z V^2 + P_I V + P_P \tag{10.31}$$

$$Q_{ZIP} = Q_Z V^2 + Q_I V + Q_P \tag{10.32}$$

$$\textbf{ADN:} \qquad P = P_a + P_{ZIP} - P_s \tag{10.33}$$

$$Q = Q_a + Q_{ZIP} - Q_s \tag{10.34}$$

Eqs. (10.19)–(10.21) and (10.23)–(10.25) are the third-order models of the asynchronous machine and the SG, respectively. The dynamics of terminal voltages V_a and V_s are defined in Eqs. (10.22) and (10.26), respectively. In the converter-connected cases, they follow the simplified first-order dynamics with time constants τ_a and τ_s [s]. In the directly connected case, they are equal to the bus voltage V. By Eq. (10.25), in the SG model, the generator angular velocity ω_s [p.u.] is synchronous with the grid electric angular velocity ω in the directly connected case, and decoupled in the converter-connected case.

Eqs. (10.27)–(10.32) define the power exchanged by the three components of the EDM: P_a [W] and Q_a [var] are the active and reactive powers absorbed (if positive) or exported (if negative) by the asynchronous machine; P_s [W] and Q_s [var] are the active and reactive powers exported by the SG; and P_{ZIP} and Q_{ZIP} are the active and reactive powers absorbed by the ZIP load [6]. The total active and reactive powers P and Q absorbed (if positive) or exported (if negative) by the represented ADN are finally defined by Eqs. (10.33), (10.34).

In Eqs. (10.27)–(10.32): S_a^{nom} and S_s^{nom} [VA] are the nominal apparent powers of the asynchronous machine and the SG, respectively. $P_{Z/I/P}$ and $Q_{Z/I/P}$ are the ZIP model load component with constant impedance, constant current, and constant power, respectively; γ_a is a scaling factor. This last parameter is used to include the reactive power exchange of the equivalent asynchronous machine. Details about the use of this scaling factor are provided in Section 3.2.1.

As mentioned before, models (10.19)–(10.34) are an extension of the one in [25, 26], which is the same adopted in [27] and reported in Section 2. Comparing the two models, it can be noted that Eqs. (10.19)–(10.21) coincide with Eqs. (10.1)–(10.3) and (10.23), (10.24) coincide with Eqs. (10.6), (10.7). Also, the ZIP model is equal to the one in Eqs. (10.4), (10.5). Whereas the interfaces with the grid modeled by Eq. (10.22), Eqs. (10.25), (10.26) are different from the one assumed in [25–27].

3.1.2 Modeling and operating constraints

The parameters of the EDM Eqs. (10.19)–(10.34) are collected in the vector $\theta = [\theta_a \ \theta_{ZIP} \ \theta_s]$, where

$$\theta_a = [T'_{da} \ X_a \ X'_a \ H_a \ T_{ma} \ \tau_a \ S_a^{\text{nom}} \ \gamma_a] \tag{10.35}$$

$$\theta_{ZIP} = [P_I \ P_Z \ P_P \ Q_I \ Q_Z \ Q_P] \tag{10.36}$$

$$\theta_s = [T'_{ds} \ E_f \ X_s \ X'_s \ H_s \ T_{ms} \ D \ \tau_s \ S^{\text{nom}}] \tag{10.37}$$

Modeling and operational constraints on the parameters in θ are introduced in the following.

For the asynchronous machine, a first model assumption is $X_a > X'_a$. Under this condition, with $V = 1$ and $\delta_a^0 \neq 0$, E'^0_a results to be lower than 1. Therefore, the following operating condition can be assumed:

$$E'^0_{a,\min} \leq E'^0_a < 1, \quad E'^0_{a,\min} < 1 \tag{10.38}$$

As previously introduced and indicated in Fig. 10.2, the asynchronous machine can work as motor (AM) or as generator (AG). For selecting between these two different operating modes, the mechanical torque T_{ma} is set alternatively using the following two constraints:

$$\text{AM: } 0 \leq T_{ma} \leq 1 \tag{10.39}$$

$$\text{AG:} \ -1 \leq T_{ma} \leq 0 \tag{10.40}$$

It is indeed clear from Eq. (10.20) that a positive T_{ma} is a decelerating (load) torque and, therefore, the steady-state model is obtained with $\delta_a^0 < 0$. Under this condition and recalling that $E_a'^0 < 1$, from Eqs. (10.27), (10.28), it follows that (with $V = 1$) $P_a > 0$ and $Q_a >> 0$. Differently, a negative T_{ma} is an accelerating (generation) torque and, therefore, the steady-state model is obtained with $\delta_a^0 > 0$. In this case, from Eqs. (10.27), (10.28), it follows that (always with $V = 1$) $P_a < 0$ and $Q_a > 0$. Therefore, the following operating conditions can be assumed:

$$\text{AM:} \ \delta_{a,\min}^0 \leq \delta_a^0 < 0, \ \ \delta_{a,\min}^0 \in (-90, 0)\text{degrees} \tag{10.41}$$

$$\text{AG:} \ 0 < \delta_a^0 \leq \delta_{a,\max}^0, \ \ \delta_{a,\max}^0 \in (0, 90)\text{degrees} \tag{10.42}$$

Finally, the following set of constraints on the model parameters in θ_a can be defined:

$$0 < T'_{da} \leq T'_{da,\max} \tag{10.43}$$

$$0 < X'_a < X_a \tag{10.44}$$

$$0 < H_a \leq H_{a,\max} \tag{10.45}$$

$$0 \leq S_a^{\text{nom}} \tag{10.46}$$

$$\gamma_{a,\min} \leq \gamma_a \leq \gamma_{a,\max} \tag{10.47}$$

$$0 \leq \tau_a \leq \tau_{a,\max} \tag{10.48}$$

$$\text{AM:} \begin{cases} 0 \leq T_{ma} \leq 1 \\ 0 \leq X_a\left(\cos\left(\delta_{a,\min}^0\right) - E_{a,\min}'^0\right) - X'_a \cos\left(\delta_{a,\min}^0\right) \\ 0 \leq -X'_a T_{ma} - E_{a,\min}'^0 \sin\left(\delta_{a,\min}^0\right) \end{cases} \tag{10.49}$$

$$\text{AG:} \begin{cases} -1 \leq T_{ma} \leq 0 \\ 0 \leq X_a\left(\cos\left(\delta_{a,\max}^0\right) - E_{a,\min}'^0\right) - X'_a \cos\left(\delta_{a,\max}^0\right) \\ 0 \leq T_{ma} X'_a + E_{a,\min}'^0 \sin\left(\delta_{a,\max}^0\right) \end{cases} \tag{10.50}$$

Constraints (10.43)–(10.48) are modeling assumptions. All parameters, excepting for T_{ma}, are indeed forced to be positive, since they are defined as positive quantities. Moreover, T'_{da} in Eq. (10.43), τ_a in Eq. (10.48), and H_a in Eq. (10.45) are limited by the maximum values $T'_{da,\max}$, $\tau_{a,\max}$, and $H_{a,\max}$, respectively. Constraints (10.49), (10.50) are obtained by combining the operating constraints (10.38), (10.41), (10.42) with the steady-state equations of system (10.19)–(10.21).

Notice that constraints (10.43)–(10.50) are all linear with respect to model parameters in θ_a, excepting for the last ones in Eqs. (10.49), (10.50). However, they can be easily transformed into linear inequalities by replacing X_a and X'_a with the auxiliary parameters $\alpha_a = X_a/X'_a$ and $\alpha'_a = 1/X'_a$.

The unique requirement assumed for the ZIP model is that all load components are positive, that is,

$$P_{Z/I/P} \geq 0, \quad Q_{Z/I/P} \geq 0 \tag{10.51}$$

For the SG, the excitation voltage should be large enough to obtain that $E'^0_s > 1$. For obtaining that $P_s > 0$, it is required that $\delta_s > 0$. Therefore, the following operating condition can be assumed:

$$0 < \delta^0_s \leq \delta^0_{s,\max}, \quad \delta^0_{s\max} \in (0, 90)\text{degrees} \tag{10.52}$$

Assuming $V_s = 1$, $X_s > X'_s$, from Eq. (10.30) it follows that $Q_s > 0$ if $E'^0_s \geq 1/\cos(\delta^0_{s,\max})$, leading to the operating constraint

$$E'^0_{s,\min} \leq E'^0_s \leq E'^0_{s,\max}, \quad E'^0_{s,\max} > 1 \tag{10.53}$$

where $E'^0_{s,\min}$ must satisfy $E'^0_{s,\min} \geq 1/\cos(\delta^0_{s,\max}) > 1$. Finally, the following set of constraints on the model parameters in θ_s can be defined:

$$0 < T'_{ds} \leq T'_{ds,\max} \tag{10.54}$$

$$0 < X'_s < X_s \tag{10.55}$$

$$0 < H_s \leq H_{s,\max} \tag{10.56}$$

$$0 \leq S^{\text{nom}}_s \tag{10.57}$$

$$0 < D \tag{10.58}$$

$$0 < \tau_s \leq \tau_{s,\max} \tag{10.59}$$

$$0 \leq T_{ms} \leq 1 \tag{10.60}$$

$$0 < X_s(\cos(\delta^0_{s,\max}) - E'^0_{s,\min}) + X'_s E_f - X'_s \cos(\delta^0_{s,\max}) \tag{10.61}$$

$$0 \leq X_s(E'^0_{s,\max} - 1) - X'_s E_f + X'_s \tag{10.62}$$

$$0 \leq -X'_s T_{ms} + E'^0_{s,\max} \sin(\delta^0_{s,\max}) \tag{10.63}$$

Constraints (10.54)–(10.60) are modeling assumptions. All parameters are indeed forced to be positive, since they are defined as positive quantities. Moreover, as for the asynchronous machine, T'_{sa} in Eq. (10.43), τ_s in Eq. (10.48), and H_s in Eq. (10.45) are limited by the maximum values $T'_{sa,\max}$, $\tau_{s,\max}$, and $H_{s,\max}$, respectively. Constraints (10.61)–(10.63) are obtained by combining the operating constraints (10.52), (10.53) with the steady-state equations of system (10.23)–(10.26).

Notice that constraints (10.54)–(10.60) are all linear with respect to the model parameters in θ_s. Whereas Eqs. (10.61)–(10.63) are quadratic but they can be easily transformed into linear inequalities by replacing X_s and X'_s with the auxiliary parameters $\alpha_s = X_s/X'_s$ and $\alpha'_s = 1/X'_s$.

3.1.3 EDM

To summarize, the EDM is defined by the state and output Eqs. (10.19)–(10.34) and constraints (10.43)–(10.48), (10.49) or (10.50), (10.51) and (10.54)–(10.63). The directly connected or converter-connected model versions are obtained by selecting the suitable state equations in Eq. (10.22) for the asynchronous machine and in Eqs. (10.25), (10.26) for the SG.

3.2 Identification procedure

The objective of the identification procedure is to set the values of the parameters in vector θ that defines the EDM. Such values should be defined in order to make the EDM reproduce the profiles of active and reactive powers P and Q absorbed by the modeled ADN.

The proposed identification procedure consists of two phases: *definition* and *estimation*. In the definition phase (Section 3.2.1), a starting version of the EDM is defined based on the available information on the ADN to be modeled. In the estimation phase (Section 3.2.2), the parameters values are tuned using a training set.

3.2.1 Model definition

The initial version of the EDM is defined based on basic information about the composition of the ADN, which are normally available to the DSO [1]. The required data are

- maximum amount of dynamic load S_{AM}^{nom} [VA];
- maximum amount of asynchronous generation S_{AG}^{nom} [VA];
- maximum amount of converter-connected synchronous generation S_{cSG}^{nom} [VA]; and
- maximum amount of directly connected synchronous generation S_{dSG}^{nom} [VA].

Further optional useful information are

- average percentage of constant impedance, constant current, and constant power load components, denoted as p.u. quantities by $\overline{P}_{Z/I/P}$ and $\overline{Q}_{Z/I/P}$;
- mechanical inertia time constants $H_{a,i}$ and $H_{s,i}$ and associated nominal apparent powers $S_{a,i}^{\text{nom}}$ and $S_{s,i}^{\text{nom}}$ of the larger generation plants within the ADN; and
- maximum amount of static active and reactive power loads P_{ZIP}^{nom} and Q_{ZIP}^{nom}.

The initial version of the EDM is thus defined using the following rules.

(a) If $S_{AM}^{\text{nom}} \geq S_{AG}^{\text{nom}}$ and AGs absorb reactive power, select the motor mode of the asynchronous machine model, and set the model parameters $S_a^{\text{nom}} = S_{AM}^{\text{nom}}$ and $\gamma_a = 1 + S_{AG}^{\text{nom}}/S_{AM}^{\text{nom}}$, with $\gamma_{a,\min} = 1$ and $\gamma_{a,\max} = 2$.

(b) If $S_{AM}^{\text{nom}} \geq S_{AG}^{\text{nom}}$ and AGs deliver reactive power, select the motor mode of the asynchronous machine model, and set the model parameters $S_a^{\text{nom}} = S_{AM}^{\text{nom}}$ and $\gamma_a = 1 - S_{AG}^{\text{nom}}/S_{AM}^{\text{nom}}$, with $\gamma_{a,\min} = 1$ and $\gamma_{a,\max} = 1$.

(c) If $S_{AG}^{\text{nom}} > S_{AM}^{\text{nom}}$ and AGs absorb reactive power, select the generator mode of the asynchronous machine model, and set the model parameters $S_a^{\text{nom}} = S_{AG}^{\text{nom}}$ and $\gamma_a = 1 + S_{AM}^{\text{nom}}/S_{AG}^{\text{nom}}$, with $\gamma_{a,\min} = 1$ and $\gamma_{a,\max} = 2$.

(d) If $S_{AG}^{\text{nom}} > S_{AM}^{\text{nom}}$ and AGs deliver reactive power, select the generator mode of the asynchronous machine model, and set the model parameters $S_a^{\text{nom}} = S_{AG}^{\text{nom}}$ and $\gamma_a = -1 + S_{AM}^{\text{nom}}/S_{AG}^{\text{nom}}$, with $\gamma_{a,\min} = -1$ and $\gamma_{a,\max} = 1$.

(e) If $S_{cSG}^{\text{nom}} \geq S_{dSG}^{\text{nom}}$, select the converter-connected model of the SG.

(f) If $S_{dSG}^{\text{nom}} > S_{cSG}^{\text{nom}}$, select the directly connected model of the SG.

(g) Set the model parameter $S_s^{\text{nom}} = S_{cSG}^{\text{nom}} + S_{dSG}^{\text{nom}}$.

(h) If no additive information is available, set $P_Z = P_I = P_P = P_{ZIP}^0/3$ and $Q_Z = Q_I = Q_P = Q_{ZIP}^0/3$, where P_{ZIP}^0 and Q_{ZIP}^0 is the load working point.

(i) If the average percentage of constant impedance, constant current, and constant power loads is known, set $p_{Z/I/P}$ and $q_{Z/I/P}$, set $P_{Z/I/P} = \overline{p}_{Z/I/P} P_{ZIP}^0$ and $Q_{Z/I/P} = \overline{q}_{Z/I/P} Q_{ZIP}^0$.

(j) If the mechanical inertia time constants and the associated nominal apparent powers of the larger generation plants are available, set $H_a = \sum_i H_{a,i} S_{a,i}^{\text{nom}} / \sum_i S_{a,i}^{\text{nom}}$ (in the generator mode case) and $H_s = \sum_i H_{s,i} S_{s,i}^{\text{nom}} / \sum_i S_{s,i}^{\text{nom}}$.

All other parameters which cannot be set following rules (a)–(j) can be initialized using typical standard values (as provided, e.g., in [34, 35]). They are obviously required to satisfy constraints (10.43)–(10.48), (10.49) or (10.50), (10.51), and (10.54)–(10.63).

If additional information is available, it is possible to include additional constraints on the model parameters. For example, if the maximal amount of static active and reactive power loads P_{ZIP}^{nom} and Q_{ZIP}^{nom} are known, the following constraints can be added:

$$P_Z + P_I + P_P \leq P_{ZIP}^{\text{nom}} \tag{10.64}$$

$$Q_Z + Q_I + Q_P \leq P_{ZIP}^{\text{nom}} \tag{10.65}$$

Finally, a percentage confidence interval can be associated to each parameter value. Given a parameter p and its value $\tilde{p}$ defined in the definition phase, the confidence interval is defined as follows:

$$\tilde{p}\left(1 - \frac{\Delta^{\%} p}{100}\right) \leq p \leq \tilde{p}\left(1 + \frac{\Delta^{\%} p}{100}\right) \tag{10.66}$$

where $\Delta^{\%}p$ is the percentage confidence interval. These constraints are optional and they can be used when the confidence about any a priori knowledge of the parameter is available.

Notice that rules (a)–(j) are provided as "guideline" for a correct definition of the initial version of the EDM. However, they should not be considered as rigorous formal conditions, since they are based on heuristic modeling assumptions.

For example, rules (a) and (d) follow from the hypothesis that the asynchronous machine is representing the aggregate of dynamical loads (motors) and asynchronous generators in the ADN. As discussed before, when the asynchronous machine works as a motor, it absorbs active and reactive power, whereas, when it works as generator, it exports active power and absorbs reactive power. However, thanks to inverter interfaces, AGs can deliver reactive power. This is, for example, the case of Wind Doubly Fed Induction Generators (DFIGs). However, the representation of the overall inverter-generator dynamics will result in the augmentation of the model complexity and of the number of parameters. For these reasons, for the purpose of this chapter, in rules (a) and (d), the sign of the reactive power exchange is determined using a simple scaling factor γ_a.

Based on this modeling assumption, in rules (a)–(d), the nominal apparent power S_a^{nom} is set equal to the maximum within the two aggregates S_{AM}^{nom} and S_{AG}^{nom}, whereas the reactive power balance is corrected using the scaling factor γ_a in order to obtain the sum of the two contributions.

3.2.2 Model estimation

Assuming to have a training set composed by a set of N measurements of the inputs V and ω and of the outputs P and Q, the model parameters in vector θ are tuned by solving the following constrained nonlinear optimization problem:

$$\begin{aligned} &\min_{\theta} J_\theta \ \ s.t. \text{ Eqs. } (10.43)-(10.48),\ (10.49) \text{ or } (10.50),\ (10.51) \\ &\quad \text{and } (10.54)-(10.63) \end{aligned} \tag{10.67}$$

where J_θ is defined as in Eq. (10.18). Optional constraints, such as Eqs. (10.64)–(10.66) can be added to the optimization problem.

It is important to remark that constraints have a crucial role. Indeed, if no constraint is imposed, parameters can lose their physical-based meaning, which is a key point for a gray-box identification approach. Moreover, without operational constraints, the estimated combination of parameters can define an unstable model.

3.3 Extension to microgrids

The method earlier introduced has been developed for ADNs. In this section, the modification adopted in [31] to migrate the approach from ADNs to microgrids are presented. The general frame of the microgrid EDM, shown in Fig. 10.3, includes a ZIP load, an SG, and a static source, which covers inverter-driven generation, and inverter-interfaced loads. Differently from models (10.19)–(10.34), there is no asynchronous

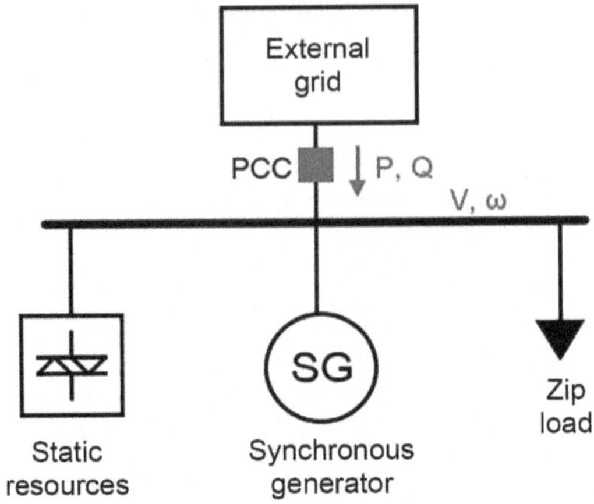

Fig. 10.3 EDM scheme for microgrids.

machine, since it is not installed in the experimental test site considered in this work; therefore, it has not been included in this specific application.

The SG is modeled by Eqs. (10.23)–(10.26), (10.29), (10.30) in the directly connected version. The ZIP load is modeled by Eqs. (10.31), (10.32). The static source active and reactive power exchanges are modeled as follows:

$$P_{static} = R_P V + D_P(\omega - 1) \tag{10.68}$$

$$Q_{static} = R_Q V + D_Q(\omega - 1) \tag{10.69}$$

where parameters R_P, D_P, R_Q, and D_Q, are defined to catch the most significant behavior of static sources, equipped with fast droop-based controllers [31, 36–38].

The active and reactive power exchange at the PCC is obtained by the sum of the three contributions of SG, ZIP load, and static source:

$$P = P_s + P_{ZIP} + P_{static} \tag{10.70}$$

$$Q = Q_s + Q_{ZIP} + Q_{static} \tag{10.71}$$

The identification procedure adopted for microgrids is the same introduced for ADNs in Section 3.2, obviously without considering the constraints for the asynchronous machine, which is not included in the model.

4 Simulation and experimental results

4.1 Linear EDM simulation results

In order to study the effectiveness of the method proposed in Section 2, a detailed ADN model has been implemented on the MATLAB/Simulink Power System blockset environment. The test network has been chosen starting from the Medium

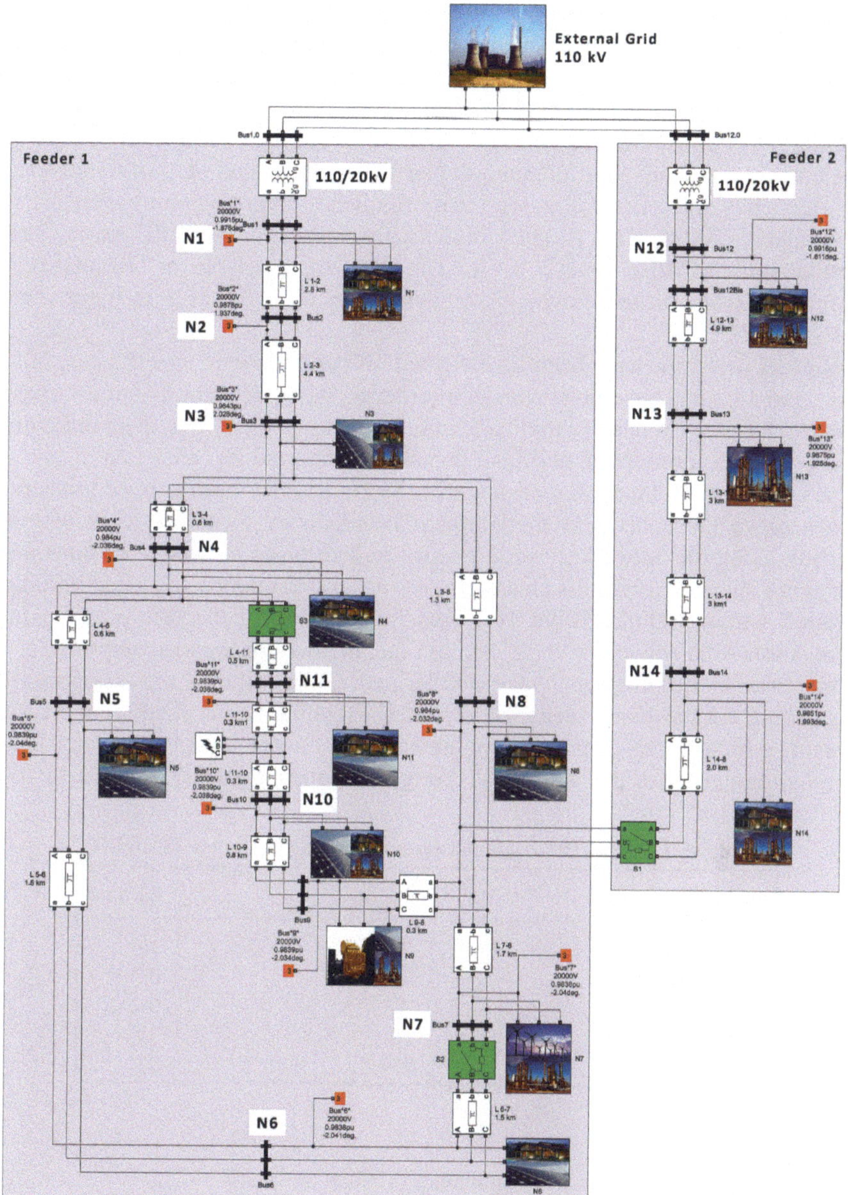

Fig. 10.4 Cigré Task Force C6.04.02 Benchmark Medium Voltage Distribution Network: Simulink implementation diagram.

Voltage (MV) Distribution Network Benchmark developed by Cigré Task Force C6.04.02 [39]. As reported in Fig. 10.4, which directly represents the Simulink diagram, it is composed of 14 secondary substations and an equivalent HV network, connected through 17 lines. Four main feeders are supplied by two 110/20 kV transformers (rated 25 MVA each).

All parameters are specified in [39]. The network has been modeled taking into consideration the dynamic behavior of the different components. Loads are assumed with variable profiles set according to the residential and commercial load specifications. DERs data come from Cigré task force case study, including PV panels, wind turbines (WTs), and combined heat and power (CHP) units. The HV network has been modeled using an equivalent plant equipped with a steam turbine prime mover, primary frequency droop controller, secondary frequency controller, and automatic voltage regulator (AVR). The adopted models are based on the IEEE steam turbine governor and the IEEE type 2 AVR. The data are available in [40]. DERs are implemented using different dynamic models, whereas loads are implemented based on classical ZIP [34].

Different scenarios have been investigated, in particular two specific case studies (2 a.m. and 12 a.m.) have been chosen in order to show different dynamic behavior. These two cases are related to night and noon conditions; therefore, there are two different generation production and load demand as reported in Table 10.1.

The identification procedure presented in Section 2.2 has been applied to Feeder 1.

To achieve the equivalent modeling, it is necessary to define a set of events occurring along the network, outside Feeder 1. Two types of events are considered: three-phase short-circuit faults cleared in 500 ms and temporary load step variations, recovered within 500 ms. Table 10.2 lists the location of the five possible faults events. Table 10.3 shows the locations and the amounts of the load steps.

Two different training sets have been defined. One for the 12 a.m. scenario (high load) and one for the 2 a.m. scenario (low load). Both the training sets were measured at Bus 1, which is the connection point between Feeder 1 and the external grid. The measurements are the voltage V, the grid frequency f [Hz] (equivalent to the

Table 10.1 Linear EDM simulation scenarios: Operating points [MW].

Scenario	2 a.m.	12 a.m.
Wind turbine	0.3	1.3
PV production	0	0.5
Diesel genset	0.3	0.2
Aggregated load	10	33
Import	9.4	31

Table 10.2 Linear EDM simulations: Faults locations.

ID	Location
F1	Bus 13
F2	Bus 12
F3	Bus 14
F4	In the middle between bus 13 and bus 14
F5	2 km after bus 14

Table 10.3 Linear EDM simulations: Load steps' locations and magnitudes.

ID	Location	Active power [MW]	Reactive power [MVAr]
E1	Bus 12	0.75	0.25
E2	External grid	1	0.5
E3	Bus 12	1.2	0.24
E4	External grid	10	5
E5	External grid	20	10
E6	External grid	30	15
E7	External grid	40	20
E8	External grid	50	25

angular velocity ω), the active power P, and the reactive power Q absorbed by Feeder 1. All measurements were collected with a sampling time $\delta_s = 10^{-2}$ s.

Each training set lasts 90 s during which a sequence of 10 events between faults and load variations have been simulated. Each event occurred after 9 s from the previous one. In both the scenarios, the event sequence was F1–F3–F5–E1–E2–E3–E4–E5–E6–E8.

The two training sets have been processed using the identification algorithm described in Section 2.2. In this way, two different realizations of the equivalent model (10.1)–(10.17) were identified: the 12 a.m. EDM and the 2 a.m. EDM.

Fig. 10.5 shows the result of the identification procedure for three faults and four load step variations in the 12 a.m. scenario (a portion of the whole training set). In

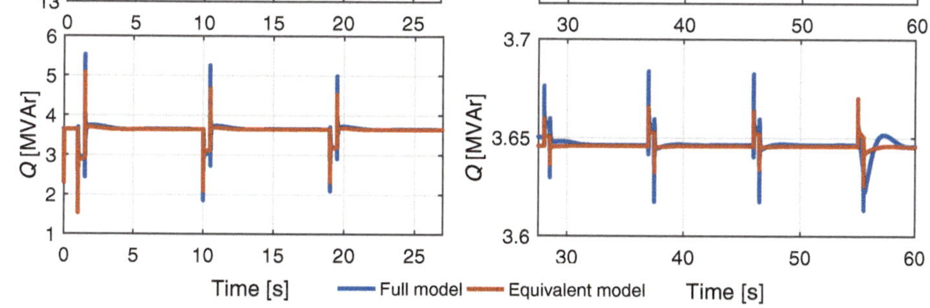

Fig. 10.5 Linear EDM simulations. Identification results: active and reactive power absorbed by Feeder 1 in the 12 a.m. scenario. *Left*: faults F1–F3–F5. *Right*: load steps E1–E2–E3–E4.

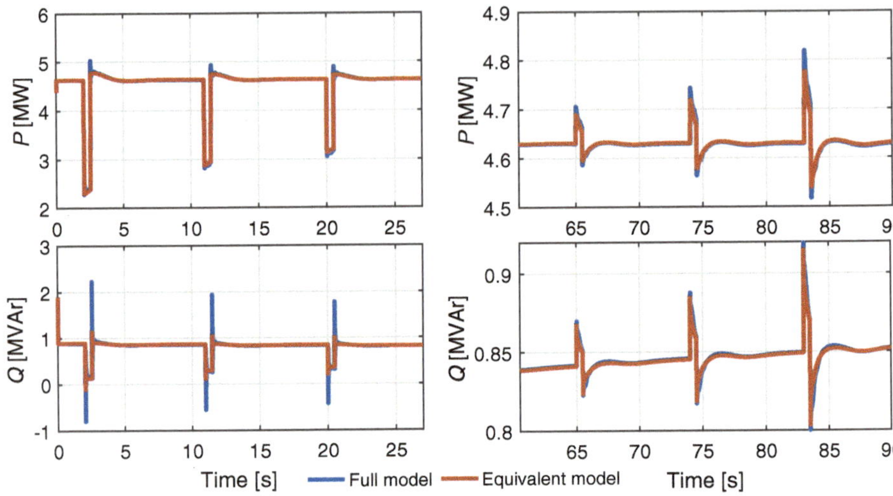

Fig. 10.6 Linear EDM simulations. Identification results: active and reactive power absorbed by Feeder 1 in the 2 a.m. scenario. *Left*: faults F1–F3–F5. *Right*: load steps E5–E6–E8.

particular, the active and reactive power absorbed by Feeder 1 simulated by the full Simulink model is compared with the outputs of the 12 a.m. equivalent model, given the measured voltage V and grid frequency ω as input signals. Fig. 10.6 shows the identification results obtained for three faults and three load steps in the 2 a.m. scenario (again a portion of the whole training set). In both cases, the identified models succeed in reproducing the power exchange profiles of Feeder 1.

To get a quantitative evaluation of the identification procedure, the root mean square errors (RMSEs) between the trajectories of active and reactive powers simulated with the full Simulink model and reproduced by the EDM using the two relevant training sets have been computed. The RMSEs are defined as

$$\sigma_P = \sqrt{\frac{1}{N}\sum_{k=0}^{N-1}(P_k - \hat{P}_{k,\theta^*})^2}, \quad \sigma_Q = \sqrt{\frac{1}{N}\sum_{k=0}^{N-1}(Q_k - \hat{Q}_{k,\theta^*})^2} \tag{10.72}$$

where P_k and Q_k are the measured active and reactive powers, $\hat{P}_{k,\theta^*}$ and $\hat{Q}_{k,\theta^*}$ are the ones reproduced by the EDM given θ^*, which is the identified set of parameters. The resulting RMSEs are reported in Table 10.4.

To evaluate the effectiveness of the approach, the 12 a.m. and 2 a.m. equivalent models have been validated by replacing the detailed model of Feeder 1 in the

Table 10.4 Linear EDM simulations: Identification RMSEs on the training sets.

	12 a.m. Model	**2 a.m. Model**
σ_P [MW]	0.035	0.012
σ_Q [MVAr]	0.033	0.032

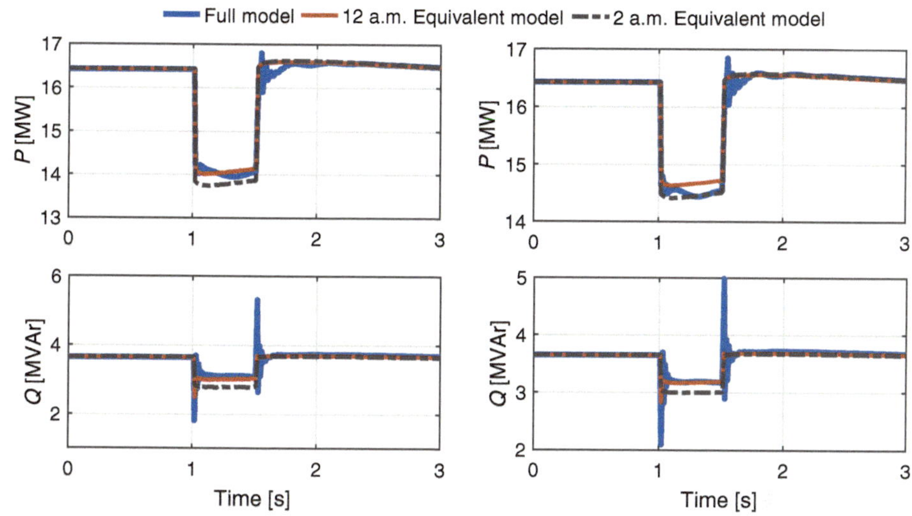

Fig. 10.7 Linear EDM simulations. Validation results: Active and reactive power absorbed by Feeder 1 in the 12 a.m. scenario. *Left*: fault F2. *Right*: fault F4.

Simulink implementation with a dynamic three-phase load driven by a state-space system block, which implements the equivalent models.

The active and reactive power profiles obtained with the full and the equivalent models by simulating faults F2 and F4 and the load step E7, all belonging not to the training set sequences, are therefore compared. Figs. 10.7 and 10.8 show the profiles obtained for the two faults in the 12 a.m. and 2 a.m. scenarios, respectively. In order to evaluate the robustness and the scalability of the equivalent models, both

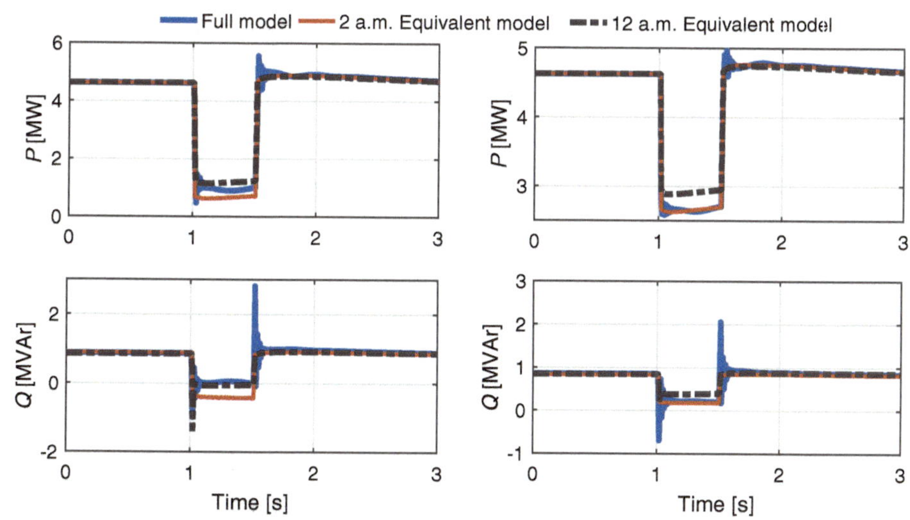

Fig. 10.8 Linear EDM simulations. Validation results: Active and reactive power absorbed by Feeder 1 in the 2 a.m. scenario. *Left*: fault F2. *Right*: fault F4.

Table 10.5 Validation RMSEs.

Scenario	**12 a.m.**	**2 a.m.**	**12 a.m.**	**2 a.m.**
Equivalent model	**12 a.m.**	**2 a.m.**	**2 a.m.**	**12 a.m.**
σ_P [MW]	0.046	0.064	0.055	0.064
σ_Q [MVAr]	0.059	0.086	0.085	0.065

the two identified model realizations have been applied within the two different scenarios.

In both cases, the modeling accuracy is comparable, even if the coherent models generally produce more precise power profiles. This highlights the potential scalability of the equivalent models.

In order to quantify this validation analysis, the RMSEs between the active and reactive power profiles obtained within the two scenarios using the detailed Simulink model and the two equivalent models have been computed and reported in Table 10.5.

It is worth noticing first that the obtained RMSEs are all comparable with a magnitude order of tens of kW and kVAr. This means that, in general, the equivalent modeling provides a good resolution in representing the ADN dynamics. As expected, using the coherent equivalent models returns a better resolution. However, such a resolution is not a severely deterioration when a different equivalent model is applied.

4.2 Nonlinear constrained EDM simulation results

4.2.1 Test network model

A benchmark network model has been developed in order to reproduce the layout of an ADN connected to a primary substation (PSUB) of the HV transmission system. Each component has been characterized with the aim of simulating the dynamical behavior of DERs and loads, according to the real characteristics of the components. The network model has been implemented on the DIgSilent PowerFactory [41] simulation platform.

The distribution system layout is an extended version of the Cigré European MV distribution network model [39]. The proposed test system is composed by four feeders, obtained by replicating the original layout of the Cigré model with a custom characterization of the two new feeders. In particular, DERs size and lines length have been updated, while the topological configuration and the secondary substations (SSUBs) layout have maintained the original structure. Fig. 10.9 shows a geographical representation of the test network. The figure reports the scheme for four types of SSUBs: Load, PV, Wind, and Diesel. Notice that DERs are all localized in Feeders 1 and 2, which are both connected to the equivalent HV transmission network through transformer TR_0_1.

The external grid represents the equivalent HV transmission network. It is modeled using a 1000 MVA synchronous machine driven by a steam turbine, with a 400 MW ($\cos\phi = 0.95$) equivalent load connected at the HV side. A step-up transformer

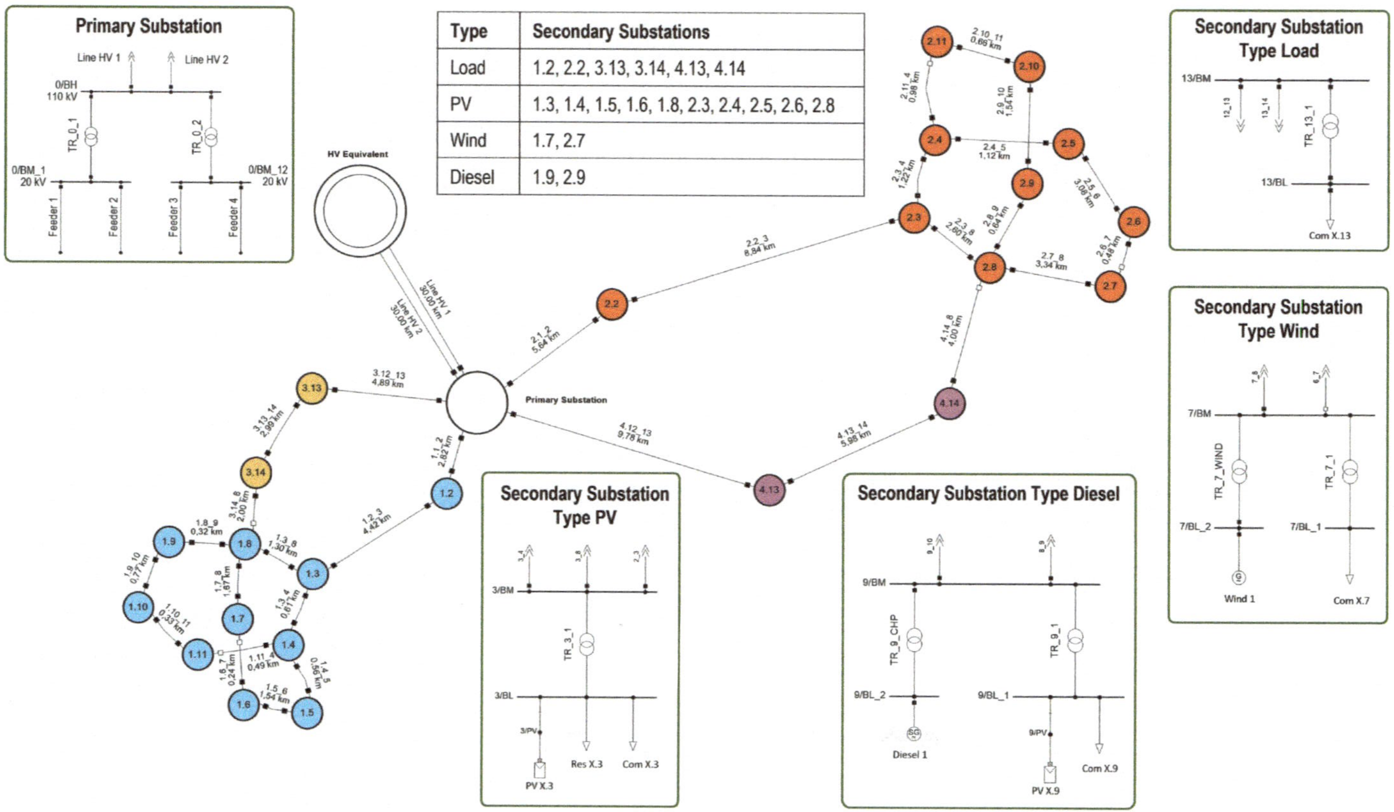

Type	Secondary Substations
Load	1.2, 2.2, 3.13, 3.14, 4.13, 4.14
PV	1.3, 1.4, 1.5, 1.6, 1.8, 2.3, 2.4, 2.5, 2.6, 2.8
Wind	1.7, 2.7
Diesel	1.9, 2.9

Fig. 10.9 Nonlinear constrained EDM simulations. Geographical representation of the test network with the detailed scheme of the primary substation (PSUB) and secondary substation (SSUB) types: Load, PV, Wind, and Diesel.

Table 10.6 Nonlinear constrained EDM simulations: Load models' parameters.

Load type	e_{aP}	e_{aQ}	k_{pf} [s]	k_{qf} [s]	t_d [s]
Residential	1.50	3.20	1.00	−1.50	0.01
Commercial	0.99	3.50	1.50	−1.10	0.01

connects the MV plant busbar with the HV node, which supplies the primary substation of the distribution network through transmission lines. The TGOV1 governor and the IEEE type 2 AVR models [41] complete the external grid model.

In all, 28 aggregated load models have been distributed among the network busses. They are divided into two classes: *residential* and *commercial*, with a constant power factor $\cos\phi$ equal to 0.97 and 0.85, respectively. All loads have been implemented with the following frequency-voltage-dependent model [41]:

$$P_\ell = P_\ell^0 V^{e_{aP}} \left(1 + \frac{k_{pf}}{1 + s t_d} \Delta f\right), \quad Q_\ell = Q_\ell^0 V^{e_{aQ}} \left(1 + \frac{k_{qf}}{1 + s t_d} \Delta f\right) \tag{10.73}$$

where P_ℓ and Q_ℓ are the active and reactive power absorbed by the load, P_ℓ^0 and Q_ℓ^0 are the nominal values, and Δf is the p.u. frequency variation from the nominal value f^{nom}. The implemented load parameter values are reported in Table 10.6.

The set of implemented DERs includes PV plants, WTs, and CHP generators. The PV plant model includes an inverter controlled through direct and quadrature current signals provided by a Maximum Power Point Tracker (MPPT). The solar radiation drives each PV generator, connected to the DC side of the inverter with a shunt capacitor. Controllers regulate the DC side voltage magnitude and the AC side power factor, set to a constant value, according to the majority of European grid codes.

The WT generator is modeled as a DFIG. It includes the asynchronous generator model driven by the mechanical model of the WT, which converts the wind speed into mechanical torque. The Maximum Power Tracker generates the reference signal for the pitch control, while a power controller provides the reference signal to the rotor converter interface [40].

The CHP generator is modeled as a synchronous machine driven by a Diesel prime mover. The DEGOV1 governor and the EXAC1A AVR models have been implemented. The dynamics behavior of the generating units corresponds to a G2 performance class [42]. Frequency droop is set equal to 4%.

The list of DERs installed in the test network is provided in Table 10.7.

4.2.2 Scenarios

The identification procedure introduced in Section 3.2 is applied to the benchmark network described in Section 4.2.1. The considered ADN portion is composed by Feeders 1 and 2, which are connected to the main grid by transformer TR_0_1. Nine

Table 10.7 Nonlinear constrained EDM simulations.

Name	Feeder	SSUB	Type	S^{nom} [kVA]
PV 1.9	1	1.9	PV	60
PV 1.8	1	1.8	PV	60
PV 1.6	1	1.6	PV	60
PV 1.5	1	1.5	PV	60
PV 1.4	1	1.4	PV	40
PV 1.3	1	1.3	PV	40
PV 1.11	1	1.11	PV	20
PV 1.10	1	1.10	PV	80
PV 2.9	2	2.9	PV	60
PV 2.8	2	2.8	PV	60
PV 2.6	2	2.6	PV	60
PV 2.5	2	2.5	PV	60
PV 2.4	2	2.4	PV	40
PV 2.3	2	2.3	PV	40
PV 2.11	2	2.11	PV	20
PV 2.10	2	2.10	PV	80
Wind 1	1	1.7	DFIG	1500
Wind 2	2	2.7	DFIG	1500
CHP diesel 1	1	1.9	Diesel CHP	450
CHP diesel 2	2	2.9	Diesel CHP	1900

Benchmark network: DER units locations and nominal powers.

different simulation scenarios are considered. Table 10.8 reports the working points, in terms of apparent power S and power factor ($\cos\phi$), of the aggregates of loads and DERs. The table also shows the working points of the three aggregates of DERs (PV, Wind, and Diesel) and the relevant percentages with respect to the total generation, indicated with $S_{\%}$. In all the scenarios, $\cos\phi = 1$ for the PV aggregate, $\cos\phi = 0.95$ for the Wind aggregate, and $\cos\phi = 0.85$ for the Diesel aggregate.

In each scenario, a different set of DER units is in service, yielding to different values of aggregated nominal powers, which are listed in Table 10.9. The relevant percentages $S_{\%}^{\mathrm{nom}}$ with respect to the entire generation aggregate are also reported. From a dynamical point of view, this information is of particular interest. Indeed, given a DER aggregate, its weight in the dynamical response of the overall portion of ADN to any disturbance is obviously related to its nominal power. Table 10.10 lists the state of service of the AGs and the SGs included in the ADN.

Among the scenarios, four different DER settings can be distinguished: Scenarios 1 and 2, with prevailing wind generation; Scenarios 3 and 4, with prevailing PV generation; Scenarios 5 and 6, with prevailing Diesel generation; and Scenarios 7–9, with a balanced mix of generation, with a slight prevalence of Wind and Diesel. For Scenarios 1–8, two levels of load demand are considered: the high load cases, with $S = 6.28$ MVA and the low load cases, with $S \in [3.55, 4.26]$ MVA. Notice that in all

Table 10.8 Nonlinear constrained EDM simulation scenarios: Working points.

	Import		Load		Generation		PV		Wind		Diesel	
Sc.	S **[MVA]**	$\cos\phi$	S	$\cos\phi$	S	$\cos\phi$	S **[MVA]**	$S_{\%}$	S **[MVA]**	$S_{\%}$	S **[MVA]**	$S_{\%}$
1	3.27	0.90	6.28	0.95	3.05	0.98	0.32	10	2.46	80	0.29	10
2	1.17	0.78	3.55	0.95	2.48	0.98	0.32	13	2.10	84	0.07	3
3	5.48	0.93	6.28	0.95	0.83	1.00	0.76	90	0	0	0.08	10
4	2.06	0.90	2.84	0.95	0.83	1.00	0.76	90	0	0	0.08	10
5	4.34	0.97	6.28	0.95	1.97	0.88	0.26	13	0	0	1.74	87
6	2.34	0.98	4.26	0.95	1.97	0.88	0.26	13	0	0	1.74	87
7	3.51	0.93	6.28	0.95	2.78	0.96	0.75	26	1.13	40	0.97	34
8	0.87	0.68	3.55	0.95	2.82	0.98	0.75	26	1.13	40	0.96	34
9	−1.01	0.96	4.26	0.95	5.28	0.95	0.75	14	2.70	50	1.93	36

Table 10.9 Nonlinear constrained EDM simulation scenarios: Nominal apparent power of the aggregated DER units in service.

	Generation	PV		Wind		Diesel	
Sc.	S^{nom} [MVA]	S^{nom} [MVA]	$S^{nom}_{\%}$	S^{nom} [MVA]	$S^{nom}_{\%}$	S^{nom} [MVA]	$S^{nom}_{\%}$
1	3.85	0.40	8	3.00	80	0.45	12
2	3.85	0.40	8	3.00	80	0.45	12
3	1.29	0.84	62	0.00	0	0.45	38
4	1.29	0.84	62	0.00	0	0.45	38
5	2.67	0.32	12	0.00	0	2.35	88
6	2.67	0.32	12	0.00	0	2.35	88
7	6.19	0.84	12	3.00	49	2.35	39
8	6.19	0.84	12	3.00	49	2.35	39
9	6.19	0.84	12	3.00	49	2.35	39

Table 10.10 Nonlinear constrained EDM simulation scenarios: Operational state of AGs and SGs.

Scenario	1	2	3	4	5	6	7	8	9
Wind 1	1	1	0	0	0	0	1	1	1
Wind 2	1	1	0	0	0	0	1	1	1
CHP diesel 1	1	1	1	1	1	1	1	1	1
CHP diesel 2	0	0	0	0	1	1	1	1	1

1: in service, 0: out of service.

scenarios the ADN is importing power from the external grid, excepting for Scenario 9, where generation is higher than load and thus the ADN is exporting power.

The information listed in Tables 10.8–10.10 can be used to define the Scenario Dataset, composed by the following information:

(1) lists of generators in service (Table 10.10);
(2) working points and nominal powers of the generators aggregates (Tables 10.8 and 10.9); and
(3) working point of the load aggregate (Table 10.8).

4.2.3 Events

In order to identify the EDM, a set of events occurring along the network, outside Feeders 1 and 2, are defined. Two types of events are considered: 0.5 Ohm purely resistive three-phase short-circuit faults and temporary load step variations. Table 10.11 (a) lists the location of the eight possible faults events. Table 10.11

Table 10.11 Nonlinear constrained EDM simulations.

(a)		(b)			
ID	**Location**	**ID**	**Location**	***P* [MW]**	***Q* [Mvar]**
F1	In the middle between SSUB 3.13 and SSUB 3.14	E1	External grid	20.00	6.50
F2	In the middle between SSUB 3.12 and SSUB 3.13	E2	External grid	30.00	9.75
F3	In the middle between SSUB 4.13 and SSUB 4.14	E3	External grid	40.00	13.00
F4	2 km after SSUB 3.14	E4	External grid	32.00	10.00
F5	SSUB 4.13				
F6	SSUB 3.14				
F7	In the middle between SSUB 4.12 and SSUB 4.13				
F8	2 km after SSUB 4.14				

Faults locations (a) and load steps' locations and magnitudes (b).

(b) shows the locations and the amounts of the four load steps. All events are cleared after 500 ms, excepting for fault F2, which is cleared after 300 ms.

4.2.4 Simulation procedure

Fig. 10.10 depicts the adopted simulation scheme, described in the following.

For each scenarios introduced in Section 4.2.2, a *training set* and a *validation set* are generated using the DIgSilent Full Model of the benchmark network described in Section 4.2.1. Both the sets are composed by the following measurements, collected at the MV side of transformer TR_0_1 with a granularity of 0.01 s: the voltage V [p.u.], the grid frequency f [Hz] (equivalent to the electrical angular velocity ω [rad/s]), the active power P [MW], and the reactive power Q [Mvar]. Training sets last 35 s during which the event sequence F1–F2–F3–F4 is simulated. Validation sets last 60 s during which the event sequence E1–E2–E3–F5–F6–F7–F8–E4 is simulated. In both the sets, each event occurs 8 s after the previous one. Notice that, in the sequel, figures have a time scale corresponding to the simulation time or relative to the proposed zoom. For instance, the training set of Scenario 6 is reported in Fig. 10.11.

Given a Scenario j, the basic information about the considered portion of ADN, combined with the Scenario Dataset, are used to define the initial version of the EDM as specified in the rules of Section 3.2.1. After the model definition, the EDM is identified by applying the procedure described in Section 3.2, using the relevant scenario training sets. Identification is not carried out for Scenario 9.

To evaluate the accuracy of the equivalent models, the active and reactive power profiles obtained with the DIgSilent Full Model in the validation sets are compared with those obtained by the *off-line* simulation of the EDMs, given same the voltage

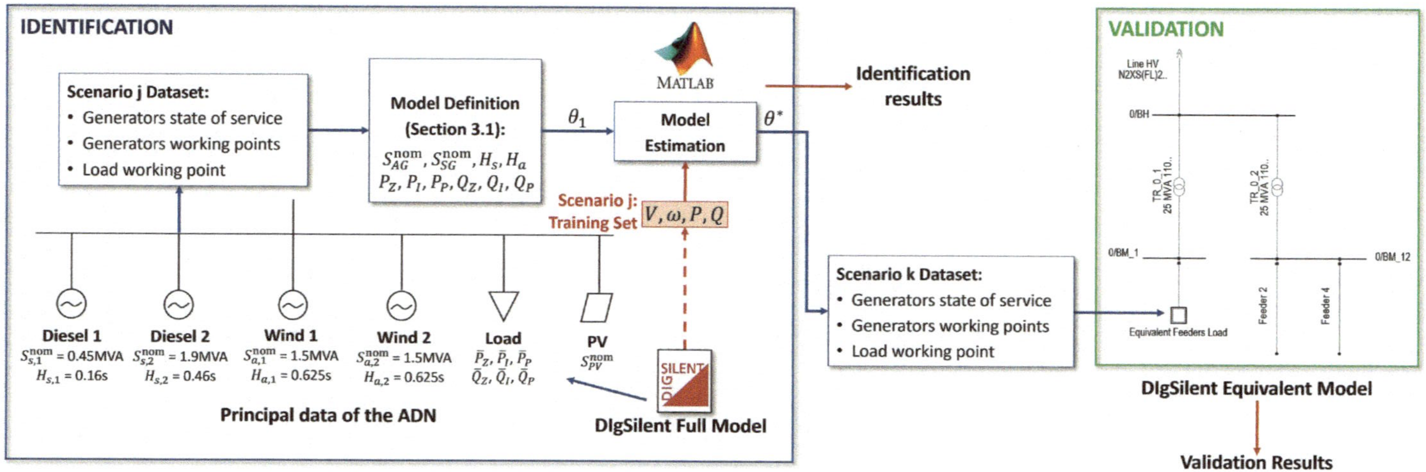

Fig. 10.10 Nonlinear constrained EDM simulations. Simulation scheme.

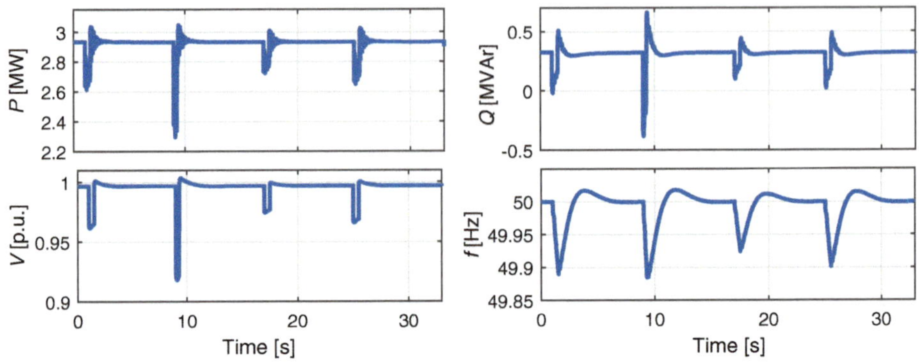

Fig. 10.11 Nonlinear constrained EDM simulations. Training set of Scenario 6.

and frequency input profiles. Such comparisons are referred to as *identification results*.

To finally validate the proposed method, a given EDM is implemented on the DIgSilent platform replacing the full model of Feeders 1 and 2, with a controlled load, which, at each integration step, reproduces the active and reactive power absorption, starting from the frequency and voltage magnitudes measured at the MV side of transformer TR_0_1. Then, the validation sets event sequence is simulated and the resulting power profiles are compared with those in the training sets. These results are referred to as *validation results*.

Identification and validation results are evaluated in terms of RMSEs, defined as in Eq. (10.72). A second performance index is the percentage RMSEs, defined with respect to the nominal power of the overall portion of the ADN generation aggregate $S_{GEN}^{nom} = 6.19$ MVA:

$$\sigma_P^{\%} = \sigma_P / S_{GEN}^{nom} \cdot 100, \quad \sigma_Q^{\%} = \sigma_Q / S_{GEN}^{nom} \cdot 100 \tag{10.74}$$

4.2.5 Models identification

As shown in Fig. 10.10, the basic information assumed to be available are the nominal powers and the mechanical inertia time constants of the AGs and the SGs included in the ADN, and the average percentage of constant impedance, current, and power components of loads.

In particular, the ADN includes two SGs (CHP Diesel 1 and CHP Diesel 2) and two AGs (Wind 1 and Wind 2). The SGs have nominal powers, $S_{s,1}^{nom} = 0.45$ MVA and $S_{s,2}^{nom} = 1.90$ MVA, and inertia time constants $H_{s,1} = 0.16$ s and $H_{s,2} = 0.46$ s, respectively. The AGs are identical with $S_{a,i}^{nom} = 1.50$ MVA and $H_{a,i} = 0.625$ s. The load components are computed by applying a curve fitting procedure between the load model used in the DIgSilent implementation (10.73) and the ZIP models (10.31), (10.32). The resulting values are $\overline{P}_Z = 0.0652$, $\overline{P}_I = 0.8593$, $\overline{P}_P = 0.0744$, $\overline{Q}_Z = 1$, and

Table 10.12 Nonlinear constrained EDM simulations: Scenario parameters.

Parameter	Description
S_a^{nom} [MVA]	Total nominal power of the AGs in service
S_s^{nom} [MVA]	Total nominal power of the SGs in service
T_{ma} [p.u.]	Asynchronous generation working point
T_{ms} [p.u.]	Synchronous generation working point
$P_{Z/I/P}$ [MW]	Load active power ZIP components
$Q_{Z/I/P}$ [Mvar]	Load reactive power ZIP components

$\overline{Q}_I = \overline{Q}_P = 0$. Consider that, in a real scenario, this procedure can be substituted by a preliminary estimate or by an a priori knowledge of the ZIP load components.

The identification procedure is applied to all the scenarios, excepting for Scenario 9. For each one, the basic information data are combined with the Scenario Dataset to set the *EDM scenario parameters* listed in Table 10.12, using the rules introduced in Section 3.2.1. For all scenarios, since there are large asynchronous generators (Wind 1 and Wind 2), the AG mode of the asynchronous machine model is selected. Because these generators are DFIG, which deliver reactive power, rule (d), with $S_{AM}^{\text{nom}} = 0$, is applied. Moreover, since the two diesel generators are directly connected to the grid, rule (f) is applied with $S_{cSG}^{\text{nom}} = 0$.

The rest of the parameters are set using typical standard values (e.g., from [34, 35]). For all scenarios, the functional limits used to define the optimization constraints are those reported in Table 10.13.

Moreover, a percentage confidence constraint (as defined in Eq. 10.66) is set for all model parameters. For the scenario parameters listed in Table 10.12 and for the inertia time constants H_a and H_s, the percentage confidence is $\Delta^{\%}p = 10\%$; for the remaining ones it is $\Delta^{\%}p = 40\%$. These different choices are due to the fact that the scenario parameters and H_s and H_a are computed from the ADN basic information and from the scenario working points.

Remark. The idea is that such data must have a decisive weight in the identification procedure. Indeed, a low percentage confidence will limit the variation of the scenario parameters. In this way, the dynamical behavior of the identified EDM will be strongly related to the values of such parameters returning an effective gray-box representation of the ADN, with the possibility to adapt the model to different configurations.

The identification procedure has been implemented on the MATLAB platform. The constrained optimization problem (10.67) has been solved using the *interior-point* algorithm and percentage stop threshold equal to 1%. The average computational time required to identify an EDM on a PC with an Intel Xeon CPU 3.40 GHz has been equal to 100 s.

By applying the proposed identification procedure to the training sets of Scenarios 1–8, eight different EDMs are obtained. Figs. 10.12 and 10.13 depict the values of the

Table 10.13 Nonlinear constrained EDM simulations: Functional limits.

$T'_{da,\max}$	$H_{a,\max}$	$E'^{0}_{a,\min}$	$\delta_{a,\max}$	$\gamma_{a,\min}$	$\gamma_{a,\max}$	$\tau_{a,\max}$	$T'_{ds,\max}$	$H_{s,\max}$	$E'^{0}_{s,\min}$	$E'^{0}_{s,\max}$	$\delta'^{0}_{s,\max}$
2 s	5 s	0.6	10 degrees	−1	0	1 s	2 s	5 s	1.0353	1.3	15 degrees

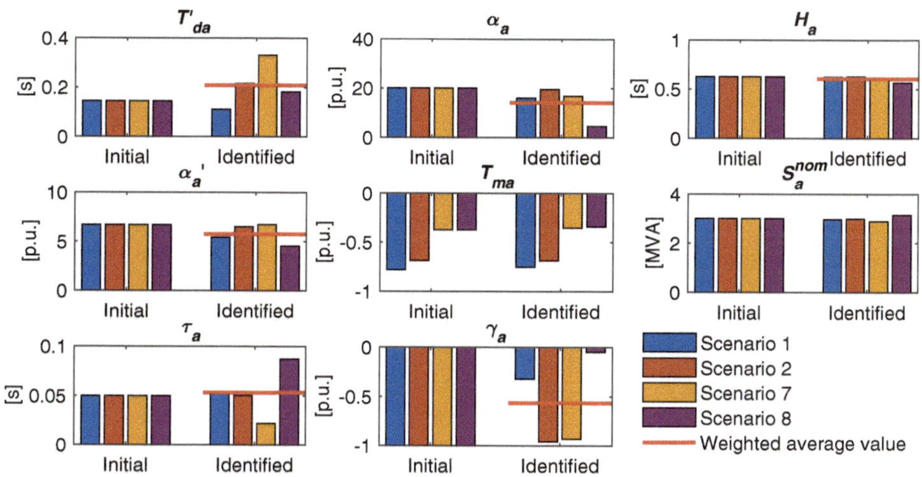

Fig. 10.12 Nonlinear constrained EDM simulations. Initial and identified parameters of the equivalent asynchronous generator.

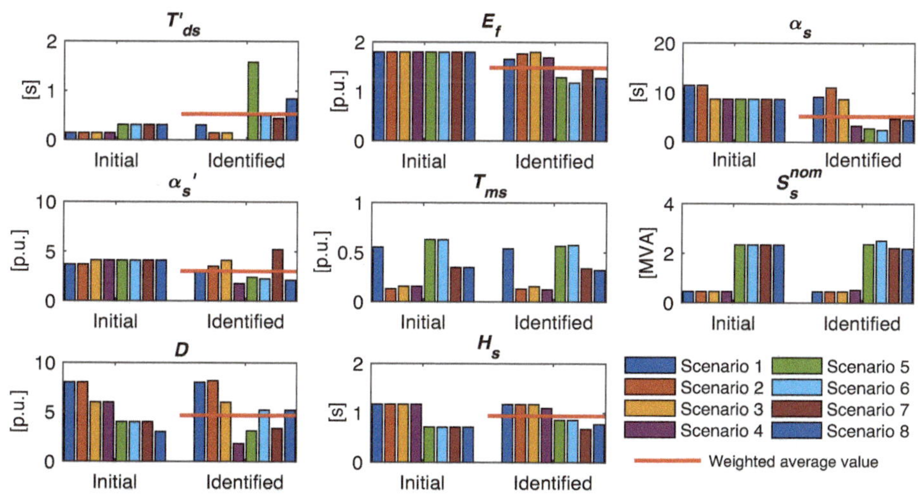

Fig. 10.13 Nonlinear constrained EDM simulations. Initial and identified parameters of the equivalent synchronous generator.

parameters of the equivalent AG and SG, respectively, before and after the identification procedure. In Fig. 10.12, Scenarios 3–5 are not considered since, in such cases, there is no asynchronous generation and, thus, the nominal value S_a^{nom} is forced to be zero (with a 0% confidence percentage).

It can be noted that the initial values are modified by the identification procedure in order to obtain the correct dynamic behavior. As expected, the variation of EDM scenario parameters is limited, thanks to the low confidence percentage (10%) used in the optimization.

The figures report (as solid red lines) the *weighted average values* of the parameters identified within the eight scenarios, i.e., $\overline{p} = \sum_{i=1}^{8} w_p^{(i)} p^{(i)}$, where $p^{(i)}$ is the value of parameter p computed in Scenario i and $w_p^{(1)} + w_p^{(2)} + \cdots + w_p^{(8)} = 1$. Weights $w_p^{(i)}$ are set larger for the values computed within the scenarios where a certain type of generation is prevalent. Specifically, for the equivalent SG, the weights assigned to the parameter values computed in Scenarios 1–4, where synchronous generation is less significant (only Diesel 1 [0.45 MVA] is in service), is half of the ones identified in Scenarios 5–8, where synchronous generation is more significant (both Diesel 1 [0.45 MVA] and Diesel 2 [1.9 MVA] are service). For the equivalent AG, weights are all the same for Scenarios 1–2 and 7–8, and zero for Scenarios 3–6.

Notice that no average value has been computed for the mechanical torques T_{ma} and T_{ms} and for the nominal powers S_a^{nom} and S_s^{nom}, which are EDM scenario parameters (and thus there is no solid red line reported in the figures). This is due to the fact that the average values are computed for defining a single *average EDM* (a-EDM). This model should offer the possibility to be readily adapted to represent all possible settings of the ADN. Therefore, the idea is to leave the mentioned EDM scenario parameters (T_{ma}, T_{ms}, S_a^{nom}, and S_s^{nom}) free to be modified in coherence with the scenario to be reproduced by the a-EDM.

In order to better evaluate the simulation results, a black-box identification technique has been applied. In particular, recurrent artificial neural networks (ANNs) have been used to identify eight different models, using the same eight 35 s training sets earlier described. Similar to [9], two hidden layers ANNs with 40 and 15 neurons, respectively, and a recursion with 10 delayed samples, has returned best results.

4.2.6 Results

Figs. 10.14 and 10.15 show a sketch of the obtained identification and validation results, respectively. In particular, the power profiles are referred to three simulated events, two faults and one load step belonging to the validation sets, in the cases of the high load scenarios (1, 3, 5, and 7). Notice that results obtained in the low load scenarios (2, 4, 6, 8), and for the other events are similar.

Considering the identification results in Fig. 10.14, in general, it can be noted that the identified EDMs are able to represent the dynamics of the ADN with a good accuracy. Better performances are obtained in the fault cases. However, in the load steps cases, the error scale is smaller since the power variations are smaller than those registered when a fault occurs.

The power profiles obtained with the EDMs and the ANNs are similar. In some cases, ANNs have a slight better accuracy, but in other cases, especially with load steps, the EDMs clearly outperform ANNs.

The validation power profiles shown in Fig. 10.15 are almost coincident with the ones reported in Fig. 10.14. Moreover, even if not showed, consider that frequency and voltage profiles obtained using the DIgSilent Equivalent Model are almost coincident to the ones obtained with the full model (with a difference smaller than 0.01 Hz and

0.001 p.u., respectively). This means that the EDM can be readily integrated in a power system simulation platform.

Fig. 10.15 also reports the power profiles obtained using the a-EDM, suitably adapted to the different scenarios using the EDM scenario parameters (Table 10.12). These profiles are very close to the ones computed with the single

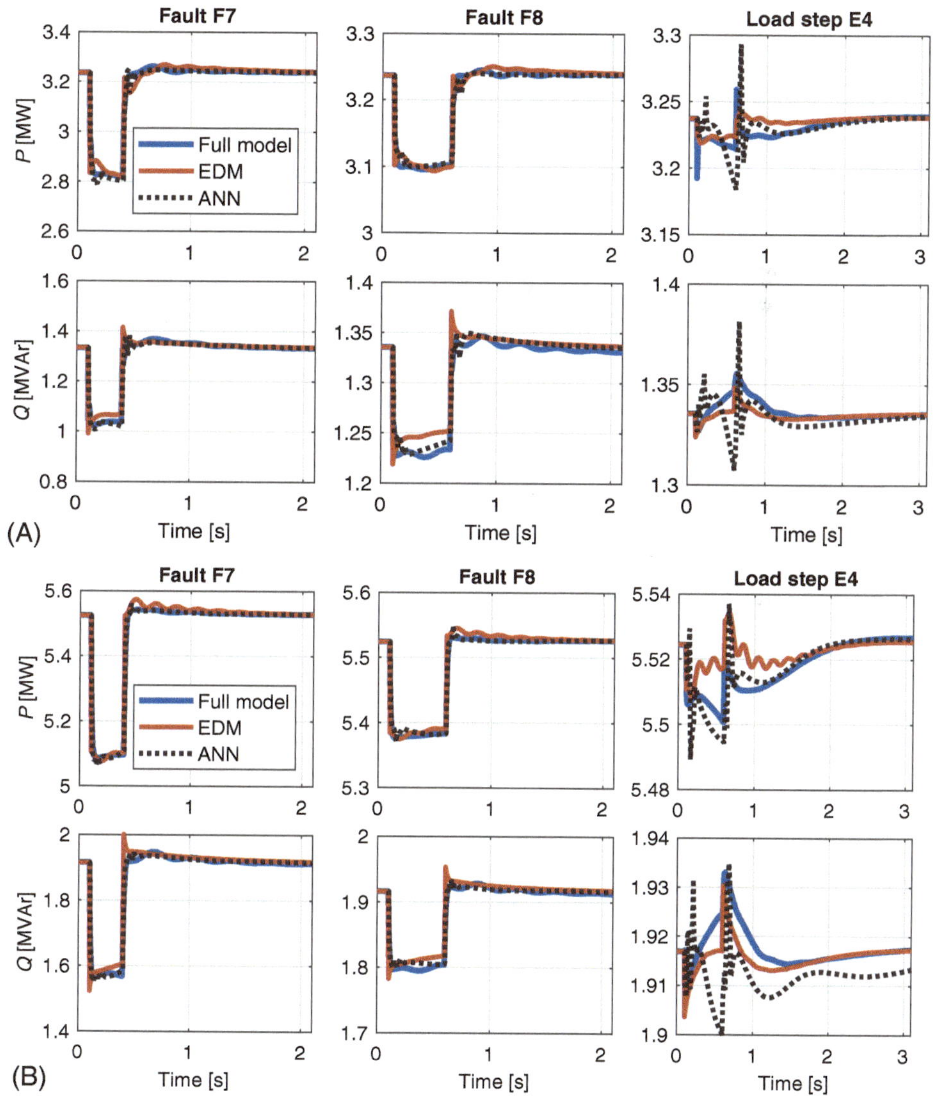

Fig. 10.14 Nonlinear constrained EDM simulations. Identification results: Active and reactive power profiles obtained for the high load scenarios, in the case of events F7, F8, and E4 with the DIgSilent Full Model, with the off-line simulated EDMs and ANNs models. (A) Scenario 1; (B) Scenario 3;

(Continued)

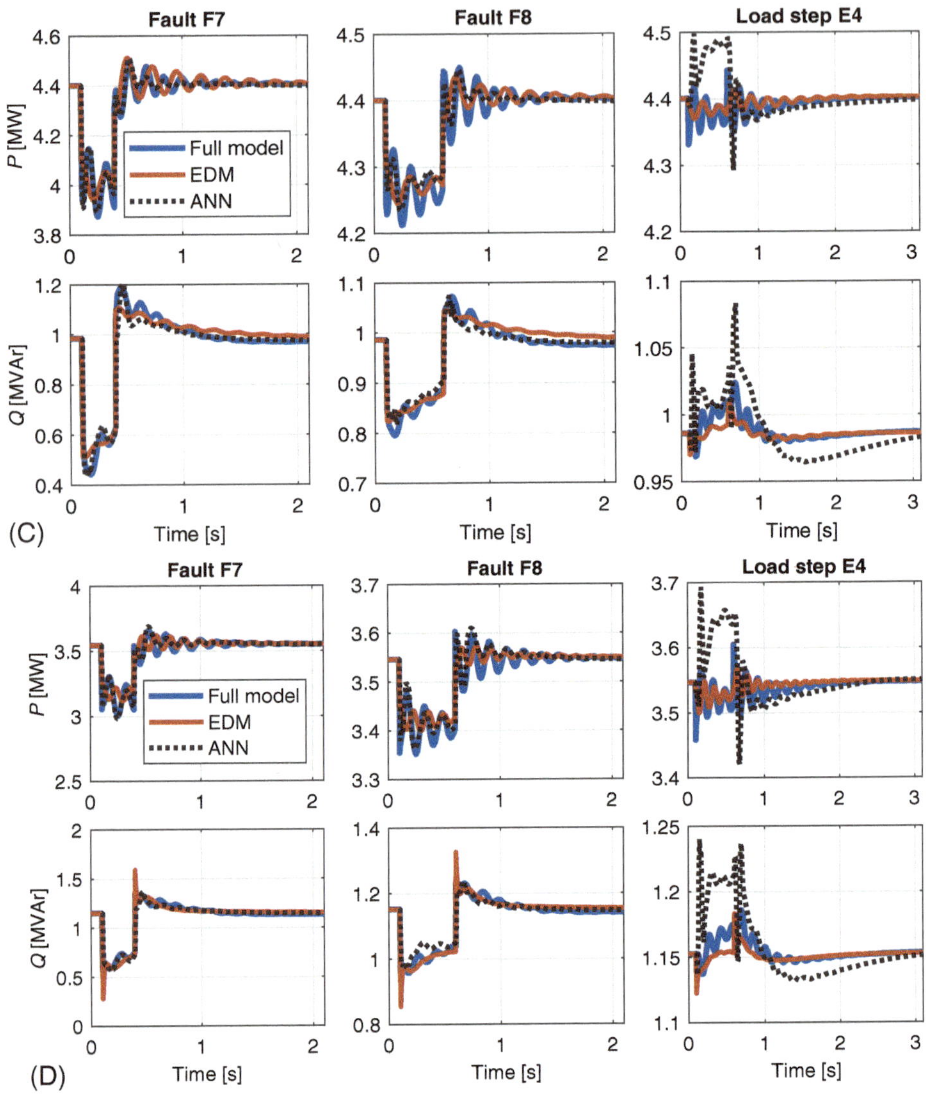

Fig. 10.14, Cont'd (C) Scenario 5; and (D) Scenario 7.

scenario EDMs, identified using the relevant training sets. This result is significant since it proves that the proposed EDM has a good scalability and it can be readily adapted to different settings of a given portion of ADN.

The numerical results reported in Table 10.14 confirm the considerations deduced from the analysis power profiles depicted in Figs. 10.14 and 10.15. The active power RMSEs are all lower than 50 kW, whereas the reactive power ones are all lower than 40 kVAr. In percentage terms (with respect to S_{GEN}^{nom}), these two values correspond to

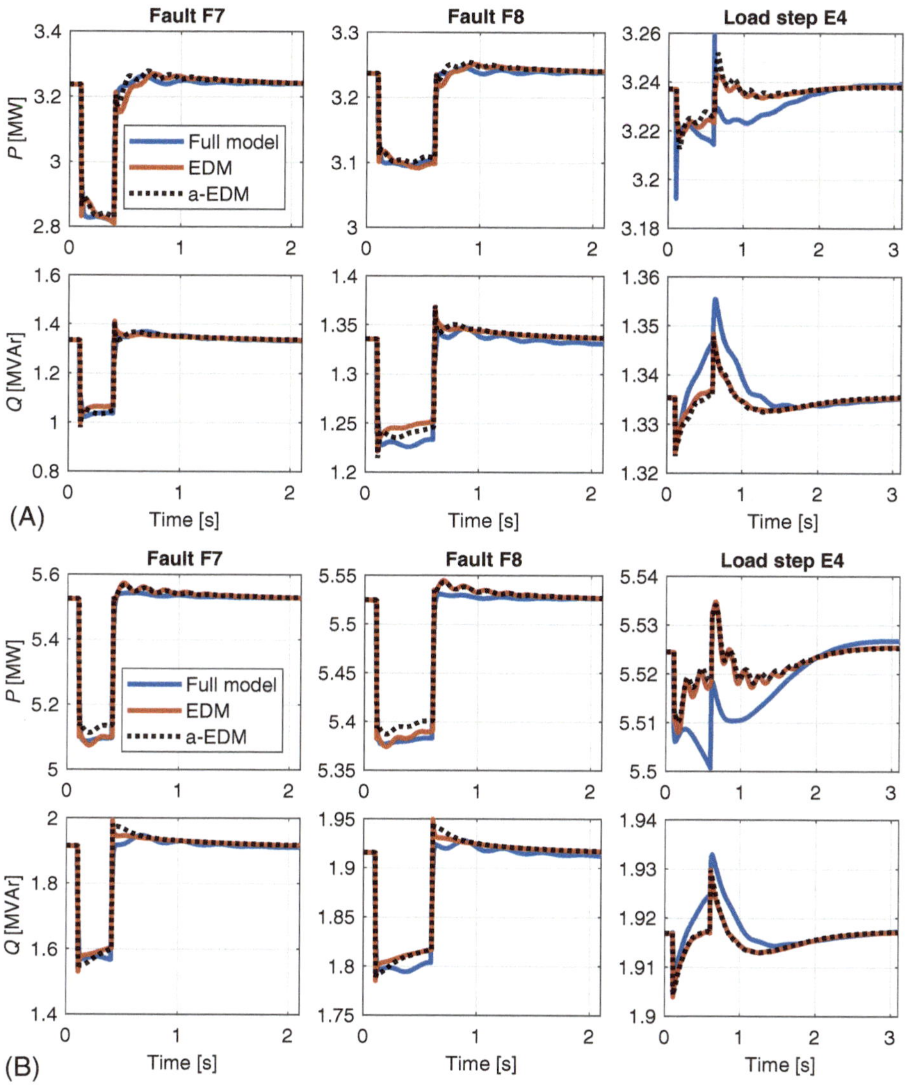

Fig. 10.15 Nonlinear constrained EDM simulations. Validation results: active and reactive power profiles obtained for the high load scenarios, in the case of events F7, F8, and E4 with the DIgSilent Full Model and the DIgSilent Equivalent Model implementing the EDMs computed using the relevant scenario training sets and the a-EDM. (A) Scenario 1; (B) Scenario 3;

(Continued)

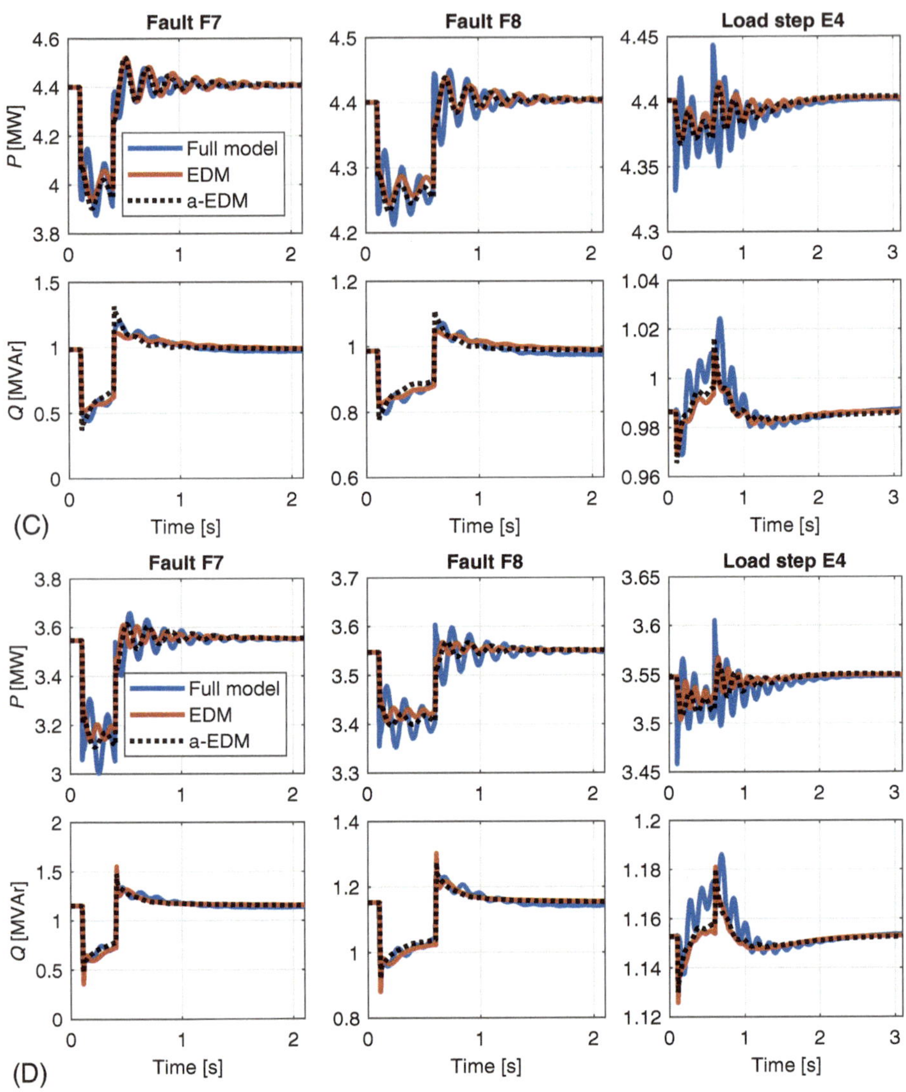

Fig. 10.15, Cont'd (C) Scenario 5; and (D) Scenario 7.

$\sigma_P^\% = 0.8\%$ and the $\sigma_Q^\% = 0.7\%$, respectively. A better accuracy is obtained in Scenarios 1–4, especially in the low load cases (Scenarios 2 and 4), where the maximum RMSEs are equal to 11–12 kW, corresponding to $\sigma_P^\% = 0.2\%$, for active power, and lower than 10 kVAr, corresponding to $\sigma_Q^\% = 0.16\%$, for reactive power.

The differences between the RMSEs obtained using the single scenario EDM and the a-EDM are particularly small, in general lower the 10 kW and 10 kVAr,

Table 10.14 Nonlinear constrained EDM simulations: Numerical results.

Scenario	1	2	3	4	5	6	7	8	9
Identification results: EDMs									**Id. res.: a-EDM**
σ_P [kW]	23.1	12.8	26.4	13	38.9	30.6	43.9	38.5	27.9
σ_Q [kVAr]	11.1	8.2	15.5	5.8	21.7	18.1	24.9	25.2	11.6
$\sigma_P^{\%}$	0.37	0.21	0.43	0.21	0.63	0.49	0.71	0.53	0.45
$\sigma_Q^{\%}$	0.18	0.13	0.25	0.09	0.35	0.29	0.40	0.30	0.19
Identification results: ANNs									**Id. res.: a-ANN**
σ_P [kW]	31.4	26.9	22.2	16.6	28.4	50.5	48.1	47.4	55.8
σ_Q [kVAr]	19.2	14.5	11.2	9.8	17.9	24.0	24.1	23.3	57.8
$\sigma_P^{\%}$	0.51	0.43	0.36	0.27	0.46	0.81	0.78	0.77	0.90
σ_P [kW]	0.31	0.23	0.18	0.16	0.29	0.39	0.39	0.38	0.93
Validation results									
σ_P [kW]	29.9	21.3	20.5	13.6	46.6	35	53.2	41.5	–
σ_Q [kVAr]	18.2	15.1	10.8	5.8	32.1	29.3	40.7	39.1	–
$\sigma_P^{\%}$	0.48	0.35	0.33	0.22	0.75	0.57	0.86	0.67	–
$\sigma_Q^{\%}$	0.29	0.24	0.17	0.09	0.52	0.47	0.66	0.63	–
a-EDM validation results									
σ_P [kW]	24.5	18.8	18.2	12.4	43.7	40.2	43.5	36.9	28.5
σ_Q [kVAr]	18.0	13.9	11.9	8.8	32.7	31.1	37.0	35.4	12.7
$\sigma_P^{\%}$	0.40	0.30	0.29	0.20	0.71	0.65	0.70	0.60	0.46
$\sigma_Q^{\%}$	0.29	0.22	0.19	0.14	0.53	0.50	0.60	0.57	0.20

corresponding to $\sigma_P^{\%} = \sigma_Q^{\%} = 0.16\%$. This confirms the suitability of the proposed model to be adapted to any given different setting of the ADN.

The identification results obtained with the ANNs models are also reported in Table 10.14. The performance indexes' values are similar to the one obtained with the EDMs, even if, in general, EDMs result to be more accurate than ANNs. It is well known that ANNs are particularly suitable in representing complex nonlinear systems, and potentially they may have a high accuracy in reproducing the system outputs. However, this can be carried out when suitable and complete training sets are available. In these simulations, training sets are relatively small and they cannot display all possible dynamical behaviors of the ADN. In the EDMs, this lack of information is compensated by the gray-box definition of the model.

Finally, Table 10.14 shows the numerical results obtained for Scenario 9. This setting is of particular interest because of two motivations: first, no data about this scenario have been used to identify the equivalent models; second, differently from all other scenarios, the ADN is exporting power. Since the training set of this scenario has not been used, no relevant EDMs has been identified. Therefore, the a-EDM has been tested both with off-line simulations (identification results) and with the DIgSilent Equivalent Model (validation results). Fig. 10.16 depicts some examples of the obtained power profiles. The performance indexes reported in Table 10.14 are comparable with those returned by the other EDMs suitably identified for each scenario, and with those returned by the a-EDM applied to the first eight scenarios. This further confirms the capability of the proposed model to be adapted to any given different setting of the ADN, both if it is exporting or importing power.

In order to provide a comparison also for Scenario 9, an average ANN (a-ANN) model has been computed by processing together all the eight training sets of Scenarios 1–8 and distinguishing them using the EDM scenario parameters, as additive constant inputs. In load modeling and forecasting, this is a technique usually used to allow an ANN model to take into account known parameters and to be adapted to different scenarios [43]. Both the numerical results reported in Table 10.14 and the power profiles depicted in Fig. 10.16 show the larger difficulties of the a-ANN in being adapted to settings of the ADN different from those used in the training stage, if compared with the proposed a-EDM.

4.3 Microgrid EDM experimental results

The modeling and identification method described in Section 3.3 is here applied to microgrids and validated with experimental data. Measurements are collected from a real LV test facility.

4.3.1 Test facility

The network is composed of a MV/LV transformer (800 kVA) and six LV feeders that can be extended using line segments (100, 150, and 200 m length). It is designed and organized as a LV microgrid to perform studies and experimental tests on DERs and

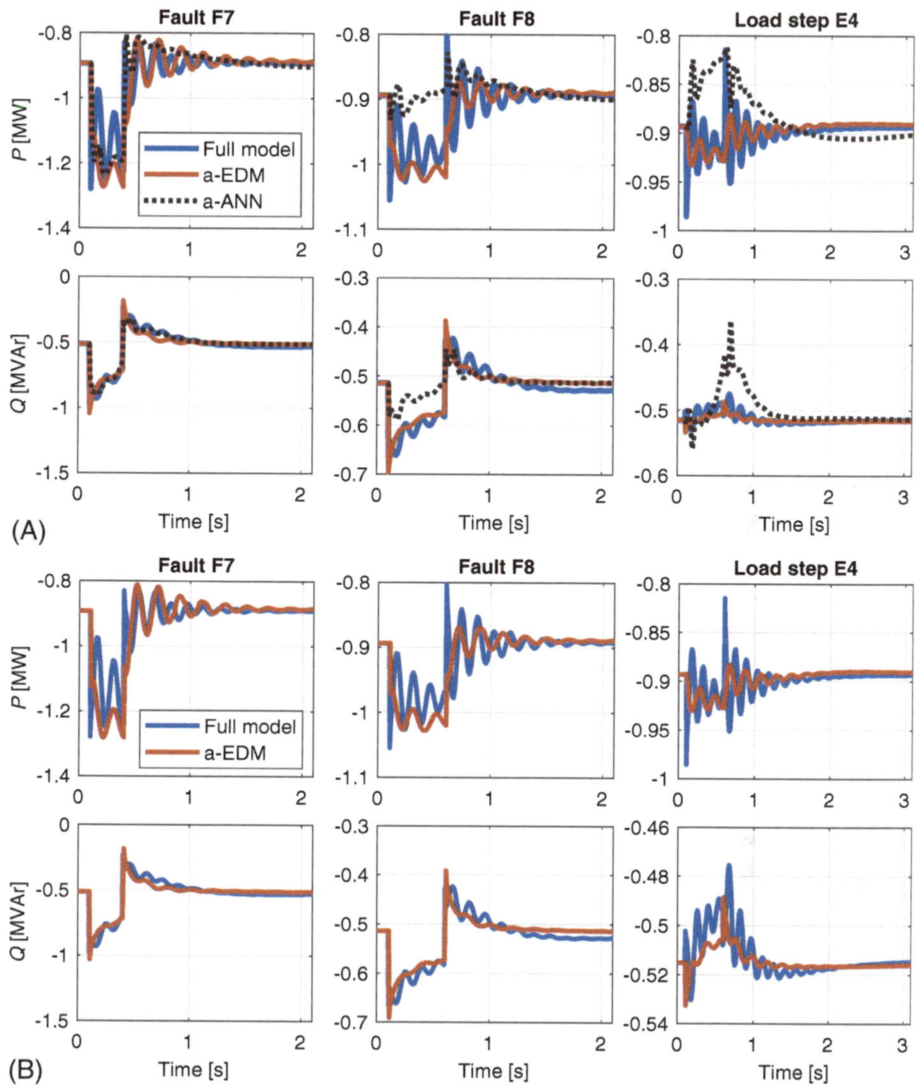

Fig. 10.16 Nonlinear constrained EDM simulations. Scenario 9: active and reactive power profiles in the case of events F7, F8, and E4. (A) Profiles obtained with the DIgSilent Full Model and the off-line simulations of the a-EDM and of the a-ANN; (B) profiles obtained with the DIgSilent Full Model and with the DIgSilent Equivalent Model implementing the a-EDM. (A) Identification; (B) validation.

smart grid methodologies [44]. The following generators are installed on the test facility (TF): different types of PV plant (total of 25 kWp), a microwind generator (2 kWp), and a CHP with a synchronous generator (50 kW). Moreover, there are BESSs (lithium, SoNick, vanadium redox, and lead acid), programmable loads (resistive, inductive, and capacitive), and a 400 V–100 kW AC/DC interface with a DC grid. Fig. 10.17

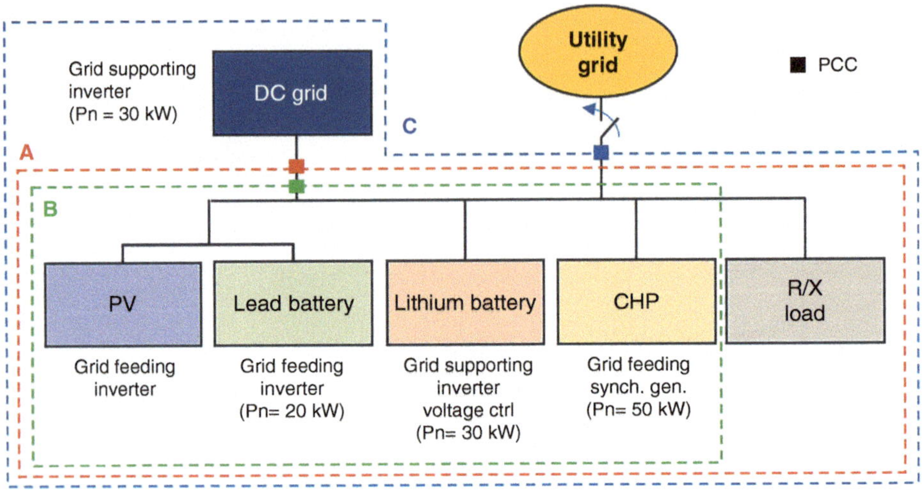

Fig. 10.17 Experimental tests. Diagram of the portion of the test facility. *Dashed lines* define the aggregate which is represented by the EDM for each configuration.

shows the block diagram and the composition of the portion of the TF, and the three configurations adopted to generate the experimental datasets. In all the configurations, inverters are voltage or current controlled and implement a fast equivalent droops control action [45].

4.3.2 Experimental scenarios

Three classes of configurations have been considered:

1. *Configuration A*: TF in islanded mode with perturbations determined by the AC/DC converter. The portion of TF considered as the system to be identified is the one within the red box in Fig. 10.17, the PCC is the point of connection with the DC grid, who plays the role of external grid. The AC/DC inverter operates in "grid forming" mode [45] and imposes frequency and voltage.
2. *Configuration B*: TF in islanded mode with load variations. The portion of TF considered as the system to be identified is the one within the green box in Fig. 10.17, the PCC is the point of connection with the DC grid.
3. *Configuration C*: TF connected to the main grid. The portion of TF considered as the system to be identified is the one within the blue box in Fig. 10.17, the PCC is the point of connection with the main grid.

In all, 10 measurement datasets have been collected in 10 different scenarios, using a PMU with sampling time of 0.02 s. Scenarios 1–3 adopt Configuration A, Scenarios 4–8 adopt Configuration B, and Scenarios 9 and 10 adopt Configuration C. Further than the FT configuration, scenarios differ each others in the inverters droop parameters, and the load and generators set points.

The identification algorithm described in Section 2.2 has been applied to the datasets collected in Scenarios 1–8. In all these cases, only the first half (in terms of time) of measurements has been used for the EDM parameters identification. Two possible initial values of parameters have been used to initialize the optimization:

- *CHP out of service (Scenarios 1, 2, and 5–8)*: The initial conditions for the nominal power of the SG is equal to zero with a null confidence interval.
- *CHP in service (Scenarios 3 and 4)*: The nominal power of the SG is initialized with 50 kVA, which is the nominal power of the CHP. In this case, the confidence interval is 0 since the information is exactly known. Moreover, based on an estimate of the owner of the TF, the inertia time constant of the CHP is within the interval [0.2 0.5] s. Thus, H_s is initialized with 0.35 s with 40% of confidence.

To study the capability of the identification procedure to provide an estimate of the inertia time constant, a test on Scenario 4 has been carried out using an erroneous initialization of H_s: 5 s with a 100% confidence interval.

4.3.3 Identification and validation results

Fig. 10.18 shows the results of identification and validation obtained in Scenarios 3 and 4. The measured profiles of P and Q are compared with the ones reproduced by the EDM. We recall that the first half (in terms of time) of the measurements is used in the identification procedure, whereas the rest is used for validation. Observing the figures, in both the scenarios, the EDM appears to be able to represent microgrid dynamics with a good accuracy.

Power profiles in Fig. 10.18 are provided as examples of identification and validation results. Table 10.15 reports the RMSEs, computed as in Eq. (10.72), obtained in Scenarios 1–8. Under the item "Identification," we have the RMSEs obtained with the first time half of datasets; under the item "Validation," we report the RMSEs obtained with the second time half of datasets, not used in the identification. In the table, and hereafter in the chapter, results obtained with the erroneous initialization of H_s are indicated with "4e."

In Scenarios 3–8, RMSEs are always lower than 0.3 kW for P and 0.2 kVAr for Q. In Scenarios 1 and 2, RMSEs are lower than 0.5 kW for P and 0.6 kVAr for Q. This consideration holds true both for identification and validation. In all the scenarios, the order of magnitude of P and Q variations are lower than 1 kW and 1 kVAr. Therefore, errors can be considered sufficiently low, meaning that the EDM is able to reproduce the microgrid active and reactive power responses with a good accuracy.

Notice also that the difference between the RMSEs obtained in identification and validation is lower than 0.1 kW and 0.1 kVAr in Scenarios 1 and 2 and of the order of centimes of kW and kVAr in Scenarios 3–8. Finally, observe that RMSEs obtained in Scenarios 4 and 4e are very close each other.

4.3.4 Cross-validation results

In this final part of the analysis, we evaluate the consistency and the scalability of the proposed EDM. The two EDMs identified with the dataset of Scenario 4, with correct and incorrect initializations (models 4 and 4e) have been applied to the other scenarios

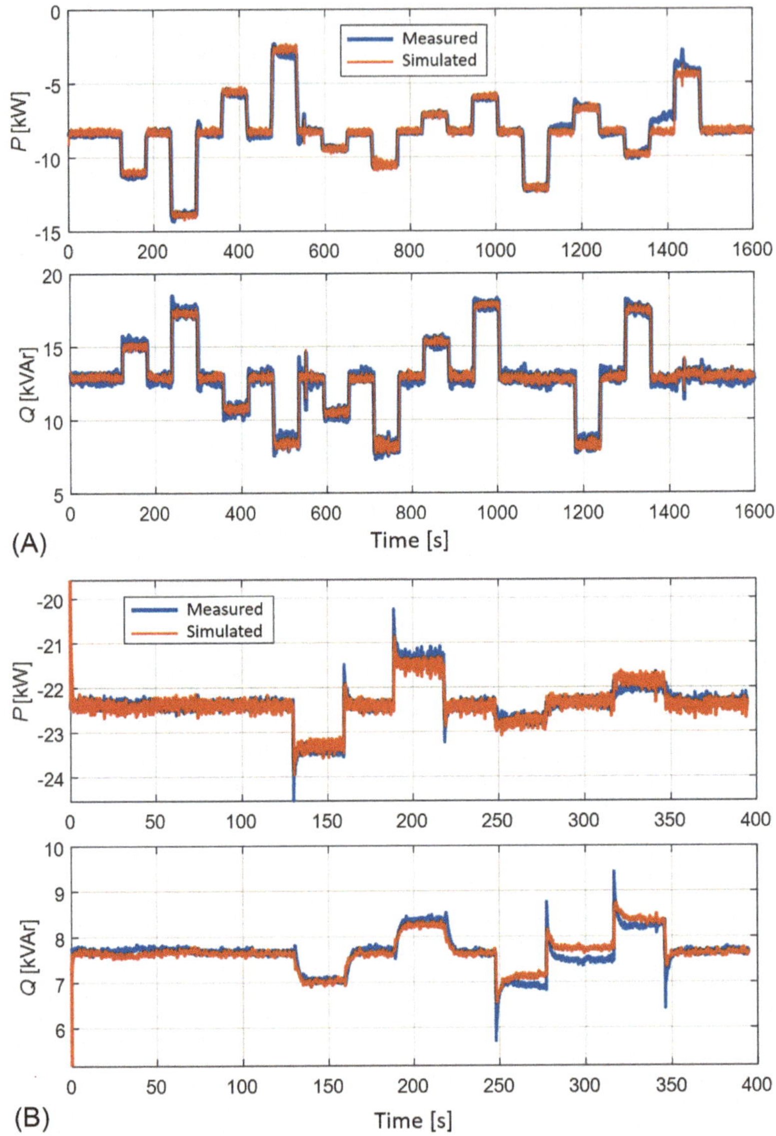

Fig. 10.18 Experimental tests. Identification and validation results: (A) Configuration A and (B) Configuration B.

where the CHP is in service (i.e., Scenarios 3, 9, and 10). Notice that in these two last cases, differently from all other scenarios, the TF is connected to the main grid (Configuration C).

Fig. 10.19 shows two examples of the obtained results. In Fig. 10.19A, the power profiles measured in Scenario 3 are reproduced by the EDM 4. By comparing such

Table 10.15 Experimental tests: RMSEs for identification and validation.

	Identification		Validation	
Sc.	σ_P **[kW]**	σ_Q **[kVAr]**	σ_P **[kW]**	σ_Q **[kVAr]**
1	0.388	0.437	0.510	0.352
2	0.227	0.630	0.238	0.534
3	0.290	0.210	0.295	0.223
4	0.173	0.146	0.181	0.149
4e	0.180	0.145	0.183	0.152
5	0.101	0.079	0.125	0.078
6	0.109	0.114	0.171	0.125
7	0.162	0.141	0.169	0.126
8	0.221	0.134	0.212	0.126

profiles with the ones in Fig. 10.18A, where the EDM 3 is employed, no significant differences can be appreciated. Fig. 10.19B reports the results obtained with model EDM 4e, applied to Scenario 10. Also in this case, the EDM appears to be capable to correctly reproduce the microgrid dynamic response.

Table 10.16 reports the cross-validation RMSEs. These values are not significantly different from the ones obtained for identification and validation (see Table 10.15). We can, therefore, conclude that the EDM shows a good accuracy of representation and the approach is consistent and scalable.

5 Conclusions

The chapter has considered the problem of the equivalent dynamic modeling for ADNs, starting from a *gray-box linear modeling* method, where a nonlinear model is initially defined, then it is linearized, and the parameters of the resulting linear model are identified. Subsequently, a *gray-box nonlinear* modeling method is introduced, including system operating constraints and avoiding any linearization of the model. This approach highlights an effective relation with the actual configuration of the ADN to be considered. In this case, all model parameters maintain a physical meaning, and the utilization of the prior knowledge of the network improves the identification procedure robustness. The result is a model that can be effectively adapted to different configurations and operating conditions of the ADN. The methodology is tested on a simulation environment, where detailed models of the target distribution networks are implemented. Simulations clearly show the capability of the equivalent model to accurately reproduce the dynamical behavior of a given ADN. In particular, results prove the suitability of the model to be adapted to different configurations of the network by modifying a limited set of scenario parameters. In the last part of the chapter, the nonlinear modeling method is adapted and validated with experimental data on a real microgrid facility, made up with a low-voltage distribution network,

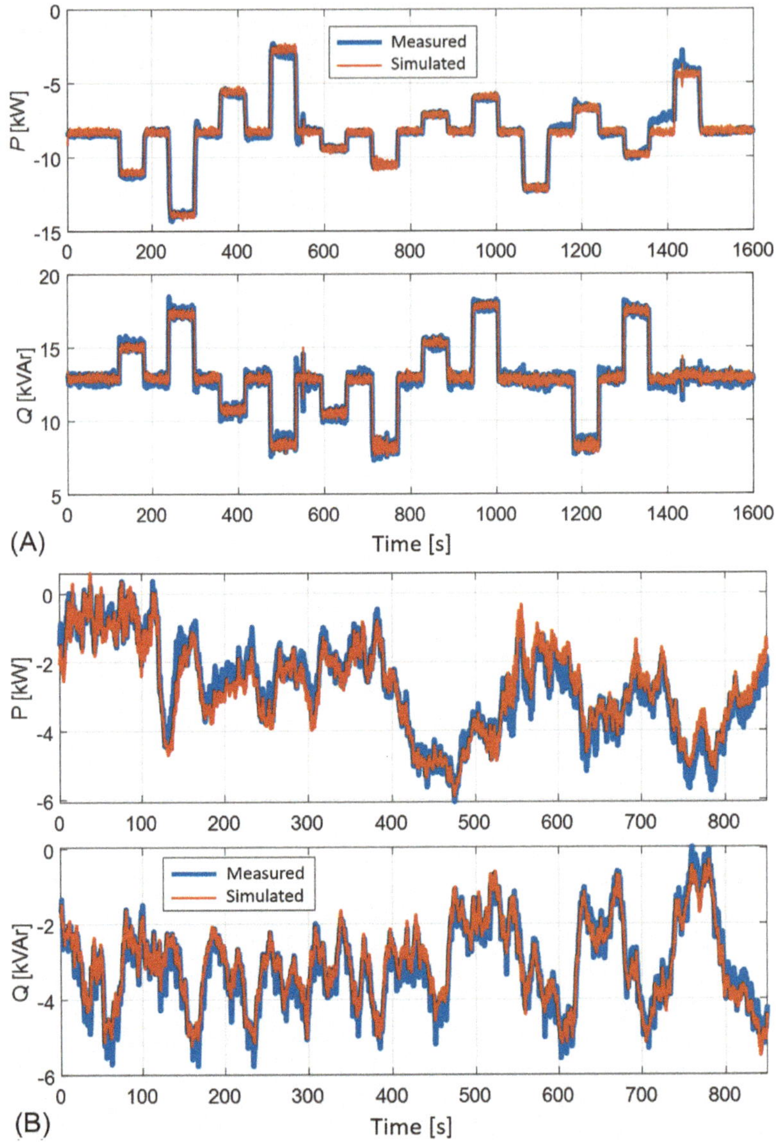

Fig. 10.19 Experimental tests. Cross-validation results. (A) Active and reactive power measured in Scenario 3 and reproduced with the EDM identified with the dataset of Scenario 4 (EDM 4). (B) Active and reactive power measured in Scenario 10 and reproduced with the EDM identified with the dataset of Scenario 4 with erroneous initialization (EDM 4e).

a synchronous generator, two batteries, a photovoltaic (PV) plant, and static loads. Results show that the proposed EDM is able to accurately reproduce the dynamic response of the microgrid to external disturbances. In particular, given a variation on voltage and frequency at the PCC, the EDM correctly returns the variations of active and reactive power exchanged by the microgrid.

Table 10.16 Experimental tests: RSMEs for cross-validation.

Scenario	Model	σ_P [kW]	σ_Q [kVAr]
3	4	0.290	0.210
3	4e	0.290	0.211
9	4	0.488	0.226
9	4e	0.488	0.233
10	4	0.359	0.236
10	4e	0.356	0.235

References

[1] F. Pilo, G. Mauri, B.B. Jensen, J. Taylor, E. Kampf, F. Silvestro, Control and automation functions at the TSO and DSO interface—impact on network planning, in: CIRED, 2017. Glasgow, UK.

[2] Control and Automation Systems for Electricity Distribution Networks (EDN) of the future, Technical Report, Joint Working Group (JWG), 2017. C6/B5.25/CIRED.

[3] P. Kotsampopoulos, N. Hatziargyriou, B. Bletterie, G. Lauss, Review, analysis and recommendations on recent guidelines for the provision of ancillary services by distributed generation, in: 2013 IEEE International Workshop on Intelligent Energy Systems (IWIES), 2013, pp. 185–190, https://doi.org/10.1109/IWIES.2013.6698583.

[4] A.O. Rousis, D. Tzelepis, Y. Pipelzadeh, G. Strbac, C.D. Booth, T.C. Green, Provision of voltage ancillary services through enhanced TSO-DSO interaction and aggregated distributed energy resources, IEEE Trans. Sustain. Energy 12 (2) (2021) 897–908, https://doi.org/10.1109/TSTE.2020.3024278.

[5] F. Adinolfi, F. Baccino, F. D'Agostino, S. Massucco, M. Saviozzi, F. Silvestro, An architecture for implementing state estimation application in distribution management system (DMS), in: IEEE PES ISGT Europe 2013, 2013, pp. 1–5, https://doi.org/10.1109/ISGTEurope.2013.6695289.

[6] J.V. Milanovic (convener), Modelling and aggregation of loads in flexible power networks, Cigre Working Group C4.605, 2014. Technical Report.

[7] F.O. Resende, J. Matevosyan, J.V. Milanovic, Application of dynamic equivalence techniques to derive aggregated models of active distribution network cells and microgrids, in: 2013 IEEE PowerTech, 2013. Grenoble, France.

[8] A.M. Azmy, I. Erlich, Identification of dynamic equivalents for distribution power networks using recurrent ANNs, in: IEEE PES Power Systems Conference and Exposition, 2004, pp. 348–353.

[9] A.M. Azmy, I. Erlich, P. Sowa, Artificial neural network-based dynamic equivalents for distribution systems containing active sources, IEE Proceedings—Generation, Transmission and Distribution 151 (6) (2004) 681–688.

[10] G.H. Shakouri, H.R. Radmanesh, Identification of a continuous time nonlinear state space model for the external power system dynamic equivalent by neural networks, Int. J. Electr. Power Energy Syst. 31 (7–8) (2009) 334–344.

[11] F. Calero, C.A. Cañizares, K. Bhattacharya, Aggregated BESS dynamic models for active distribution network studies, IEEE Trans. Smart Grid 12 (3) (2021) 2077–2088, https://doi.org/10.1109/TSG.2020.3048648.

[12] C. Changchun, W. Min, D. Lihua, D. Zhixiang, Z. Jianyong, Micro-grid dynamic modeling based on RBF artificial neural network, in: 2014 International Conference on Power System Technology, 2014, pp. 3348–3353, https://doi.org/10.1109/POWERCON.2014.6993926.
[13] C. Cai, H. Liu, Y. Tao, Z. Deng, W. Dai, J. Chen, Microgrid equivalent modeling based on long short-term memory neural network, IEEE Access 8 (2020) 23120–23133, https://doi.org/10.1109/ACCESS.2020.2966238.
[14] S.M. Zali, J.V. Milanovic, Dynamic equivalent model of distribution network cell using Prony analysis and nonlinear least square optimization, in: 2009 IEEE Bucharest PowerTech, 2009.
[15] P.N. Papadopoulos, T.A. Papadopoulos, P. Crolla, A.J. Roscoe, G.K. Papagiannis, G.M. Burt, Measurement-based analysis of the dynamic performance of microgrids using system identification techniques, IET Gen. Transm. Distrib. 9 (1) (2015) 90–103.
[16] A. Ishchenko, J.M.A. Myrzik, W.L. Kling, Dynamic equivalencing of distribution networks with dispersed generation using Krylov methods, in: 2007 IEEE PowerTech, 2007, pp. 2023–2028, https://doi.org/10.1109/PCT.2007.4538629.
[17] A.M. Miah, Study of a coherency-based simple dynamic equivalent for transient stability assessment, IET Gen. Transm. Distrib. 5 (4) (2011) 405–416.
[18] E.O. Kontis, S.P. Dimitrakopoulos, A.I. Chrysochos, G.K. Papagiannis, T.A. Papadopoulos, Dynamic equivalencing of active distribution grids, in: IEEE Manchester PowerTech, 2017, https://doi.org/10.1109/PTC.2017.7981103.
[19] G. Mitrentsis, H. Lens, A dynamic active distribution network equivalent for enhancing the generalization capability of the exponential recovery model in stability studies, IEEE Trans. Power Syst. 36 (3) (2021) 2709–2712, https://doi.org/10.1109/TPWRS.2021.3053137.
[20] G. Mitrentsis, H. Lens, Data-driven dynamic models of active distribution networks using unsupervised learning techniques on field measurements, IEEE Trans. Smart Grid 12 (4) (2021) 2952–2965, https://doi.org/10.1109/TSG.2021.3057763.
[21] U.D. Annakkage, N.K.C. Nair, Y. Liang, A.M. Gole, V. Dinavahi, B. Gustavsen, T. Noda, H. Ghasemi, A. Monti, M. Matar, R. Iravani, J.A. Martinez, Dynamic system equivalents: a survey of available techniques, IEEE Trans. Pow. Deliv. 27 (1) (2012) 411–420.
[22] N. Hatziargyriou (convener), Contribution to bulk system control and stability by distributed energy resources connected at distribution network, 2013. IEEE PES Technical Report PES-TR22.
[23] G. Chaspierre, P. Panciatici, T. Van Cutsem, Aggregated dynamic equivalent of a distribution system hosting inverter-based generators, in: 2018 Power Systems Computation Conference (PSCC), 2018, pp. 1–7, https://doi.org/10.23919/PSCC.2018.8442968.
[24] G. Chaspierre, G. Denis, P. Panciatici, T. Van Cutsem, An active distribution network equivalent derived from large-disturbance simulations with uncertainty, IEEE Trans. Smart Grid 11 (6) (2020) 4749–4759, https://doi.org/10.1109/TSG.2020.2999114.
[25] J.V. Milanovic, S.M. Zali, Validation of equivalent dynamic model of active distribution network cell, IEEE Trans. Power Syst. 28 (3) (2013) 2101–2110.
[26] S.M. Zali, J.V. Milanovic, Generic model of active distribution network for large power system stability studies, IEEE Trans. Power Syst. 28 (3) (2013) 3126–3133.
[27] F. Conte, F. D'Agostino, S. Massucco, G. Palombo, F. Silvestro, C. Bossi, M. Cabiati, Dynamic equivalent modelling of active distribution networks for TSO-DSO interactions, in: IEEE ISGT Europe 2017, 2017. Torino, Italy.
[28] F.O. Resende, J.A.P. Lopes, Development of dynamic equivalents for microgrids using system identification theory, in: 2007 IEEE PowerTech, 2007, pp. 1033–1038. Lousanne, Switzerland.

[29] G. Chaspierre, P. Panciatici, T.V. Cutsem, Dynamic equivalent of a distribution grid hosting dispersed photovoltaic units, in: IREP '17 Symposium, 2017. Espinho, Portugal.
[30] F. Conte, F. D'Agostino, F. Silvestro, Operational constrained nonlinear modeling and identification of active distribution networks, Electr. Power Syst. Res. 168 (2019) 92–104.
[31] F. Conte, F. D'Agostino, S. Massucco, F. Silvestro, C. Bossi, M. Cabiati, Experimental validation of a dynamic equivalent model for microgrids, IEEE Trans. Ind. Appl. 57 (3) (2021) 2202–2211.
[32] W.J. Rugh, Linear System Theory, Prentice Hall, Inc., 1996.
[33] D.W. Marquardt, An algorithm for least-squares estimation of nonlinear parameters, SIAM J. Appl. Math. 11 (2) (1963) 431–441.
[34] P. Kundur, Power System Stability and Control, McGraw-Hill, 1994.
[35] P.M. Anderson, A.A. Fouad, Power System Control and Stability, Wiley, 2003.
[36] K. De Brabandere, B. Bolsens, J. Van den Keybus, A. Woyte, J. Driesen, R. Belmans, A voltage and frequency droop control method for parallel inverters, IEEE Trans. Power Electron. 22 (4) (2007) 1107–1115, https://doi.org/10.1109/TPEL.2007.900456.
[37] L. Meng, A. Luna, E.R. Díaz, B. Sun, T. Dragicevic, M. Savaghebi, J.C. Vasquez, J.M. Guerrero, M. Graells, F. Andrade, Flexible system integration and advanced hierarchical control architectures in the microgrid research laboratory of Aalborg University, IEEE Trans. Ind. Appl. 52 (2) (2016) 1736–1749, https://doi.org/10.1109/TIA.2015.2504472.
[38] N. Soni, S. Doolla, M.C. Chandorkar, Analysis of frequency transients in isolated microgrids, in: 2016 IEEE Industry Applications Society Annual Meeting, 2016, pp. 1–9, https://doi.org/10.1109/IAS.2016.7731879.
[39] Benchmark systems for network integration of renewable and distributed energy resources, Cigre Task Force C6.04.02, 2009. Technical Report.
[40] F. Baccino, F. Conte, S. Grillo, S. Massucco, F. Silvestro, An optimal model-based control technique to improve wind farm participation to frequency regulation, IEEE Trans. Sustain. Energy 6 (3) (2015) 993–1003.
[41] DIgSILENT GmbH, DIgSILENT GmbH, Gomaringen, Germania, 2017. DIgSILENT PowerFactory User's Manual, version 2017.
[42] ISO, 2005. International Standard ISO 8528-2-2005. Reciprocating internal combustion engine driven alternating current generating sets—Part 5: Generating sets.
[43] F. Adinolfi, F. D'Agostino, A. Morini, M. Saviozzi, F. Silvestro, Pseudo-measurements modeling using neural network and Fourier decomposition for distribution state estimation, in: IEEE PES Innovative Smart Grid Technologies, Europe, 2014, pp. 1–6, https://doi.org/10.1109/ISGTEurope.2014.7028770.
[44] F. D'Agostino, S. Massucco, F. Silvestro, C. Bossi, A. Guagliardi, C. Sandroni, Implementation of a distribution state estimation algorithm on a low voltage test facility with distributed energy resources, in: 2016 IEEE PES Innovative Smart Grid Technologies Conference Europe (ISGT-Europe), 2016, pp. 1–6.
[45] J. Rocabert, A. Luna, F. Blaabjerg, P. Rodríguez, Control of power converters in AC microgrids, IEEE Trans. Power Electron. 27 (11) (2012) 4734–4749.

Transactive control for residential demand-side management **11**

Lessons learned from noncooperative game theory

Luciana Marques[a,b], *Miguel Heleno*[c], *and Wadaed Uturbey*[a]
[a]Universidade Federal de Minas Gerais, Belo Horizonte, Brazil, [b]Flemish Institute for Technological Research (VITO), Mol, Belgium, [c]Lawrence Berkeley National Laboratory, Berkeley, CA, United States

Abbreviations

BNE	Bayesian Nash equilibrium
BRD	best-response dynamics
DER	distributed energy resource
NE	Nash equilibrium
PoA	price-of-anarchy
PTC	proportional-to-consumption billing
PTS	per-time-slot billing
TC	transactive control
TCL	thermostatically controlled load

1 Introduction

Demand-side management (DSM) is a key element of the grid of the future, allowing consumers to participate in the system operation through automatic variations of demand that can be offered as flexibility services to improve efficiency and decrease grid costs. The majority of DSM methodologies developed over the last decades rely on centralized control of load devices, in which a single entity (system operator, retailer, or aggregator) is responsible to gather information about the consumption, perform a consolidated calculation of the control actions, and communicate set points to a large number of devices via central communication systems. This centralized form of control has the advantage of guaranteeing optimality of the solutions, but it raises important privacy concerns, especially in residential applications. Alternatively, traditional decentralized approaches rely on economic incentives and price signals with the objective of inducing optimal control actions taken independently by the consumers, without intrusive control orders. However, these price signals are defined a priori, and they do not internalize the response of the consumers in the price formation, which may lead to suboptimal load scheduling solutions.

Decentralized Frameworks for Future Power Systems. https://doi.org/10.1016/B978-0-323-91698-1.00006-6

To address the drawbacks of centralized and decentralized approaches, transactive control (TC) models have emerged as a way of orchestrating multiple consumers without harming their autonomy. In DSM applications, the idea is to optimize the overall use of flexible demand by enabling participants to interact with each other, exchange information about consumption, and iteratively schedule their loads, under individual constraints and comfort preferences, until an equilibrium solution is reached. These decentralized characteristics, together with the transparent communication and decision-making process, make TC an attractive solution for practical implementation of DSM technologies in the residential sector, with the potential to unleash the flexibility of millions of consumers and provide a new flexible resource to the system operation. To achieve this, the recent theoretical advancements were made in TC, particularly in the field of noncooperative game theory, which enables a formal modeling of consumers' preferences, priorities, and conflicting interests in a decentralized manner, while providing important insights about the economic rationality and the equilibrium conditions of the decentralized DSM programs.

This chapter provides a review and a structured analysis of the TC applications to residential DSM, in particular the use of noncooperative games and the key theoretical aspects that affect the equilibrium conditions, and therefore the success of transactive load control implementations. First, a literature review is conducted to show how noncooperative game models can be applied to manage the flexible load of a population of consumers. Then, given different deployment environments, cost functions, and billing methods, the theoretical load control aspects are analyzed in a form of existence, multiplicity, and optimality of Nash equilibria, fairness of billing methods, users' engagement, cheating behavior, and applicability to different contexts and types of load devices. Although the subject of this analysis is predominantly theoretical, the nature of this chapter is practical, focusing on the implications of TC design for the actual deployment of residential DSM. To illustrate each step of the analysis, numerical results are provided using a medium-scale residential community of consumers.

2 Literature review

As discussed in the previous section, TC has emerged as a form of coordinating different agents in power systems (consumers, producers, DSOs (distribution system operators), TSOs (transmission system operators), and aggregators), while considering their particularities, priorities, interests, and autonomy [1]. The idea is to optimize the allocation of resources (e.g., generation, storage devices, and loads) by enabling actors to interact with each other and exchange information about consumption, generation, constraints, and preferences until an equilibrium solution is reached [2]. This market-based control is naturally decentralized, which makes it attractive to applications in residential sector, in which privacy is a main concern [3]. When applied to control of domestic loads, the implementation of TC is able to capture the load scheduling interactions between mid- to small-size consumers, allowing demand flexibility to be offered via home management systems and cleared in a form of an electronic local market [4]. In this context, noncooperative game theory allows to translate the implicit dialectic between agents [5] and inform the control and market design process. This section presents main literature on the topic of noncooperative games applied to TC in a form of Table 11.1 that condenses the following features of the models.

- *Application*: For what reason the noncooperative game was designed (e.g., optimal residential energy management). The classification proposed in Hu et al. [4] is followed, but with the addition of more detailed information for some references.
- *Implementing method*: What algorithm is used to solve the game (e.g., best-response dynamics [BRD]). This determines if the method is one-time or iterative information exchange based [4].
- *Load model*: What variables are used to model the resources (continuous, integer), including if the study uses simple equations to represent the appliances operations; and if the load is energy variant (e.g., thermostatically controlled loads [TCLs]), or fixed/constant (e.g., shiftable loads as washing machines).
- *Utility function*: What are the payoffs for participating consumers. It is defined by the total cost of the TC (e.g., quadratic function), and a billing mechanism that divides this total cost among participants (e.g., per-time-slot [PTS] billing). In some references, the utility for each consumer is defined directly, without considering a total cost function with a billing method. For those, the total cost column is the utility function, and the billing is filled with "according to the total cost."
- *Aspects analyzed*: What features of the games are studied on the reference. Eight aspects are possible: equilibrium existence; equilibrium uniqueness; equilibrium optimality; solution fairness; cheating behavior; price-of-anarchy (PoA); information losses; and consumers' privacy. Equilibrium states for: Nash equilibrium (NE), correlated equilibrium, ϵ-equilibrium, Bayesian Nash equilibrium (BNE), and equilibrium (in general).

3 Noncooperative games for the coordination of residential loads

As discussed in the previous sections, noncooperative games can be used to optimally manage residential consumption. To model the coordination problem as a game, some basic blocks must be defined together with an information exchange structure. Fig. 11.1 presents a general framework with those building blocks, including the communication notion behind TC.

In the following sections, the building blocks and the key aspects of a noncooperative game formulation are discussed in detail. Later, some aspects of these games can be further analyzed, such as the existence, optimality, and uniqueness of equilibria, applicability of solution algorithms (including communication aspects), fairness, and strategy proof of billing methods, among others.

It is important to note that the group of consumers can be an energy community, a virtual power plant, or any other structure allowing the implementation of the model.

Table 11.1 Literature review about TC models applying noncooperative game approaches.

			Load model		Utility function		
Reference	Application	Implementing method	Variables	Energy variant	Total cost	Billing	Aspects analyzed
Mohsenian-Rad et al. [6]	Residential optimal energy management	BRD	Continuous (simple)	No	Quadratic	PTC	NE existence
Baharlouei et al. [7]	Residential optimal energy management	Not specified	Continuous (simple)	No	Quadratic	PTS vs. PTC	NE existence Fairness
Baharlouei and Hashemi [8]	Residential optimal energy management	BRD (modified)	Continuous (simple)	No	Increasing strictly convex	PTS vs. PTC vs. proposition	NE existence Fairness Cheating
Chen et al. [9]	Residential optimal energy management	Proximal point, synchronous agreement, asynchronous gossip	Continuous (simple)	No	Polynomial	PTS	NE existence NE uniqueness NE optimality Fairness Cheating
Chakraborty and Khargonekar [10]	Optimal management of demand response	None (theoretical demonstration only)	Continuous (simple with limits)	No	Positive, concave, differentiable, monotonically increasing	PTS	NE existence PoA
Bahrami and Wong [11]	Electric vehicles optimal charging coordination	Subgradient method	Continuous (charging rates)	No	Quadratic	PTS and dissatisfaction	NE existence NE uniqueness
Rahman et al. [12]	Residential optimal energy management	BRD (secure)	Continuous (simple)	No	Quadratic	PTS vs. PTC	NE existence Cheating Info losses Privacy

Liang et al. [13]	Residential optimal energy management	Projected gradient	Continuous (simple)	No	Quadratic	PTC	NE existence NE optimality Info losses Privacy
Baharlouei et al. [14]	Residential optimal energy management	Fast convergent BRD model	Continuous (simple)	No	Increasing strictly convex	PTC	NE existence NE uniqueness NE optimality
Fernandez et al. [15]	Optimal energy management of neighborhood	Round-robin	Continuous (simple)	No	Quadratic	PTC	NE existence NE uniqueness NE optimality
Noor et al. [16]	Optimal energy management of microgrids	BRD	Continuous (simple with limits)	No	Quadratic cost with users' discomfort	PTC	NE existence Privacy
Wang et al. [17]	Social welfare optimization with electric vehicles charging response	Spatial adaptive play	Continuous	No	Benefits of DR minus generation costs	According to the total cost	NE existence NE optimality
Collins and Middleton [18]	Peak demand response	Iterative synchronous BRD	Continuous (simple)	No	Compound function	According to the total cost	NE existence PoA Privacy
Zhou et al. [19]	Electric vehicles online optimal charging coordination	Queues (Lyaponuv method)	Continuous (simple)	No	Convex functions	PTC	ϵ-NE existence ϵ-NE optimality Privacy
Jacquot et al. [20]	Offline/online demand response optimization	Cycling BRD, projected gradient descent	Continuous	Yes	Affine, positive, increasing	PTS	NE existence NE uniqueness NE optimality PoA

Continued

Table 11.1 Continued

Reference	Application	Implementing method	Load model		Utility function		Aspects analyzed
			Variables	Energy variant	Total cost	Billing	
Gong et al. [21]	Coordination of flexible devices to offer energy reserve	Agent-based model (Lyaponuv method)	Continuous (simple with limits)	No	Electricity costs, reserve revenue, EVs' discomfort	Revenue minus costs	NE existence NE optimality
Tang et al. [22]	Optimal energy management of building	BRD with Nikaido-Isoda analysis	Continuous	Yes	Peak pricing	Constant factor for each consumer	NE existence
Tang et al. [23]	Optimal energy management of building	Closed formula given by Stackelberg solution	Continuous	No	Leader: profit, Follower: bill, dissatisfaction, mismatching	According to the total cost	NE existence NE uniqueness NE optimality
De Paola et al. [24]	Optimal management of TCLs to provide frequency support	Distributed price-based coordination	Continuous	Yes	Quadratic costs	Revenue minus costs	Equilibrium existence
Bhatti and Broadwater [25]	Optimal energy management of microgrids	Extremum seeking approach	Continuous (generic)	No	Nonquadratic	Revenue minus costs	NE existence NE uniqueness NE optimality Fairness
Eksin et al. [26–28]	Residential optimal energy management with renewables	BRD	Continuous (preferences described by constant)	No	Quadratic	PTS	BNE existence BNE uniqueness
Wang et al. [29]	Optimal energy management of microgrids	BRD	Continuous and integer	Yes	Benefits of DR minus generation costs	According to the total cost	NE existence

Wang et al. [30]	Social welfare optimization	Spatial adaptive play under network anomaly	Continuous and integer	No	Benefits of DR minus generation costs	According to the total cost	NE existence NE optimality Info losses
Zhu et al. [31]	Residential optimal energy management	BRD	Integer (simple)	No	Quadratic	PTC	NE existence NE optimality
Barbato et al. [32]	Residential optimal energy management	BRD	Integer (simple)	No	Regular functions	PTS	NE existence
Barbato et al. [33]	Residential optimal energy management	Reinforcement learning	Integer (simple)	No	Increasing functions	PTS	NE existence Fairness
Yaagoubi and Mouftah [34]	Residential optimal energy management	Reinforcement learning	Integer (simple)	No	Quadratic cost with users' discomfort	PTS	Correlated equilibrium existence
Rottondi et al. [35]	Residential optimal energy management	Round-robin	Integer (simple)	No	Regular functions	PTS	NE existence Privacy
Zeng et al. [36]	Optimal energy management of microgrids	BRD	Integer	Yes	Sum of utilities	Depending on DER	NE existence NE optimality Info losses

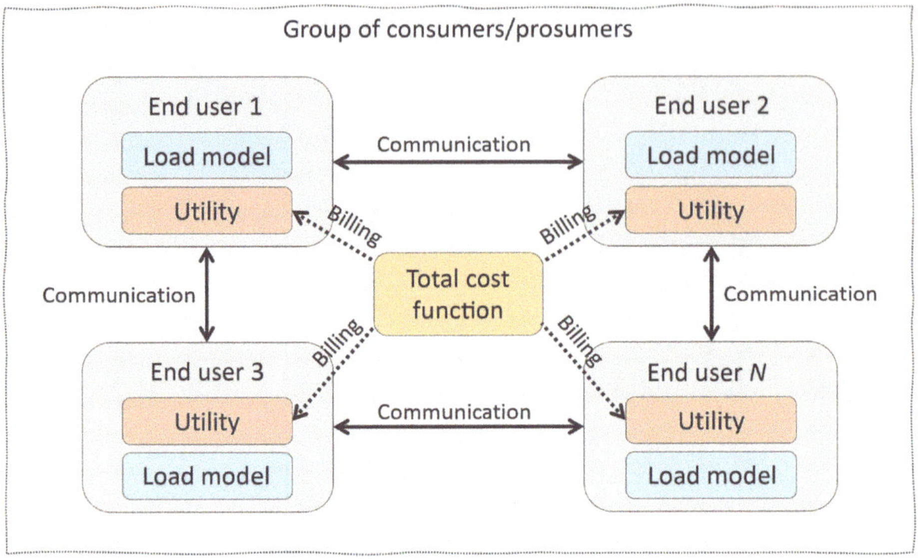

Fig. 11.1 General framework to model the DSM problem using noncooperative games.

3.1 Load modeling

In the literature, different load models are used to represent the operation of end users' appliances and the possible scheduling strategies. One general mathematical formulation often adopted considers simple load models to represent shiftable (e.g., washing machines) and/or interruptible (e.g., electric vehicles) appliances. Consider $x_{n,a}^t$ as a continuous decision variable on the amount of energy appliance a of consumer n will be using at time slot t. Following the preferences of this consumer for the use of the load, his/her constraints are defined as [6]

$$\sum_{t=\alpha_{n,a}}^{\beta_{n,a}} x_{n,a}^t = E_{n,a}^{tot} \quad \forall n \in \mathcal{N} \ \forall a \in \mathcal{I}_n \tag{11.1}$$

in which $\alpha_{n,a}$ is the earliest starting time of this appliance, $\beta_{n,a}$ is the latest ending time, $E_{n,a}^{tot}$ is the total energy consumption of the appliance, and $\mathcal{I}_n$ is the set of interruptible loads of consumer n. A constraint for consumption on standby mode can also be added:

$$\gamma_{n,a}^{\min} \leq x_{n,a}^t \leq \gamma_{n,a}^{\max} \quad \forall n \in \mathcal{N} \ \forall a \in \mathcal{I}_n \tag{11.2}$$

in which $\gamma_{n,a}^{\min}$ and $\gamma_{n,a}^{\max}$ are minimum and maximum consumption, respectively, of the appliance. If the load has an on/off operation mode, then Eqs. (11.1), (11.2) can be expressed as [32]

$$\sum_{t=\alpha_{n,a}}^{\beta_{n,a}} x_{n,a}^{t} = \zeta_{n,a} \quad \forall n \in \mathcal{N} \ \forall a \in \mathcal{I}_n \tag{11.3}$$

$$\gamma_{n,a}^{\min} \leq E_{n,t} x_{n,a}^{t} \leq \gamma_{n,a}^{\max} \quad \forall n \in \mathcal{N} \ \forall a \in \mathcal{I}_n \tag{11.4}$$

In this case, the variables $x_{n,a}^{t}$ are binary and $\zeta_{n,a}$ is the duration of the appliances cycle. In addition, $E_{n,a}$ is the power rate of the load (i.e., $E_{n,a}^{tot} = E_{n,a}\zeta_{n,a}$).

In some cases, shiftable appliances have a predefined consumption schedule. Therefore, their load model must be represented by binary variables representing the time they are turned on [31]

$$\sum_{t=\alpha_{n,a}}^{\beta_{n,a}-d_{n,a}+1} y_{n,a}^{t} = 1 \quad \forall n \in \mathcal{N} \ \forall a \in \mathcal{U}_n \tag{11.5}$$

in which $y_{n,a}^{t}$ is the binary decision variable representing the starting time of the appliance. Although many literature models use those simple equations to describe any type of appliances, more complex load models can also be included. For example, to represent the operation of thermal loads with variables $x_{n,a}^{t}$, the following constraints can be used [37]:

$$\theta_{n,a}^{t} \geq \theta_{n,a}^{\min,t} \quad \forall n \in \mathcal{N} \ \forall a \in \mathcal{H}_n \ \forall t \in \mathcal{T} \tag{11.6}$$

$$\theta_{n,a}^{t} \leq \theta_{n,a}^{\max,t} \quad \forall n \in \mathcal{N} \ \forall a \in \mathcal{H}_n \ \forall t \in \mathcal{T} \tag{11.7}$$

$$\theta_{n,a}^{t} = f(x_{n,a}^{t}, \theta_{n,a}^{t-1}, \Delta) \tag{11.8}$$

in which $\theta_{n,a}^{t}$ is the temperature related to the thermal load (e.g., room temperature for air conditioners [ACs], water temperature for water heaters), $\theta_{n,a}^{\min,t}$ and $\theta_{n,a}^{\max,t}$ are the minimum and maximum temperature preferences, respectively, of the appliance owner, $f(.)$ is the operation model of the thermal load, and $\mathcal{H}_n$ is the set of thermal appliances of consumer n. For example, ACs with on/off operation model can be described by

$$\begin{aligned} \theta_{n,a}^{t} &= f(x_{n,a}^{t}, \theta_{n,a}^{t-1}, TH_{n,a}, R_{n,a}, \eta_{n,a}, E_{n,a}, \theta_{t}^{et}, \delta) \\ &= \theta_{n,a}^{t-1} - \frac{\delta}{TH_{n,a}R_{n,a}}(\theta_{n,a}^{t-1} - \theta_{t}^{et} + \eta_{n,a}R_{n,a}E_{n,a}x_{n,a}^{t}) \end{aligned} \tag{11.9}$$

in which δ is the time slot size (in hours), $TH_{n,a}$ is the AC thermal capacity, $R_{n,a}$ its thermal resistance, $E_{n,a}$ its power rate, $\eta_{n,a}$ its performance, and θ_{t}^{et} is the external temperature. The variables $x_{n,a}^{t}$ are binary.

Uncontrollable loads are added to the model as a deterministic parameter w_n defined for each end user n. In a day-ahead implementation, this value can be forecasted based on past consumption.

Depending on the model used to define consumers' preferences and constraints, an energy scheduling vector $l_{n,t}$ for each consumer can be defined. For example, if integer variables are used to represent the operation of shiftable and thermal loads, the total consumption vector of the scheduling horizon is

$$l_{n,t} = \delta\left(w_{n,t} + \sum_{a \in \mathcal{U}_n} E_{n,a} y_{n,a}^{t} + \sum_{a \in \mathcal{H}_n} E_{n,a} x_{n,a}^{t}\right) \tag{11.10}$$

In summary, a *feasible* energy consumption scheduling set $\mathcal{S}_n$ for each user n can be defined, which includes all possible scheduling vectors respecting their preferences:

$$\mathcal{S}_n = \{\mathbf{l}_n = [l_{n,1}, l_{n,2}, \ldots, l_{n,T}] \in \mathbb{R}^T : \text{users' preferences; decision variables' domain}\} \tag{11.11}$$

The energy consumption vector of the group of participants can be defined as

$$L_t = \sum_{n \in \mathcal{N}} l_{n,t} \quad \forall t \in \mathcal{T} \tag{11.12}$$

It is worth mentioning that other distributed energy resources (DERs) as distributed generation or storage can also be added to the model. However, this chapter is focused on the control of consumers' load.

3.2 Total cost function

Another core building block to define the optimization of residential loads is the cost function. It is applied to the group of consumers and drives load control in order to achieve specific objectives (e.g., reduce the group's peak). Cost function defines consumers' bills and determines their willingness to participate in TC. Two cost models are studied in this chapter: the common quadratic cost function and a peak pricing function.

3.2.1 Quadratic cost function

Extensively used in the literature of TC, quadratic cost functions are strictly convex, allowing the application of potential games and the study of theoretical aspects. They define the total community cost in each scheduling time slot $t \in \mathcal{T}$ as

$$C_t^Q(L_t) = a_t (L_t)^2 + b_t L_t \tag{11.13}$$

Thus, the total community cost of the day-ahead operations planning is

$$C^Q(\mathbf{L}) = \sum_{t \in \mathcal{T}} C_t^Q(L_t) \tag{11.14}$$

where $a_t > 0$ and $b_t \geq 0$ are constants and can be time varying, that is, they have different values for different time slots of the day. It is worth mentioning that the fixed component on the quadratic cost function (11.13) is not considered, because the scheduling of controllable appliances only affects the volumetric part of electricity costs [38]. Fixed costs and/or demand rates do not depend on the amount consumed, and therefore they can be charged independently of this function.

In practice, Eq. (11.14) can represent real energy costs associated with thermal generation or power losses or specific tariffs contracted with aggregators or retailers. For example, two-step tariffs used to encourage consumers to reduce their energy load, as applied by British Columbia (BC) Hydro in Canada [39], are piecewise linear and can be approximated by quadratic functions.

The solution to the problem described earlier that minimizes the total system cost for a group of consumers $\mathcal{N}$ can be calculated as the following (mixed-integer) quadratic optimization model,[a] with the decision variables constrained by the scheduling set defined in Eq. (11.11):

$$C_{\mathcal{N}}^{Q*} = \min_{\substack{\mathbf{l}_n \in \mathcal{S}_n \\ n \in \mathcal{N}}} C^Q(\mathbf{L}) \tag{11.15}$$

3.2.2 Peak pricing function

Peak pricing models represent energy costs by a combination of a retail energy price and a demand charge, which is a more common structure in residential tariffs [40, 41]. Assuming the volumetric energy component as variant in time, for example, a time-of-use (TOU) tariff, the total cost of a community of consumers can be written as

$$C^P(\mathbf{L}) = \sum_{t \in \mathcal{T}} c_t L_t + d \max_{t \in \mathcal{T}} \frac{L_t}{\delta} \tag{11.16}$$

In which $c_t > 0 \; \forall \, t \in \mathcal{T}$ are the TOU tariffs for the energy consumption, and $d > 0$ is the peak demand charge for the community. It is important to note that this function is convex, as it results from the sum of two convex functions [42]. The solution that minimizes the costs of a community composed by a group of consumers $\mathcal{N}$ can be calculated as the following (mixed-integer) linear program,[b] in which ρ_t are auxiliary variables for computing the peak load:

[a] If the variables are only continuous, the model is a QP; if integer load models are added, the model is a MIQP.

[b] Again, if only continuous control is considered, the model is an LP; if integer loads are added, the model becomes a MILP.

$$
\begin{aligned}
C_{\mathcal{N}}^{P*} &= \min \sum_{t\in\mathcal{T}} c_t L_t + d\rho_t \\
\rho_t &\geq \frac{L_t}{\delta} \quad \forall t\in\mathcal{T} \\
L_t &= \sum_{n\in\mathcal{N}} l_{n,t} \quad \forall t\in\mathcal{T} \\
\mathbf{l}_n &\in \mathcal{S}_n \quad \forall n\in\mathcal{N}
\end{aligned}
\tag{11.17}
$$

3.3 Billing functions for defining consumers' utilities

The problem of dividing the total community costs between individual consumers is determined by the billing function. The billing function is key to design TC and many alternatives have been studied in the literature. In this chapter, two different billing mechanisms for the TC are presented: (1) PTC and (2) PTS.

3.3.1 Proportional to consumption

Proportional to consumption (PTC) is a very popular billing mechanism used in the literature for sharing the cost among participants of a scheduling game. Consumers' utilities are defined by a constant share f_n multiplied by the total cost of the energy provision:

$$u_n^C(\mathbf{l}_n, \mathbf{l}_{-n}) = -f_n C(\mathbf{l}_n, \mathbf{l}_{-n}) \tag{11.18}$$

$$f_n = \frac{\Sigma_{t\in\mathcal{T}} l_{n,t}}{\Sigma_{t\in\mathcal{T}} L_t} \tag{11.19}$$

In this billing setting, if two participants n and m have total load $\sum_{t\in\mathcal{T}} l_{n,t} = \beta \sum_{t\in\mathcal{T}} l_{m,t}$ after the scheduling game is solved, then consumer n will pay β times consumer m's bill. Moreover, the sum of all consumers' payments will be equal to the community's total cost: note that $\sum_{n\in\mathcal{N}} \sum_{t\in\mathcal{T}} l_{n,t} = \sum_{t\in\mathcal{T}} L_t$.

3.3.2 Per time slot

This mechanism shares the total cost of the transactive solution according to consumers' energy use at each time slot. Therefore, their utility function is written as:

$$u_n^S(\mathbf{l}_n, \mathbf{l}_{-n}) = -\sum_{t\in\mathcal{T}} \frac{l_{n,t}}{L_t} C_t(l_{n,t}, l_{-n,t}) \tag{11.20}$$

This equation depends on consumers' load in each time slot. If applied to divide a quadratic total cost, the previous equation can be rewritten as Eq. (11.21), where $P_t(l_{n,t}, l_{-n,t}) = a_t L_t + b_t$ are linear prices arising from the game.

$$u_n^S(\mathbf{l}_n,\mathbf{l}_{-n}) = -\sum_{t\in\mathcal{T}} l_{n,t}P_t(l_{n,t},l_{-n,t}) \tag{11.21}$$

3.4 Noncooperative scheduling game

Given a load model for consumers' appliances, a total cost function, and a billing mechanism, the noncooperative scheduling game can be designed to coordinate their consumption and optimize their bills. Using the strategic form, the game is a tuple $\Gamma = \langle \mathcal{N}, (\mathcal{S}_n)_{n\in\mathcal{N}}, \{u_n\}_{n\in\mathcal{N}} \rangle$, where

$\mathcal{N} = \{1,2,\ldots,N\}$ is the set of players.
$\mathcal{S}_n = \{\mathbf{l}_n\}_{\mathbf{l}_n\in\mathcal{S}_n}$ denotes the action space for consumer $n\in\mathcal{N}$. This set is composed by feasible scheduling vectors $\mathbf{l}_n$ that respect users' preferences—see Eq. (11.11). Moreover, $\mathcal{S} = \times_{n=1}^{N}\mathcal{S}_n$ is the joint action space.
u_n: $\mathcal{S} \mapsto \mathbb{R}$ is the utility function that defines user $n\in\mathcal{N}$ payoff. It can be written as a function of the actions chosen by all players $(\mathbf{l}_n,\mathbf{l}_{-n})\in\mathcal{S}$, where $\mathbf{l}_n$ is the scheduling vector of player n, and $\mathbf{l}_{-n} = [\mathbf{l}_m]_{m\neq n}$ are the scheduling vectors of all players except n.

As mentioned in the last section, the utility function depends on the billing mechanism used to share the total cost of the load scheduling among consumers. Some examples were described in Eqs. (11.18), (11.21).

4 Game aspects

After defining the noncooperative scheduling game, the following aspects can be analyzed:

- existence of equilibrium;
- optimality of equilibrium;
- uniqueness of equilibrium;
- billing fairness;
- billing strategy proof; and
- efficiency of equilibria (PoA).

These aspects influence the efficiency and the applicability of the TC. Moreover, depending on those characteristics, different algorithms are suitable for solving the game. Finally, those aspects impact practical applications as end users' engagement, infrastructure requirements, and regulation.

4.1 Existence of Nash equilibria

The existence of NE for a noncooperative scheduling game depends on its nature. For the readers who are unfamiliar with this concept, an NE is a solution concept for games and is defined as a strategy profile $(\mathbf{l}_n^*,\mathbf{l}_{-n}^*)$ of scheduling vectors from which no player has incentive to deviate:

$$u_n(\mathbf{l}_n^*,\mathbf{l}_{-n}^*) \geq u_n(\mathbf{l}_n,\mathbf{l}_{-n}^*) \quad \forall\, \mathbf{l}_n\in\mathcal{S}_n \tag{11.22}$$

Different games proposed in the literature apply multiple theoretical frameworks to verify if a game has an equilibrium (which is the solution of the problem)—see Table 11.1. One widely used concept to verify the existence of NE is potential games:

Definition 11.1 Exact potential games [43]. The game Γ is an exact potential game, if there exists a potential function ϕ: $\mathcal{S} \mapsto \mathbb{R}$ such that, for every player $n \in \mathcal{N}$, for every opponents' strategy $\mathbf{l}_{-n} \in \mathcal{S}_{-n}$, and for every two strategies of player n, $\mathbf{l}_n$, $\mathbf{q}_n \in \mathcal{S}_n$, it holds that

$$u_n(\mathbf{l}_n, \mathbf{l}_{-n}) - u_n(\mathbf{q}_n, \mathbf{l}_{-n}) = \phi(\mathbf{l}_n, \mathbf{l}_{-n}) - \phi(\mathbf{q}_n, \mathbf{l}_{-n}) \tag{11.23}$$

Exact potential games always have at least one NE, which can be easily calculated using one-sided best-response algorithms (as shown in the following sections). Examples of games with and without an NE are given next.

4.1.1 Game with PTC billing to schedule energy invariant loads

When the game is modeled with a PTC billing—see Eq. (11.18), the existence of NE depends on the nature of the load model. For instance, it is guaranteed when only energy invariant loads are considered—for example, shiftable/interruptible appliances following Eqs. (11.1), (11.3), (11.5). Those loads have a fixed amount of energy to be scheduled in the time horizon $E_{n,a}^{tot}$.

In addition, if the variables are continuous and the total cost function is strictly convex, the NE is unique and minimizes the total system cost [14, 19]. Moreover, if the variables are integer, the game is an exact potential with potential function equal to the total system cost [31], which means that consumers try to minimize the total system cost.

The proof is straightforward for all cases of variables types and total cost functions with minimum values. Because f_n in Eq. (11.18) is a constant for energy invariant loads, consumers will try to optimize $C(\mathbf{l}_n, \mathbf{l}_{-n})$ to minimize their utilities. First, if the variables are continuous and the total cost function is strictly convex, as in the case of the quadratic total cost $C^Q(\mathbf{l}_n, \mathbf{l}_{-n})$ defined in Eq. (11.13), this function has a unique optimum value, thus the solution of the game will reach this minimum. Moreover, as f_n is fixed for all $n \in \mathcal{N}$, consumers' bills will be unique (and minimum)—see Example 11.1. Second, if the total cost is convex, as in the peak pricing total cost function $C^P(\mathbf{l}_n, \mathbf{l}_{-n})$ defined in Eq. (11.16), it also has an attainable minimum (both global and local are optimum) that will be the solution of the game. Third, if the variables are integer, the total cost $C(\mathbf{l}_n, \mathbf{l}_{-n})$ will be the exact potential function ϕ of Definition 11.1. However, in this last case, suboptimal solutions of the game can be the NE, as can be seen in Example 11.2.

Example 11.1. Consider a scheduling game between three consumers $\mathcal{N} = \{1,2,3\}$. Each participant has one appliance to schedule in two time slots. The total energy to be scheduled is $E_1^{tot} = 1$ kWh, $E_2^{tot} = 2$ kWh, and $E_3^{tot} = 3$ kWh. The total cost of the resulting schedule is a quadratic function of the form

$\phi = \sum_{t=1}^{2} a_t L_t^2$, in which $L_t = l_{1,t} + l_{2,t} + l_{3,t}$, $l_{n,t}$ is the energy consumer n places in time slot t, $a_1 = 1$ and $a_2 = 2$. The cost for n is proportional to his/her consumption: $u_n = \frac{E_n^{tot}}{\sum_{m \in \mathcal{N}} E_m^{tot}} \phi$. This is a potential game with potential function equal to ϕ. If the strategies $l_{n,t}$ are continuous, the set of possible scheduling vectors can be defined as $\mathcal{S}_n = \{[l_{n,1}, E_n^{tot} - l_{n,1}]\}$. The total cost function can be rewritten as $\phi = (l_{1,1}+l_{2,1}+l_{3,1})^2 + 2(6-l_{1,1}-l_{2,1}-l_{3,1})^2$. Each consumer seeks to minimize his/her utility. Thus, by taking the derivative $\frac{\partial u_n}{\partial l_{n,1}} = 0$, any solution $(l_{1,1}, l_{2,1}, l_{3,1})$ that satisfies $l_{1,1} + l_{2,1} + l_{3,1} = 4$ and $0 \le l_{n,1} \le E_n^{tot}$, $n = \{1, 2, 3\}$, is an NE. Even though there are infinity possible combinations, all of them have total cost equal to 24 and in all of them consumer 1's payoff is $u_1 = -4$, consumer 2 is $u_2 = -8$, and consumer 3 is $u_2 = -12$.

Example 11.2. Let us assume the game in Example 11.1 is integer and the power rate of consumers' appliances is equal to E_n^{tot}. Thus, the set of possible strategies for each consumer is reduced to turning it on at one of the two time slots: $\mathcal{S}_n = \{[E_n^{tot}, 0]; [0, E_n^{tot}]\}$. This new version of the scheduling game in matrix form is represented in Table 11.2. All possible combinations of consumers' strategies and the resulting cost for each of them are shown in the table. The best responses of each player to the opponents' strategies are underlined in the table. This game has three NEs: one equal to the continuous version (and with optimal total cost), which strategies are $sNE_1 = \{l_1 = [1, 0], l_2 = [0, 2], l_3 = [3, 0]\}$; and two suboptimal with total cost equal to 27, which strategies are $sNE_2 = \{l_1 = [0, 1], l_2 = [2, 0], l_3 = [3, 0]\}$ and $sNE_3 = \{l_1 = [1, 0], l_2 = [2, 0], l_3 = [0, 3]\}$, and utilities are $u_1 = -4.5$, $u_2 = -9$, and $u_3 = -13.5$.

Table 11.2 Example 11.2 payoff matrix of the game with integer and energy invariant loads, considering a PTC billing.

l_3 = [3, 0]	**l_2 = [2, 0]**	**l_2 = [0, 2]**
l_1 = [1, 0]	−6.0, −12.0, −18.0	<u>−4.0</u>, <u>−6.0</u>, <u>−12.0</u>
l_1 = [0, 1]	<u>−4.5</u>, <u>−9.0</u>, <u>−13.5</u>	−4.5, <u>−9.0</u>, <u>−13.5</u>
l_3 = [0, 3]	l_2 = [2, 0]	l_2 = [0, 2]
l_1 = [1, 0]	<u>−4.5</u>, <u>−9.0</u>, <u>−13.5</u>	<u>−8.5</u>, −17.0, −25.5
l_1 = [0, 1]	−6.0, <u>−12.0</u>, −18.0	−12.0, −24.0, −36.0

Underlined values represent best response strategies.

4.1.2 Game with PTC billing to schedule energy variant loads

For energy variant loads with a multiperiod scheduling characteristic, as thermal loads in Eq. (11.8), the total energy to be scheduled is no longer fixed. This energy variant nature results from the fact that thermal loads are not purely shiftable and entail the so-called "energy payback" [44, 45]. In other words, shifting thermal loads in time while maintaining comfort standards implies overall energy variations in relation to a base consumption. In fact, since the comfort constraints of thermal loads' owners are related to temperature targets, different scheduling solutions result in a different overall energy consumption. Thus, looking at Eq. (11.18), the factor multiplying the total cost function is no longer constant, depending on the variables results, and consumers no longer seek to minimize the total cost. Under this scenario, the game might not be potential, as can be seen in Theorem 11.1 for the quadratic total cost function.

Theorem 11.1. *The noncooperative scheduling game* $\boldsymbol{\Gamma} = \langle \mathcal{N}, (\mathcal{S}_n)_{n\in\mathcal{N}}, \{u_n^C\}_{n\in\mathcal{N}}\rangle$ *with* $\mathcal{S}_n$ *defined in Eq. (11.11), relaxing the integer constraints, that is,* $0 \le y_{n,t} \le 1$, *and* u_n^C *equals to Eq. (11.18) with* $C(\boldsymbol{l}_n, \boldsymbol{l}_{-n}) = C^Q(\boldsymbol{l}_n, \boldsymbol{l}_{-n})$ *(PTC billing to divide a total quadratic cost) is not a potential game.*

Proof. Consider a two-player game $\mathcal{N} = \{n, -n\}$ (one player against his/her opponents), the strategies of player n as $\mathbf{l}_n = \{l_{n,1}, l_{n,2}, \ldots, l_{n,t}\}$, and the coupled strategies of opponents $-n$ as $\mathbf{q}_{-n} = \{q_{-n,1}, q_{-n,2}, \ldots, q_{-n,t}\}$, that is, $q_{-n,t} = \sum_{n\in\mathcal{N}\setminus n} l_{n,t}$. To simplify the notation, consider $\sum_{t\in\mathcal{T}}(a_t(l_{n,t}+q_{-n,t})^2 + b_t(l_{n,t}+q_{-n,t})) = C^Q(\mathbf{l}_n, \mathbf{q}_{-n})$. The first derivatives of player n's utility with respect to variables $\mathbf{q}_{-n}$ are

$$\frac{\partial u_n^C(\mathbf{l}_n, \mathbf{q}_{-n})}{\partial \mathbf{q}_{-n}} = \frac{\partial}{\partial \mathbf{q}_{-n}}\left[-\frac{\sum_{t\in\mathcal{T}} l_{n,t}}{\sum_{t\in\mathcal{T}}(l_{n,t}+q_{-n,t})} C^Q(\mathbf{l}_n, \mathbf{q}_{-n})\right]$$

$$= \left[\frac{\sum_{t\in\mathcal{T}} l_{n,t}}{\left(\sum_{t\in\mathcal{T}} l_{n,t}+q_{-n,t}\right)^2} C^Q(\mathbf{l}_n, \mathbf{q}_{-n}) - \frac{\sum_{t\in\mathcal{T}} l_{n,t}}{\sum_{t\in\mathcal{T}}(l_{n,t}+q_{-n,t})}(2a_v(l_{n,v}+q_{-n,v})+b_v)\right]_{v\in\mathcal{T}} \tag{11.24}$$

The first derivatives are a vector with values defined for each time slot $v \in \mathcal{T}$. Considering $c_v = 2a_v(l_{n,v} + q_{-n,v}) + b_v$ for all $v \in \mathcal{T}$, and $c_h = 2a_h(l_{n,h} + q_{-n,h}) + b_h$ for all $h \in \mathcal{T}$, $h \neq v$, the second derivatives of player n's utility with respect to variables $\mathbf{l}_n$ are

$$\left.\frac{\partial^2 u_n^C(\mathbf{l}_n,\mathbf{q}_{-n})}{\partial \mathbf{l}_n \partial \mathbf{q}_{-n}}\right|_{h \neq v} = \left[\frac{1}{\left(\sum_{t\in\mathcal{T}} l_{n,t}+q_{-n,t}\right)^2} C^Q(\mathbf{l}_n,\mathbf{q}_{-n}) - 2\frac{\sum_{t\in\mathcal{T}} l_{n,t}}{\left(\sum_{t\in\mathcal{T}} l_{n,t}+q_{-n,t}\right)^3} C^Q(\mathbf{l}_n,\mathbf{q}_{-n}) + \frac{\sum_{t\in\mathcal{T}} l_{n,t}}{\left(\sum_{t\in\mathcal{T}} l_{n,t}+q_{-n,t}\right)^2} c_h - \frac{1}{\sum_{t\in\mathcal{T}}(l_{n,t}+q_{-n,t})} c_v + \frac{1}{\sum_{t\in\mathcal{T}}(l_{n,t}+q_{-n,t})} c_v\right] \tag{11.25}$$

$$\left.\frac{\partial^2 u_n^C(\mathbf{l}_n,\mathbf{q}_{-n})}{\partial \mathbf{l}_n \partial \mathbf{q}_{-n}}\right|_{h = v} = \left[\frac{1}{\left(\sum_{t\in\mathcal{T}} l_{n,t}+q_{-n,t}\right)^2} C^Q(\mathbf{l}_n,\mathbf{q}_{-n}) - 2\frac{\sum_{t\in\mathcal{T}} l_{n,t}}{\left(\sum_{t\in\mathcal{T}} l_{n,t}+q_{-n,t}\right)^3} C^Q(\mathbf{l}_n,\mathbf{q}_{-n}) + \frac{\sum_{t\in\mathcal{T}} l_{n,t}}{\left(\sum_{t\in\mathcal{T}} l_{n,t}+q_{-n,t}\right)^2} c_v - \frac{1}{\sum_{t\in\mathcal{T}}(l_{n,t}+q_{-n,t})} c_v + \frac{1}{\sum_{t\in\mathcal{T}}(l_{n,t}+q_{-n,t})} c_v - 2\frac{\sum_{t\in\mathcal{T}} l_{n,t}}{\sum_{t\in\mathcal{T}}(l_{n,t}+q_{-n,t})} a_v\right] \tag{11.26}$$

Therefore, $\frac{\partial^2 u_n^C(\mathbf{l}_n,\mathbf{q}_{-n})}{\partial \mathbf{l}_n \partial \mathbf{q}_{-n}}$ is a $T \times T$ matrix with diagonal values equal to Eq. (11.26) and off-diagonal values equal to Eq. (11.25).

On the other hand, the first derivatives of player $-n$'s utility with respect to variables $\mathbf{q}_{-n}$ are

$$\frac{\partial u^C_{-n}(\mathbf{l}_n,\mathbf{q}_{-n})}{\partial \mathbf{q}_{-n}} = \frac{\partial}{\partial \mathbf{q}_{-n}}\left[-\frac{\sum_{t\in\mathcal{T}} q_{-n,t}}{\sum_{t\in\mathcal{T}}(l_{n,t}+q_{-n,t})}C^Q(\mathbf{l}_n,\mathbf{q}_{-n})\right]$$

$$= \left[-\frac{1}{\sum_{t\in\mathcal{T}}(l_{n,t}+q_{-n,t})}C^Q(\mathbf{l}_n,\mathbf{q}_{-n}) + \frac{\sum_{t\in\mathcal{T}} q_{-n,t}}{\left(\sum_{t\in\mathcal{T}} l_{n,t}+q_{-n,t}\right)^2}C^Q(\mathbf{l}_n,\mathbf{q}_{-n})\right.$$

$$\left. -\frac{\sum_{t\in\mathcal{T}} q_{-n,t}}{\sum_{t\in\mathcal{T}}(l_{n,t}+q_{-n,t})}(2a_v(l_{n,v}+q_{-n,v})+b_v)\right]_{v\in\mathcal{T}} \tag{11.27}$$

and the second derivatives of player $-n$'s utility with respect to variables l_n are

$$\left.\frac{\partial^2 u^C_{-n}(\mathbf{l}_n,\mathbf{q}_{-n})}{\partial \mathbf{l}_n \partial \mathbf{q}_{-n}}\right|_{h\neq v} = \left[\frac{1}{\left(\sum_{t\in\mathcal{T}} l_{n,t}+q_{-n,t}\right)^2}C^Q(\mathbf{l}_n,\mathbf{q}_{-n}) - \frac{1}{\sum_{t\in\mathcal{T}}(l_{n,t}+q_{-n,t})}c_h\right.$$

$$-2\frac{\sum_{t\in\mathcal{T}} q_{-n,t}}{\left(\sum_{t\in\mathcal{T}} l_{n,t}+q_{-n,t}\right)^3}C^Q(\mathbf{l}_n,\mathbf{q}_{-n}) + \frac{\sum_{t\in\mathcal{T}} q_{-n,t}}{\left(\sum_{t\in\mathcal{T}} l_{n,t}+q_{-n,t}\right)^2}c_h,$$

$$\left.+\frac{\sum_{t\in\mathcal{T}} q_{-n,t}}{\sum_{t\in\mathcal{T}}(l_{n,t}+q_{-n,t})}c_v\right] \tag{11.28}$$

$$\left.\frac{\partial^2 u_{-n}^C(\mathbf{l}_n,\mathbf{q}_{-n})}{\partial \mathbf{l}_n \partial \mathbf{q}_{-n}}\right|_{h=v} = \left[\frac{1}{\left(\sum_{t\in\mathcal{T}} l_{n,t}+q_{-n,t}\right)^2} C^Q(\mathbf{l}_n,\mathbf{q}_{-n}) - \frac{1}{\sum_{t\in\mathcal{T}}(l_{n,t}+q_{-n,t})} c_v \right.$$
$$-2\frac{\sum_{t\in\mathcal{T}} q_{-n,t}}{\left(\sum_{t\in\mathcal{T}} l_{n,t}+q_{-n,t}\right)^3} C^Q(\mathbf{l}_n,\mathbf{q}_{-n}) + \frac{\sum_{t\in\mathcal{T}} q_{-n,t}}{\left(\sum_{t\in\mathcal{T}} l_{n,t}+q_{-n,t}\right)^2} c_v$$
$$\left. + \frac{\sum_{t\in\mathcal{T}} q_{-n,t}}{\sum_{t\in\mathcal{T}}(l_{n,t}+q_{-n,t})} c_v - 2\frac{\sum_{t\in\mathcal{T}} q_{-n,t}}{\sum_{t\in\mathcal{T}}(l_{n,t}+q_{-n,t})} a_v \right] \tag{11.29}$$

Again, $\frac{\partial^2 u_{-n}^C(\mathbf{l}_n,\mathbf{q}_{-n})}{\partial \mathbf{l}_n \partial \mathbf{q}_{-n}}$ is a $T \times T$ matrix with diagonal values equal to Eq. (11.29) and off-diagonal values equal to Eq. (11.28). One can notice that the derivatives for player n are different from the derivatives of player $-n$, because consumers' own variables on the fraction numerator are different. Therefore, Theorem 4.5 of Monderer and Shapley [43] is not respected for every n, $-n \in \mathcal{N}$, meaning the game with PTC billing, a quadratic cost function, and energy variant loads is not potential. □

Gopalakrishnan et al. [46, Th. 1] prove that, in cost sharing games,[c] a potential formulation is necessary to ensure the existence of a pure-strategy NE. Therefore, as a potential game formulation is no longer guaranteed for the game with continuous variables, the game does not have pure-strategy equilibria.

On the other hand, for the integer version, there is no theorem to generally demonstrate a game is not potential. Therefore, the integer load scheduling game with PTC billing and energy variant loads may or may not have Nash equilibria. This notion is discussed with the following example.

Example 11.3. Consider a game with PTC billing, any convex total cost function (e.g., quadratic function, peak pricing function, or another one), and energy variant loads. A consumer n has the choice between two total consumption levels: one equal to κ and the other equal to μ. The rest of the community has fixed total load

[c] The scheduling game analyzed here can be rewritten as a cost sharing game, considering each time slot as a resource and the appliances power consumption as different demands of those resources.

equal to ζ, which is equally split among the consumers. Consider that two total costs are possible for this game: $C_1 > C_2$. As the rest of the community has a fixed consumption, it prefers the smaller total cost C_2. However, consumer n can have a better utility when choosing the strategy leading to the total cost C_1, say strategy κ, and a worst utility when choosing strategy μ leading to C_2. Thus,

$$\frac{\kappa}{\kappa+\zeta}C_1 < \frac{\mu}{\mu+\zeta}C_2 \tag{11.30}$$

And consumer n chooses κ, enhancing the cost to C_1. As a response, the rest of the community manages to bring the total cost back to C_2, choosing ζ:

$$\begin{aligned} \frac{\zeta}{\kappa+\zeta}C_1 &> \frac{\zeta}{\kappa+\zeta}C_2 \\ C_1 &> C_2 \end{aligned} \tag{11.31}$$

For both Eqs. (11.30), (11.31) to be true, it is necessary that

$$\begin{aligned} \kappa(\mu+\zeta)C_1 &< \mu(\kappa+\zeta)C_2 \\ C_1 &< \frac{\mu(\kappa+\zeta)}{\kappa(\mu+\zeta)}C_2 \\ \frac{\mu(\kappa+\zeta)}{\kappa(\mu+\zeta)} &> 1 \\ \mu &> \kappa \end{aligned} \tag{11.32}$$

Therefore, an NE does not exist if consumer n has a strategy κ with total consumption smaller than μ, leading to a higher total cost to the community (C_1) in a way that this higher total cost does not surpasses $\frac{\mu(\kappa+\zeta)}{\kappa(\mu+\zeta)}C_2$. It is important to notice that, for this NE to not exist, both community and consumer n must have the possibility to choose a strategy to bring the cost back to C_1 or C_2 when the opponent changed it. That is the reason why the game is not potential with an infinite possibility of actions to change the total cost (continuous variables). However, with integer loads, it depends on the strategies the players have access to define if an NE exists.

4.1.3 Game with PTS billing and quadratic total cost

For continuous games, the PTS billing has been shown to have Nash equilibria [8, 20]. In the case in which the decision variables in the vectors $\mathrm{x}_{n,\,a}$ are binary, or shiftable loads are included (with decision variables $\mathrm{y}_{n,\,a}$), the existence proof must use the concept of exact potential games instead of derivatives. Function (11.33) is introduced to prove it is the potential function for the integer scheduling game with PTS to divide the total quadratic cost in Theorem 11.2. With this proof, the finite game is guaranteed to have at least one pure-strategy Nash equilibria [47]. A simple example in Example 11.4 is also shown to illustrate the application of the PTS billing to divide a total quadratic cost with integer and energy variant loads.

$$\phi(\mathbf{s}) = -\sum_{t\in\mathcal{T}}\left[\sum_{j\in\mathcal{N}} b_t l_{j,t} + \sum_{j\in\mathcal{N}} a_t (l_{j,t})^2 + a_t \sum_{j\in\mathcal{N}} \sum_{\substack{i\in\mathcal{N}\\ i<j}} l_{j,t} l_{i,t}\right] \tag{11.33}$$

Theorem 11.2. *The noncooperative scheduling game $\Gamma = \langle \mathcal{N}, (\mathcal{S}_n)_{n\in\mathcal{N}}, \{u_n^S\}_{n\in\mathcal{N}}\rangle$ with $\mathcal{S}_n$ defined in Eq. (11.11), and u_n^S equal to Eq. (11.21) (PTS billing to divide a total quadratic cost) is an exact potential game.*

Proof. To prove the theorem, it is necessary to show that for any two strategies $\mathbf{l}_n$ and $\mathbf{q}_n$, and for every player $n\in\mathcal{N}$, Eq. (11.23) holds. Starting from the left-hand side of Eq. (11.23), for a fixed player $n\in\mathcal{N}$:

$$\begin{aligned}
u_n^S(\mathbf{l}_n, \mathbf{l}_{-n}) - u_n^S(\mathbf{q}_n, \mathbf{l}_{-n}) &= -\sum_{t\in\mathcal{T}}\left(b_t + a_t \sum_{m\in\mathcal{N}} l_{m,t}\right) l_{n,t} \\
&\quad + \sum_{t\in\mathcal{T}}\left[b_t + a_t\left(q_{n,t} + \sum_{\substack{m\in\mathcal{N}\\ m\neq n}} l_{m,t}\right)\right] q_{n,t} \\
&= -\sum_{t\in\mathcal{T}}\left[b_t l_{n,t} + a_t (l_{n,t})^2 + a_t l_{n,t} \sum_{\substack{m\in\mathcal{N}\\ m\neq n}} l_{m,t}\right] \\
&\quad + \sum_{t\in\mathcal{T}}\left[b_t q_{n,t} + a_t (q_{n,t})^2 + a_t q_{n,t} \sum_{\substack{m\in\mathcal{N}\\ m\neq n}} l_{m,t}\right]
\end{aligned} \tag{11.34}$$

Consider $\phi_t(\mathrm{s})$ as the potential of a strategy profile s for a specific time slot $t \in \mathcal{T}$ of Eq. (11.33). The difference in the potential function when this same player changes his/her strategy from $\mathbf{l}_n$ to $\mathbf{q}_n$, which means the game strategy profile changes from $(\mathbf{l}_n, \mathbf{l}_{-n})$ to $(\mathrm{q}_n, \mathbf{l}_{-n})$, is

$$\begin{aligned}\phi_t(\mathbf{l}_n,\mathbf{l}_{-n}) - \phi_t(\mathbf{q}_n,\mathbf{l}_{-n}) &= -b_t l_{n,t} - b_t \sum_{\substack{m\in\mathcal{N}\\ m\neq n}} l_{m,t} - a_t (l_{n,t})^2 - a_t \sum_{\substack{m\in\mathcal{N}\\ m\neq n}} (l_{m,t})^2 \\ &\quad - a_t l_{n,t} \sum_{\substack{m\in\mathcal{N}\\ m\neq n}} l_{m,t} - a_t \sum_{\substack{m\in\mathcal{N}\\ m\neq n}} \sum_{\substack{j\in\mathcal{N}\\ j<m\\ j\neq n}} l_{j,t} l_{m,t} \\ &\quad + b_t q_{n,t} + b_t \sum_{\substack{m\in\mathcal{N}\\ m\neq n}} l_{m,t} + a_t (q_{n,t})^2 + a_t \sum_{\substack{m\in\mathcal{N}\\ m\neq n}} (l_{m,t})^2 \\ &\quad + a_t q_{n,t} \sum_{\substack{m\in\mathcal{N}\\ m\neq n}} l_{m,t} + a_t \sum_{\substack{m\in\mathcal{N}\\ m\neq n}} \sum_{\substack{j\in\mathcal{N}\\ j<m\\ j\neq n}} l_{j,t} l_{m,t} \\ &= -b_t l_{n,t} - a_t (l_{n,t})^2 - a_t l_{n,t} \sum_{\substack{m\in\mathcal{N}\\ m\neq n}} l_{m,t} \\ &\quad + b_t q_{n,t} + a_t (q_{n,t})^2 + a_t q_{n,t} \sum_{\substack{m\in\mathcal{N}\\ m\neq n}} l_{m,t}\end{aligned} \tag{11.35}$$

By summing up both sides of the earlier equation with respect to $t \in \mathcal{T}$, the difference in the potential function $\phi(\mathrm{s})$ is equal to Eq. (11.34). Therefore, Definition 11.1 holds, which means that the game proposed is an exact potential game with potential function equal to Eq. (11.33) and the proof is complete. □

Example 11.4. Consider a scheduling game between two consumers $\mathcal{N} = \{1,2\}$. Each participant has one AC to schedule in four time slots. The total cost of the resulting group's schedule is a quadratic function of the form $\phi = \sum_{t=1}^{4} a_t L_t^2$, in which $L_t = l_{1,t} + l_{2,t}$, $l_{n,t}$ is the energy consumer n places in time slot t, $a_1 = 1$, $a_2 = 1$, $a_3 = 2$, and $a_4 = 2$. To respect their temperature preferences, each consumer has three possible scheduling strategies (in kWh), that is, $\mathcal{S}_1 = \{[1,1,0,1],[0,1,1,0],[1,1,1,0]\}$ and $\mathcal{S}_2 = \{[0,0,2,2],[2,2,0,2],[0,2,2,0]\}$. One can notice that, by summing the consumption of the possible strategies, the flexible load is energy variant. Using a PTS billing, the game in matrix form is presented in Table 11.3. All possible combinations of consumers' strategies and the resulting cost for each of them are shown in the table. The best responses of each player to the opponents' strategies are underlined. This game

Table 11.3 Example 11.4 payoff matrix of the game with integer and energy variant loads, considering a PTS billing.

	$l_2 = [0, 0, 2, 2]$	$l_2 = [2, 2, 0, 2]$	$l_2 = [0, 2, 2, 0]$
$l_1 = [1, 1, 0, 1]$	$-8, -20$	$-12, -24$	$\underline{-6}, \underline{-14}$
$l_1 = [0, 1, 1, 0]$	$\underline{-7}, -20$	$\underline{-5}, \underline{-18}$	$-9, \underline{-18}$
$l_1 = [1, 1, 1, 0]$	$-8, -20$	$-8, -20$	$-10, \underline{-18}$

Underlined values represent best responses.

has two NEs: one with total cost equal to 20, in which consumers' utilities are $u_1 = -6$ and $u_2 = -14$; one with total cost equal to 23, in which consumers' utilities are $u_1 = -5$ and $u_2 = -18$.

4.2 Solution algorithm to solve potential games

Exact potential games have the finite improvement property, which assures the convergence to the equilibrium set of any myopic learning method based on the one-sided better reply dynamic [43]. Therefore, the BRD can be applied to solve some of the games present in the literature and discussed in the last sections—in Sections 4.1.1 and 4.1.3. This learning method consists of a sequential decision model in which consumers take turn and best respond to opponents' last strategies (appliances schedules). Consumers communicate the total group's load, instead of their individual profiles, which leads to less messages exchanges and reduces the privacy issue [12]. A brief overview of the algorithm is provided below, together with a pseudo-code in Algorithm 11.1.

Assuming that each consumer's home energy management system keeps track of opponents' total consumption vector ($\mathbf{L}^k_{-n}$), and the total load of all players is initialized with zeros $\mathrm{L}^0 = [0]$, each iteration of the gameplay consists of three steps:

1. Consumers receive the total load vector and calculate the load of opponents $\mathbf{L}^k_{-n}$ using their previous best strategies.
2. They update their strategies $\mathbf{l}^{k+1}_n$ as a response to $\mathbf{L}^k_{-n}$, by solving the local (mixed-integer) linear/quadratic[d] program (11.36)

$$\mathbf{l}^{k+1}_n = BR_n(\mathbf{L}^k_{-n}) = \underset{\mathbf{q}_n \in \mathcal{S}_n}{\operatorname{argmax}}\, u_n(\mathbf{q}_n, \mathbf{L}^k_{-n}) \tag{11.36}$$

3. Consumers add their local strategy $\mathbf{l}^{k+1}_n$ to the opponents' total consumption vector $\mathbf{L}^k_{-n}$ and send this new aggregated consumption profile L^{k+1} to the next player.

[d] If the total cost function is the peak pricing and the scheduling variables are continuous, the model to be solved is an LP; if the variables are integer, it is a MILP; if the total cost is quadratic and the variables are continuous, the model is a QP, and if the variables are integer, a MIQP.

Algorithm 11.1 Best-response dynamics

Input: Local data about consumers' appliances and preferences $(\mathcal{N}, \mathcal{S})$; market data related to the total cost function which defines consumers' utility (u_n)
Output: load schedules $\mathbf{s} = (\mathbf{l}_1, \mathbf{l}_2, \ldots, \mathbf{l}_N)$ which forms the NE

$k \leftarrow 0$;
Initialize $\mathbf{l}_n^k = [\mathbf{0}]\ \forall n \in \mathcal{N}$ and $\mathbf{L}^k = \sum_{n \in \mathcal{N}} \mathbf{l}_n^k$;
count $\leftarrow 1$;
while *count* > 0 **do**
  count $\leftarrow 0$;
  for all $n \in \mathcal{N}$ **do**
    Calculate $\mathbf{L}_{-n}^k$ using L^k received: $\mathbf{L}_{-n}^k = \mathbf{L}^k - \mathbf{l}_n^k$;
    Solve local problem (11.36): $\mathbf{l}_n^{k+1} = BR_n(\mathbf{L}_{-n}^k)$, to calculate the new schedule;
    Update and broadcast new $\mathbf{L}^{k+1} = \mathbf{l}_n^{k+1} + \mathbf{L}_{-n}^k$ to the next consumer $(n+1)$;
    if $u_n(\mathbf{l}_n^{k+1}, \mathbf{L}_{-n}^k) < u_n(\mathbf{l}_n^k, \mathbf{L}_{-n}^k)$ **then**
      count $\leftarrow$ *count* + 1
    end if
    $k \leftarrow k + 1$;
  end for
end while
return s;

As BRD converges to an NE for all exact potential games [43], the process continues until an equilibrium is reached and consumers can no longer reduce their bills when changing their schedules. It is worth mentioning that this algorithm is decentralized and each consumer has autonomy to schedule locally his/her TCLs, which are one of the pillars of TC.

The convergence of this algorithm to an NE is guaranteed for potential games, for example, for games applying the PTC billing to schedule energy invariant loads, and for games with the PTS billing to schedule any type of loads. In the case of the PTC applied to energy variant loads, the game was proved to not be potential in Theorem 11.1, which means the existence of NE is not ensured, and thus the algorithm may not converge. In this case, other solution methods are needed to analyze the game. Finally, it is important to stress that, in this version of the BRD, the consumers communicate the group's total load. This design guarantees users' privacy, as they do not share individual preferences, strategies, or load profiles: their personal consumption is integrated in the total load vector and no participant knows opponents' individual values. Other methods to enhance data privacy and secrecy can be added to the algorithm, for example, multiparty computation protocol (MPC), as in Rahman et al. [12].

4.3 Multiplicity of Nash equilibria

In this section, the aspect of uniqueness/multiplicity of possible Nash equilibria (solutions) is analyzed, as well as its implications to the implementation of load scheduling games. To have a unique NE, a game must respect two conditions: players' utility functions must be strictly concave and the strategy sets must be convex [48, 49]. To explain why those conditions are necessary for a unique equilibrium, exact potential games theory is used. As shown in Definition 11.1, a game is exact potential if it has a function $\phi: \mathcal{S} \mapsto \mathbb{R}$ that maps the joint strategy set to the set of real numbers. When a game has an exact potential function, each consumer's change in strategy generates a change in his/her payoff that equals the change in the potential function, while opponents' strategies are kept fixed. For example, if a consumer n shifts from strategy $\mathbf{l}_n^1$ to $\mathbf{l}_n^2$, when opponents are playing a joint strategy profile $\mathbf{l}_{-n}$, his/her payoff change will be equal to $u_n(\mathbf{l}_n^2, \mathbf{l}_{-n}) - u_n(\mathbf{l}_n^1, \mathbf{l}_{-n})$, which will be the same as the change in the potential function $\phi(\mathbf{l}_n^2, \mathbf{l}_{-n}) - \phi(\mathbf{l}_n^1, \mathbf{l}_{-n})$. If the potential function is strictly concave, it has a unique global maximum. Moreover, if the strategy set is convex and bounded, the unique maximum will be always reached, as every consumer will try to unilaterally ameliorate his/her utility (thus the potential function), until its maximum is reached.

For example, in the game with the PTC billing and energy invariant loads, the total cost function will be the exact potential function of the game—see Section 4.1.1. This function is strictly convex for the total quadratic cost function.[e] If the variables are continuous, each consumer seeks to minimize this convex total cost, reachable by the continuous control, the NEs have the same minimum total cost, and even if they are different in terms of schedules of appliances, they allocate the same bills to consumers, which make the result unique—the reader is referred to Baharlouei et al. [14] for the complete proof. Even with the PTS billing, if the variables are continuous, those conditions (potential function strictly concave and convex strategy sets) are respected [20].

However, if the strategy sets are no longer convex (cases with integer variables), one of the conditions is no longer respected. This means that pure-strategy NEs with suboptimal total cost may also exist, corresponding to "local minima" [43]. The consequences of this are twofold. First, the total cost to be shared can vary among equilibria, which means that the result of the play can be suboptimal for the group of consumers (in terms of potential function). In Example 11.2, two out of the three equilibria have total cost 27, instead of the minimum 24.

Second, multiple equilibria mean dispersion of individual payments: in one equilibrium a specific consumer can pay more than in another. As seen in Example 11.4, one equilibrium is better for consumer 1 ($l_1 = [0, 1, 1, 0]$ and $l_2 = [2, 2, 0, 2]$) and the other is better for consumer 2 ($l_1 = [1, 1, 0, 1]$ and $l_2 = [0, 2, 2, 0]$). Thus, when solving the load scheduling problem with integer potential games, choosing one equilibrium out of the possible ones is an important concern. If some participants are always benefited at the end of the gameplay, and this process is transparent, unfairness can be raised by the other participants.

[e] Remember that maximize a strictly concave function is the same as minimize a strictly convex function.

One question remains: what determines which NE will be reached? This will depend on the solution algorithm to solve the problem. In the case of the BRD—described in the pseudo-code (Algorithm 11.1), consumers' playing order and the initial solution will determine the finite improvement path generated for the gameplay: at each iteration k, one consumer $n \in \mathcal{N}$ gathers the total load information of the past iteration ($k-1$), and best responds to it. If $u_n(\mathbf{l}_n^k, \mathbf{l}_{-n}^{k-1}) > u_n(\mathbf{l}_n^{k-1}, \mathbf{l}_{-n}^{k-1})$, this consumer is the "deviator." The process continues until there is no more "deviators," which means that a terminal point (NE) is reached. Moreover, at each round of the game, all consumers play once to verify if there is no more "deviators" in the community. Also, the process starts from load schedules $\mathbf{l}_n^k = [\mathbf{0}]\ \forall\, n$. Therefore, it is the consumers playing order defined in the set $\mathcal{N}$, and the algorithm's starting point that determines which finite improvement path will be generated, thus what terminal point will be reached. In Example 11.4, if the BRD starts with strategies $l_1 = [1, 1, 0, 1]$ and $l_2 = [0, 0, 2, 2]$, and the playing order is first consumer 1 and then consumer 2, the NE attained is the one better for consumer 1 ($l_1 = [0, 1, 1, 0]$ and $l_2 = [2, 2, 0, 2]$). However, if the starting point is $l_1 = [1, 1, 0, 1]$ and $l_2 = [2, 2, 0, 2]$, and the playing order is first consumer 2 and then consumer 1, the NE reached is the one better for consumer 2 ($l_1 = [1, 1, 0, 1]$ and $l_2 = [0, 2, 2, 0]$).

4.4 Fairness of different billings

In order to engage end users to participate in DSM programs as the coordination model analyzed in this chapter, it is important that the total scheduling cost is shared in a fair way. The design of a utility function depends on the billing mechanism chosen, which should consider the contributions of each consumer to the total cost. As stated by Baharlouei et al. [7] and Baharlouei and Hashemi [8], consumers with more flexibility to schedule their loads and who intend to consume less energy should have better payoffs.

Shapley value (SV) can be used to define the best sharing among consumers [50]. It is based on the contribution of each player n to all possible coalitions that he/she does not participate in $\mathcal{G} \subseteq \mathcal{N} \backslash \{n\}$. The utility for player n can be calculated by Eq. (11.37), in which $C_{\mathcal{G}}^*$ is the minimum cost of the scheduling problem for a coalition $\mathcal{G}$ and $C_{\mathcal{G}\cup\{n\}}^*$ is the minimum cost when consumer n is added to that coalition. Both costs are calculated by solving the total cost problem in Eq. (11.15) or (11.17)—that is, $C_{\mathcal{G}}^*$ can be either $C_{\mathcal{G}}^{Q*}$ or $C_{\mathcal{G}}^{P*}$, depending on the game under analysis. One should notice that the complexity of calculating the SV is $o(2^n)$, as long as it implies the total cost calculation of all possible subsets of $\mathcal{N}$ [51].

$$SV_n = \sum_{\mathcal{G} \subseteq \mathcal{N} \backslash \{n\}} \frac{|\mathcal{G}|!(|\mathcal{N}| - |\mathcal{G}| - 1)!}{|\mathcal{N}|!} [C_{\mathcal{G}\cup\{n\}}^* - C_{\mathcal{G}}^*] \tag{11.37}$$

Then, a fairness index can be calculated by the distance between the vector of the consumers' utilities after the game and the utilities they should have when applying the

SV [7, 8]. This can be done using Eq. (11.38), where u_n is the utility of consumer n at the NE of the game and u_n^* is his/her utility calculated by the SV.

$$F = \sum_{n \in \mathcal{N}} \left| \frac{u_n}{\sum_{m \in \mathcal{N}} u_m} - \frac{SV_n}{\sum_{m \in \mathcal{N}} SV_m} \right| \tag{11.38}$$

An example to compare the fairness of both the PTS and PTC mechanisms when applying the SV is given in Example 11.5.

Example 11.5. To compare the fairness of PTS and PTC mechanisms, an example with energy invariant loads and a quadratic total cost function is given. Consider a scheduling game between two participants $\mathcal{N} = \{1,2\}$ for the next four time slots $\mathcal{T} = \{1,2,3,4\}$. Consumer 1 has base load w_1 = [6, 2, 1, 1] and consumer 2 has w_2 = [3, 1, 2, 2]. Each consumer has three possible strategies for scheduling a flexible load: n = 1 has $\mathbf{x}_1^1 = [2,2,0,0]$, $\mathbf{x}_1^2 = [0,2,2,0]$, and $\mathbf{x}_1^3 = [0,0,2,2]$; n = 2 has $\mathbf{x}_2^1 = [0,6,0,0]$, $\mathbf{x}_2^2 = [0,0,6,0]$, and $\mathbf{x}_2^3 = [0,0,0,6]$. The final scheduling vectors for each consumer are $\mathbf{l}_n = \mathbf{w}_n + \mathbf{x}_n$. Considering $a_t = 1$ and $b_t = 0$ for all $t \in \mathcal{T}$, the payoff matrix of this game using the PTS billing is shown in Table 11.4, and with the PTC billing in Table 11.5. In both matrices, u_1, u_2 is the utility of consumers 1 and 2, respectively. They are calculated using Eq. (11.21) for the PTS billing and Eq. (11.18) for the PTC, considering the total cost as a quadratic function. The best responses of each player to the opponents' strategies are underlined.

Therefore, it is easily verified that this game has two NE in the PTS game: $(\mathbf{l}_1^3, \mathbf{l}_2^1)$ and $(\mathbf{l}_1^2, \mathbf{l}_2^3)$. The first one is the global optimum because it maximizes the potential

Table 11.4 Payoff matrix of the fairness example using the PTS billing.

	$\mathbf{l}_2^1 = [3,7,2,2]$	$\mathbf{l}_2^2 = [3,1,8,2]$	$\mathbf{l}_2^3 = [6,2,3,3]$
$\mathbf{l}_1^1 = [8,4,1,1]$	−138, −122	−120, <u>−116</u>	−120, <u>−116</u>
$\mathbf{l}_1^2 = [6,4,3,1]$	−116, −120	−110, −126	<u>−98</u>, <u>−114</u>
$\mathbf{l}_1^3 = [6,2,3,3]$	<u>−102</u>, <u>−110</u>	<u>−108</u>, −128	−108, −128

Underlined values represent best responses.

Table 11.5 Payoff matrix of the fairness example using the PTC billing.

	$\mathbf{l}_2^1 = [3,7,2,2]$	$\mathbf{l}_2^2 = [3,1,8,2]$	$\mathbf{l}_2^3 = [6,2,3,3]$
$\mathbf{l}_1^1 = [8,4,1,1]$	−130, −130	<u>−118</u>, <u>−118</u>	−118, <u>−118</u>
$\mathbf{l}_1^2 = [6,4,3,1]$	−118, −118	<u>−118</u>, −118	<u>−106</u>, <u>−106</u>
$\mathbf{l}_1^3 = [6,2,3,3]$	<u>−106</u>, <u>−106</u>	<u>−118</u>, −118	−118, −118

Underlined values represent best responses.

Table 11.6 Minimum cost of the possible coalitions of Example 11.5.

$\mathcal{G}$	User 1's strategy	User 2's strategy	Value ($C^*_{\mathcal{G}}$)
{∅}	$\mathbf{l}_1^0 = [0,0,0,0]$	$\mathbf{l}_2^0 = [0,0,0,0]$	0
{1}	$\mathbf{l}_1^3 = [6,2,3,3]$	$\mathbf{l}_2^0 = [0,0,0,0]$	−58
{2}	$\mathbf{l}_1^0 = [0,0,0,0]$	$\mathbf{l}_2^1 = [3,7,2,2]$	−66
{1, 2}	$\mathbf{l}_1^3 = [6,2,3,3]$	$\mathbf{l}_2^1 = [3,7,2,2]$	−212
	$\mathbf{l}_1^2 = [6,4,3,1]$	$\mathbf{l}_2^3 = [3,1,2,8]$	

(equals to -168), and the second NE is a local optimum, having potential equal to -176. Both potentials were calculated using Eq. (11.33). Even though both equilibria minimize the total cost, the first NE is better for consumer 2 and the second for consumer 1. On the other hand, in the case of the PTC billing, there are three NE: $(\mathbf{l}_1^3, \mathbf{l}_2^1)$, $(\mathbf{l}_1^2, \mathbf{l}_2^3)$, and $(\mathbf{l}_1^1, \mathbf{l}_2^2)$. The first two are the same as the PTS game and are Pareto efficient. However, the third NE is a local optimum. Even though it harms both consumers, it can be attained depending on the BR sequence and the tie-break rule when two strategies lead to the same payoff.

The total cost of the load scheduling for each coalition $\mathcal{G} \subseteq \mathcal{N}$ is calculated in Table 11.6. Therefore, the SV for each consumer is $SV_1 = -102$ and $SV_2 = -110$. One should notice that consumer 1 has more flexibility than consumer 2 as long as he/she can split his/her load among more time slots. This explains why consumer 1 pay less using the SV than consumer 2, even though they have the same total load of 14. For each NE and billing mechanism, the fairness index is calculated. When applying the PTS billing, for the point $(\mathbf{l}_1^3, \mathbf{l}_2^1)$, the consumers' utilities are the same as the SV and the fairness index equals 0.00%. For the NE $(\mathbf{l}_1^2, \mathbf{l}_2^3)$, the consumers' utilities are $(-98, -114)$ and the fairness index equals 3.77%. In the case of the PTC, the utilities are $(-106, -106)$ or $(-118, -118)$. Both lead to a fairness index equals to 3.77%. Therefore, one of the solutions of the PTS is fairer than the NEs of the PTC.

4.5 Strategy proof of different billings

A noncooperative game can open space for players to cheat. Depending on the utility function design, one participant can have a better payoff by communicating false information to opponents. In this section, this idea is explained using the two billing functions described in Section 3.3.

4.5.1 Proportional to consumption

The PTC mechanism prevents cheating behaviors when only energy invariant loads are considered, as the potential function is equal to the total cost. This means that lying about consumption increases the total cost, which increases players own bills as a consequence [8]. However, for energy variant loads, the potential function is no longer the total cost, as shown in Section 4.1.2. Therefore, the strategy proof of this mechanisms is no longer guaranteed. The following example illustrates that:

Example 11.6. Consider an integer game with PTC billing, any convex total cost function (e.g., quadratic function, peak pricing function, or another one), and energy variant loads. A consumer n has the choice between two total consumption values: one equal to κ and the other equal to μ. The rest of the community has fixed total load equal to ζ, which is equally split among the consumers. Consider that two total costs are possible for this game: $C_1 > C_2$. Consumer n can have a better utility when choosing the strategy κ leading to the total cost $C_1 = C(\kappa, \zeta)$, and a worst utility when choosing strategy μ leading to $C_2 = C(\mu, \zeta)$. Consider that consumers best respond to the last total consumption vector of the opponent, which is updated after each best response. Starting with consumer n playing against strategy κ, the following will happen:

$$\frac{\kappa}{\mu+\zeta}C(\mu,\zeta) < \frac{\kappa}{\kappa+\zeta}C(\kappa,\zeta) \tag{11.39}$$

So he chooses the strategy μ, because it leads to a smaller total cost. In the following iteration, f_n is updated to μ, leading to

$$\frac{\mu}{\mu+\zeta}C(\mu,\zeta) < \frac{\mu}{\mu+\zeta}C(\kappa,\zeta) \tag{11.40}$$

Therefore, consumer n keeps the strategy μ, because of the minimization of the total cost. Even though this pseudo-version of BRD converges to an equilibrium, it is not an NE of the original game, because the following is true:

$$\frac{\mu}{\mu+\zeta}C(\mu,\zeta) > \frac{\kappa}{\kappa+\zeta}C(\kappa,\zeta) \tag{11.41}$$

Thus, consumer n can cheat after reaching this equilibrium by changing strategy μ to κ, which increases the total community cost but decreases his own utility. One can notice that this behavior will harm the other consumers because the additional cost will be divided among them (according to their consumption).

4.5.2 Per time slot

The PTS billing has been shown to lead to untruthful behavior of the players for continuous games [8, 12]. Both studies claim that a consumer can take advantage of informing opponents a different consumption than his/her true value, which would force other users to schedule their loads apart from the cheater's preferred time interval.

Here, a theoretical proof that the cheating behavior is more likely to arise when the PTS billing is done over the total cost realized after real consumption (ex post), as considered in Baharlouei and Hashemi [8] and Rahman et al. [12], because cheating can be a dominant strategy in this scenario. However, if prices $P_t(L_t)$ are obtained as a result of the game, as shown in Eq. (11.21), consumers have no incentives to lie. The prices are defined the day before consumption (ex ante), following the output of the BRD algorithm, and they are used to calculate consumers' payoffs. Therefore, informing opponents a higher load would increase the prices of the cheater's preferred time interval, which would increase his/her own cost. Example 11.7 is used to illustrate this idea.

Theorem 11.3. *In a noncooperative scheduling game* $\Gamma = \langle \mathcal{N}, (S_n)_{n\in\mathcal{N}}, \{u_n\}_{n\in\mathcal{N}} \rangle$ *with PTS billing Eq. (11.21), a consumer has incentive to declare false information about his/her consumption if prices are defined ex post consumption. On the other hand, if prices are defined ex ante consumption, consumers have no incentive to lie.*

Proof. If prices are ex ante consumption, the amount of energy a cheater would have to add to a time slot to make an opponent move his/her consumption away from it would increase its price, which would decrease his/her utility, resulting in no benefit for him/her. In the ex post case, the prices do not increase with the cheater's lie. To prove that, consider a group of consumers participating in the scheduling game with equal preferences and AC parameters. Assume that $\delta = 1$ and $\mathbf{w}_i = [0]\ \forall\, i \in \mathcal{N}$. Assume that consumers only need to turn on their AC once during the time horizon to keep the temperature inside the feasible region. At a certain stage of the game k, a consumer n schedules his/her AC at time slot t. His/her payoff is

$$u_n(l_n, l_{-n}) = -l_{n,t}P(l_{n,t} + l_{-n,t}) \tag{11.42}$$

which means the energy price at time slot t is $p_t = a_t(l_{n,t} + l_{-n,t}) + b_t$. Now consider that n is a dishonest consumer and he/she declares a consumption $l_{n,t} + \Delta l_{n,t}$ on the time slot t he/she wants to schedule his/her AC. As a response, opponents will play a strategy $l'_{-n,t} = l_{-n,t} - \Delta l_{-n,t}$, which means they will move $\Delta l_{-n,t}$ from time slot t. Therefore, if prices are ex post consumption, they will be calculated without considering consumer n's lie ($\Delta l_{n,t}$), resulting in $p''_t = a_t(l_{n,t} + l_{-n,t} - \Delta l_{-n,t}) + b_t$. The utility of consumer n is then $u''_n(l_n, l'_{-n}) = -l_{n,t}p''_t$. Thus, $u''_n(l_n, l'_{-n}) \geq u_n(l_n, l_{-n})$ if

$$-l_{n,t}[a_t(l_{n,t} + l_{-n,t} - \Delta l_{-n,t}) + b_t] \geq -l_{n,t}[a_t(l_{n,t} + l_{-n,t}) + b_t] \tag{11.43}$$

As a result, it is always interesting to declare a higher consumption during the time slot player n intends to turn on the AC, because $p''_t \leq p_t$. One can see that if no opponent moves his/her consumption from t, then $\Delta l_{-n,t} = 0$ and $p''_t = p_t$, which means the cheater would pay at most the same amount as Eq. (11.42). Therefore, in this framework, declaring a higher consumption for the time slot t is a dominant strategy.[f]

[f] There are other frameworks where this would not be a dominant strategy and consumer n could in fact pay more. For instance, if he/she cheats in more time slots and opponents' constraints make them move consumption to time slots consumer n uses energy; or if there are more cheaters in the game. However, the discussed framework shows that there exists the possibility of benefiting from cheating in the ex post scenario.

If prices are ex ante consumption (as a result of the game), they can be calculated as $p'_t = a_t(l_{n,t} + \Delta l_{n,t} + l_{-n,t} - \Delta l_{-n,t}) + b_t$. The utility of consumer n is then $u'_n(l_n, l'_{-n}) = -l_{n,t} p'_t$. Thus, $u'_n(l_n, l'_{-n}) \geq u_n(l_n, l_{-n})$ if

$$-l_{n,t}[a_t(l_{n,t} + \Delta l_{n,t} + l_{-n,t} - \Delta l_{-n,t}) + b_t] \geq -l_{n,t}[a_t(l_{n,t} + l_{-n,t}) + b_t] \quad (11.44)$$

Thus, consumer n takes benefit from cheating only if

$$\Delta l_{n,t} \leq \Delta l_{-n,t} \quad (11.45)$$

Now, it is necessary to analyze how opponents respond ($\Delta l_{-n,t}$) to an increase of $\Delta l_{n,t}$ during a time slot they scheduled their AC. As long as consumer n only uses time slot t, this depends on the consumption of opponents on time slots $\{-t\}$. Considering $a_j = a$and $b_j = b \, \forall j \in \mathcal{T}$, and given that the BRD algorithm is sequential, the response of the next player is analyzed, say opponent i. He/she will change his/her AC from time slot t to $t' \in \{-t\}$, if

$$L_t + \Delta l_{n,t} - \Delta l_{i,t} > L_{t'} + \Delta l_{i,t} \quad (11.46)$$

where $L_t = l_{n,t} + l_{-n,t}$. Therefore, there are three scenarios for the relation between L_t and $L_{t'}$ which would impact the behavior of opponent i. If $L_t = L_{t'}$, then $\Delta l_{n,t} > 2\Delta l_{i,t}$ to make consumer i change from t to t', which contradicts the necessary condition (11.45) for consumer n to take benefit from cheating.

If $L_t > L_{t'}$, then $\Delta l_{n,t} > 2\Delta l_{i,t} - f_1, f_1 = L_t - L_{t'} > 0$. Thus, if $f_1 > 2\Delta l_{i,t}$, condition (11.45) is respected. However, one can easily notice that this would never be true, because consumer i would have chosen to place $\Delta l_{i,t}$ at time slot t' before, at round $k - 1$. On the other hand, if $f_1 \leq 2\Delta l_{i,\ t}$, condition (11.45) is not respected.

Finally, if $L_t < L_{t'}$, then $\Delta l_{n,t} > 2\Delta l_{i,t} - f_2, f_2 = L_t - L_{t'} < 0$. As a result, consumer n would have to declare a value greater than $2\Delta l_{i,t}$ plus a positive value to make i moves $\Delta l_{i,t}$ to time slot t', which again, contradicts (11.45). □

Example 11.7. Consider the example shown in Rahman et al. [12, Figs. 3 and 4] (necessary information is contained in Tables 11.7 and 11.8 herein), but with binary variables.[g] There are three users playing the scheduling game. User A plays first, followed by user B and user C. The PTS billing mechanism to divide a total quadratic cost is applied, and $a_t = 1$ and $b_t = 0$ for all $t \in \mathcal{T}$. Therefore, in the first round, user A declares the consumption vector $\mathrm{l}_A = [5, 6, 4]$, for which user B best responds with $\mathrm{l}_B = [2, 4, 6]$ and user C with $\mathrm{l}_C = [5, 5, 5]$. After some rounds, the NE is reached, and the results are shown in Table 11.7 herein. The final prices are $\mathbf{P} = [28, 28, 28]$ and consumers' utilities are calculated using Eq. (11.21). Suppose that user B decides to cheat in an attempt to have less opponents consuming energy in time slot 3, provided that he/she wants to consume more in it. Thus, at the first round, he/she declares the double of

[g] For the purpose of this example, it is not necessary to detail users' preferences. In Rahman et al. [12], integer values are used and the comparison is possible.

Table 11.7 Example 11.7 without cheating behavior.

	First round			Nash equilibrium			Billing
Consumer	**S 1**	**S 2**	**S 3**	**S 1**	**S 2**	**S 3**	**Per time slot**
A	5	6	4	6	4	5	420
B	2	4	6	3	3	6	336
C	5	5	5	5	7	3	420

Source: From M.A. Rahman, M.H. Manshaei, E. Al-Shaer, M. Shehab, Secure and private data aggregation for energy consumption scheduling in smart grids, IEEE Trans. Dependable Secure Comput. 14 (2) (2017) 221–234.

Table 11.8 Example 11.7 when user B cheats and the prices are defined ex post or ex ante.

	First round			Nash equilibrium			Real consumption		
Consumers	**S 1**	**S 2**	**S 3**	**S 1**	**S 2**	**S 3**	**S 1**	**S 2**	**S 3**
A	5	6	4	7	6	2	7	6	2
B	2	4	12	3	3	12	3	3	6
C	5	5	5	6	7	2	6	7	2

his/her real consumption, leading to the results in Table 11.8. Therefore, if the billing is ex post, and the prices are defined after the consumption, then $\mathbf{P} = [32, 32, 20]$ and user B benefits from lying, because he/she pays \$312 instead of \$336. His/her attitude also harms the other players, adding \$36 to their bills. However, if the prices are defined ex ante, right after the end of the game, then $\mathbf{P} = [32, 32, 32]$ and user B pays more than he/she would have paid being honest (\$384 instead of \$336). The other users are also damaged (they pay \$480 instead of \$420).

4.6 Price of anarchy

PoA is a concept in algorithmic game theory related to the question: how inefficient is the equilibrium reached by selfish rational players in comparison to an idealized situation in which the agents would collaborate with a common goal? [52]. It measures the amount of damage suffered by the agents due to the absence of a central authority [47]. More specifically, it is computed as a ratio between the socially Pareto optimal outcome and the equilibrium outcome from the distributed interaction among selfish players. In a transactive coordination problem, the PoA can be calculated as a ratio between the solution of the centralized optimization problem, for example, in Eq. (11.15) or (11.17), and the sum of the utilities of players in the decentralized game.

Mathematically, consider a *welfare function* $\Theta(\mathrm{s})$ measuring the efficiency of a strategy profile $\mathbf{s} = (\mathbf{l}_n, \mathbf{l}_{-n})$. The *utilitarian* welfare function is used, as defined in Lã et al. [47].

Definition 11.2. Utilitarian welfare function. Given a strategy profile $\mathbf{s} = (\mathbf{l}_n, \mathbf{l}_{-n}) \in \mathcal{S}$, the utilitarian welfare function $\Theta: \mathcal{S} \mapsto \mathbb{R}$ is the sum of the utilities (or costs) of all players:

$$\Theta(\mathbf{s}) = \sum_{n \in \mathcal{N}} u_n(\mathbf{s}) \tag{11.47}$$

Let $\mathbf{s}^*$ be the optimal socially point with respect to the welfare function, for example, the solution of Eq. (11.15) or (11.17), depending on the game type. Let $\mathbf{s}'$ be an NE of a coordination game. PoA is defined as

$$PoA = \frac{\Theta(\mathbf{s}')}{\Theta(\mathbf{s}^*)} \tag{11.48}$$

Therefore, the optimal PoA is 1, the minimum value a game can achieve. Other results indicate how many times the decentralized approach is worse than the centralized one.

5 An application of noncooperative games to coordinate thermal loads

To demonstrate the aspects of modeling a DSM program using noncooperative game, a case study using real data collected from a real community of consumers in the South of Spain is used. Hourly active power consumption from June 2019 is collected and averaged to build a daily consumption curve for each of the 201 consumers of the community.

Consumers are considered to have one thermal load, more specifically, one AC. Physical parameters of these ACs (power, thermal resistance, capacity, etc.) are calculated according to Heleno et al. [37], while each household occupation is estimated considering the real data. Only consumers with high enough consumption are assumed to have an AC installed. As a result, 70 consumers out of the 201 have ACs. Their AC load is simulated using Eq. (11.9), and consumers inflexible loads are calculated by subtracting the AC load from the data collected. The other 131 consumers without AC are kept in the simulations, as their inflexible load impacts the community's consumption and the total cost. This initial scenario is called base case (BAS) and is shown in Fig. 11.2 and the system parameters in Table 11.9.

A quadratic total cost function is used to analyze the game aspects with different billings (PTS and PTC), and its parameters are calculated from a three-step piecewise function. The parameters are defined based on the following tiers prices: 5, 15, and 30 ¢/kWh, and the thresholds between tiers are adopted as 60% and 75% of the group's peak load on the base scenario. From these values, a quadratic curve without intercept is adjusted, resulting in parameters $a_t = 0.065$¢/kWh^2 and $b_t = -0.858$¢/kWh for all $t \in \mathcal{T}$.

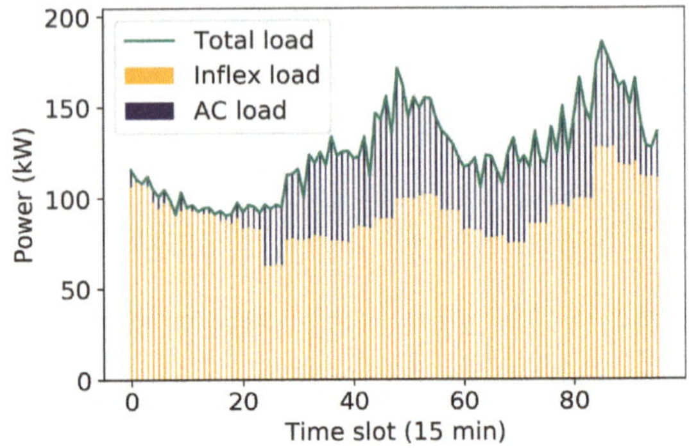

Fig. 11.2 Total energy consumption of the community with ACs in the base scenario (without energy management).

Table 11.9 System parameters.

N	201	$\theta_{n,t}^{\min}$ (°C)	[18, 22]
T	96	$\theta_{n,t}^{\max}$ (°C)	[20, 26]
δ	0.25 (15 min)	$\theta_{n,0}$ (°C)	$[\theta_{n,t}^{\min}, \theta_{n,t}^{\max}]$
R_n (°C/kW)	[5.0, 8.0]	θ_t^{et} (°C)	35
E_n (kW)	[1.5, 3.5]	a_t (¢/kWh^2)	0.065
C_n (kWh/°C)	[0.5, 4.2]	b_t (¢/kWh)	−0.858
η_n	[3.0, 3.2]		

The BRD is implemented in Python 3.7.3 and the optimization model (11.36) of each iteration is solved using Gurobi 8.1.1. Because the PTC billing for energy variant loads might not have equilibrium, a modified version of the BRD is implemented, in which a consumer's load is updated after each best response (m-BRD).

5.1 Existence of Nash equilibria

As mentioned before, the NE of the PTS billing is calculated using the BRD, and the equilibrium of the PTC billing is calculated using the m-BRD. Both are solved with the study case data. In addition, the solution to problem (11.15) is calculated to have a quality measurement of the equilibria. Other aspects are analyzed as the ability of the TC models to flatten the community's load curve and if the equilibrium of the PTC using the m-BRD is an NE of the game with energy variant loads.

Results are shown in Table 11.10. Both total cost and peak-to-average (PAR) ratio—which measures how flat the load curve is [19]—are optimized with the TC approaches. Moreover, the PTC with m-BRD is able to reach an optimal total cost, while the solution of the PTS with BRD is very close to optimal. This leads to a PoA close to 1.0 for both cases. Being close to 1.0 indicates that the game solution with PTC/PTS can reach the optimal value, and the community as a whole has no loss when giving consumers the autonomy to decide their own schedules. In addition, the PARs are also reduced, indicating that the load curves with TC are flattened. One can

Table 11.10 Solution of each scenario.

Scenario	Total cost ($)	PAR	Total energy (kWh)
Base scenario	37.387	1.492	2991.062
Centralized	36.579	1.254	2996.737
PTS (BRD)	36.594	1.233	3002.512
PTC (m-BRD)	36.577	1.260	2996.287

Table 11.11 Bills some consumers can have if they choose the schedule of the base case instead of the equilibrium solution reached by the m-BRD for the PTC approach (in $).

n	0	5	28	48	63	92	100
m-BRD	0.811	0.205	0.486	0.258	1.007	1.432	0.292
BAS	0.802	0.199	0.478	0.244	1.000	1.424	0.285

notice that, to reduce the total cost and PAR, the final schedules turn-on the ACs more times than the original BAS case. The sum of consumers' total load in the base scenario is 2991.062 kWh. After the PTS with BRD, this value is 3002.512 kWh, and the PTC with m-BRD, 2996.287. The additional kWh in both cases confirms the energy variant nature of TCLs and their "payback" characteristic: there is a trade-off between energy consumption and total cost with the TC approaches, especially when the quadratic parameter of Eq. (11.13) is high, inducing consumers to flatten the load curve.

Finally, it is demonstrated that the equilibrium reached by the m-BRD for the scheduling game with energy variant loads and the PTC does not reach a NE. Even though the algorithm converges to the solution discussed above, it is not an NE of the game with PTC: fixing consumers' energy share f_n at each iteration can lead participants to optimize the community's total cost, but they can have better payoffs if they change it. Table 11.11 shows that seven consumers can have a smaller bill if they play the solution in the base case instead of the equilibrium solution reached by the m-BRD, while the other consumers are playing the m-BRD equilibrium. Therefore, this equilibrium solution is not an NE of the game with the original utility function (11.18). Consumers can in fact have a better payoff when increasing the community's total cost, because their shares are a function of the results. If one of those seven consumers plays the BAS schedule instead of the equilibrium reached, the total community cost can increase from \$36.580 to \$36.607 depending on which of them deviates. This demonstrates that, when energy variant loads are present and the PTC is applied, consumers' goal disconnects with the community's goal, the game is no longer potential, and consumers can cheat.

5.2 *Fairness of the PTS and PTC*

In this section, the fairness of both billing mechanisms applied to integer and energy variant loads is compared. As the PTC may not have an NE when energy variant loads are considered, the equilibrium solution of the m-BRD is used.

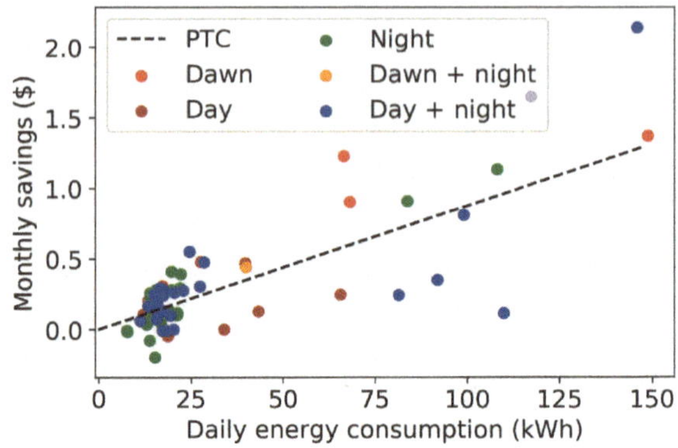

Fig. 11.3 Consumers' savings from BAS scenario with the TC approach and the PTS billing. The *dashed line* shows the same consumers' savings when the PTC approach is used.

Consumers' savings in each TC solution are calculated by comparing their payoffs with the BAS solution. Results are shown for an entire month. Only active participants (i.e., with AC loads) are analyzed. They are classified into three groups according to their preferences (day periods they want their AC to be operating): dawn for use between midnight and 7 h; day from 7 to 17 h; and night from 17 h to midnight. Consumers can choose to use their AC in more than one period of the day. To illustrate the impact on consumers, monthly individual savings are plotted against consumers' total energy consumption in Fig. 11.3. In PTS case, consumers using their AC during peak periods (specially night) have less savings when comparing with those using the AC during valley periods (e.g., dawn). This time-domain differentiation feature is not observed in PTC (see dotted line in Fig. 11.3), where savings only depend on the total consumption.

To analyze the fairness of payoffs/savings distribution, the SV and fairness index presented in Section 4.4 are used on subcases of the data. Six subcases with 10 consumers are constructed: the first three by picking them randomly; one for those with the largest total energy consumption; one for those with the largest AC load as a percentage of their total energy use; and one for those with the smallest total energy consumption. The BRD applying the PTS and the m-BRD for the PTC billing are run for all subcases, followed by the calculation of the SV and fairness index. Results are shown in Table 11.12: for all six subcases, this index is smaller when the PTS is applied, which means that its solution is closer to the SV than when using the PTC, thus the PTS is fairer.

Table 11.12 Fairness index comparison between PTS and PTC.

Case	Random 01	Random 02	Random 03	Highest load	Highest AC	Smallest load
PTS	3.15%	1.86%	0.55%	0.53%	1.87%	3.38%
PTC	7.92%	6.78%	12.45%	7.26%	3.91%	4.76%

5.3 Strategy proof of the billing mechanisms

5.3.1 Per time slot

As this billing might have the problem of cheating—see Section 4.5.2, ex ante prices to discourage this behavior are tested. To verify its efficiency, a large load consumer ($n = 14$) is chosen and the best-response algorithm with the PTS billing is run five times, varying the amount of lie this consumer adds to his/her consumption vector. It is considered that he/she declares to opponents, at the time slots he/she turns on the AC, a load value α times the real one ($l_{n,t} \times \alpha$), with $\alpha = [1.0, 1.1, 1.5, 2.0, 3.0]$. Results in Table 11.13 show that if the prices are ex post consumption, the cheater has a real incentive to lie, adding a large load to his preferred time interval, in an attempt to make opponents move as much load as possible away from it. For instance, if he/she declares to opponents 3.0 times the real AC load, opponents move their consumption away from the cheater's preferred time slots, reducing cheater's bill from the NE \$1.883–\$1.864. Moreover, the other participants are harmed because of this nonequilibrium solution, with a bill increase, for example, consumer $n = 96$ has a bill of \$1.602 instead of \$ 1.578. However, if the prices are ex ante, the more load the cheater adds to the consumption vector, the bigger the prices are at those time slots, decreasing his utility. In the $\alpha = 3.0$ scenario, his/her bill with an ex ante price increases to \$1.977. Therefore, ex ante prices are an effective way to discourage cheating behavior when applying the PTS billing.

5.3.2 Proportional to consumption

The m-BRD is able to reach an equilibrium solution in which consumers seek to reduce the community's total cost. However, as shown in Section 5.1, the final solution is not an NE of the game with the original PTC utility. It is shown that cheating is possible in this scenario. Table 11.14 shows that, when consumer $n = 0$ plays the BAS solution instead of the equilibrium reached by the m-BRD, while all the other consumers play the equilibrium, his/her bill decreases (from \$0.811 to \$0.802). Moreover, seven opponents are harmed with this change, seeing a slightly increase in their bills—three harmed participants are shown in Table 11.14. Even though the bill differences are small, this example illustrates the cheating possibility when applying the PTC to

Table 11.13 Cheater's and opponent's bills when a big consumer with AC cheats (in \$).

	Cheater $n = 14$		Opponent with AC $n = 96$	
α	**Ex ante**	**Ex post**	**Ex ante**	**Ex post**
1.0	1.883		1.578	
1.1	1.890	1.884	1.586	1.584
1.5	1.908	1.878	1.595	1.585
2.0	1.931	1.875	1.610	1.593
3.0	1.977	1.864	1.639	1.602

Table 11.14 Cheater's and opponents' bills when a consumer changes the PTC equilibrium solution to the BAS schedule (in $).

Cheater plays	Cheater $n = 0$	Opponent $n =$ 14	Opponent $n =$ 68	Opponent $n =$ 96
Equilibrium	0.811	1.781	1.341	1.808
BAS solution	0.802	1.783	1.342	1.809

All opponents play the equilibrium solution in both cases.

energy variant loads, using the BAS solution. Consumers could find even better strategies to play after the equilibrium was reached, which would increase more the community's total cost and cause greater harms to opponents.

6 Conclusions

This chapter has shown how TC may be utilized in DSM contexts to explore the potential of residential flexibility to reduce energy bills and achieve efficiency in a decentralized manner. A noncooperative game theoretic framework was used to model consumers interacting to schedule their loads under individual constraints and comfort preferences. The framework comprises a total cost and a billing function that defines consumers utility. A general mathematical formulation of loads that can accommodate shiftable and interruptible residential appliances and/or represent the operation of thermal loads together with temperature preferences was adopted. Important game aspects related to existence and multiplicity of Nash equilibria, fairness and strategy proof of the billing mechanisms, and the efficiency of equilibria were analyzed in the context of potential games. Two popular billing mechanisms, the PTC and the PTS, were discussed when scheduling both variant and invariant loads. It is shown that the existence of NE in some cases depends on the nature of the load model. Simple examples were built to illustrate the main characteristics of the games, including fairness and strategy proof of the billing mechanisms, using a study with data collected from a real energy community of consumers in Spain.

References

[1] K. Kok, S. Widergren, A society of devices: integrating intelligent distributed resources with transactive energy, IEEE Power Energ. Mag. 14 (3) (2016) 34–45.
[2] D. Hammerstrom, T. Oliver, R. Melton, R. Ambrosio, Standardization of a hierarchical transactive control system, Grid Interop 9 (2009) 35–42.
[3] GWAC, GridWise Transactive Energy Framework, GridWise Architecture Council, 2015.
[4] J. Hu, G. Yang, K. Kok, Y. Xue, H.W. Bindner, Transactive control: a framework for operating power systems characterized by high penetration of distributed energy resources, J. Mod. Power Syst. Clean Energy 5 (3) (2017) 451–464.

[5] W. Saad, Z. Han, H.V. Poor, T. Basar, Game-theoretic methods for the smart grid: an overview of microgrid systems, demand-side management, and smart grid communications, IEEE Signal Process. Mag. 29 (5) (2012) 86–105.
[6] A.-H. Mohsenian-Rad, V.W.S. Wong, J. Jatskevich, R. Schober, A. Leon-Garcia, Autonomous demand-side management based on game-theoretic energy consumption scheduling for the future smart grid, IEEE Trans. Smart Grid 1 (3) (2010) 320–331.
[7] Z. Baharlouei, M. Hashemi, H. Narimani, H. Mohsenian-Rad, Achieving optimality and fairness in autonomous demand response: benchmarks and billing mechanisms, IEEE Trans. Smart Grid 4 (2) (2013) 968–975.
[8] Z. Baharlouei, M. Hashemi, Efficiency-fairness trade-off in privacy-preserving autonomous demand side management, IEEE Trans. Smart Grid 5 (2) (2014) 799–808.
[9] H. Chen, Y. Li, R.H.Y. Louie, B. Vucetic, Autonomous demand side management based on energy consumption scheduling and instantaneous load billing: an aggregative game approach, IEEE Trans. Smart Grid 5 (4) (2014) 1744–1754.
[10] P. Chakraborty, P.P. Khargonekar, A demand response game and its robust price of anarchy, in: 2014 IEEE SmartGridComm, IEEE, 2014, pp. 644–649.
[11] S. Bahrami, V.W.S. Wong, A potential game framework for charging PHEVs in smart grid, in: 2015 IEEE Pacific Rim Conference on Communications, Computers and Signal Processing (PACRIM), IEEE, 2015, pp. 28–33.
[12] M.A. Rahman, M.H. Manshaei, E. Al-Shaer, M. Shehab, Secure and private data aggregation for energy consumption scheduling in smart grids, IEEE Trans. Dependable Secure Comput. 14 (2) (2017) 221–234.
[13] Y. Liang, F. Liu, C. Wang, S. Mei, Distributed demand-side energy management scheme in residential smart grids: an ordinal state-based potential game approach, Appl. Energy 206 (2017) 991–1008.
[14] Z. Baharlouei, H. Narimani, M. Hashemi, On the convergence properties of autonomous demand side management algorithms, IEEE Trans. Smart Grid 9 (6) (2018) 6713–6720.
[15] E. Fernandez, M.J. Hossain, M.S.H. Nizami, Game-theoretic approach to demand-side energy management for a smart neighbourhood in Sydney incorporating renewable resources, Appl. Energy 232 (2018) 245–257.
[16] S. Noor, W. Yang, M. Guo, K.H. van Dam, X. Wang, Energy demand side management within micro-grid networks enhanced by blockchain, Appl. Energy 228 (2018) 1385–1398.
[17] S. Wang, D. Sun, L. Du, J. Ye, Noncooperative distributed social welfare optimization with EV charging response, in: IECON 2018—44th Annual Conference of the IEEE Industrial Electronics Society, IEEE, 2018, pp. 2097–2102.
[18] L.D. Collins, R.H. Middleton, Distributed demand peak reduction with non-cooperative players and minimal communication, IEEE Trans. Smart Grid 10 (1) (2019) 153–162.
[19] Y. Zhou, R. Kumar, S. Tang, Incentive-based distributed scheduling of electric vehicle charging under uncertainty, IEEE Trans. Power Syst. 34 (1) (2019) 3–11.
[20] P. Jacquot, O. Beaude, S. Gaubert, N. Oudjane, Analysis and implementation of an hourly billing mechanism for demand response management, IEEE Trans. Smart Grid 10 (4) (2019) 4265–4278.
[21] X. Gong, A. De Paola, D. Angeli, G. Strbac, A game-theoretic approach for price-based coordination of flexible devices operating in integrated energy-reserve markets, Energy 189 (2019) 116153.
[22] R. Tang, H. Li, S. Wang, A game theory-based decentralized control strategy for power demand management of building cluster using thermal mass and energy storage, Appl. Energy 242 (2019) 809–820.

[23] R. Tang, S. Wang, H. Li, Game theory based interactive demand side management responding to dynamic pricing in price-based demand response of smart grids, Appl. Energy 250 (2019) 118–130.
[24] A. De Paola, V. Trovato, D. Angeli, G. Strbac, A mean field game approach for distributed control of thermostatic loads acting in simultaneous energy-frequency response markets, IEEE Trans. Smart Grid 10 (6) (2019) 5987–5999.
[25] B.A. Bhatti, R. Broadwater, Distributed nash equilibrium seeking for a dynamic microgrid energy trading game with non-quadratic payoffs, Energy 202 (2020) 117709.
[26] C. Eksin, H. Deliç, A. Ribeiro, Distributed demand side management of heterogeneous rational consumers in smart grids with renewable sources, in: 2014 IEEE International Conference on Acoustics, Speech and Signal Processing (ICASSP), IEEE, 2014, pp. 1100–1104.
[27] C. Eksin, H. Deliç, A. Ribeiro, Demand response management in smart grids with heterogeneous consumer preferences, IEEE Trans. Smart Grid 6 (6) (2015) 3082–3094.
[28] C. Eksin, H. Deliç, A. Ribeiro, Demand response with communicating rational consumers, IEEE Trans. Smart Grid 9 (1) (2018) 469–482.
[29] Q. Wang, J. Zeng, Z. Wang, J. Liu, Distributed game-theoretic interactive algorithm for microgrid optimization, in: 2018 37th Chinese Control Conference (CCC), IEEE, 2018, pp. 8758–8763.
[30] S. Wang, L. Du, J. Ye, L. He, Noncooperative social welfare optimization with resiliency against network anomaly, IEEE Trans. Ind. Inf. 16 (4) (2020) 2403–2412.
[31] Z. Zhu, S. Lambotharan, W.H. Chin, Z. Fan, A game theoretic optimization framework for home demand management incorporating local energy resources, IEEE Trans. Ind. Inf. 11 (2) (2015) 353–362.
[32] A. Barbato, A. Capone, L. Chen, F. Martignon, S. Paris, A distributed demand-side management framework for the smart grid, Comput. Commun. 57 (2015) 13–24.
[33] A. Barbato, A. Capone, L. Chen, F. Martignon, S. Paris, Distributed demand-side management in smart grid: how imitation improves power scheduling, in: 2015 IEEE International Conference on Communications (ICC), IEEE, 2015, pp. 6163–6168.
[34] N. Yaagoubi, H.T. Mouftah, User-aware game theoretic approach for demand management, IEEE Trans. Smart Grid 6 (2) (2015) 716–725.
[35] C. Rottondi, A. Barbato, L. Chen, G. Verticale, Enabling privacy in a distributed game-theoretical scheduling system for domestic appliances, IEEE Trans. Smart Grid 8 (3) (2017) 1220–1230.
[36] J. Zeng, Q. Wang, J. Liu, J. Chen, H. Chen, A potential game approach to distributed operational optimization for microgrid energy management with renewable energy and demand response, IEEE Trans. Ind. Electron. 66 (6) (2019) 4479–4489.
[37] M. Heleno, M.A. Matos, J.A.P. Lopes, Availability and flexibility of loads for the provision of reserve, IEEE Trans. Smart Grid 6 (2) (2015) 667–674.
[38] Study on tariff design for distribution systems, AF-Mercados, EMI, Technical Report, 2015. *https://ec.europa.eu/energy/sites/ener/files*.
[39] B.C. Hydro, Residential rates, 2020. Accessed 10 July 2020 https://www.bchydro.com/accounts-billing/rates-energy-use/electricity-rates/residential-rates.html.
[40] G. Fridgen, M. Kahlen, W. Ketter, A. Rieger, M. Thimmel, One rate does not fit all: an empirical analysis of electricity tariffs for residential microgrids, Appl. Energy 210 (2018) 800–814.
[41] A. Gholian, H. Mohsenian-Rad, Y. Hua, Optimal industrial load control in smart grid, IEEE Trans. Smart Grid 7 (5) (2016) 2305–2316.

[42] S. Boyd, S.P. Boyd, L. Vandenberghe, Convex Optimization, Cambridge University Press, 2004.
[43] D. Monderer, L.S. Shapley, Potential games, Games Econom. Behav. 14 (1) (1996) 124–143.
[44] R.F. Bischke, R.A. Sella, Design and controlled use of water heater load management, IEEE Trans. Power Apparatus Syst. PAS-104 (6) (1985) 1290–1293.
[45] D.-C. Wei, N. Chen, Air conditioner direct load control by multi-pass dynamic programming, IEEE Trans. Power Systems 10 (1) (1995) 307–313.
[46] R. Gopalakrishnan, J.R. Marden, A. Wierman, Potential games are necessary to ensure pure Nash equilibria in cost sharing games, Math. Oper. Res. 39 (4) (2014) 1252–1296.
[47] Q.D. Lã, Y.H. Chew, B.-H. Soong, Potential Game Theory: Applications in Radio Resource Allocation, Springer, 2016.
[48] A. Neyman, Correlated equilibrium and potential games, Int. J. Game Theory 26 (2) (1997) 223–227.
[49] J.B. Rosen, Existence and uniqueness of equilibrium points for concave N-person games, Econometrica 33 (3) (1965) 520–534.
[50] L.S. Shapley, Contributions to the Theory of Games (AM-28), in: A Value for n-Person Games, vol. 2, Princeton University Press, 1953, pp. 307–317.
[51] O. Shehory, S. Kraus, Coalition formation among autonomous agents: strategies and complexity (preliminary report), in: European Workshop on Modelling Autonomous Agents in a Multi-Agent World, Springer, 1993, pp. 55–72.
[52] N. Nisan, T. Roughgarden, E. Tardos, V.V. Vazirani, Algorithmic Game Theory, Cambridge University Press, 2007.

Distributed dynamic algorithm for energy management in smart grids

12

Shailesh Wasti[a], *Pablo Macedo*[a], *Shahab Afshar*[a], *James Griffin*[a], *Vahid R. Disfani*[a], *and Pierluigi Siano*[b]
[a]ConnectSmart Research Laboratory, Department of Electrical Engineering, University of Tennessee at Chattanooga, Chattanooga, TN, United States, [b]Department of Management and Innovation Systems, University of Salerno, Fisciano (SA), Italy

1 Introduction

Economic dispatch (*ED*) is a typical resource allocation problem in a power system, where each generator finds its optimum strategy to ensure power balance in the network. In ED problems, generators share their cost variables and generator limits to a global control center, which then implements the centralized dispatch algorithms and sends back dispatching information to all the generating units. However, microgrids and distributed energy resources (DER) owned and managed by private sectors urge the ED problem to be solved differently. In a decentralized ED, there is a need for a master node(s) for tight coordination of all agents. To remove this coordinator, a dynamic average consensus algorithm is leveraged to achieve completely distributed versions of subgradient algorithms. Each agent is empowered with the ability to update the time-varying estimate of average power mismatch and the market price in a distributed fashion and find consensus on those variables. In other words, in a *distributed economic dispatch* (DED) each agent finds its optimal strategy based on its own and that of its directly connected neighbors' information [1,2]. Fig. 12.1 shows the structure of the centralized and distributed structure of ED in power grids.

DERs managed by private sectors raise the concern of sharing cost parameters and generator limits with a central, global control center. Such information is private and, if shared, could be used to manipulate the power market. Distributed solutions offer a way for an agent (in this study, an agent is considered as a bus with its connected generator(s) and load(s)) to maintain the privacy of information about local cost functions, local generation level, and local loads. The only information shared is each agent's perception of the average system-wide marginal cost and the perception of the average global power mismatch (total generation minus total load). Iterative consensus drives these shared values to global equality, with the global power mismatch being driven to zero.

A few assumptions should be noted at the outset. It is assumed that

- The cost functions are quadratic, which is true for most of the IEEE-published test grids.

Decentralized Frameworks for Future Power Systems. https://doi.org/10.1016/B978-0-323-91698-1.00008-X

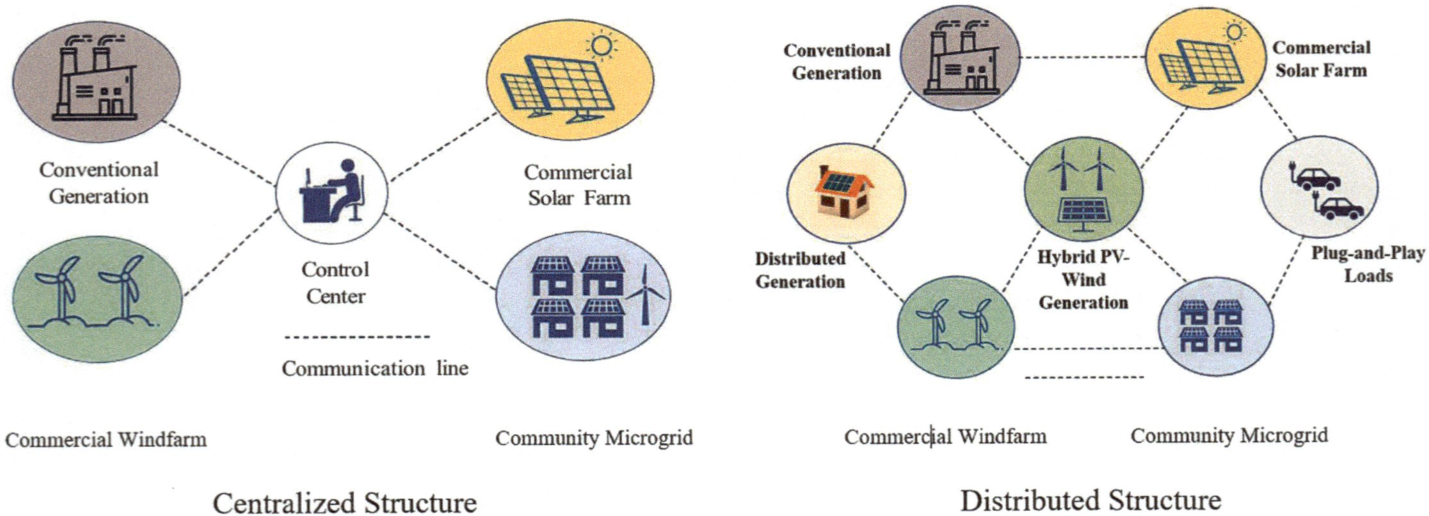

Fig. 12.1 Different approaches of economic dispatch in power grids.

- The agents are cooperative, meaning that no agent attempts to manipulate the market but, instead, makes its optimal calculations based on the equations defined herein.
- There is perfect communication between agents.
- Agents have knowledge of the total generation and load demand for themselves only.

In this chapter, to investigate the stability and convergence of the distributed subgradient algorithm, the problem is modeled as a discrete-time linear system.

1.1 Review of literature

The algorithms on decentralized and distributed architecture are inspired by the work on the multiagent systems in control theory [3], and parallel computation [4]. In the recent decade, several solutions have been proposed as an alternative approach for energy management in smart power grids including analytical target cascading (ATC), proximal message passing (PMP) [5], auxiliary proximal message passing (APMP) [6], and alternating direction method of multipliers (ADMM) [7–11]. In these methods, the central coordinator gathers the optimal decision variables from all the agents to compute the global dual variable and distributes the necessary information back to the agents. In economic dispatch problems and many other optimization problems on resource allocation, the dual variable is the market-clearing price that is updated to eliminate the power mismatch of the network. ADMM algorithm with model predictive control is implemented in Ref. [12] to control and schedule DERs in real-time in a micro-grid. This and all the decentralized algorithms, however, are susceptible to single-point failure and have data-privacy issues.

In order to achieve a fully distributed platform, it is important to design an algorithm that relies on no global variable. In Ref. [13], a distributed subgradient-based solution has been proposed to coordinate among distributed renewable generators where the local frequency measurements are used as a proxy for the power balance constraint. Although the frequency measurements are local, a central coordinator is still required to determine the local measurements.

In Ref. [14], an average consensus algorithm is developed for decentralized economic dispatch with the need of a master node(s) to obtain the primal and dual variables of the system in order to address the power imbalance of the entire network. The authors in Ref. [15] eliminate the need for a master node in the consensus algorithm and introduce an iterative algorithm called Consensus+Innovations (C+I) to address power mismatch. Two major shortcomings are identified in their proposed solutions. First, the convergence of the algorithm depends on the trade-off between the update coefficients chosen for C+I. Second, the update coefficients introduced in these references tend to zero after enough iterations, and thus the algorithm is not a suitable choice for economic dispatch to meet dynamically changing power demands.

Also, in Ref. [16], the DED problem is realized using a projected gradient and finite-time average consensus method. In this method, the update coefficients of agents are defined based on the eigenvalues of their Laplacian graph. Such a choice for updating coefficients is challenging since it simply contradicts the fact that the design of distributed algorithms requires all agents to determine their updating coefficients based on their own and, at most, their neighbors' shared information. The need for every agent to know the Laplacian graph, calculate all eigenvalues in the same order, and determine what eigenvalue to use at each iteration makes the algorithm too challenging for large networks.

1.2 Contribution

In this chapter, the algorithmic solution to distribute the ED problem is investigated, where the total demand of the network fluctuates in the granularity of seconds because of the high penetration of variable renewable energy (VRE) sources. This work leverages average consensus theory and its dynamic aspect in that agents communicate through the underlying communication network in a distributed fashion. The network is distributed if the agents can communicate only with their directly connected neighbors. Even though consensus problems can handle communication imperfections such as network splitting and time delays [17], this study is restricted to perfect communication. The agents are assumed to be cooperative with the ability to compute and communicate the information to their directly connected neighbors.

In this chapter, the mathematical foundation for the distributed algorithm is explained in detail. A consensus-based algorithm for the ED problem is then proposed. In a departure from the existing literature where all the decision variables are coupled to meet certain time-invariant constraint sets, the proposed online (real-time) algorithm finds consensus on the time-varying estimate of the average power mismatch in a purely distributed setting via dynamic average consensus algorithm detailed in Section 2, thus ensuring power balance in the network. First, the equivalent dual problem of the Lagrange relaxed problem is formulated and solved using consensus theory and the subgradient method. It is further shown that the iterative solution provably goes to optimal points using the well-known Karush-Kuhn-Tucker (KKT) conditions. As the algorithm tracks the real-time changes in demand using a dynamic average consensus algorithm, it is agnostic to any initialization vector. In other words, optimization can start from any random points without any apology.

In addition, the algorithm ensures privacy of data as agents communicate their own estimate of the average power mismatch to their neighbors, which quickly goes to zero during iteration, instead of the generated power and the demand of the node. In this approach, agents are only aware of their own cost functions [18,19]. Finally, the algorithm is modeled as a discrete dynamic system to investigate the stability and convergence of the algorithm detailed in Sections 3 and 4. The optimum gain parameter is calculated based on the study of the state matrix and modal analysis, thus ensuring stability and convergence. The efficacy of the proposed algorithm is demonstrated using the IEEE 39-bus test network.

2 Preliminaries

This section introduces some basic concepts on graph theory, convex optimization, consensus theory, optimality conditions, and stability of the discrete dynamic system. While the intent is not to detail the concepts that are quite basic, this background serves as the foundation to the algorithms presented in the following sections.

2.1 Distributed consensus algorithms

2.1.1 Graph theory

Notation: Define a graph with a set of vertices $=\{1,\ldots,N\}$ and the edges $\mathcal{V}\times\mathcal{V}$ that is restricted to a *simple*, *undirected* graph without multiedges and self-edges. For a set, its cardinality is represented by $|\mathcal{V}|$. Define $\mathcal{N}_i=\{j\in\mathcal{V}|\ (j,i)\in\mathcal{E}, j\neq i\}$ as the set of all the neighbors of the agent i and d_i as the degree of any node i; $d_i=|\{j:(i,j)\in\mathcal{E}\}|$. $\mathbb{R}$ denotes the set of all real numbers, $\in\mathbb{R}^{N\times N}$ is a null matrix, $\mathbf{I}\in\mathbb{R}^{N\times N}$ is an identity matrix, $\mathbf{1}\in\mathbb{R}^{N}$ is a column vector of all ones, γ is the eigenvalue, and $(\cdot)^T$ denotes the transpose of $(\cdot)$. Matrices and vectors are written in bold throughout the chapter.

Spectral graph theory is the study of eigenvalues and eigenvectors to understand the interesting properties of the graph. Let $\mathbf{A}$ be the *adjacency matrix* of graph $\mathcal{G}$, and $\mathbf{D}$ be the degree matrix with the vertex degree along its diagonal. Then, the graph Laplacian $\mathbf{L}\in\mathbb{R}^{N\times N}$ can be expressed in matrix form as

$$L=D-A \tag{12.1}$$

In full form:

$$l_{ij}=\begin{cases} d_i & i=j, \\ -1 & i\neq j \text{ and there is an edge } (i,j) \\ 0 & \text{otherwise} \end{cases} \tag{12.2}$$

The Laplacian consensus dynamics is given by the following equation:

$$\dot{\mathbf{x}}=-\mathbf{L}\mathbf{x} \tag{12.3}$$

where $\mathbf{x}\in\mathbb{R}^{N}$ denotes the values of corresponding vertices of the network. The spectral analysis on the Laplacian graph shows that

$$0=\gamma_1(\mathbf{L})\leq\gamma_2(\mathbf{L})\leq\cdots\leq\gamma_N(\mathbf{L}) \tag{12.4}$$

where γ_i denotes eigenvalue.

Observe that L is a symmetric positive semidefinite matrix and $\mathbf{L}\cdot\mathbf{1}=0$. For a connected graph, the connectivity is given by $\gamma_2(\mathbf{L})$, also called algebraic connectivity. The consensus is achieved if and only if $\gamma_2(\mathbf{L})$ is greater than zero. The convergence speed to consensus is governed by $\gamma_2(\mathbf{L})$, i.e., the slowest mode. With these conditions, any initial value leads to a consensus and any consensus is an equilibrium, i.e.,

$$\frac{1}{N}\mathbf{1}^T\mathrm{x}(0)=\frac{1}{N}\sum_{i}^{N}x_i(0) \tag{12.5}$$

where $x_i(0)$ denotes the initial value and $\mathbf{x^0}=(x_1(0),\ldots,x_N(0))$ stands for the vector of the initial values of the network. Below are some lemmas on a communication network, which are employed in later sections.

Lemma 12.1. *Communication topology between all the agents $A(t)$ is connected at all times $t \geq 0$. i.e., $\gamma_2(\mathbf{L}) \geq 0$. Where $\mathbf{A}$ is the adjacency matrix of graph $\mathcal{G}$.*

Lemma 12.2. *Any bus i can exchange information only with its neighboring agents, i.e., $\mathcal{N}_i = \{j \in \mathcal{V} | (j, i) \in \mathcal{E}$ and $\mathrm{j} \neq \mathrm{i}\}$.*

Lemma 12.3. *Laplacian matrix* L *is positive semidefinite, $\sum_j \mathbf{L_{ij}} = 0$, and $\gamma_2(\mathbf{L})$ is the algebraic connectivity of the network, which is the second-smallest eigenvalue of the Laplacian matrix. The speed of convergence to reach the consensus in the iterative process is governed by $\gamma_2(\mathbf{L})$ and it represents the convergence rate of the slowest mode* [20].

2.1.2 Dynamic average consensus

With $\mathbf{x^0} \in \mathbb{R}^N$ denoting the initial value of the ith element of the vector $\mathbf{x}$, i.e., $\mathbf{x^0} = [x_1(0), \ldots, x_N(0)]^T$, we discuss the computation of average $\frac{1}{N}\sum_{i=1}^{N} x_i^0$ in a distributed fashion below, where each node in the graph communicates with only its neighbors. Let us consider *distributed linear iteration* of the form

$$x_i^{k+1} = w_{ii} x_i^k + \sum_{j \in \mathcal{N}_i} w_{ij} x_j^k, \quad i = 1, \ldots, N \tag{12.6}$$

where iteration $k = 0, 1, 2, \ldots$, w_{ij} is the weight on x_j at node i, and $\mathcal{N}_i$ denotes the set of all neighbors of agent i. Setting $w_{ij} = 0$ for $j \notin \mathcal{N}_i$, this iteration can be written as

$$\mathbf{x}^{k+1} = \mathbf{W}\mathbf{x}^k \tag{12.7}$$

or equivalently as

$$\mathbf{x}_i^{k+1} = \mathbf{W}_\mathbf{i}^\mathbf{T} \mathbf{x}^k \tag{12.8}$$

where $\mathbf{W} = [W_1, \ldots, W_N] \in \mathbb{R}^{N \times N}$ with $W_i \in \mathbb{R}^N$. $\mathbf{W}$ is sparse since $W_{ij} = 0$ if agents i and j are not neighbors. Eq. (12.7) can be written as

$$\mathbf{x}^k = \mathbf{W}^k \mathbf{x}^0, \quad k = 0, 1, 2, \ldots \tag{12.9}$$

A matrix $\mathbf{W}$ is sought such that

$$\lim_{k \to \infty} \mathbf{W}^k = \frac{\mathbf{1} \cdot \mathbf{1}^T}{N} \tag{12.10}$$

The following Lemma is stated on weight matrix $\mathbf{W}$. The following conditions are necessary to guarantee the convergence [21]

$$\begin{cases} \mathbf{1}^T\mathbf{W} = \mathbf{1}^T \\ \mathbf{W}\mathbf{1} = \mathbf{1} \\ \zeta\left(\mathbf{W} - \frac{\mathbf{1}\mathbf{1}^T}{N}\right) < 1 \end{cases} \tag{12.11}$$

where $\zeta(\cdot)$ is the spectral radius of the matrix, $\mathbf{1} = [1, \ldots, 1]^T$ is the eigenvector associated with weight matrix $\mathbf{W}$, and $\mathbf{1}^T$ is the transpose of $\mathbf{1}$.

The choice of the coefficients of matrix $\mathbf{W}$ determines the speed of the convergence. Constant edge weight and local degree weight matrices are discussed in Ref. [21]. The local degree weight matrix $\mathbf{W}$ with coefficients depending only on the degree of the incident node is

$$w_{ij} = \begin{cases} \frac{1}{\max(d_i, d_j)} & i, j \in \mathcal{E} \\ 1 - \sum_{j \in \mathcal{E}} \frac{1}{\max(d_i, d_j)} & i = j \\ 0 & \text{otherwise} \end{cases} \tag{12.12}$$

where d_i and d_j are the number of edges that are incident to the vertex i and j, respectively. This method is implied from the Metropolis-Hastings algorithm and is often called the *Metropolis* method. The improved *Metropolis* called *Mean Metropolis* is proposed in Ref. [22] where

$$w_{ij} = \begin{cases} \frac{2}{\{d_i + d_j + \epsilon\}} & \{i, j\} \in \mathcal{E} \\ 1 - \sum_{j \in \mathcal{E}} \frac{2}{\{d_i + d_j + \epsilon\}} & i = j \\ 0 & \text{otherwise} \end{cases} \tag{12.13}$$

where ϵ denotes a very small number. The average consensus described by Eq. (12.7) can be extended to a consensus on a general time-varying signal, called dynamic average consensus [17].

If the average of a dynamically changing signal $\mathbf{z}$ is desired, some modifications in the algorithm are required. Let $\Delta\mathbf{z}$ be the input bias (the difference of z in two consecutive time steps) applied to the average consensus system. The following modification to Eq. (12.7) makes the dynamic consensus algorithm track the time-varying average consensus:

$$\mathbf{x}^{k+1} = \mathbf{W}\mathbf{x}^k + \Delta\mathbf{z}^{\mathbf{k+1}} \tag{12.14}$$

where the bias is $\Delta\mathbf{z}^{\mathbf{k+1}} = \mathbf{z}^{k+1} - \mathbf{z}^k$. The extension in Eq. (12.14) is still fully distributed as each agent needs to obtain the information from its directly connected neighbors.

2.2 Convex optimization

2.2.1 Constrained optimization

Constrained optimization is of the form:

$$\text{minimize}_{x \in R^N} f(x) \tag{12.15}$$

Subject to,

$h_i(x) = 0, \forall i \in$ E, where E stands for equality
$g_i(x) \leq 0, \forall i \in$ I, where I stands for inequality

Preliminarily, the Lagrange function of the problem (12.15) is

$$\mathcal{L}(x, \lambda, \mu) = f(x) + \sum_{i \in E} \lambda_i h_i(x) + \sum_{i \in I} \mu_i g_i(x) \tag{12.16}$$

2.2.2 Convex functions

A set $S \in \mathbb{R}^N$ is a convex set if the straight line segment connecting two points in S lies entirely inside S. Mathematically,

$$\alpha x + (1 - \alpha) y \in S \forall \alpha \in [0, 1]$$

The function f is a convex function in its convex set domain S if it satisfies:

$$f(\alpha x + (1 - \alpha) y) \leq \alpha f(x) + (1 - \alpha) f(y) \forall \alpha [0, 1], \quad \forall x, y \in S \tag{12.17}$$

A continuously differential function $f: \mathbb{R}^N \rightarrow$ is smooth if it has a globally Lipschitz gradient, $\exists L > 0 \ni \nabla f(y) - \nabla f(x) \leq L y - x \forall x,\ y \in \mathbb{R}^N$. Readers can refer to Ref. [23] for more insight on convexity.

2.2.3 Optimality conditions: KKT conditions

The first-order necessary conditions for the optimality, also known as KKT conditions, are defined as follows. Suppose x^* is a local solution of Eq. (12.15), the functions f, h_i, g_i are continuously differentiable, and the linearly independent constraint qualification holds at x^*. Then there is a vector of Lagrange multipliers $\lambda_i \in \mathcal{E}, \mu_i \in \mathcal{I}$, such that the following conditions hold at (x^*, λ^*, μ^*) [24]:

- Stationarity: $\nabla(x^*, \lambda^*, \mu^*) = \nabla f(x) + \sum_{i \in \mathcal{E}} \lambda_i \nabla h_i(x) + \sum_{i \in \mathcal{I}} \mu_i \nabla g_i(x) = 0$
- Primal feasibility: $h_i(x^*) = 0, \forall i \in \mathcal{E}\ g_i(x^*) \leq 0, \forall i \in \mathcal{I}$
- Dual feasibility: $\mu_i^* \geq 0, \forall i \in \mathcal{I}$
- Complementary slackness: $\sum_{i \in \mathcal{I}} \mu_i g_i(x^*) = 0$

For convex problems, KKT conditions are sufficient for optimality. For any convex optimization problem with the differentiable objective and constraint functions, any points that satisfy the KKT conditions are primal and dual optimal and have zero duality gap, and vice versa [24].

2.3 Distributed optimization

2.3.1 Primal and dual decomposition

Some optimization problems are inherently decomposed. The optimization problem is decomposable if the decision variables are decoupled for each subproblem. The problem is nondecomposable if the subproblems share some decision variables [25,26].

- Decomposable problem:

$$\min_{x,y} f(x) + g(y) = \min_{x} f(x) + \min_{y} g(y) \tag{12.18}$$

- Nondecomposable problem:

$$\min_{x,y,z} f(x,z) + g(y,z) \neq \min_{x,z} f(x,z) + \min_{y,z} g(y,z) \tag{12.19}$$

Eq. (12.19), however, can be decomposed using primal and dual decomposition.

(i) *Primal decomposition:*

$$\min_{x,y} f(x,z) + g(y,z) = \min_{x} f(x,z) + \min_{y} g(y,z)$$

Algorithm:

1. Solve subproblems for a fixed value of z
2. Update z

(ii) *Dual decomposition:*

Let us rewrite the problem in Eq. (12.19)

$$\min_{x,y,z} f(x,z) + g(y,z) = \mathbf{p}^* \tag{12.20}$$

where $\mathbf{p}^*$ is the optimal solution for the nonempty feasible domain of Eq. (12.20). Correspondingly,

$$\min_{x,y,z_1,z_2} f(x,z_1) + g(y,z_2) \text{ subject to}, z_1 = z_2 \tag{12.21}$$

The equality constraint can be relaxed using Lagrange relaxation. The Lagrange function for Eq. (12.21) is

$$(x,y,z_1,z_2;\lambda) = f(x,z_1) + g(y,z_2) + \lambda(z_1 - z_2) \tag{12.22}$$

Corresponding Lagrangian dual is [8]

$$\begin{aligned} D(\lambda) &= \min_{x,z}[f(x,z)+\lambda z] + \min_{y,z}[g(y,z)-\lambda z] \\ &= \max_{\lambda}\left\{\min_{x,z}[f(x,z)+\lambda z] + \min_{y,z}[g(y,z)-\lambda z]\right\} \end{aligned} \tag{12.23}$$

Readers can refer to Ref. [25] for a detailed explanation of decomposition. Note that Eq. (12.23) is an unconstrained problem. Constructing a dual problem from the original problem, sometimes called a primal problem, can provide more insight such as finding the lower bound (upper bound) for a minimization (maximization) problem. Once the constrained problem is transformed to the equivalent unconstrained dual problem, all optimization algorithms for unconstrained can easily be applied. Section 3 employs this method to solve the DED problem.

2.3.2 Distributed gradient algorithm

For a system with N agents in a network, a typical unconstrained optimization problem can be written as

$$\text{minimize}_{x\in R^N} \sum_{i\in V} f_i(x) \tag{12.24}$$

Any agent i can find consensus in the underlying communication network with a descent step along with the local (sub)gradient direction of its convex objective function [27]. Mathematically,

$$x_i^{k+1} = \mathbf{W}_\mathbf{i}^\mathbf{T}\mathbf{x}^k - \rho(k)g_i(k) \tag{12.25}$$

where $x_i^k \in \mathbb{R}$ is agent i's estimate of the optimal solution at iteration k, $\mathbf{x}^k$ is the estimate vector comprising all agents' estimates of x_i^k, $\rho(k)$ is a diminishing update factor, $\mathbf{W}$ is a stochastic weight matrix, and $g_i(k)$ is the (sub)gradient of the local objective function $f_i(x)$ which is convex.

2.3.3 Algorithm stability: Discrete dynamic systems

A system is discrete if the time variables have been quantized. The state-space representation of a discrete-dynamic system is

$$\mathbf{x}^{k+1} = \mathbf{A}\mathbf{x}^k + \mathbf{B}\mathbf{u}^k = \mathbf{f}(\mathbf{x}^k, \mathbf{u}^k) \tag{12.26}$$

where $\mathbf{x}^k$ and $\mathbf{u}^k$ are state and input vectors, respectively, at iteration k. The natural response of state equation Eq. (12.26) is

$$\mathbf{x}^k = \mathbf{A}^k\mathbf{x}^k \tag{12.27}$$

where $\mathbf{x}^0$ is the initial condition for state matrix $\mathbf{x}$. The stability of the system exclusively depends on matrix $\mathbf{A}$.

Theorem 12.1. *Let $\gamma_1, \ldots, \gamma_m$ $m \leq n$ be the eigenvalues of $\mathbf{A} \in \mathbb{R}^{N \times N}$. The system is* [28]

- *asymptotically stable iff $|\gamma_i| < 1, \forall i = \{1, \ldots, m\}$*
- *stable if $|\gamma_i| \leq 1, \forall i = \{1, \ldots, m\}$*
- *unstable if $\exists$ i such that $|\gamma_i| > 1$*

The spectral radius of a matrix $\mathbf{A}$ is defined as the maximal modulus of all of its real and complex numbers; $\zeta(\mathbf{A}) = \max\{\gamma_1, \ldots, \gamma_N\}$.

Correspondingly,

$$(1 = |\gamma_1| \geq |\gamma_2| \geq \cdots \geq |\gamma_N| \tag{12.28}$$

The stability is guaranteed if at any iteration, and the distance between the states $\mathbf{x}$ and the optimal point $\mathbf{x}^*$ diminishes over iterations. Mathematically, a fixed point $\mathbf{x}$ of an iterative system $\mathbf{x}^{k+1} = \mathbf{A}\mathbf{x}^k$ is called stable if for every $\epsilon > 0$ there exists a $\delta > 0$ such that whenever $||\mathbf{x}^0 - \mathbf{x}^*|| < \delta$, then the resulting iterates satisfy $||\mathbf{x}^k - \mathbf{x}^*|| < \epsilon$ for all k (see Chapter 9 in [29] for more details).

3 Application of distributed algorithms in economic dispatch problem

This section presents a privacy-preserving consensus algorithm for ED for distributed optimization. Each agent is empowered with the ability to update the time-varying estimate of average power mismatch and the market price in a distributed fashion and find consensus on those variables.

3.1 Economic dispatch problem

Economic dispatch is a typical resource allocation problem in a power system, where each generator finds its optimum strategy to ensure power balance in the network.

Let each agent $i \in \mathcal{V} = \{1, \ldots, N\}$ be the bus of the power network. The economic dispatch is, then, modeled as

$$\underset{\mathbf{p_g^k}}{minimize} \sum_{i \in \mathcal{V}} \underbrace{a_i \left(p_{gi}^k\right)^2 + b_i p_{gi}^k + c_i}_{C_i\left(p_{gi}^k\right)} \tag{12.29a}$$

$$\text{Subject to,} \sum_{i \in \mathcal{V}} p_{gi}^k = \sum_{i \in \mathcal{V}} p_{di}^k \tag{12.29b}$$

$$p_{gi}^{\min} \leq p_{gi}^{k} \leq p_{gi}^{\max} \forall i \in \mathcal{V} = \{1, \ldots, N\} \tag{12.29c}$$

where c_i is the cost function of agent i, given in a generic quadratic function form by the coefficients a_i, b_i and c_i. Note the following details

(i) $\mathbf{p_g^k} = \{p_{gi}^k, \ldots, p_{gN}^k\}$ denotes the power generation of all buses at the time-step k and is the set of decision variables.

(ii) Eq. (12.29a) represents the objective function as minimization of the total generation cost at any time-step k with $C_i(\cdot)|\rightarrow$ equal to the generation cost function of agent i. To preserve the convexity, it is assumed that $C_i(p_{gi})$ is Lipschitz continuous and convex, defined as $C_i(\cdot) = a_i p_{gi}^2 + b_i p_{gi} + c_i$ if agent i has a generating unit, and 0 otherwise; the coefficients a_i, b_i, c_i are the characteristic cost parameters of generator i and p_{gi} is its generation.

(iii) Eq. (12.29b) represents the power balance constraint; it ensures that the network's total generation and total demand are equal at any time, where p_{di} denotes the demand at bus i.

(iv) Eq. (12.29c) guarantees all generators operate within the minimum and maximum generation capacities at any time.

The optimization problem in Eqs. (12.29a), (12.29b), and (12.29c) has the local objective function (Eq. 12.29a) and local constraint (12.29c), coupled by the global constraint (Eq. (12.29b) on the network's total generation and total demand. In this chapter, the problem is modeled as a completely distributed optimization problem that allows each agent to determine the value of its decision variables $p_{gi} \in \mathbf{p_g}$.

Let us define agent i's power mismatch by $p_i = p_{gi} - p_{di}$ and its feasible set by

$$\Omega_i = \left[p_i^{\min}, p_i^{\max}\right] = \left[p_{gi}^{\min} - p_{d_i}, p_{gi}^{\max} - p_{d_i}\right]$$

Substituting $p_{gi} = p_i + p_{di}$ in the cost function $C_i(p_{gi})$ is defined in Eq. (12.30), where the cost function of each agent modifies to $f_i(p_i)$.

$$f_i(p_i) = a_i p_i^2 + (2a_i p_{di} + b_i)p_i + a_i p_{di}^2 + b_i p_{di} + c \tag{12.30}$$

Then, Eq. (12.21) can be written as

$$\underset{\mathbf{p}\in\mathbf{\Omega}}{minimize} \; f(\mathbf{p}) := \sum_{i\in\mathcal{V}} f_i(p_i) \tag{12.31}$$

$$\text{Subject to, } \sum_{i\in\mathcal{V}} p_i = 0 \tag{12.32}$$

where $f_i : \mathbb{R} \rightarrow \mathbb{R}$ is a local convex objective function of agent $\mathbf{\Omega} = [\Omega_1; \ldots; \Omega_N] \in \mathbb{R}^N$ is compact and convex constraint sets, and $\mathbf{p} \in \mathbb{R}^N$ is the set of the decision variables. Eq. (12.32) represents a global equality constraint, and it is assumed that the set of feasible points is nonempty. These assumptions are formalized in the following Lemma.

Lemma 12.4. *The function* f_i *and thus* $\mathbf{f}: \mathbb{R}^N \to \mathbb{R}$ *is closed, proper, and convex. The interior of the feasible region is nonempty. This is known by Slater's condition for convex optimization problems* [30].

3.2 Dual decomposition of economic dispatch

The dual representation of Eq. (12.31) is

$$\max_{\lambda \in} \sum_{i \in \mathcal{V}} \underbrace{\left\{ \min_{\Omega_i} f_i(p_i) - \lambda p_i \right\}}_{\Phi_i^{\dagger}(\lambda)}$$

where the optimal solution of the minimization problem for given λ can be explicitly defined as

$$\Phi_i^{\dagger}(\lambda) := f_i\left(p_i^{\dagger}\right) \tag{12.33}$$

with

$$p_i^{\dagger}(\lambda) := \left[MC_i^{-1}(\lambda)\right]_{p_i^{\min}}^{p_i^{\max}} \tag{12.34}$$

where $MC_i(p_i) = \frac{\partial f_i}{\partial p_i}$ is the marginal cost corresponding to bus i, MC_i^{-1} denotes the inverse of the marginal cost of power output, and $[\cdot]_a^b$ denotes $\max\{\min\{\cdot, b\}, a\}$ for a, $b \in \mathbb{R}$, $a \leq b$.

The dual variable λ is updated as

$$\lambda^{k+1} = \lambda^k - \alpha \sum_{i \in \mathcal{V}} p_i^k = \lambda^k - \alpha N \overline{p}^k = \lambda^k - \rho \overline{p}^k \forall i \tag{12.35}$$

to ensure that the power balance constraint is met where α and ρ are the feedback gain parameters for the average mismatch, and N is the total number of agents of the network. ρ is chosen below certain critical values to ensure the stability of convergence, which will be discussed later in the chapter. The price update reflects the fact that with a demand excess, the market price must increase to attract more generators to supply power, and the opposite otherwise (Fig. 12.2).

Let $\chi_g \subset \mathcal{V}$ be the set of buses with the generating units. For the cost function defined in Eq. (12.35), the marginal cost of generation is

$$\frac{df_i(p_i)}{dp_i} = 2a_i p_i + 2a_i p_{di} + b_i \doteq \lambda_i \tag{12.36}$$

in its feasible region. The feasible region of any agent i is bounded by the lower and the upper limit of its power mismatch. Consequently,

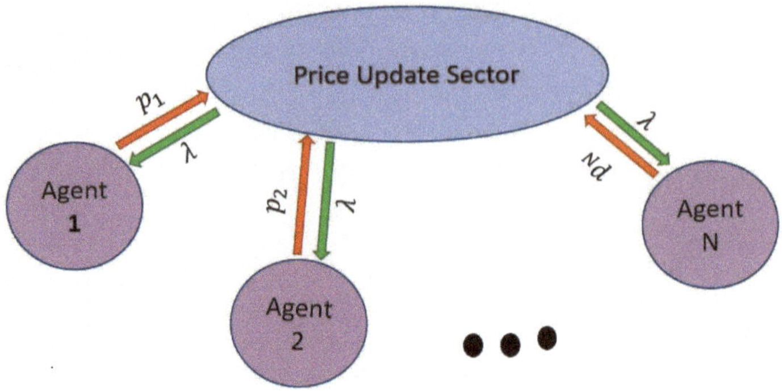

Fig. 12.2 Information flow in a decentralized platform.

$$p_i^* = \left[\frac{\lambda^* - b_i - 2a_i p_{di}^k}{2a_i}\right]_{p_i^{\min}}^{p_i^{\max}} \tag{12.37}$$

where superscript p_i* and λ* denote the corresponding parameter values at the optimal solution. Thus, Eq. (12.10) can be written in a generalized form as

$$\mathbf{p}^{k+1} = \left[\mathbf{\Lambda}\boldsymbol{\lambda}^k - \mathbf{p}_d^k - \mathbf{\Lambda}\mathbf{b}\right]_{\mathbf{p}^{\min}}^{\mathbf{p}^{\max}} \tag{12.38}$$

where $\mathbf{a} = \text{diag}\,\{a_i\}$, $\mathbf{b} = \text{diag}\,\{b_i\}$, $\mathbf{\Lambda} = \text{diag}\,\{1/(2a_i)\}$. For $i \notin \chi_g$, $a_i = b_i = \Lambda_i = 0$.

3.3 Decentralized algorithm for economic dispatch

The algorithm for all the agents in the network can be summarized in vector format as

$$\begin{cases} \mathbf{p}^{k+1} = \left[\Lambda\boldsymbol{\lambda}^k - \mathbf{p}_d^k - \mathbf{\Lambda}\mathbf{b}\right]_{\mathbf{p}^{\min}}^{\mathbf{p}^{\max}} \\ \lambda^{k+1} = \lambda^k - \rho\overline{p}^k \end{cases} \tag{12.39}$$

In a decentralized framework, a central coordinator (price updater) tightly coordinates between all agents to ensure supply and demand balance in the network. In the semantics of optimization, λ is a dual variable in Lagrange relaxed function associated with equality constraint and is updated based on the information from all the participating agents. It should be noted that in an optimization problem with equality constraint(s) binding all the agents, the feasible region of an individual agent's decision variable is influenced by other agents' decision variables. In addition to the concerns on data privacy and vulnerability to single-point failure, the central question is, "Who adjusts the market price?"

Remark. The price adjustment process in general equilibrium theory is known as the tatonnement process, which is a simple way to model the supply and demand of any market. The fundamental idea is that all generators find an optimum strategy to adjust their output based on their cost function and local constraints for the given market price. With all agents in perfect coordination, the market goes to equilibrium after some trial and error process. That is, the price updater increases the market price when total generation falls short and decreases for a surplus in supply. The equilibrium price is in locally stable equilibrium if the initial price in the small neighborhood of λ^* converges to optimum price, and it is globally stable if any initial value of market price λ converges to λ^* [31]. This raises a very important question of who can possibly be that benign updater for price adjustment. The following section presents the formulation without the price updater.

3.4 Distributed economic dispatch

The objective function $\Phi_i^\dagger(\lambda)$ is not separable across buses $i \in \mathcal{V}$ since it has a scalar variable λ which is common between all $i \in \mathcal{V}$. The variables λ_i are constrained to be equal for all $i \in \mathcal{V}$ at the optimal solution, i.e., $\lambda_i = \lambda_j \forall i, j \in \mathcal{V}$. In a connected graph, it suffices to define the constraint $\lambda_i = \lambda_j$ just across the neighboring agents, i.e., $\forall (i,j) \in \mathcal{E}$. With these modifications, the DED problem can be written as below.

$$\max_{\lambda} \Phi(\lambda) := \sum_{i \in \mathcal{V}} \Phi_i^\dagger(\lambda_i) \text{ subject to } \lambda_i = \lambda_j \forall (i,j) \in \mathcal{E} \tag{12.40}$$

The following two results suggest that solving the distributed optimization problem in Eq. (12.40) is equivalent to solving Eqs. (12.29a), (12.29b), and (12.29c). The unique optimal point (p_i^*) will then be recovered from the unique dual optimal λ^*. The objective function Φ of DED in Eq. (12.41) is strictly concave over R^N. Eq. (12.40) has a unique optimal point (λ^*) with $\lambda_i^* = \lambda_j^* = \lambda^*$ for all $(i,j) \in \mathcal{E}$. DED has a unique optimal point p^* where $p_i^* = p_i^\dagger(\lambda^*)$.

Similar to Eqs. (12.33) and (12.34), any agent i can find its optimal solution point given λ_i as

$$\Phi_i^\dagger(\lambda_i) := f_i\left(p_i^\dagger\right) \tag{12.41}$$

with

$$p_i^\dagger(\lambda_i) := \left[MC_i^{-1}(\lambda_i)\right]_{p_i^{\min}}^{p_i^{\max}} \tag{12.42}$$

In order to realize this, all the market prices should reach one consensus value while power balance constrained is ensured in the network. Based on Lemmas 12.1–12.3, all participating agents reach consensus on the market price given as

$$\lambda_i^{k+1} = W_i^T \boldsymbol{\lambda}^k - \rho \overline{p}_i^k \tag{12.43}$$

where $\boldsymbol{\lambda} = [\lambda_1; \ldots; \lambda_N] \in \mathbb{R}^N$, and $W_i \in \mathbb{R}^{N\times 1}$ is the vector of agent i's update weights as defined in Eq. (12.13). Notice that the global variable average power mismatch $\overline{p}$ is replaced with the local variable $\overline{p_i}$, defined as the agent i's estimate of the average power mismatch (Fig. 12.3). For this, any agent i estimates the network's average mismatch using a dynamic average consensus algorithm in Eq. (12.14).

$$\overline{p}_i^{k+1} = W_i^T \overline{\mathbf{p}}^k + \left(p_i^{k+1} - p_i^k\right) \tag{12.44}$$

where $\overline{\mathbf{p}} = [\overline{p}_1; \ldots; \overline{p}_N] \in^N$. Consequently, the fully distributed form of Eq. (12.35) is

$$p_i^\dagger(\lambda_i) := \left[MC_i^{-1}(\lambda_i)\right]_{p_i^{\min}}^{p_i^{\max}} = \left[\frac{\lambda_i^* - b_i - 2a_i p_{di}^k}{2a_i}\right]_{p_i^{\min}}^{p_i^{\max}} \tag{12.45}$$

The distributed algorithm for all the agents in the network can be summarized in vector format as

$$\begin{cases} \mathbf{p}^{k+1} \in \left[\boldsymbol{\Lambda}\boldsymbol{\lambda}^k - \mathbf{p}_d^k - \boldsymbol{\Lambda}\mathbf{b}\right]_{\mathbf{p}^{\min}}^{\mathbf{p}^{\max}} \\ \boldsymbol{\lambda}^{k+1} = \mathbf{W}\boldsymbol{\lambda}^k - \rho\overline{\mathbf{p}}^k \\ \overline{\mathbf{p}}^{k+1} = \mathbf{W}\overline{\mathbf{p}}^k + \left(\mathbf{p}^{k+1} - \mathbf{p}^k\right) \end{cases} \tag{12.46}$$

where $\boldsymbol{\lambda} = [\lambda_1; \ldots; \lambda_N] \in \mathbb{R}^N$.

With reference to Lemma 12.4, the following lemma investigates the optimality of the DED algorithm in Eq. (12.46).

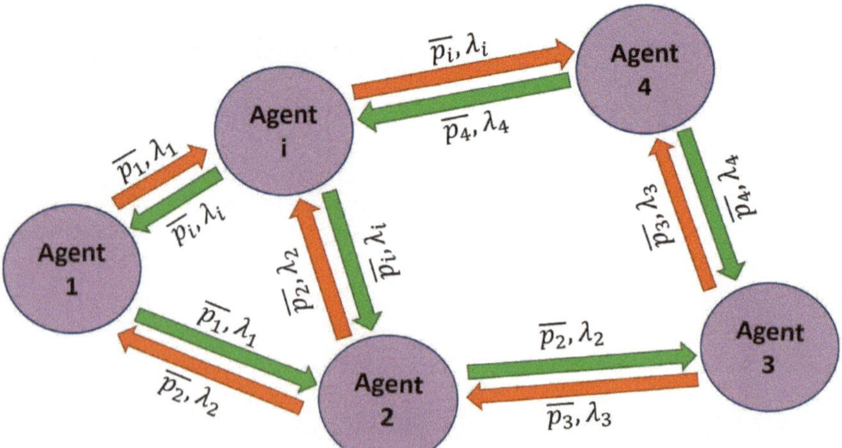

Fig. 12.3 Information flow in a distributed platform.

Lemma 12.5. *For any convex optimization problem with differentiable objective and constraint functions, any points that satisfy the Karush-Kuhn-Tucker (KKT) conditions are primal and dual optimal and have zero duality gap, and vice versa* [30].

Let us reintroduce the DED optimization problem to investigate the optimality. The Lagrange relaxation of the DED for any agent i is

$$\begin{aligned}\inf_{p_i\in\Omega_i}\sup_{\lambda_i\in\mathbb{R}}\Phi_i^\dagger(p_i,\lambda_i) &:= \inf_{p_i\in\Omega_i}\sup_{\lambda_i\in\mathbb{R}}\left\{f_i(p_i)-\lambda_i\left(p_i-p_i^k+N_p^{-k}\right)\right\}\\ &:= \inf_{p_i\in\Omega_i}\sup_{\lambda_i\in\mathbb{R}}\{f_i(p_i)-\lambda_i p_i\}\end{aligned} \tag{12.47}$$

where $p_i-p_i^k+N\overline{p}^k=\mathbf{1}^T\mathbf{p}=0$ is the power balance constraint of the network. Consequently, based on the strong duality of the problem

$$\underbrace{\inf_{\mathbf{p}\in\boldsymbol{\Omega}}\sup_{\boldsymbol{\lambda}\in\mathbb{R}^N}\boldsymbol{\Phi}^\dagger(\mathbf{p},\boldsymbol{\lambda})}_{\text{primalproblem}}=\underbrace{\sup_{\boldsymbol{\lambda}\in\mathbb{R}^N}\inf_{\mathbf{p}\in\boldsymbol{\Omega}}\boldsymbol{\Phi}^\dagger(\mathbf{p},\boldsymbol{\lambda})}_{\text{dualproblem}} \tag{12.48}$$

where $\boldsymbol{\Phi}^\dagger=[\Phi_1^\dagger;\cdots;\Phi_N^\dagger]$.

It must be shown that a fixed point $(\mathbf{p},\boldsymbol{\lambda})$ is the solution of Eq. (12.48). In other words, $(\mathbf{p},\boldsymbol{\lambda})$ is the KKT points. Using stationary conditions for in vector format

$$\nabla_{\mathbf{p}}\boldsymbol{\Phi}\in\partial f(\mathbf{p}^*)-\boldsymbol{\lambda}^*\in\mathbf{0} \tag{12.49}$$

$$\nabla_{\boldsymbol{\lambda}}\boldsymbol{\Phi}\in\mathbf{p}^*-\mathbf{p}^k+\mathbf{p}^k\in\mathbf{0} \tag{12.50}$$

where ∇ is the partial differentiation operator. The update process of market price $\boldsymbol{\lambda}$ in Eq. (12.46) can be written as $\mathbf{W}\boldsymbol{\lambda}^k-\boldsymbol{\lambda}^{k+1}=\rho\overline{\mathbf{p}}^k$. It is pretty straightforward that each market price finds a stable point once the mismatch term goes to zero. Then, the consensus on λ is guaranteed based on the properties of $\mathbf{W}$ in Eq. (12.13). So, for a fixed point $(\mathbf{p},\boldsymbol{\lambda})$, $\mathbf{p}^*-\mathbf{p}^k=0$ and $\overline{\mathbf{p}}^k=0$, the left-hand side of Eq. (12.50) goes to zero, and all market prices settle in their optimum points. This can now be readily interpreted as a price adjustment process in general equilibrium theory but in a distributed fashion. All players find the optimum strategy to adjust their output based on their objective function and local sets of constraints for their estimated market price, which in turn depends on its estimated mismatch of the network.

3.5 Distributed algorithm for economic dispatch

The proposed DED in Eq. (12.46) now can be implemented as an iterative procedure. Each agent after receiving the information on price, an estimate of the average generation and demand of the network from its directly connected neighbors N_i finds the optimum value of generation. It then updates its price and the estimate of the network's power generation and demand and broadcasts to its connected neighbors. Algorithm 12.1 elaborates the iterative processes for the consensus-based economic dispatch problem.

Algorithm 12.1. The iterative processes for the consensus-based economic dispatch problem

Algorithm 1 Distributed-subgradient Algorithm from the perspective of agent i

Input: cost-coefficients $\{a_i, b_i\}$, and control parameter ρ

Initialization: $\lambda_i{}^0, p_i{}^0 = \overline{p_i}^0 = -p_d{}^0$

for $k = 0$ to ∞ *do*

Receive $\overline{\mathbf{p}}$ and $\boldsymbol{\lambda}$. Each agent receives data from its neighboring agents only.

Compute power mismatch p^{k+1} as $p^{k+1} \in [\Lambda_i \lambda_i{}^k - p_{di}^k - \Lambda_i b_i]_{p^{min}}^{p^{max}}$

Update $\lambda_i^{k+1} = W_i^T \boldsymbol{\lambda}^k - \rho \overline{p}_i^k$

Update $\overline{p}_i^{k+1} = W_i^T \overline{\mathbf{p}}^k + (p_i^{k+1} - p_i^k)$

Broadcast $\overline{p_i}^{k+1}, \lambda_i{}^{k+1}$. Each agent broadcasts its data to its neighboring agents only.

4 Numerical stability and convergence

A system is discrete if the time variables have been quantized. The optimization problem in Eq. (12.46) is recast into a discrete state-space problem to investigate the stability and convergence of the proposed DED algorithm. The state-space representation of a discrete-dynamic system is defined in Eqs. (12.26) and (12.27). The state equation for the market price vector is

$$\boldsymbol{\lambda}^{k+1} = \mathbf{W}\boldsymbol{\lambda}^k - \rho\overline{\mathbf{p}}^k \tag{12.51}$$

Similarly, the state equation of power output is

$$\mathbf{p}^{k+1} = \boldsymbol{\Lambda}\boldsymbol{\lambda}^k - \mathbf{p}_d^k - \boldsymbol{\Lambda}\mathbf{b} \tag{12.52}$$

The state equation for average power mismatch is

$$\overline{\mathbf{p}}^{k+1} = \mathbf{W}\overline{\mathbf{p}}^{k} + \left(\mathbf{p}^{k+1} - \mathbf{p}^{k}\right) \tag{12.53}$$

Consequently,

$$\overline{\mathbf{p}}^{k+1} = \mathbf{W}\overline{\mathbf{p}}^{k} + \left(\mathbf{\Lambda}\boldsymbol{\lambda}^{k} - \mathbf{p}_{d}^{k} - \mathbf{\Lambda}\mathbf{b} - \mathbf{p}^{k}\right) \tag{12.54}$$

The state equations Eqs. (12.51), (12.52), and (12.54) can be written in matrix form as

$$\begin{bmatrix} \boldsymbol{\lambda}^{k+1} \\ \mathbf{p}^{k+1} \\ \overline{\mathbf{p}}^{k+1} \end{bmatrix} = \underbrace{\begin{bmatrix} \mathbf{W} & \mathbb{O} & -\rho\mathbf{I} \\ \mathbf{\Lambda} & \mathbb{O} & \mathbb{O} \\ \mathbf{\Lambda} & -\mathbf{I} & \mathbf{W} \end{bmatrix}}_{\mathbf{A}} \underbrace{\begin{bmatrix} \boldsymbol{\lambda}^{k} \\ \mathbf{p}^{k} \\ \overline{\mathbf{p}}^{k} \end{bmatrix}}_{\mathbf{x}} + \underbrace{\begin{bmatrix} \mathbb{O} & \mathbb{O} \\ -\mathbf{I} & -\mathbf{b}\mathbf{\Lambda} \\ -\mathbf{I} & -\mathbf{b}\mathbf{\Lambda} \end{bmatrix}}_{\mathbf{B}} \underbrace{\begin{bmatrix} \mathbf{P_d}^{k} \\ \mathbf{1}_{N} \end{bmatrix}}_{\mathbf{u}} \tag{12.55}$$

Output Matrix:

$$\begin{bmatrix} \boldsymbol{\lambda}^{k} \\ \mathbf{P}_{\mathbf{g}}^{k} \end{bmatrix} = \underbrace{\begin{bmatrix} \mathbf{I} & \mathbb{O} & \mathbb{O} \\ \mathbb{O} & \mathbf{I} & \mathbb{O} \end{bmatrix}}_{\mathbf{C}} \underbrace{\begin{bmatrix} \boldsymbol{\lambda}^{k} \\ \mathbf{p}^{k} \\ \overline{\mathbf{p}}^{k} \end{bmatrix}}_{\mathbf{x}} + \underbrace{\begin{bmatrix} \mathbb{O} & \mathbb{O} \\ \mathbf{I} & \mathbb{O} \end{bmatrix}}_{\mathbf{D}} \underbrace{\begin{bmatrix} \mathbf{P_d}^{k} \\ \mathbf{1}_{N} \end{bmatrix}}_{\mathbf{u}} \tag{12.56}$$

Comparing Eqs. (12.29a), (12.29b), (12.29c) and (12.30) with the canonical form in Eq. (12.24), $\mathbf{A} \in \mathbb{R}^{3N\times 3N}$ is the state matrix, $\mathbf{B} \in \mathbb{R}^{3N\times 2N}$ is the input matrix, $\mathbf{C} \in \mathbb{R}^{2N\times 3N}$ is the output matrix, and $\mathbf{D} \in \mathbb{R}^{2N\times 2N}$ is the feed-forward matrix.

Theorem 12.2. *Let $\gamma_1, \ldots \gamma_m$ $m \leq n$ be the eigenvalues of $\mathbf{A} \in \mathbb{R}^{3N\times 3N}$. The system Eq. (12.26) is [28]*

- *Asymptotically stable iff $|\gamma_i| < 1, \forall i = \{1, \ldots, m\}$*
- *Stable if $|\gamma_i| \leq 1, \forall i = \{1, \ldots, m\}$*
- *Unstable if $\exists$ i such that $|\gamma_i| > 1$*

The spectral radius of a matrix **A** is defined as the maximal modulus of all of its real and complex numbers; $\zeta(\mathbf{A}) = \max\{\gamma_1, \ldots, \gamma_{3N}\}$. Based on Theorem 12.2, the stability can be examined by the study of the spectral radius. Since the stability of a nonzero fixed point is sought

$$\zeta(\mathbf{A}) = 1 \tag{12.57}$$

Correspondingly,

$$1 = \gamma_1 \geq \gamma_2 \geq \cdots \geq \gamma_{3N} \tag{12.58}$$

Lemma 12.6. *A fixed point* $\mathbf{x}$ *of an iterative system* $\mathbf{x}^{k+1} = \mathbf{A}\mathbf{x}^k$ *is called stable if for every* $\epsilon > 0$ *there exists a* $\delta > 0$ *such that whenever* $\mathbf{x}^0 - \mathbf{x}^* < \delta$*, then the resulting iterates satisfy* $\mathbf{x}^0 - \mathbf{x}^* < \epsilon$ *for all* k [29].

The solution of the natural response (Eq. 12.27) is

$$\mathbf{x}(k) = \sum_{i=1}^{3N} c_i \gamma_i^k \xi_i \tag{12.59}$$

where c_i is the scalar prescribed by initial condition $\mathbf{x}(0)$, γ_i is the eigenvalue of $\mathbf{A}$, and ξ_i is a linearly independent eigenvector. Consequently,

$$\lim_{k \to \infty} \mathbf{x}(k) = c_1 \xi_1 \tag{12.60}$$

Note that the stability means the stability of all the fixed points, which in fact are the elements of eigenspace corresponding to $\gamma = 1$.

Corollary 12.1. *All solutions of linear iterative solution* $\mathbf{x}^{k+1} = \mathbf{A}\mathbf{x}^k$ *converges to a vector* ξ *that lies in the* $\gamma_1 = 1$ *eigenspace provided Eq.* (12.31) *holds true. Moreover, the rate of convergence of the solution is governed by the modulus* γ_2 *of the subdominant eigenvalue* [29].

The correlation of gain parameter ρ with the convergence speed of the algorithm should be considered in the optimization. Fig. 12.4 shows the plot of the second-highest eigenvalues of the state matrix $\mathbf{A}$ for different input of ρs. The algorithm becomes unstable beyond the critical value ρ_{cr}. As stated in Corollary 12.1, the smaller the second largest eigenvalue, the faster the convergence. As depicted in Fig. 12.4, the subdominant eigenvalue quickly decreases as ρ is decreased. The system has oscillatory behavior at the margin of stability, and the oscillation is the controller with a lower gain. Recall that the update process of the market price λ has a consensus part and a gradient part. With the higher ρ, the mismatch of the network goes to an acceptable limit before the state variables settle to a fixed point. A trade-off can be sought by choosing ρ in between ρ_{cr} and ρ corresponding to minimum subdominant eigenvalue.

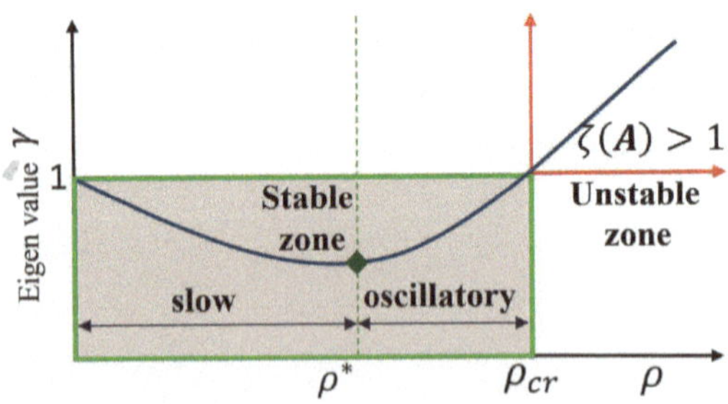

Fig. 12.4 Plot showing the locus of eigenvalues for different feedback gain parameter $|\gamma_j|, j \notin i$; $|\gamma_i| = 1$.

5 Results and discussions

5.1 Simulation setup

To test the efficacy of the optimization problem, Algorithm 12.1 is implemented for the IEEE-39 bus network. The cost coefficients a_i and b_i, generator limits, and the initial values of power demand at each bus are adopted from Ref. [32]. The power demands of all buses are artificially shifted four times during the iterative process to mimic the dynamic nature of the load. This is accomplished by uniform distribution functions used to randomize the drop percentages for each load.

In all cases, as the algorithm is robust enough to drive any initial value to convergence, power generations are initialized at zero and prices at a uniform distribution around the initial optimal value. The Mean Metropolis algorithm is used with $\epsilon = 1$ to set up the weight matrix **W** assuming that each bus of the network is an agent and the communication topology follows the electrical connection between buses. The case study results are benchmarked against MATPOWER 7.0. Based on the discussion in Section 4, the optimal gain parameter ρ is calculated to be $8.2e-04$.

5.2 Algorithm performance

Fig. 12.5A shows the convergence of market price λ (dual variable) of all the nodes. All agents reach one consensus value, and any stable consensus is the optimal dual variable. Fig. 12.5B depicts the magnified version of the initial consensus process, where all agents take $0.0084781s$ (or 500 iterations) to find the settling point. In the iterative process, agents calculate their iterates based on their estimation of average and gradient descent to find the market equilibrium. In Fig. 12.5C all agents increase λ based on their estimation as the network had a sudden deficit of power because of the increment in demand.

Fig. 12.6A shows that all generating units in the IEEE 39-bus network collectively ensure power balance in the network while finding their optimal strategies. Fig. 12.6B is the magnified version showing the convergence at $0.0084781s$, and Fig. 12.6C shows the magnified version when all the generating units move from one equilibrium point to another. The demand fluctuates every 0.2 s, and the generating units quickly find their optimal points based on the demand of the network. While the number of iterations is higher, the time for each iteration is quite small because all nodes participate in the update process and the optimization problem does not require sophisticated computation.

The algorithm proposed in this chapter preserves the privacy of the agents participating in the market. Fig. 12.7A shows the estimate of the total power mismatch by each agent. With the change in demand in the network in different nodes, all agents quickly estimate the power imbalance of the network to adjust their market price. Fig. 12.7B and C is the enlarged versions of the individual estimates of price mismatch. In the algorithm presented here, any agent i shares just two pieces of information (λ_i and $\overline{p_i}$) to their directly connected neighbors.

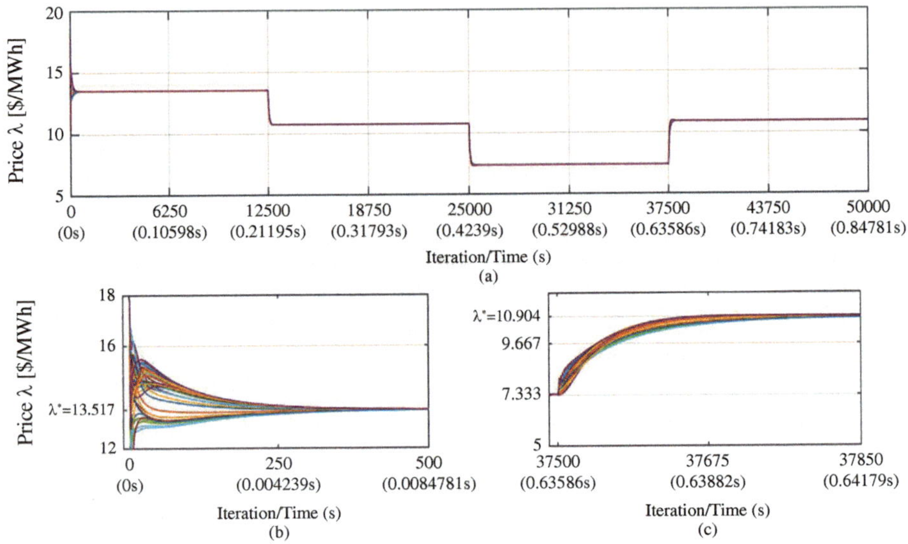

Fig. 12.5 Results for IEEE 39-bus case; (A) displays the convergence process for the market price (λ); (B) shows the detail of the initial convergence process; and (C) shows the detail of the last convergence process.

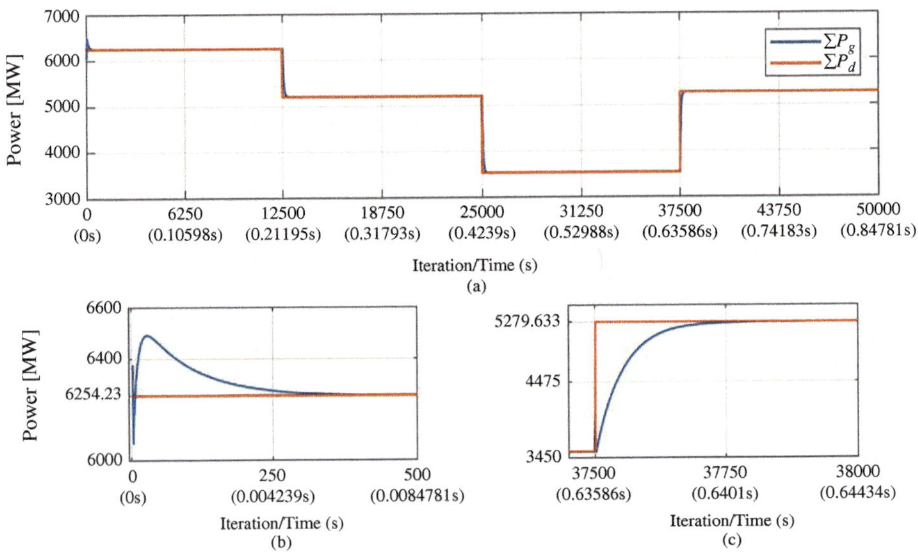

Fig. 12.6 Results for IEEE 39-bus case: (A) displays the total generated power and total demand of the network, (B) shows the magnified version of the convergence process when all the units start from 0 MW, and (C) shows the generating units moving together from one equilibrium point to another after the change in demand.

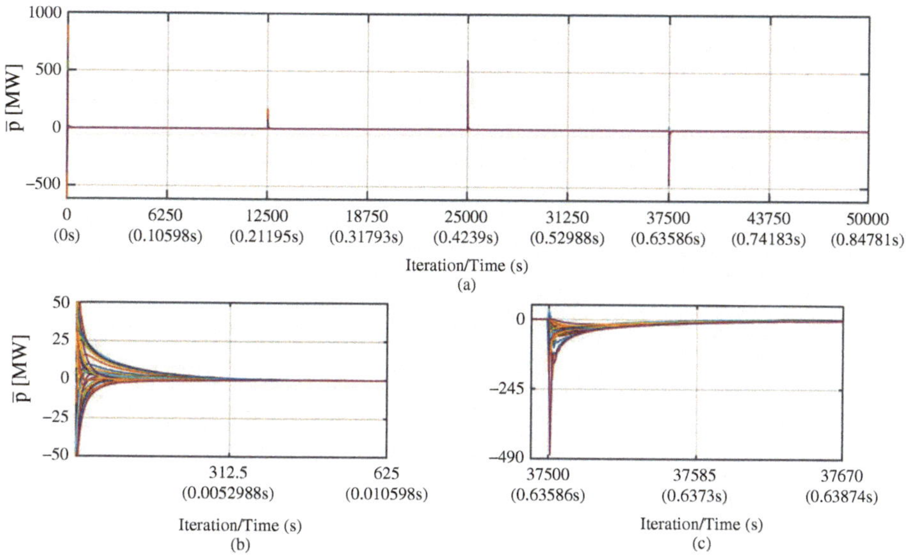

Fig. 12.7 Results for IEEE 39-bus case; (A) displays the estimate of the average power mismatch $\overline{P_i}$ converging to zero; (B) details of the initial convergence process; and (C) details of the last convergence process.

6 Conclusions

In this chapter, a fully distributed algorithm is proposed for the economic dispatch problem in which all agents' computations are based solely on their own data and their direct neighbors' shared information. The algorithm employs dual decomposition and dynamic average consensus algorithms to develop the update procedures. The use of dynamic average consensus has enabled the algorithm to track the time-varying constraint set of the optimization problem, which translates to tracking of demand change in the network. Eigensystem analysis is presented to speed up the convergence to a fixed point solution. The performance of the proposed solution is tested against different IEEE 39 bus test cases. Simulations demonstrate promising results for the algorithm to solve real-time economic dispatch problems with dynamic load variations. It would be interesting to study the communication imperfection in data transfer as future work.

References

[1] M. Dolatabadi, P. Siano, A scalable privacy preserving distributed parallel optimization for a large-scale aggregation of prosumers with residential PV-battery systems, IEEE Access 8 (2020) 210950–210960.

[2] S.S. Kia, B. Van Scoy, J. Cortes, R.A. Freeman, K.M. Lynch, S. Martinez, Tutorial on dynamic average consensus: the problem, its applications, and the algorithms, IEEE Control. Syst. Mag. 39 (3) (2019) 40–72.

[3] R. Olfati-Saber, J.A. Fax, R.M. Murray, Consensus and cooperation in networked multi-agent systems, Proc. IEEE 95 (1) (2007) 215–233.
[4] J.N. Tsitsiklis, Problems in Decentralized Decision Making and Computation, Massachusetts Inst of Tech Cambridge Lab for Information and Decision Systems, 1984.
[5] A. Kargarian, J. Mohammadi, J. Guo, S. Chakrabarti, M. Barati, G. Hug, et al., Toward distributed/decentralized DC optimal power flow implementation in future electric power systems, IEEE Trans. Smart Grid 9 (4) (2016) 2574–2594.
[6] S. Chakrabarti, R. Baldick, Look-ahead scopf (lascopf) for tracking demand variation via auxiliary proximal message passing (APMP) algorithm, Int. J. Electr. Power Energy Syst. 116 (2020) 105533.
[7] G. Chen, Q. Yang, An ADMM-based distributed algorithm for economic dispatch in islanded microgrids, IEEE Trans. Ind. Inf. 14 (9) (2017) 3892–3903.
[8] S. Boyd, N. Parikh, E. Chu, Distributed Optimization and Statistical Learning Via the Alternating Direction Method of Multipliers, Now Publishers Inc., 2011.
[9] S. Afshar, S. Wasti, V. Disfani, Coordinated EV aggregation management via alternating direction method of multipliers, in: 2020 International Conference on Smart Grids and Energy Systems (SGES), IEEE, 2020, pp. 882–887.
[10] S. Afshar, V. Disfani, A distributed EV charging framework considering aggregators collaboration, in: 2021 IEEE Madrid PowerTech, IEEE, 2021, pp. 1–6.
[11] S. Afshar, S. Wasti, V. Disfani, An ADMM-based MIQP platform for the EV aggregation management, 2020. arXiv preprint arXiv:2010.03475.
[12] T. Wang, D. O'Neill, H. Kamath, Dynamic control and optimization of distributed energy resources in a microgrid, IEEE Trans. Smart Grid 6 (6) (2015) 2884–2894.
[13] Y. Xu, W. Zhang, W. Liu, X. Wang, F. Ferrese, C. Zang, H. Yu, Distributed subgradient-based coordination of multiple renewable generators in a microgrid, IEEE Trans. Power Syst. 29 (1) (2013) 23–33.
[14] Z. Zhang, X. Ying, M.Y. Chow, Decentralizing the economic dispatch problem using a two-level incremental cost consensus algorithm in a smart grid environment, in: 2011 North American Power Symposium, IEEE, 2011, pp. 1–7.
[15] G. Hug, S. Kar, C. Wu, Consensus+ innovations approach for distributed multiagent coordination in a microgrid, IEEE Trans. Smart Grid 6 (4) (2015) 1893–1903.
[16] F. Guo, C. Wen, J. Mao, Y.D. Song, Distributed economic dispatch for smart grids with random wind power, IEEE Trans. Smart Grid 7 (3) (2015) 1572–1583.
[17] R.O. Saber, R.M. Murray, Consensus protocols for networks of dynamic agents, 2003.
[18] S. Wasti, P. Ubiratan, S. Afshar, V. Disfani, Distributed dynamic economic dispatch using alternating direction method of multipliers, 2020. arXiv preprint arXiv:2005.09819.
[19] S. Wasti, Distributed online algorithms for energy management in smart grids, 2020.
[20] L.C. Kempton, G. Herrmann, M. Di Bernardo, Distributed optimisation and control of graph Laplacian eigenvalues for robust consensus via an adaptive multilayer strategy, Int. J. Robust Nonlinear Control 27 (9) (2017) 1499–1525.
[21] L. Xiao, S. Boyd, Fast linear iterations for distributed averaging, Syst. Control Lett. 53 (1) (2004) 65–78.
[22] Y. Xu, W. Liu, Novel multiagent based load restoration algorithm for microgrids, IEEE Trans. Smart Grid 2 (1) (2011) 152–161.
[23] S. Boyd, L. Vandenberghe, Convex Optimization, Cambridge University Press, 2004. Google Scholar Google Scholar Digital Library Digital Library.
[24] J. Nocedal, S. Wright, Numerical Optimization, Springer Science & Business Media, 2006.

[25] S. Boyd, L. Xiao, A. Mutapcic, J. Mattingley, Notes on decomposition methods, Notes for EE364B, vol. 635, Stanford University, 2007, pp. 1–36.
[26] P. Siano, D. Mohammad, MILP optimization model for assessing the participation of distributed residential PV-battery systems in ancillary services market, CSEE J. Power Energy Syst. 7 (2) (2020) 348–357.
[27] T. Yang, X. Yi, J. Wu, Y. Yuan, D. Wu, Z. Meng, et al., A survey of distributed optimization, Annu. Rev. Control. 47 (2019) 278–305.
[28] W.L. Brogan, Modern Control Theory, Pearson Education India, 1991.
[29] P.J. Olver, C. Shakiban, C. Shakiban, Applied Linear Algebra, vol. 1, Prentice Hall, Upper Saddle River, NJ, 2006.
[30] S. Boyd, S.P. Boyd, L. Vandenberghe, Convex Optimization, Cambridge University Press, 2004.
[31] J. Tuinstra, Price Dynamics in Equilibrium Models: The Search for Equilibrium and the Emergence of Endogenous Fluctuations, vol. 16, Springer Science & Business Media, 2001.
[32] R.D. Zimmerman, C.E. Murillo-Sánchez, D. Gan, MATPOWER: a MATLAB power system simulation package, vol. 1, Manual, Power Systems Engineering Research Center, Ithaca, NY, 1997, pp. 10–17.

Decentralized power exchange control methods among subsystems in future power network

13

Mahdi Zolfaghari[a] *and Gevork B. Gharehpetian*[b]
[a]Power System Secure Operation Research Centre, Amirkabir University of Technology, Tehran, Iran, [b]Department of Electrical Engineering, Amirkabir University of Technology, Tehran, Iran

1 Introduction

Electricity indisputably plays the most important role in human activities. For a long time, DC and AC resources have been extensively used to obviate the needs for lighting, heating, mechanical movements, etc., due to the main advantages of electrical energy in transforming to other types of energies [1]. The century-long power systems have had a distinctive portrayal in interconnecting the resources and loads through transmission and distribution networks. At the beginning, power systems consisted of independent local DC islands, i.e., each DC generator was used to supply local loads and this island acted as a DC subsystem. Many of these subsystems were in use at those times, forming a power system with separate DC subsystems. However, as electricity became more popular, power demand started to increase, and actually it is ever-increasing now. This inevitable increasing demand along with reliability requirements, and by revealing the advantages of AC systems, the interconnection of AC subsystems and some of the DC subsystems, which were in use for some time, became necessary. In fact, the power system later became a complex power grid interconnected by many AC branches, meshes, and loops, which enjoyed power exchange benefits among different branches when necessary [2,3]. This resulted in reliability increment and benefited the customers and utilities. The topology of such a power grid, which is currently in use in many countries, mainly consists of four sections: generation, transmission, subtransmission, and distribution. The fossil fuels have been used to produce electricity in the generation section. Thermal power plants have been responsible as the major source of supplying demand. The transmission, subtransmission, and distribution sections have intertied thousands of domestic, commercial, and industrial loads with the power generation section. The main features of the current power networks are as follows [2,4,5]:

Decentralized Frameworks for Future Power Systems. https://doi.org/10.1016/B978-0-323-91698-1.00003-0

- Synchronous generators are the most important machines in the generation side whereas the induction motors are known as the workhorses of power system and shape the lion's share part of electrical loads in current power networks.
- Employing of fossil fuels to produce electricity has resulted in air pollution and environmental issues in many parts of the world. These fuels are also decreasing and shrinking with time and could not meet the increasing demand in the near future.
- The structure of the current power networks provides one-sided power flow from the generation level to the distribution level.
- The basic infrastructure of the current grid suffers from aging. The tolerability of the system components has been reduced because of numerous stresses on the system, which come from lack of appropriate monitoring and control.
- Many parts of the current power systems do not share information with each other. Therefore, there may be miscoordination among system levels and also components.

There is an increasing trend in using cleaner energies, for example, wind, solar, hydrogen, and geothermal to name but a few. These clean resources, more commonly referred to as distributed generations (DGs) in the literature, can improve the power quality and reliability if they are appropriately located and controlled. The most commonly used DGs are wind units and photovoltaic (PV) systems. They are indisputably penetrating into the current power grids in distribution, transmission, and sub-transmission levels. More specially, the DGs are operated and controlled in the form of AC or DC subsystems. These subsystems in the distribution level are called microgrids since a microgrid operates in low voltage level, as defined by IEEE [6].

Nevertheless, from the big picture, the future power system would be defined as a system of systems with too many interconnected subsystems, as shown in Fig. 13.1. Each subsystem can be figured out as AC or DC. Mainly, AC and DC subsystems are

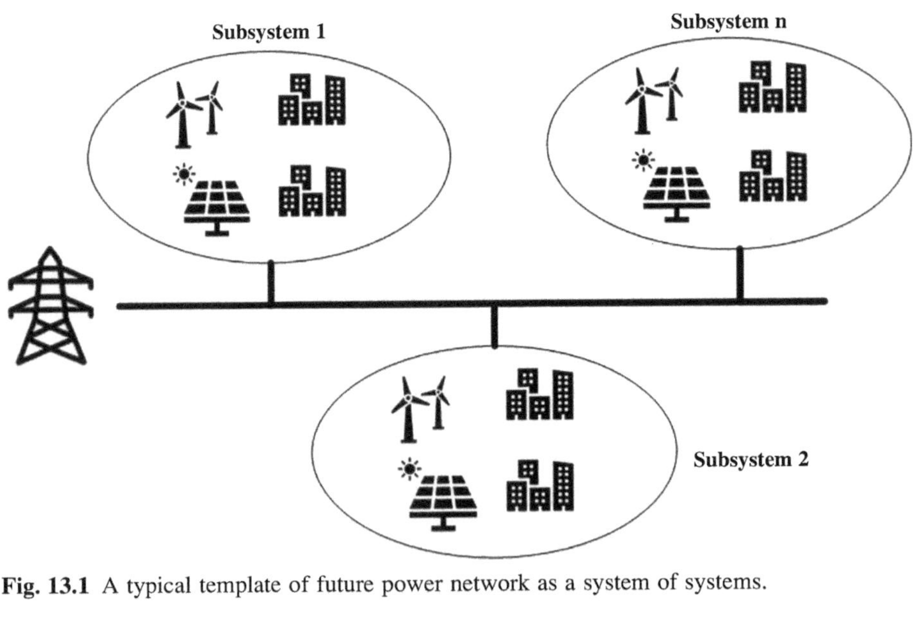

Fig. 13.1 A typical template of future power network as a system of systems.

expected to be interconnected to form a hybrid structure. In light of this heterogeneous structure, the smart grid implementation is facilitated. The AC subsystems include mainly AC sources and loads. There is a common AC bus with all loads and sources connected to it. On the other hand, in a DC subsystem the major part of sources and loads are DC. Similarly, a common DC bus exists to which all sources and load are connected. In both AC and DC subsystems, if there is a load/source which cannot be connected to the common bus directly, for example, a DC load in an AC subsystem, the DC/AC power converters are used as the interface [7,8].

According to the abovementioned definition, the main features of the future power networks can be summarized as follows [9–12]:

- Since the DGs are power electronics (PE)-based, the future power systems are converter-dominated. This may result in other issues, for example, stability, control, and protection challenges.
- Information sharing, cloud-based data storages, and communication links among different subsystems and components.
- Real-time control and monitoring of power system sections and components.
- Deregulated power market and dynamic active and reactive power pricing.
- More complex structure because of interconnection of different subsystems and infrastructure, which have various characteristics and dynamics.
- Bidirectional power flow is expected between subsystems or between prosumers with power grid.
- One main issue in the power systems with structure similar to Fig. 13.1 is the interactions between different subsystems. The interconnection of these subsystems is necessary due to the need for power exchange requirements and reliability enhancement. The subsystems are mainly interconnected by bidirectional power converters (BLPCs) or other alternatives, for example, solid-state transformers (SSTs), unified power flow controllers, etc. These interconnections are expected to provide bidirectional power flow among subsystems with different characteristics. The power exchange problem is to be resolved as centralized or decentralized.

The major purpose of this chapter is to give the reader a comprehensive view of the interconnection of the AC and DC subsystems in future power systems concentrating on the power exchange requirements. The remainder parts of this chapter are set-up as follows: Section 2 classifies the linkage topologies used to intertie subsystems. Section 3 analyzes the properties of centralized and decentralized power exchange control methods among subsystems. Section 4 investigates the decentralized control strategies for multiple BLPCs implemented in the literature. Finally, Section 5 concludes this chapter.

2 Classification of linkage topologies for AC and DC subsystems in future power networks

To intertie AC and DC subsystems in upcoming power grids, various topologies are being proposed in the literature. The main common feature of all of the linkage strategies lies in using the power electronics components to control the exchanged power

Fig. 13.2 Classification of linkage topologies for AC and DC subsystems in future power networks.

among subsystems. In fact, power converters have the main role in these strategies [12]. Fig. 13.2 indicates the classification of various linkage strategies, which have been described in the literature. As shown, BLPCs have been used in two ways: stand-alone BLPC [13,14] and multiple BLPCs [15,16]. These strategies are the mostly used methods to intertie the subsystems. SSTs [17,18] and energy routers (EROs) [19,20] have also been used to join subsystems. As another alternative methods, some form of flexible AC transmission system (FACTS) devices, i.e., UPFC, UPQC, and UIPC, have been used as the joint component for the subsystems [12,21,22]. In the remaining, we discuss these linkage strategies in more detail.

2.1 Linkage of subsystems using stand-alone BLPC

The BLPC is a power converter, which can provide bidirectional power exchange among two subsystems. The full-bridge or neutral-point-clamped structure, etc. have commonly been used to establish a BLPC.

Stand-alone BLPCs have been used as the first, and in comparison, the simplest strategy to intertie the subsystems. Fig. 13.3 shows a future power network consisting of many AC and DC subsystems, which have been linked using a stand-alone BLPC. As demonstrated, subsystem 1 and subsystem 3 are intertied using a stand-alone BLPC whereas subsystem 2 is directly connected to the power grid. It should be noted that each subsystem can be DC or AC according to various factors, for example, load/source location, circumstances, economic evaluations, etc. There are many topologies for BLPC in the literature, which add some advantages to this linkage strategy or solve a challenge in power exchange procedure. The full-bridge single-phase and three-phase structures are the most used topologies. However, there are some other topologies reported in the literature in recent years. For instance, in Ref. [13] a topology occupied with an energy storage system has been presented for BLPCs. Such a topology has been able to reduce the power and voltage fluctuations and has provided stable performance for the overall subsystems. In Ref. [14], the authors have described a

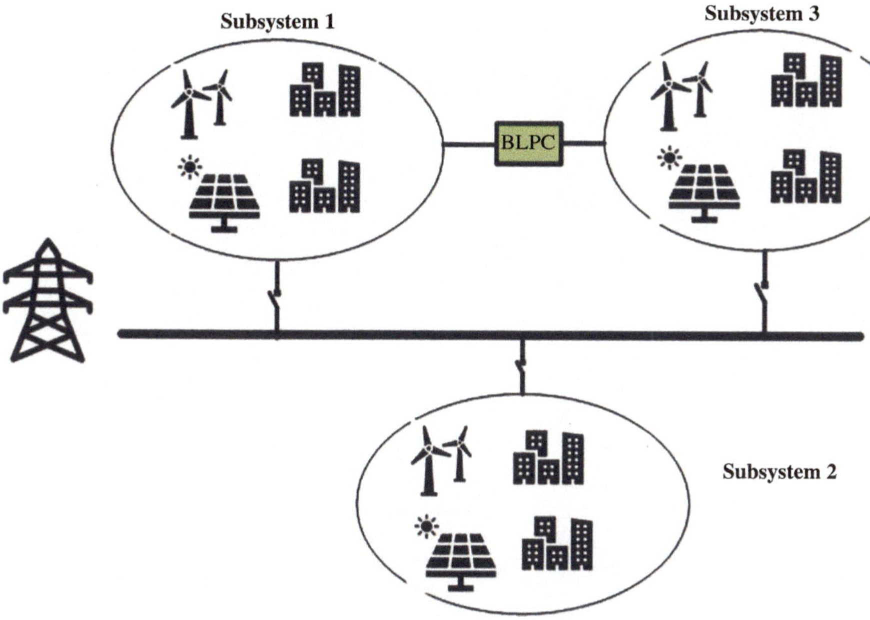

Fig. 13.3 A stand-alone BLPC interties subsystems.

novel BLPC structure by incorporating the Gamma-Impedance Source concept. The structure has enjoyed a large gain in voltage control, which has been able to increase the subsystems' reliability and overall efficiency.

2.2 *Linkage of subsystems using multiple BLPCs*

A stand-alone BLPC brings two challenges when it is implemented for joining the subsystems:

(a) the amount of exchanged power among subsystems is limited, and (b) in the case of BLPC failure, the power transfer capability is lost and this reduces the system reliability. Accordingly, many researchers have presented multiple BLPCs, operated in parallel, as shown in Fig. 13.4. This way, each BLPC acts as a controlled power junction and carries a share of exchanged power and in the case of failure, other BLPCs still exchange power. Thus, the transfer capacity limit is removed and the reliability is increased. Nevertheless, multiple BLPCs bring many challenges in operation and control. For example:

- Each BLPC must carry a power share according to its ratings.
- The power distribution among BLPCs is affected by line impedance changes, DC voltage changes, load changes, etc. The control unit of these converters must handle these issues.
- The system must be fault-tolerant as much as possible.
- The subsystems have different dynamics and characteristics. Operation of multiple power converters in such heterogeneous systems is challenging because of dynamic interactions, couplings, and different time constants.

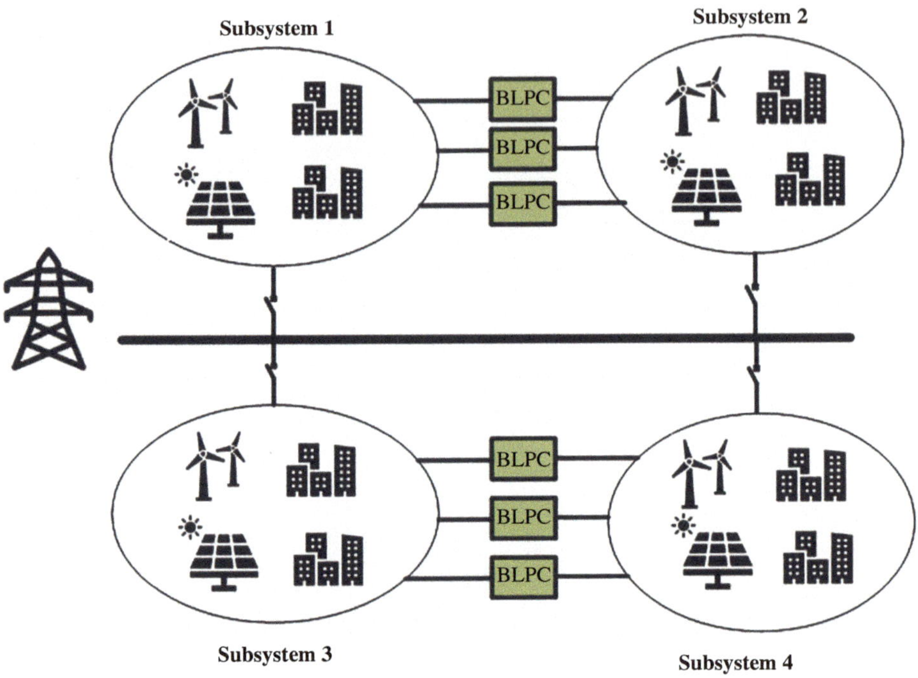

Fig. 13.4 Multiple BLPCs are used to join the subsystems.

However, until now, the structure in Fig. 13.4 is the mostly used topology because of add-on property.

2.3 Linkage of subsystems using SSTs

Traditional power transformers have been commonly used to link power system components with different voltage levels. For instance, a 20/0.4kV distribution transformer is commonly used in Iran's power grid to join the medium-voltage (MV) lines to the low-voltage (LV) system. However, these transformers are passive devices (fairly without tap changers) since they cannot control the power flow interactions. The inductive impedance of these devices imposes power losses and voltage drop especially during the peak times.

The SSTs have been designed based on PE concepts and add many properties to the power exchange control among subsystems. For the future power networks, the SSTs will be of great interest because of numerous advantages such as controllable power flow, providing a low-voltage DC side to accommodate PVs and other DC resources, reduced inductor and passive components and power losses. Various topologies for SSTs have been reported in the literature where the implementation of SSTs in future power networks has been discussed [23,24]. The SST has been used in [25] to intertie subsystems. This topology is indicated in Fig. 13.5. As shown, subsystems 1 and 3 are

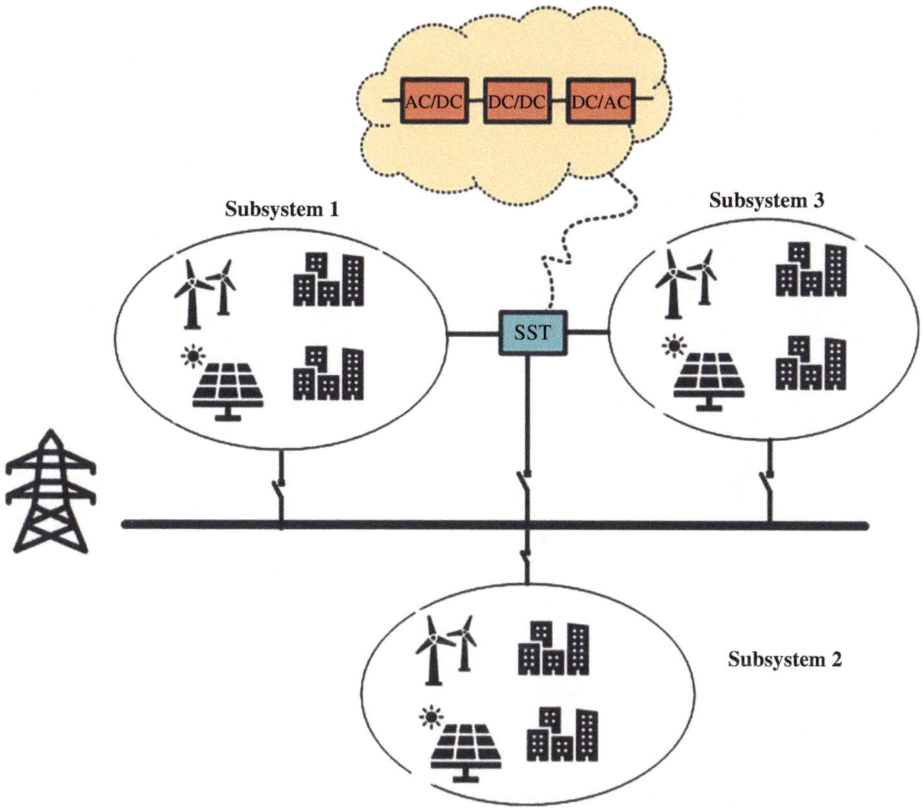

Fig. 13.5 Solid-state transformer is used to intertie multiple subsystems.

tied through an SST and can exchange power with each other or with a power grid. Other examples can be seen in Refs. [26, 27].

2.4 Linkage of subsystems using ERO

ERO has been represented in Refs. [19, 20] as a package of multiple AC/AC and DC/DC power conversion units, which have been implemented to aggregate AC and DC subsystems. This structure is shown in Fig. 13.6. As shown, this topology facilitates the uniting of multiple AC subsystems with different characteristics (namely, frequencies) and several DC subsystems with different voltage levels. However, the ERO has a bottleneck from reliability point of view, and it is the AC/DC main power conversion unit that interties the whole subsystems with the upstream grid. If failed, the utility support would be lost and the subsystems may undergo voltage or frequency stability problems if there are not preventive control tasks. Furthermore, since there are numerous power conversion units with a common DC side, the voltage oscillation and stabilization of this link are challenging. Nevertheless, all of the abovementioned

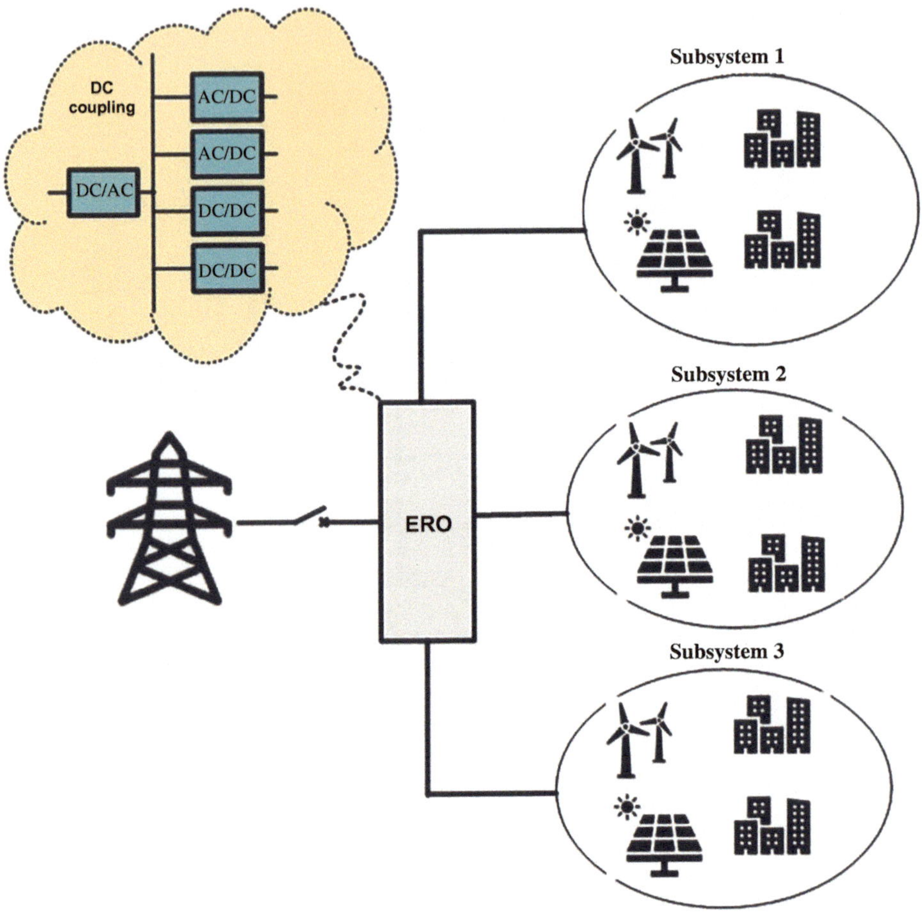

Fig. 13.6 ERO-based aggregated subsystems.

challenges could be resolved by adopting strong control and management units, which have been designed considering system cost and losses.

2.5 Linkage of subsystems using FACTS devices

In the current power networks, FACTS devices have been extensively used to power flow control and stability enhancement. These devices have been implemented in high-voltage (HV) transmission lines and substations. In addition to using FACTS in the HV level, the future power networks will also enjoy the merits of these devices in MV and LV levels. Until now, many scholars have developed some control strategies to adopt the FACTS devices operation with converter-dominated power systems. Moreover, some revised models of FACTS devices have been presented to embed with the new-structured power grids. For instance, a modified UIPC model

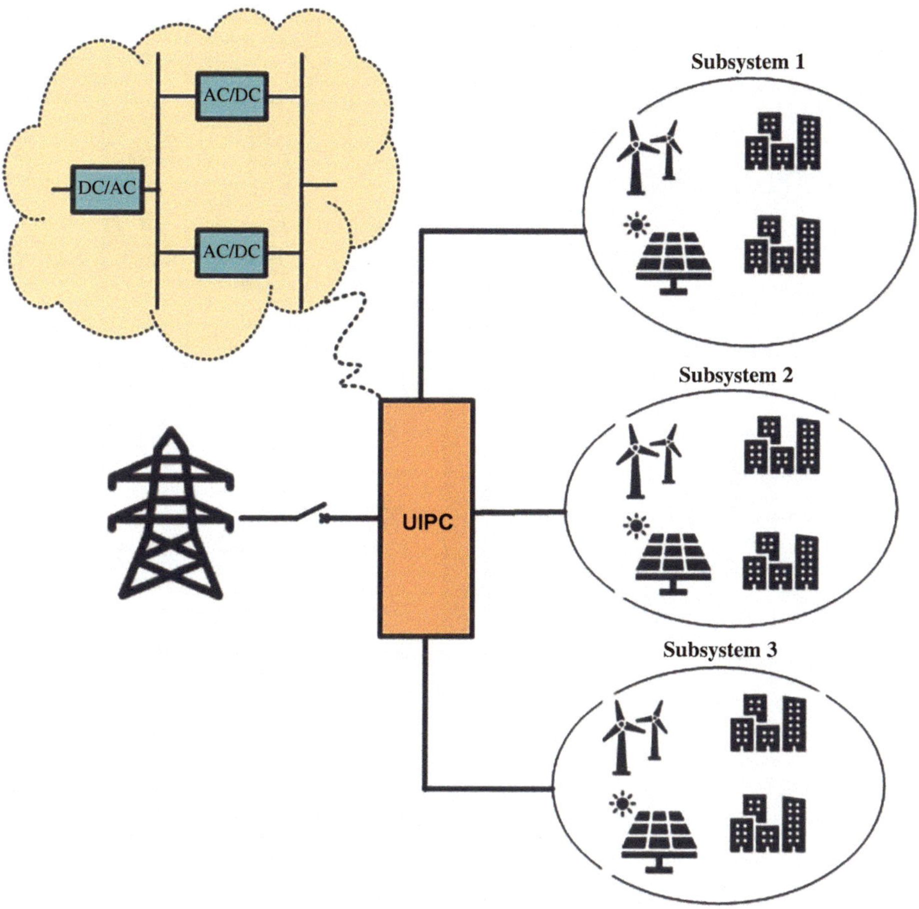

Fig. 13.7 The UIPC is used to intertie subsystems.

has been presented in Refs. [12, 28] to comply with AC and DC subsystems. Such a structure is illustrated in Fig. 13.7. The UIPC consists of two series power converters and one parallel power converter. The series ones are used to control the power exchange between subsystems whereas the parallel unit tries to regulate the voltage at the bus to which this power converter is connected.

In Ref. [29], the UPQC has been implemented to concatenate AC and DC subsystems. The topology is demonstrated in Fig. 13.8. This device has been able to improve power quality and active and reactive power control in distribution systems. According to the results provided in Ref. [29], the topology of Fig. 13.8 enables the subsystems to interact and transfer power in emergencies. The UPQC has the advantage of back-to-back connected power converters, which can provide bidirectional power flow among subsystems. Moreover, if controlled appropriately, the UPQC is capable of voltage control and power oscillation reduction.

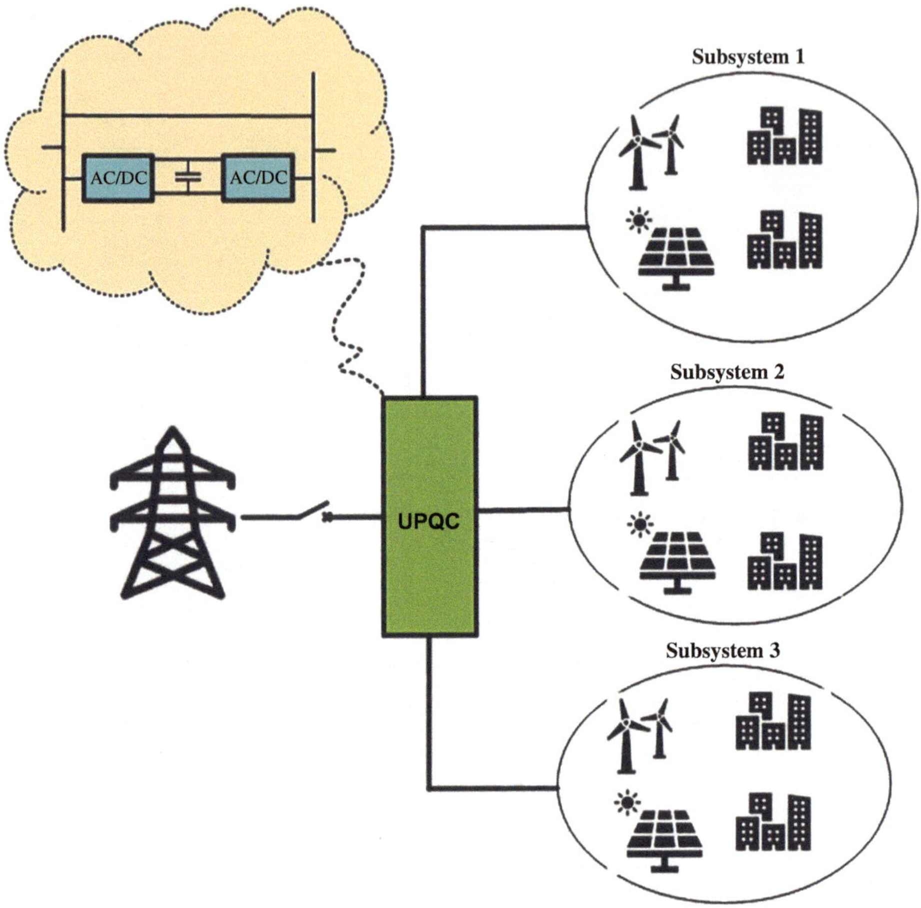

Fig. 13.8 UPQC is used to concatenate subsystems.

Further, the authors in Ref. [21] have developed a topology based on UPFC to intertie several AC and DC subsystems. The topology is shown in Fig. 13.9. As shown, three power conversion units have been implemented to accommodate AC and DC subsystems. A DC/DC conversion unit facilitates the integration of DC subsystems, provided with a sturdy control system to reduce power flow oscillations. In the conventional models of UPFC, heavy power transformers were implemented to intertie the whole subsystems with the power grid. However, in [21] these voluminous power transformers have been replaced by PE-based conversion units. This makes the system more flexible and controllable from the operating point of view.

As the concluding remark, we should note that a combination of the all of abovementioned linkage strategies are expected to be used in the future power networks. For instance, in a typical future power grid some subsystems may be intertied using stand-alone BLPCs and others may be interconnected by UIPC. It is worth mentioning that the multiple BLPC topology bring many control challenges, as mentioned

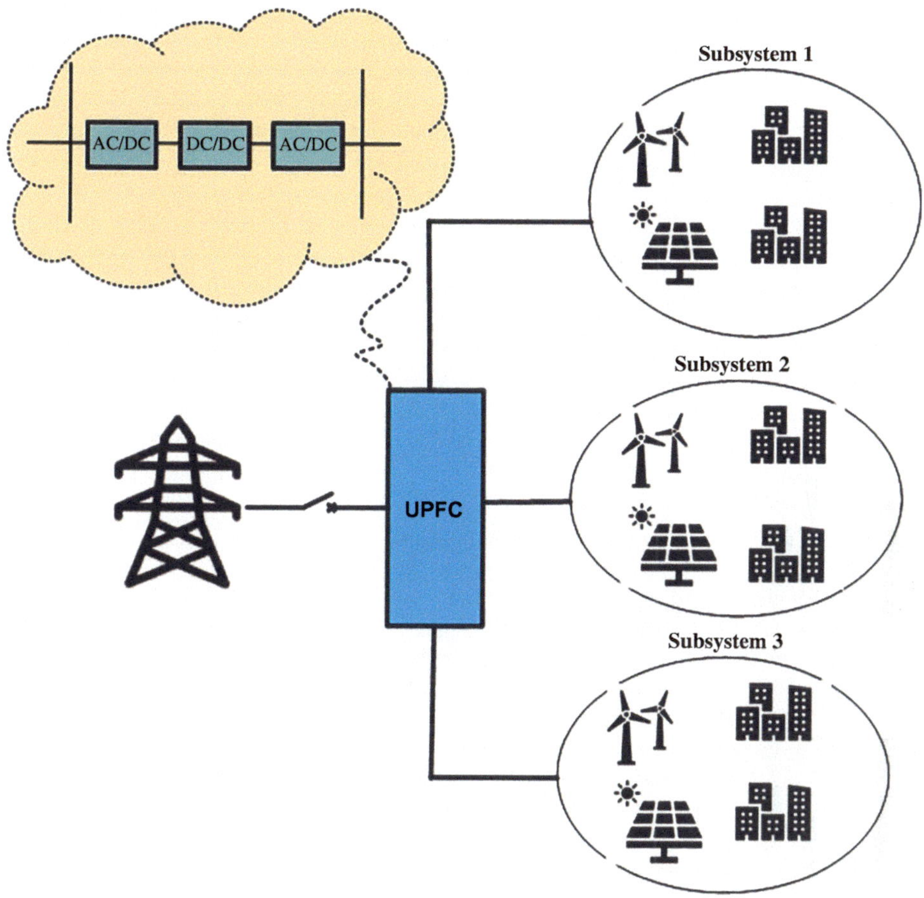

Fig. 13.9 UPFC is used to intertie subsystems.

earlier, and the research direction is to replace these BLPCs with alternative methods (e.g., the UIPC, etc.) in favor of obtaining a straightforward decentralized power exchange control among subsystems of future power networks. Also, in Ref. [30], a semi-Luenberger observer strategy has been described to handle power flow exchange among various subsystems with active DC link control.

2.6 Comparison of linkage strategies

Table 13.1 presents the major merits and problems/difficulty of the linkage strategies for subsystems in future power networks, according to classification introduced in this section.

From Table 13.1, it is obvious that each linkage strategy represents its own merits and challenges when implemented in networked subsystems. However, among them, the multiple BLPCs and FACTS devices, specially, the UIPC, are the most

Table 13.1 Major merits and problems/difficulties of linkage strategies for subsystems in future power networks.

Linkage strategy	Merits	Problems/difficulties
Stand-alone BLPC	– The system general cost is low and the control signal is not complicated – A stand-alone BLPC can be supported with an ESS to reduce DC voltage oscillation	– The amount of exchanged power is limited – The reliability is low because of loss of power exchange in case of BLPC failure – System stability and dynamic response are crucial for a stand-alone BLPC to provide a stable operation – Balancing of the DC voltage is an important problem which must be considered
Multiple BLPCs	– The reliability is improved because of the parallel interconnection – The amount of exchanged power can be increased by adding a greater number of BLPCs – The topology gives a modular-based configuration	– Overall cost is increased and the control becomes complex – Removing the circulating current is a problem – Providing unified power distribution among BLPCs is a serious difficulty – Because of multiple DC coupling, the DC voltage stabilization is a problem
SST	– The problems of BLPCs are removed – The SST brings more property for the future power network. For example, resolving power quality, smart protection, etc. – Galvanic isolation property	– The control system may be more complex – Addressing DC side fluctuation is essential – The stability of the grid-tied power converters of SST must be studied well
Energy router	– In case where there are a greater number of DC subsystems than AC ones in future power networks, the energy router is economic and feasible – In the structure of energy routers, multilevel or multiinput-multioutput power converters can be implemented to facilitate the	– Generally, the energy routers are required to adopt high input voltages and high output currents. This becomes a difficulty in design process – The energy routers may have efficiency problems if they are not designed well – A superfluous power converter must be allocated

Table 13.1 Continued

Linkage strategy	Merits	Problems/difficulties
	integration of various subsystems in the future power networks	in topology of energy router. This may increase its cost and also reduce the reliability in case of failure in this power converter – Stabilizing the DC side of energy routers is a difficulty
FACTS	– FACTS provides higher reliability – The amount of exchanged power is extended – Swift and fast power exchange among subsystems – DC voltage adjustment – The power quality problems can also be resolved	– Some versions of FACTS are costly. For instance, traditional UIPCs which use phase-shifter transformers – I some FACTS devices, for example, in UIPC and UPQC, control of series and parallel power converters is difficult when interfacing a great number of DC subsystems

sophisticated choice to intertie subsystems with reliable power control, reduced cost, and regulated DC bus voltage.

3 Power exchange control strategies among subsystems

As explained earlier, in the future power networks the power converters are dominated. Thus, the main elements in power exchange control are these power converters, which can be implemented to interconnect subsystems in the form of structures represented in the previous section. Accordingly, the power exchange control strategies among subsystems can be clustered to centralized and decentralized, as demonstrated in Fig. 13.10. As shown, the power control strategies can be clustered as centralized and decentralized.

The centralized strategies have the following properties [31,32]:

- The centralized strategies are mostly used in interconnection of multiple BLPCs, i.e., the linkage scheme in Fig. 13.4.
- There is a central compensator which generates a control signal for all BLPCs.
- The central compensator has to make all computations and then produce a unique control signal for BLPCs.
- The performance of the centralized strategies is dependent on the central compensator. Therefore, the design of this compensator is important since it affects the power exchange performance of all BLPCs.

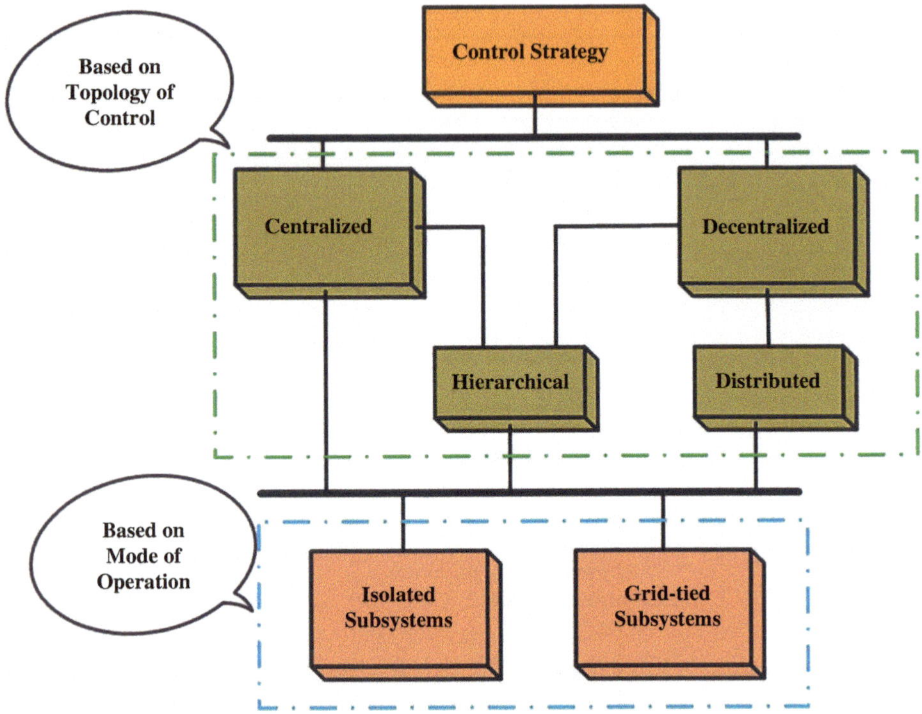

Fig. 13.10 Classification of power control strategies among subsystems.

- Since all computations are carried out in the central controller, the real time or hardware-in-the-loop implementation of these strategies may encounter problems because of communications delay and considerable computational burden.
- The centralized strategies adopt intertied communication links among all subsystems.
- The reliability of the centralized strategies is low because of loss of a communication link or the central compensator failures.
- The system operates in the optimal mode.

The decentralized strategies have the following properties [4,33,34]:

- The decentralized strategies are mostly applied in the control of multiple BLPCs or two (or more) power converters in other structures described in the previous section.
- Each power converter has its own compensator. Therefore, each compensator operates locally and there is no communication link among the compensators.
- The computation is done in each local compensator. Thus, the computation burden is distributed among local controllers. Each subsystem has its own control signal.
- The delay affects the real-time implementation much less than the centralized strategies.
- The reliability of these control strategies is improved considerably.
- The system operates in suboptimal or near optimal mode because of fine tuning of each local compensator.

Table 13.2 summarizes and compares the main properties of the centralized and decentralized strategies.

Table 13.2 comparison of main properties of centralized and decentralized strategies.

Property	Centralized	Decentralized
Communications	Yes	Fairly no
Reliability	Low	High
Power dispatch among subsystems	High	Low
Need to know the working situations of other subsystems	Highly dependent	Fairly independent
Integration of newly connected subsystems	Applicable only by altering the control configuration and communications	Add-up property
Optimal operation of subsystems	Optimal	Near optimal
Dynamic speed response of control	Low	High
Size of the control system	Big dimensions	Fairly small

According to Fig. 13.10, there are other subcategories for the centralized and decentralized strategies. The distributed strategy is directly derived from the decentralized strategies while the hierarchical strategy is a combination of the centralized and decentralized.

The main properties of the distributed strategies are as follows [35,36]:

- This strategy can be adapted to multiple converter-based topologies. For example, Figs. 13.4, 13.6, 13.7, and 13.9.
- All of the properties of the decentralized strategies exist in the distributed strategies.
- In such strategies, the compensators can communicate with each other.
- When compared to the centralized strategies, the distributed strategies have higher reliability with adequate performance.
- The optimality of the distributed strategies is higher than purely decentralized strategies because of availability of information of other subsystems.

The main properties of hierarchical strategies can be described as follows [37,38]:

- This strategy is mostly used in multiple BLPCs, the topology of Fig. 13.4.
- Three are three control levels: primary, secondary, and tertiary. The first level performs the voltage and current regulation for each BLPC. The second level is used to resolve power quality problems, and eliminating steady-state errors of the first level. Finally, the tertiary level is responsible for providing reference values and power exchange performance among the subsystems that have been intertied by the BLPCs.
- The droop control is the main control concept considered in the hierarchical strategies in literature.

The subsystems may be operated in isolated or gird-tied mode. Therefore, all categories demonstrated in Fig. 13.10 can be considered in islanded mode or grid-tied mode.

In future power networks, many of the subsystems can be tied to each other and then interconnected with the main power grid. However, this cluster of subsystems may be isolated from the power grid due to intentional islanding or even due to faults. Thus, the control objectives may change but the control topology does not change in most cases. It should be noted that a subsystem may be connected to the power grid lonely, for example, subsystem 2 in Figs. 13.1 and 13.3. In this case, when operating in isolated or grid-tied modes, the control objectives are varied according to the situations of that subsystem.

4 Decentralized control of multiple BLPCs for interlinking subsystems

For the most part, literature is enriched with many control strategies to intertie AC and DC subsystems using multiple BLPC topology, as illustrated in Fig. 13.4. These control strategies, as shown in Fig. 13.11, are classified here into five main groups: (1) droop-based strategies, (2) intelligent control strategies, (3) robust, observer-based, and optimal control strategies, (4) active power sharing strategies, and (5) instantaneous power theory-based strategies.

Each of the abovementioned strategies have some subclass (expect the last one) and are discussed in the following subsections.

4.1 Droop-based control of multiple BLPCs

In overall, the traditional droop control has been used in synchronous generators where a droop characteristic has been used to adjust the output power of each parallel generator. However, this concept has also been developed and implemented for multiple BLPCs operation to tune the voltage and power exchange control among subsystems. In fact, droop-based control is the most control strategy to control the interconnection of AC and DC subsystems [39,40]. The earlier forms of droop control strategies are called "conventional droop" here, and as shown in Fig. 13.11, other forms of these control strategies have been presented in the literature to improve its performance, for example, virtual impedance-based droop and adaptive and robust droop strategies. The latter is divided into three subgroups: (1) linear matrix inequality (LMI)-based strategy, (2) linear quadratic regulation (LQR)-based strategy, and (3) data-driven-based strategy.

The conventional droop control strategy of multiple BLPCs is based on the following equations:

$$\omega_k = \omega_{\text{nom}} - \gamma_k P_k \tag{13.1}$$

$$V_k = V_{\text{nom}} - \delta_k Q_k \tag{13.2}$$

where k is the number of the BLPC, ω is the frequency in rad/s and ω_{nom} is its nominal value, V is the output voltage of each BLPC and V_{nom} is its nominal value, P_k and Q_k

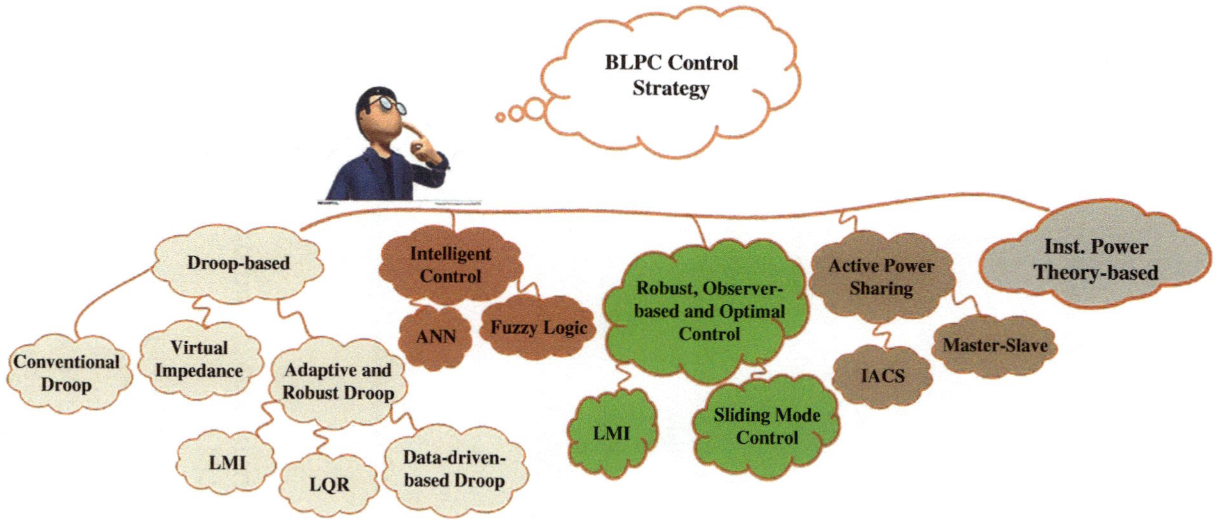

Fig. 13.11 Classification of multiple BLPCs control strategies when they are implemented to interconnect AC and DC subsystems.

are the active and reactive power of BLPC, respectively, and γ_k and δ_k are the droop factors.

It should be noted than the droop-based strategies are classified as decentralized strategies for power exchange control among subsystems in future power networks.

4.2 Intelligent control of multiple BLPCs

These kinds of strategies refer to the introduction of artificial intelligent concepts into control and management of multiple BLPCs in the interconnection of subsystems. For example, optimization algorithms such as particle swarm, genetic algorithms, etc., and fuzzy logic inference systems, and artificial neural networks can be adopted in control loops of BLPCs to provide a fast, flexible, and robust power exchange in control manner among multiple subsystems with different characteristics. As an example, in Ref. [41] a fuzzy logic inference system has been adopted to manage the interactions of AC and DC subsystems. The rule-base characteristic of fuzzy logic lets the designer to make a control parameter tunable according to the interconnected AC and DC subsystem situations. For example, the DC current I_{DC} of a BLPC can be tubed according to the changes in DC active power and state of charge of an ESS as follows:

$$\begin{cases} \text{if } \Delta P_{DC} \text{ is } PL \text{ and } \Delta SoC \text{ is } NL, \text{ then } I_{DC} \text{ is } PL \\ \text{if } \Delta P_{DC} \text{ is } PL \text{ and } \Delta SoC \text{ is } NS, \text{ then } I_{DC} \text{ is } PL \\ \text{if } \Delta P_{DC} \text{ is } PL \text{ and } \Delta SoC \text{ is } Z, \text{ then } I_{DC} \text{ is } Z \end{cases} \tag{13.3}$$

where P stands for positive, N for negative, Z for zero, L for large, and S for small.

In Refs. [10, 42], the artificial neural network has been used to control multiple BLPCs. The main purpose of the control system has been set to provide optimal power exchange management, power quality improvement, and to increase utilization factor for renewable resources.

4.3 Robust, observer-based, and optimal control of multiple BLPCs

The robust and observer-based strategies have recently been implemented in controlling multiple BLPCs to successfully adopt the changes in the heterogeneous structure of intertied subsystems. For example, in Ref. [43] a near-optimal control system has been designed for a multiple BLPC system. The Lyapunov theorem has been used to verify the stability and design the control parameters of the BLPC control loops in Ref. [44]. The authors in Ref. [35] have resolved the power quality problems of AC and DC subsystems, which have been intertied through multiple BLPCs. The solution method has been based on robust predictive control of BLPCs. The reference voltage has been determined according to the following equation:

$$E_{\text{desired}}(k+1)=E_o(k)+L_1\left(\frac{i_{L1,\text{desired}}(k+1)-i_{L1}(k)}{T_s}\right) \tag{13.4}$$

where E_{desired} is the desired output voltage of each BLPC, L_1 is the output inductance of BLPC, and T_s is the sampling time interval.

The combination of sliding mode control (SMC) with observer-based control has also been used to control multiple BLPCs, which have intertied several AC and DC subsystems as discussed in Ref. [45]. The control law can be merged with the observer concept. For example, in [30] the following control signal has been attained:

$$u=-\frac{1}{B(\hat{x}(t))}\left\{f_1(\hat{x}(t))+\frac{1+\tau_{\text{SMC}}\dfrac{\partial f_2}{\partial \hat{x}_2}}{\tau_{\text{SMC}}\dfrac{\partial f_2}{\partial \hat{x}_1}}\dot{\hat{x}}_2-\frac{\left(\tau_{\text{SMC}}\ddot{\hat{x}}_{2d}+\dot{\hat{x}}_{2d}\right)}{\tau_{\text{SMC}}\dfrac{\partial f_2}{\partial \hat{x}_1}}-\frac{\zeta}{\tau_{\text{SMC}}\dfrac{\partial f_2}{\partial \hat{x}_1}}sat(\sigma(t))\right\} \tag{13.5}$$

where $\hat{x}(t)$ is the approximated state through applying the observer concept, τ_{SMC} is the time constant of SMC, and $sat(\sigma(t))$ is the saturation function. More details can be found in Ref. [30]. The authors in Ref. [46] have also made the observer adaptive based on the weighting matrixes of Riccati's equation.

4.4 Active power sharing strategies for control of multiple BLPCs

These kinds of strategies for multiple BLPCs can be classified into two groups: (1) master-slave (MS), and (2) instantaneous active power sharing (IACS). For example, in Refs. [25–27] MS-based control strategies have been presented for multiple BLPCs. In Refs. [25, 26], a MS strategy has been represented for a network of multiple subsystems, which have been intertied through multiple BLPCs. The basic objective of the strategy is to maintain network stability. Further, in Ref. [27], the authors have tried to optimize the control law in an MS structure. Recently, in Ref. [47], an IACS strategy has been illustrated for multiple BLPCs using impedance source power converters. The controller has had an adaptive behavior because of implementing a tunable gain in the current control loop. The results indicated that these strategies have been able to handle power exchange control among different subsystems successfully.

4.5 Instantaneous power theory-based control of multiple BLPCs

The power system voltage may be unbalanced because of presence of nonlinear loads, phase unbalance, etc. The subsystems, which are connected to an unbalance power grid, may have interrupted power exchange or even they may become unstable. Accordingly, the authors in Refs. [48, 49] have investigated the effects of voltage unbalance on the performance of integrated AC and DC subsystems. To this end, the instantaneous power theory has been adopted to analyze positive and negative sequences. The authors have added a redundant power converter to the multiple BLPCs to compensate the asymmetric overcurrent resulted from faults.

5 Conclusions

The future power network composes of many elements and subsystems which have different properties. This is a heterogeneous structure which its control and management is an important challenge. The subsystems contain mainly power electronics-based elements. These power converters are mandatory to interface renewable resources with the power system and with other subsystems. The subsystems are expected to interact with each other to exchange power when necessary. To intertie the subsystems, various elements can be used. For the most part, the BLPCs have been used since they provide a modular structure and can remove power exchange limits. However, the control and management of these BLPCs is challenging and have been discussed in this chapter. To overcome the barriers of these linkage methods, many researchers have presented alternative strategies based on FACTS devices, ERO, and SST. These alternatives have attracted a lot of interests among industrialist and researchers and have also been investigated in this chapter. It is expected these alternatives receives fortune in future power networks since the power electronics technology has had drastic progress in easy mode control and management of the embedded power converters in FACTS and SSTs.

References

[1] S.A. Taher, M. Zolfaghari, Designing robust controller to improve current-sharing for parallel-connected inverter-based DGs considering line impedance impact in microgrid networks, Int. J. Electr. Power Energy Syst. 63 (2014) 625–644.

[2] S.M. Dawoud, X. Lin, M.I. Okba, Hybrid renewable microgrid optimization techniques: a review, Renew. Sustain. Energy Rev. 82 (2018) 2039–2052.

[3] A. Das, A.K. Kar, C. Kumar, Mayur, U. Kumar, A. Verma, S. Pal, A. Joseph, B. Das, P.R. Kasari, A. Chakrabarti, S. Bhattacharjee, TSBC converter with BESS for DFIG-based wind energy conversion system, IEEE Trans. Ind. Appl. 56 (2020) 6158–6173.

[4] J. Kumar, A. Agarwal, V. Agarwal, A review on overall control of DC microgrids, J. Energy Storage 21 (2019) 113–138.

[5] A. Ordono, E. Unamuno, J.A. Barrena, J. Paniagua, Interlinking converters and their contribution to primary regulation: a review, Int. J. Electr. Power Energy Syst. 111 (2019) 44–57.

[6] M. Zolfaghari, S.H. Hosseinian, S.H. Fathi, M. Abedi, G.B. Gharehpetian, A new power management scheme for parallel-connected PV systems in microgrids, IEEE Trans. Sustain. Energy 9 (2018) 1605–1617.

[7] M. Kim, B.Y. Choi, J. Park, K. Kang, K. Choo, C. Won, Selective control algorithm for N-phase switching power pole of 4-leg interlinking converter in AC/DC hybrid microgrid, in: 2019 22nd International Conference on Electrical Machines and Systems (ICEMS), 2019, pp. 1–5.

[8] I.S. Freitas Gomes, Y. Perez, E. Suomalainen, Coupling small batteries and PV generation: a review, Renew. Sustain. Energy Rev. 126 (2020) 109835.
[9] G. Ding, F. Gao, S. Zhang, P.C. Loh, F. Blaabjerg, Control of hybrid AC/DC microgrid under islanding operational conditions, J. Mod. Power Syst. Clean Energy 2 (2014) 223–232.
[10] N. Chettibi, A. Mellit, G. Sulligoi, A.M. Pavan, Adaptive neural network-based control of a hybrid AC/DC microgrid, IEEE Trans. Smart Grid 9 (2018) 1667–1679.
[11] D.R.K. Aryani, J.-S. Kim, H. Song, Interlink converter with linear quadratic regulator based current control for hybrid AC/DC microgrid, Energies 10 (2017) (1799).
[12] M. Zolfaghari, M. Abedi, G.B. Gharehpetian, Power flow control of interconnected AC–DC microgrids in grid-connected hybrid microgrids using modified UIPC, IEEE Trans. Smart Grid 10 (2019) 6298–6307.
[13] U. Bose, S. Chattopadhyay, C. Chakraborty, Topological investigation on interlinking converter in a hybrid microgrid, in: 2018 IEEE International Conference on Industrial Electronics for Sustainable Energy Systems (IESES), 2018, pp. 62–67.
[14] M. Poursmaeil, S.M. Dizgah, H. Torkaman, E. Afjei, Autonomous control and operation of an interconnected AC/DC microgrid with Γ-Z-Source interlinking converter, in: 2017 Smart Grid Conference (SGC), 2017, pp. 1–6.
[15] Z. Junliu, G. Diankui, W. Fengping, Z. Yuechang, Z. Haiyan, Control strategy of interlinking converter in hybrid AC/DC microgrid, in: 2013 International Conference on Renewable Energy Research and Applications (ICRERA), 2013, pp. 97–102.
[16] F. Nejabatkhah, Y.W. Li, Overview of power management strategies of hybrid AC/DC microgrid, IEEE Trans. Power Electron. 30 (2015) 7072–7089.
[17] X. Yu, X. She, X. Zhou, A.Q. Huang, Power management for DC microgrid enabled by solid-state transformer, IEEE Trans. Smart Grid 5 (2014) 954–965.
[18] C. Kumar, Z. Zou, M. Liserre, Smart transformer-based hybrid grid loads support in partial disconnection of MV/HV power system, in: 2016 IEEE Energy Conversion Congress and Exposition (ECCE), 2016, pp. 1–8.
[19] Y.F. Liu, Y. Fang, J. Li, Interconnecting microgrids via the energy router with smart energy management, Energies 10 (2017) 1297.
[20] A.A. Eajal, M.A. Abdelwahed, E.F. El-Saadany, K. Ponnambalam, A unified approach to the power flow analysis of AC/DC hybrid microgrids, IEEE Trans. Sustain. Energy 7 (2016) 1145–1158.
[21] F. Wang, J.L. Duarte, M.A.M. Hendrix, Grid-interfacing converter systems with enhanced voltage quality for microgrid application—concept and implementation, IEEE Trans. Power Electron. 26 (2011) 3501–3513.
[22] H. Toodeji, S.H. Fathi, G.B. Gharehpetian, Power management and performance improvement in integrated system of variable speed wind turbine and UPQC, in: 2009 International Conference on Clean Electrical Power, 2009, pp. 609–614.
[23] J. Zhou, Y. Xu, H. Sun, Y. Li, M. Chow, Distributed power management for networked AC–DC microgrids with unbalanced microgrids, IEEE Trans. Ind. Inf. 16 (2020) 1655–1667.
[24] A. Kirakosyan, E.F. El-Saadany, M.S.E. Moursi, A.H. Yazdavar, A. Al-Durra, Communication-free current sharing control strategy for DC microgrids and its application for AC/DC hybrid microgrids, IEEE Trans. Power Syst. 35 (2020) 140–151.
[25] W. Deng, W. Pei, L. Li, Active stabilization control of multi-terminal AC/DC hybrid system based on flexible low-voltage DC power distribution, Energies 11 (2018) 502.
[26] R.A. Kaushik, N.M. Pindoriya, Power flow control of hybrid AC-DC microgrid using master-slave technique, in: 2014 IEEE Conference on Energy Conversion (CENCON), 2014, pp. 389–394.

[27] N.H. Saad, A.A. El-Sattar, A.E.-A.M. Mansour, A novel control strategy for grid connected hybrid renewable energy systems using improved particle swarm optimization, Ain Shams Eng. J. 9 (2018) 2195–2214.

[28] M. Zolfaghari, M. Abedi, G.B. Gharehpetian, Power exchange control of clusters of multiple AC and DC microgrids interconnected by UIPC in hybrid microgrids, in: 2019 24th Electrical Power Distribution Conference (EPDC), 2019, pp. 22–26.

[29] P.G. Khorasani, M. Joorabian, S.G. Seifossadat, Smart grid realization with introducing unified power quality conditioner integrated with DC microgrid, Electr. Power Syst. Res. 151 (2017) 68–85.

[30] M. Zolfaghari, G.B. Gharehpetian, A. Anvari-Moghaddam, Quasi-Luenberger observer-based robust DC link control of UIPC for flexible power exchange control in hybrid microgrids, IEEE Syst. J. (2020) 1–10.

[31] D.E. Olivares, A. Mehrizi-Sani, A.H. Etemadi, C.A. Cañizares, R. Iravani, M. Kazerani, A.H. Hajimiragha, O. Gomis-Bellmunt, M. Saeedifard, R. Palma-Behnke, G.A. Jiménez-Estévez, N.D. Hatziargyriou, Trends in microgrid control, IEEE Trans. Smart Grid 5 (2014) 1905–1919.

[32] Y. Li, F. Nejabatkhah, Overview of control, integration and energy management of microgrids, J. Mod. Power Syst. Clean Energy 2 (2014) 212–222.

[33] M. Fathi, H. Bevrani, Statistical cooperative power dispatching in interconnected microgrids, IEEE Trans. Sustain. Energy 4 (2013) 586–593.

[34] Y.G. Yajie Guo, H. Sun, J. Xi, Y. Hao, Overview of improved droop control methods of hybrid AC/DC microgrid interlinking converter, Adv. Comput. Sci. Res. 74 (2017) 260–266.

[35] M. Abuhilaleh, L. Li, J. Zhu, M.J. Hossain, Distributed control and power management strategy for parallel bidirectional power converters in hybrid microgrids, in: IECON 2019—45th Annual Conference of the IEEE Industrial Electronics Society, 2019, pp. 2148–2153.

[36] S.A. Bozorgavari, J. Aghaei, S. Pirouzi, A. Nikoobakht, H. Farahmand, M. Korpås, Robust planning of distributed battery energy storage systems in flexible smart distribution networks: a comprehensive study, Renew. Sustain. Energy Rev. 123 (2020) 109739.

[37] M.M. Esfahani, H.F. Habib, O.A. Mohammed, A hierarchical power routing scheme for interlinking converters in unbalanced hybrid AC-DC microgrids, in: IECON 2018—44th Annual Conference of the IEEE Industrial Electronics Society, 2018, pp. 53–58.

[38] E. Unamuno, J.A. Barrena, Hybrid ac/dc microgrids—part II: review and classification of control strategies, Renew. Sustain. Energy Rev. 52 (2015) 1123–1134.

[39] A. Eisapour-Moarref, M. Kalantar, M. Esmaili, Power sharing in hybrid microgrids using a harmonic-based multidimensional droop, IEEE Trans. Ind. Inf. 16 (2020) 109–119.

[40] V.K. Hema, R. Dhanalakshmi, Power sharing of an autonomous hybrid microgrid with an interlinking converter, in: IET Chennai Fourth International Conference on Sustainable Energy and Intelligent Systems (SEISCON 2013), 2013, pp. 481–485.

[41] M. Hosseinzadeh, F.R. Salmasi, Power management of an isolated hybrid AC/DC microgrid with fuzzy control of battery banks, IET Renew. Power Gener. 9 (2015) 484–493.

[42] T. Manjunath, S.D. Pandi, Artificial neural network based efficient power transfers between hybrid AC/DC micro-grid, Int. J. Emerg. Tech. Comp. Sci. Electron. 20 (2016) 163–168.

[43] I.R.K. Fitri, J.-S. Kim, H. Song, A robust suboptimal current control of an interlink converter for a hybrid AC/DC microgrid, Energies 11 (2018) 1382.

[44] N.M. Dehkordi, N. Sadati, M. Hamzeh, Robust backstepping control of an interlink converter in a hybrid AC/DC microgrid based on feedback linearisation method, Int. J. Control. 90 (2017) 1990–2004.

[45] M. Zolfaghari, M. Abedi, G.B. Gharehpetian, Robust nonlinear state feedback control of bidirectional interlink power converters in grid-connected hybrid microgrids, IEEE Syst. J. 14 (2020) 1117–1124.

[46] M. Zolfaghari, M. Abedi, G.B. Gharehpetian, J.M. Guerrero, Flatness-based decentralized control of bidirectional interlink power converters in grid-connected hybrid microgrids using adaptive high-gain PI-observer, IEEE Syst. J. (2020) 1–9.

[47] M. Zolfaghari, V. Nurmanova, M. Bagheri, G.B. Gharehpetian, Adaptive gain-regulating-based control of parallel-connected Γ-Z-source power converters in hybrid microgrids, in: 2020 9th International Conference on Renewable Energy Research and Application (ICRERA), 2020, pp. 321–325.

[48] F. Nejabatkhah, Y.W. Li, K. Sun, Parallel three-phase interfacing converters operation under unbalanced voltage in hybrid AC/DC microgrid, IEEE Trans. Smart Grid 9 (2018) 1310–1322.

[49] K. Sun, X. Wang, Y.W. Li, F. Nejabatkhah, Y. Mei, X. Lu, Parallel operation of bidirectional interfacing converters in a hybrid AC/DC microgrid under unbalanced grid voltage conditions, IEEE Trans. Power Electron. 32 (2017) 1872–1884.

Peer-to-peer management of energy systems

14

Sajjad Fattaheian-Dehkordi[a,b], *Mahyar Tofighi-Milani*[b], *Mahmud Fotuhi-Firuzabad*[b], *and Fei Wang*[c]
[a]Aalto University, Espoo, Finland, [b]Sharif University of Technology, Tehran, Iran
[c]North China Electric Power University, Beijing, China

Nomenclature

Variables

$P_{m,t}^{CDG}$	power generation by generation unit m at time t
$P_{m,t}^{Ch,ESS}$, $P_{m,t}^{Dis,ESS}$	charging/discharging power of storage unit m at time t
$P_{m,t}^{RES}$	power generation by RES unit m at time t
$P_{m,i}^{Producer\_Agent}$	P2P power transaction from producer agent m to consumer agent i
$P_{i,t}^{Sell,\ Consumer\_Agent}$	preferred selling power by producer agent i at time t
$P_{i,t}^{Purchase\_Retail\_entity}$	purchased power by agent i from the retail entity at time t
$E_{m,\ t}^{ESS}$	charging/discharging power of storage/electrical-vehicle unit m at time t
$\alpha_{m,t}^{Ch,ESS}$, $\alpha_{m,t}^{Dis,ESS}$	binary variables for impeding simultaneous charging/discharging power of storage/electrical-vehicle unit m at time t
$P_{m,t}^{Demand}$	power consumption by demand unit m at time t

Parameters

$Cost_{m}^{CDG}$	operational cost of generation unit i
$\lambda_{t}^{Purchase\_Retail\_entity}$, $\lambda_{t}^{Sell\_Retail\_entity}$	prices associated with purchasing/selling power to the retail entity
$\lambda_{m,t}^{Producer\_Agent}$	selling price of producer agent m at time t
$P_{m}^{Max,\ RES}$, $P_{m}^{Min,RES}$	maximum/minimum possible power generation by RES unit m
$P_{m}^{Max,CDG}$, $P_{m}^{Min,CDG}$	maximum/minimum possible power generation by generation unit m
$\rho_{m}^{Ch,ESS}$, $\rho_{m}^{Dis,ESS}$	charging/discharging efficient of storage unit m
$E_{m}^{Min,ESS}$, $E_{m}^{Max,ESS}$	minimum/maximum energy of storage unit m
$P_{Grid}^{total\_loss}$	total active power loss in the energy system
$P_{j,\ i}^{total\_loss}$	total active power loss associated with the power transaction from agent j to agent i
$Cost_{Grid}^{total\_loss}$	cost of total active power loss in the energy system
$P_{j,i}^{line}$	share of power transaction from agent j to agent i in power flow of in the designated "line"

Decentralized Frameworks for Future Power Systems. https://doi.org/10.1016/B978-0-323-91698-1.00018-2

P_{line}^{total}	total power flow in the designated "line"
P_{line}^{loss}	total active power loss in the designated "line"
$\rho_{line}^{Congestion}$	penalty factor associated with the congestion issue in the designated "line"

Sets

$\omega^{i,CDG}$	set of all CDG units in agent i
$\omega^{i,Demand}$	set of all demand units in agent i
$\omega^{i,RES}$	set of all RES units in agent i
$\omega^{i,ESS}$	set of all ESS units in agent i
$\omega^{Producer\_Agents}$	set of all producer agents

1 Introduction

In recent years, distributed energy resources (DERs) have been considerably integrated into local energy systems, which finally has long-lasting effects on the operation and planning of energy systems. In other words, the proliferation of DERs in local energy systems has facilitated the restructuring and privatization of conventional power systems. The high integration of DERs enables the supply of demand locally, which results in decrease in the dependence of local systems on upper-level power networks to ensure the supply-demand balance [1]. On the other hand, the installation of DERs has expanded the privatization of local energy systems. In this regard, decentralized structures such as multiagent systems have been introduced in energy systems to facilitate the integration of independently operated DERs into power systems [2].

Prosumers as entities that could act as power producers/consumers in energy systems are introduced as a result of the high integration of DERs into the system. The increasing trend of expanding the prosumers in energy systems would cause a more decentralized structure in the system [3]. As a result, in fully decentralized structures with open networks, system operators would be responsible for operating the grid rather than supplying the demand. In other words, the operational scheduling in decentralized systems would not be conducted in a central manner [4].

In the current centralized operational conditions, the net-metering policy is taken into account by most of the utilities in order to pay the prosumers and incentivize the integration of DERs into power systems [3]. Nevertheless, as the power systems are undergoing restructuring and privatization, net metering seems not to be an applicable technique to manage the DERs in local energy systems. Specifically, the independent operation of DERs by prosumers would limit the accessibility of system operators over their scheduling. Therefore, a significant flexibility capacity would be required to ensure the reliable operation of energy systems in the case of high penetration of DERs in power systems. However, the decreasing operational and investment costs of DERs (i.e., renewable energy sources [RESs]) in comparison with the operational and investment costs of bulk power units as well as their construction time period have limited the available flexibility capacity in the system [5,6]. Based on the aforementioned points, new management techniques should be applied by utilities to reliably operate energy systems with decentralized structures.

The centralized operational management of energy systems requires the gathering, analysis, and optimization of the operational characteristics of local resources in a centralized manner. Nevertheless, this management technique could cause privacy concerns in restructured energy systems. On the other hand, the increasing integration of DERs in local energy systems could impede the centralized operation of the system due to the limitation over the required communication infrastructure and the optimization capabilities of central operators [7]. On the other hand, privatization and restructuring in energy systems have resulted in the development of multiagent structures. While this structure facilitates the integration of independently operated DERs into power systems, the direct access of system operators over the scheduling of local resources as the base point in centralized operational management schemes is limited. Consequently, new market models should be developed to address the new operational conditions associated with the large growing number of DER units in local energy systems.

In recent years, a number of approaches have been introduced with the aim of facilitating the decentralized operational management of energy systems [7]. As a common point, these systems have tried to enable the interaction among the independent entities, i.e., prosumers/agents, in energy systems in order to optimize their power production/consumption in a decentralized manner. In this regard, the transactive energy concept is employed in several practical and research studies to facilitate the operational management of energy systems in a distributed manner [8,9]. In this approach, transactive control signals are developed in order to coordinate the scheduling of local resources without causing security issues in the system [10–12]. The transactive control signals have economic structures in order to incentivize the independent entities in the system to contribute to the management scheme. The application of transactive energy provides a management scheme in which the exchange of information between independent entities in the system would be limited to the economic signals. This approach has been applied in several pilot projects to investigate its application in the management of energy systems with decentralized structures. It is noteworthy that based on the pilot case studies, the transactive energy concept could be utilized by utilities to activate operational services in energy systems, which would finally result in the reliable and flexible operation of the network. Accordingly, value-based transactive control signals would enable the operators of energy systems to incentivize the collaboration of independently operated local resources in activating the required operational services. As a result, in transactive systems, system operators could exploit the scheduling of local DERs without direct access over their operational scheduling, which would address the privacy concerns in future energy systems with multiagent structures.

As mentioned earlier, the transactive energy concept is primarily appropriate for the rescheduling of independently operated local resources in order to address the operational condition of energy systems. Nevertheless, this approach could not facilitate energy management among the independent agents of a smart energy system in a decentralized manner. In this regard, P2P energy trading has recently been taken into consideration by researchers and utilities to develop active power markets in local energy systems [3,13–16]. This methodology enables the energy transactions between peers based on their own decision procedures, which would benefit the prosumers and consumers in the system. In this power market model, agents could directly trade their excess energy with other agents in the system. As a result, the implementation of P2P energy markets would decrease the dependence of local energy systems on the upper-level network, which would lead to the improvement of reliability and flexibility of power systems. Furthermore, the P2P local power market facilitates the transition of conventional systems into low-carbon energy systems by enabling the small DER users to participate in the market and benefit from the power transactions with other peers. These conditions would increase the investments in DERs which finally decrease the peak net load in the power system and required investments in bulk power units.

As discussed, the high increase in DERs and expansion of multiagent systems has resulted in the development of P2P power markets. P2P power markets would improve the reliability of the system by diversifying the power generation in energy systems. This structure facilitates the decentralized operational management of energy systems, which would address the operational risks associated with the power supply chain in the centralized operation of power systems. Furthermore, by decreasing the centralization in power systems, P2P power markets would reduce the risk associated with the failure of transmission networks. Finally, P2P power markets enable supplying the consumers locally in case of cyberattacks or natural disasters that would interrupt the supply chain in a centralized power system.

Based on the above points, the P2P concept would be a valuable asset to develop local power markets that enable the power/information transactions between peers in an energy system. Consequently, this chapter aims to present a thorough snapshot of the modeling and application of P2P power markets in energy systems. In this regard, the application of the P2P approach in the development of power markets in multiagent energy systems is studied in this chapter. On the other hand, the utilization of transactive control signals in the P2P power market could enable the system operator to activate operational services in the system. In this regard, along with the study of the application of P2P power markets in local energy systems, exploiting the scheduling of agents participating in the P2P power market in order to address the congestion issue in multiagent systems is also modeled in this chapter. As a result, this chapter would provide a general overview over the implementation of the P2P power market for decentralized management of energy systems with multiagent structures.

1.1 Current trends and impacts of P2P power markets

The utility regulations such as mandatory building integrated photovoltaics (BIPV) installation for new buildings, as well as incentive-based mechanisms, have resulted in increasing the RESs integration in local energy systems. This condition would reduce the dependence of the local energy systems on power grids to ensure the supply-demand balance in their systems. As a result, the conventional procedures of production and consumption of energy in power grids have been changed by defining agents in local energy systems that could purchase/sell energy at different time intervals. The extra power production by the agents in energy systems should be stored or sold to other agents or the power grid. As a result, the P2P-based consumer-centric electricity market seems to be a practical approach that enables direct power/information transactions between agents in energy systems with multiagent structures. The P2P framework enables negotiation between agents with extra power generation and agents with energy shortage in order to decide regarding the amounts of power transactions and their associated costs. This approach along with the advances in the installation of new information and communication infrastructures in power grids facilitates the implementation of power markets in local energy systems. In this framework, the price of power transactions between peers would be generally set between the price of purchasing and selling power to the grid (i.e., retail price). That is why both groups of producer agents and consumer agents in a P2P scheme would benefit from participation in the developed power market. It is noteworthy that while the conventional power systems are designed based on the one-directional power flow from bulk generation units to end user consumers, the P2P power market enables multidirectional power trading in local energy systems. As a result, the P2P power markets would benefit the supply-demand balance in local energy systems due to facilitation of the participation of local power generators in the operational management of power systems.

Development of P2P power markets aside from the energy system would also affect the human community [3]. First of all, the active P2P power market would positively affect the lifestyle of its associated community, e.g., active participation in demand response programs. Furthermore, local job positions would be introduced by the development of P2P power markets, which would finally improve the connection of the human community with the new energy environment. Implementation of P2P power markets would improve the transparency in energy transactions, benefit the local prosumers, and increase the social trust within the community. On the other hand, the resiliency and security of the energy systems would be improved as the agents of the energy system would find themselves responsible for resolving the operational issues and keeping the normal operational conditions. In this regard, agents would actively participate in providing operational services for resolving the operational issues, e.g., network congestion, in the energy system. Moreover, the P2P local power markets would provide other power resources for consumers in case of price spikes in the wholesale market. These changes would finally facilitate in transforming the hierarchical structure of power systems. In this regard, this new concept has been employed in several research and development projects to study its effectiveness in the decentralized management of multiagent energy systems [3,17,18]. These projects primarily address the business model as well as the information and communication systems required for the development of P2P power markets. In this context, Peer Energy Cloud and Smart Watts in Germany study the appropriate communication technologies for the implementation of P2P markets. Moreover, EMPOWER11 has developed a framework based on the cloud computing concept to execute the net metering and power trading in local P2P markets. On the other hand, the Enerchain project aims to provide a P2P market platform to expand or substitute conventional wholesale power markets. It is noteworthy that, in the context of developing local business systems, several start-up projects are currently focusing on enabling the prosumers to exchange their extra energy with neighbors in a P2P framework, as well as enabling the consumers to directly choose their local renewable generation suppliers.

As mentioned earlier, the communication infrastructures that enable the information transactions among peers play a key role in the modeling of sustainable local power markets. In this regard, the advancements in the implementation of 5G technology infrastructures as well as blockchain technology have facilitated the development of P2P power markets [19]. Specifically, blockchain technology seems to be a suitable option for enabling the information exchange between system agents without causing privacy concerns. This technology as the backbone of cryptocurrency could provide new opportunities for the participation of local agents in decentralized consumer-centric power markets.

In summary, the changes in energy and the human community with the implementation of P2P power markets are as follows:

- Empowerment of local consumers and increasing trust, transparency, and openness.
- Expanding the available choices for supplying the consumers.
- Increasing the flexibility and reliability of energy systems.
- Improving the competitiveness in power markets and facilitating the democratization of energy production/consumption.
- Increasing the cooperation of consumers in providing operational services.
- Increasing the knowledge of consumers regarding the environmental effects of their energy usage.
- Providing new business positions.
- Postponing the required grid investments.
- Improving the security and privacy of local resources.
- Improving the information and communication side of the power network for implementation of P2P markets, which would facilitate the transformation to smart energy systems.

2 Modeling the P2P energy management scheme in a local energy system with a multiagent structure

As mentioned, the transformation of operational conditions in multiagent energy systems shows the importance of developing reliable management systems that ensure the supply-demand balance in the system. In this respect, an energy management scheme based on the P2P concept has been developed in this section to enable the power transaction between independent agents in the system. P2P structures are defined as decentralized management systems where all active agents (i.e., peers) could cooperate in service trading. In this framework, each agent is considered to independently operate a group of assets; e.g., demands, conventional distributed generation (CDG), energy storage systems (ESSs), and RESs. In this structure, it is assumed that the agents would benefit from energy trading within the developed P2P power market, which would provide incentives for future investments in the expansion of DERs in local energy systems. Furthermore, it is considered that a retail entity would participate in the P2P power market, which would merely purchase power from the wholesale market and sell it to agents. In this regard, the agents could optimize their scheduling for the current time interval considering the announced electricity prices by the retail entity or power transactions with the other agents in the P2P market scheme. As a result, the developed structure in this section aims to enable the P2P power transactions among the system agents, while facilitating their power transactions with the upper-level network through retail entities. Moreover, a system operator is considered as the entity responsible for resolving potential conflicts and running the P2P power market to optimize real-time scheduling in the energy system. A schematic diagram of the considered information exchange in a multiagent system to develop the P2P energy management framework is shown in Fig. 14.1.

In the proposed P2P power market, agents could act as the power producer or consumer based on their objectives to sell energy to or buy energy from other agents of the system. In this regard, after announcing the retail prices from the retail entity, agents could conduct their local optimization and define their role as consumers/producers in the P2P power market. Note that, as the P2P power market would be conducted in an iterative manner, the agents could change their role based on the operational condition of the system. In this regard, in each iteration of running the P2P market, the producer agents would announce their preferred prices to the consumer agents, while consumer agents optimize their power requests considering the received prices. Then the consumer agents announce their power requests to each producer agent and this procedure would be conducted until the developed energy management scheme converges to the optimal results. Consequently, the P2P concept-based information exchange between the agents

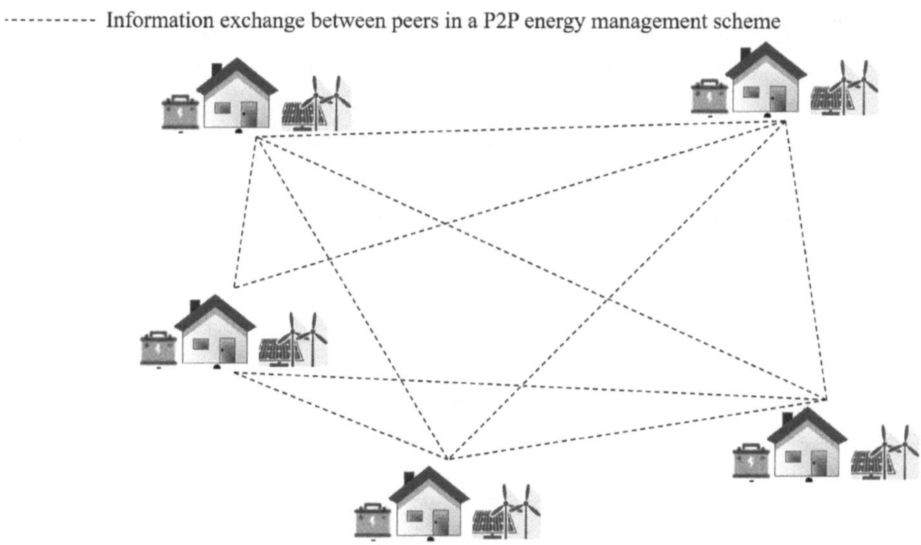

Fig. 14.1 Simplified model of the P2P information transaction in a multiagent energy system.

of the system would ensure the privacy of each agent. Moreover, while the P2P power market is conducted to determine the peers' transaction at the current time interval, agents should consider future time intervals in their optimization models to ensure management of their energy limited resources in an efficient manner. In this regard, in the proposed scheme, it is considered that agents take into account the model predictive control (MPC) technique in order to consider the $[t'+1, T]$ future time intervals while participating in the P2P power market at the current time interval (i.e., t'). Based on the above discussions, the consumer/producer agents would optimize the scheduling of their local resources in each iteration of running the P2P power market as follows:

✓ Optimization model conducted by consumer agent i:

$$
\begin{aligned}
\text{Min } Operational\_Cost_i^{Consumer\_Agent} = & \sum\nolimits_t \sum\nolimits_{m \in \omega^{i,CDG}} \left[P_{m,t}^{CDG} \cdot \left(Cost_m^{CDG} \right) \right] \\
& + \sum\nolimits_{m \in \omega^{i,Demand}} \left[P_{m,t}^{Demand} \cdot \left(P_{m,t}^{Demand} \right) \right] \\
& + \sum\nolimits_{m \in \omega^{Producer\_agent}} \\
& \left[\left(P_{m,i,t}^{Producer\_Agent} \right) \cdot \lambda_{m,t}^{Producer\_Agent} \right] \Big|_{t=t'} \\
& + \sum\nolimits_t \left[\left(P_{i,t}^{Purchase\_Retail\_entity} \right) \cdot \lambda_t^{Purchase\_Retail\_entity} \right]
\end{aligned} \tag{14.1}
$$

subject to

$$
\begin{aligned}
& P_{i,t}^{Purchase\_Retail\_entity} + \sum\nolimits_{m \in \omega^{Producer\_agent}} \left[P_{m,i,t}^{Producer\_Agent} \right] \Big|_{t=t'} \\
& = \sum\nolimits_{i \in \omega^{i,Demand}} P_{m,t}^{Demand} + \sum\nolimits_{m \in \omega^{i,ESS}} \left(P_{m,t}^{Ch,ESS} - P_{m,t}^{Dis,ESS} \right) \\
& - \sum\nolimits_{m \in \omega^{i,CDG}} P_{m,t}^{CDG} - \sum\nolimits_{m \in \omega^{i,RES}} P_{m,t}^{RES}
\end{aligned} \tag{14.2}
$$

$$
P_m^{\text{Min},CDG} \leq P_{m,t}^{CDG} \leq P_m^{\text{Max},CDG} \tag{14.3}
$$

$$
P_m^{\text{Min},RES} \leq P_{m,t}^{RES} \leq P_m^{\text{Max},RES} \tag{14.4}
$$

$$
0 \leq P_{m,t}^{Ch,ESS} \leq P_m^{\text{Max},Ch,ESS} \cdot \alpha_{m,t}^{Ch,ESS} \tag{14.5}
$$

$$
0 \leq P_{m,t}^{Dis,ESS} \leq P_m^{\text{Max},Dis,ESS} \cdot \alpha_{m,t}^{Dis,ESS} \tag{14.6}
$$

$$
\alpha_{m,t}^{Ch,ESS} + \alpha_{m,t}^{Dis,ESS} \leq 1 \tag{14.7}
$$

$$
E_{m,t}^{ESS} = E_{m,t-1}^{ESS} - \left(P_{m,t}^{Ch,ESS} \cdot \rho_m^{Ch,ESS} - P_{m,t}^{Dis,ESS} / \rho m^{Dis,ESS} \right) \tag{14.8}
$$

$$E_m^{\mathrm{Min},ESS} \leq E_{m,t}^{ESS} \leq E_m^{\mathrm{Max},ESS} \tag{14.9}$$

✓ Optimization model conducted by producer agent i:

$$\begin{aligned} Min\ Operational\_Cost_i^{Producer\_Agent} = & \sum_t \sum_{m \in \omega^{i,Demand}} \left[P_{m,t}^{CDG} \cdot \left(Cost_m^{CDG} \right) \right] \\ & + \sum_t \sum_{m \in \omega^{i,Demand}} \left[P_{m,t}^{Demand} \cdot \left(P_{m,t}^{Demand} \right) \right] \\ & - \left[\left(P_{i,t}^{Sell,Consumer\_Agent} \right) \cdot \lambda_{i,t}^{Producer\_Agent} \right] \Big|_{t=t'} \\ & - \sum_t \left[\left(P_{i,t}^{Sell\_Retail\_entity} \right) \cdot \lambda_t^{Sell\_Retail\_entity} \right] \end{aligned} \tag{14.10}$$

subject to Eqs. (14.3)–(14.9), and

$$\begin{aligned} \sum_{m \in \omega^{i,RES}} P_{m,t}^{RES} + \sum_{m \in \omega^{i,CDG}} P_{m,t}^{CDG} = & \sum_{i \in \omega^{i,Demand}} P_{m,t}^{Demand} \\ & + \sum_{m \in \omega^{i,ESS}} \left(P_{m,t}^{Ch,ESS} - P_{m,t}^{Dis,ESS} \right) \\ & + P_t^{Sell\_Retail\_entity} \\ & + \left[P_{i,t}^{Sell,Consumer\_Agent} \right] \Big|_{t=t'} \end{aligned} \tag{14.11}$$

Based on the developed mathematical optimization associated with the consumer agent i in the system, the objective function in Eq. (14.1) aims to minimize the operational costs of the agent considering the costs associated with operating the CDG units, deviation from the preferred load demand [2], as well as the costs of purchasing power from the producer agents and retail entity. Moreover, the power balance constraint of the agent is enforced by Eq. (14.2). Furthermore, the operational constraints of the CDG units, RESs, and ESSs are modeled in Eqs. (14.3), (14.4), and (14.5)–(14.9), respectively.

In the developed optimization problem associated with the producer agent i, the objective function (14.10) strives to minimize the operational costs of the agent. In this context, the operational costs of the CDG units, load demands, as well as the profit of selling power to consumer agents and the retail entity are considered in the objective function of the producer agent. Moreover, the power balance of the producer agent is modeled in Eq. (14.11) in order to ensure that the power produced by CDGs and RES units as well as the discharging of ESS units meets the power consumptions associated with demands, charging of ESS units, as well as the power sold to the retail entity and consumer agents. Note that the optimization models of the producer and consumer agents would be conducted at each iteration of running the P2P power market to optimize the real-time scheduling of local resources. That is why the agents employ the MPC technique to consider future time intervals while scheduling their local resources at the real-time interval. In this regard, the agents could utilize machine learning techniques to improve their prediction procedures associated with the required scheduling data.

As previously discussed, the consumer agents optimize their power requests from the retail entity as well as producer agents and announce the requested power to the peer agents. In this regard, producer agents could consider the total power requests by consumer agents (i.e., $\sum_{m \in \omega^{Consumer\_agent}} P_{i,m,t}^{Producer\_Agent}$) with the available power to sell to the consumer agents (i.e., $P_{i,t}^{Sell,\ Consumer\_Agent}$). In case the overall power requests by consumer agents are lower than the available selling power, the producer agent decreases its announced price to incentivize the agents to purchase more power. On the other hand, the agent would increase the announced price in the case where the overall power requests by consumers are more than the available selling power, which would incentivize the agents to minimize their power requests from the other agents. Consequently, the illustrated procedure associated with updating the selling prices by the producer i (i. e. , $\lambda_{i,t}^{Producer\_Agent}$) at iteration l could be mathematically formulated as follows:

$$\lambda_{i,t}^{Producer\_Agent}\Big|_{l} = \lambda_{i,t}^{Producer\_Agent}\Big|_{l-1} + \rho_{i,t}^{Producer\_Agent}\left(P_{i,t}^{Sell,Consumer\_Agent} - \sum_{m \in \omega^{Consumer\_agent}} P_{i,m,t}^{Producer\_Agent}\right) \tag{14.12}$$

In the developed formulation, $\rho_{i,t}^{Producer\_Agent}$ is a penalty factor that transforms the deviation between the available selling power and the overall power request by system agents to the price signal and would be determined independently by each agent. It is noteworthy that the producer agents would announce a preferred selling price based on their learning techniques at the first iteration of running the P2P power market. Nevertheless, based on the point that the proposed P2P power market is an iterative procedure, the agents would be able to update their announced prices as the proposed P2P procedure proceeds. The developed mathematical formulation for updating the prices in each iteration aims to minimize the deviation between the requested power by agents and the available selling power to agents, which facilitates the convergence of the proposed P2P structure.

The developed P2P market structure enables the agents in the system to exchange power/price information with their peers to update their operational scheduling in an iterative manner. As the developed procedure proceeds, the changes in the selling prices of the producer agents would be decreased until the converged point where the changes would be less than a negligible amount (i.e., ε). Consequently, the convergence constraint associated with the developed P2P energy management scheme in local energy systems could be formulated as follows:

$$\varepsilon \le \lambda_{i,t}^{Producer\_Agent}\Big|_{l} - \lambda_{i,t}^{Producer_{Agent}}\Big|_{l-1} \le \varepsilon, \quad i \in \omega^{Producer\_Agents} \tag{14.13}$$

As a result, the system operator as the responsible entity for running the P2P power market would check the convergence constraint in order to clear the market and

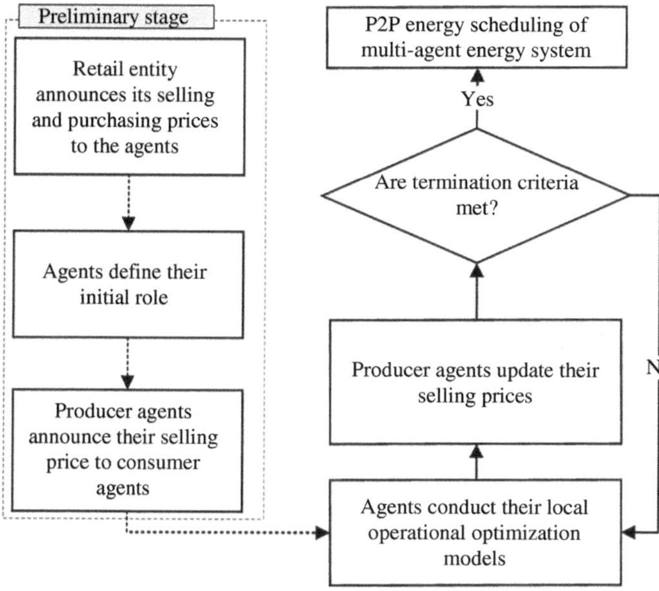

Fig. 14.2 The iterative algorithm associated with implementing the developed P2P power market.

resolve conflicts between system agents. This procedure enables direct transactions between peers during the running of the P2P management model. This provides local small power units with the chance of exchanging their extra power amounts with other peers to maximize their profits. The iterative algorithm associated with running the developed P2P power market is presented in Fig. 14.2. Note that the methodologies that could be taken into account in optimizing the scheduling of local resources as well as the improvement of the convergence of the proposed scheme are discussed in the final section of this chapter.

3 Extending the developed P2P power market in local energy systems

Network modeling has not been taken into account in the development of P2P operational management in the previous section. While most of the current works in the context of modeling P2P power markets have overlooked the operational effects associated with the network of the energy system, this section aims to extend the preliminary P2P management model to consider the power loss and operational costs of the power grid in the system. As a result, it is considered that the local energy system is operated with a radial structure, and the system operator would be responsible to announce the costs associated with the active power loss as well as operation and maintenance (O&M) of the system to the agents. One way to address the operational costs and power loss costs is to allocate the costs after finalizing the P2P power market. Nevertheless, this approach would result in a suboptimal solution, as the agents would not be able to consider these costs while scheduling their resources. Moreover, considering power loss associated with the P2P power transactions would incentivize the agents to exchange power with their neighbor agents, which would be advantageous in developing sustainable energy systems. In this regard, this section aims to develop an approach to enable the system operator to allocate the costs of power loss in the grid in a more efficient manner.

The power transactions in the P2P structure include two agents, i.e., producer and consumer agents. In this regard, the costs should be separated between both agents. Nevertheless, as the developed market model is based on the P2P concept and producer agents update their selling prices in an iterative manner, producer agents would increase their prices in the case of announcing the costs of power loss to them, which would finally be compensated by the consumer agents. Consequently, it is assumed that the system operator would announce the costs of power losses as well as the O&M of the network to the consumer agents. The system operator has to employ a management technique for allocating the power loss as well as O&M costs that align with the P2P structure of the energy system. In this regard, the transactive energy concept seems to be an applicable approach to enable the system operator to allocate the operational costs to the consumer agents.

The transactive energy concept enables the interaction between the system operator and agents in a decentralized structure. Utilizing this technique would limit the information exchange between the system operator and agents to the overall power transactions and transactive signals, which would ensure the privacy of local resources operated by the independent agents. As a result, the system operator aims to provide agents with monetary control signals in order to enable the agents to consider the power loss and O&M costs in their local optimization problems. Note that this scheme would finally result in minimizing the active power loss in the energy system, which would benefit the overall system agents. Moreover, this approach would be able to fairly allocate the O&M costs in a decentralized system based on the power transactions between agents through the energy grid. That is why transactive energy seems to be an applicable manner to be combined with the P2P structure in order to expand the preliminary power market model in a multiagent energy system.

As discussed, based on the obtained results in the P2P energy management model, the system operator would provide consumer agents with transactive signals modeling the cost of power loss associated with the P2P power transactions as well as a fixed amount of costs associated with the O&M costs of the energy grid. In this regard, the transactive signal associated with the active power loss in the energy grid (i.e., $TE_{j,i}^{loss}$) due to the power transaction from agent j to agent i would enable the consumer agent i to efficiently optimize its power transaction with agent j. In this context, after determining the active power loss in each line of the energy grid, the transactive energy associated with the power loss due to the purchasing power by agent i from agent j could be estimated as follows:

$$TE_{j,i}^{loss} = \left[Cost_{Grid}^{total\_loss} . \left(P_{j,i}^{total\_loss} \Big/ P_{Grid}^{total\_loss} \right) \right] \Big/ P_{j,i}^{Producer\_Agent} \tag{14.14}$$

$$P_{j,i}^{total\_loss} = \sum\nolimits_{line \in \omega^{j,i,line}} P_{line}^{loss} \cdot \left(P_{j,i}^{line} / P_{line}^{total} \right) \tag{14.15}$$

In the developed formulation, Eq. (14.14) models the share of the power transaction from agent j to agent i in the cost associated with the power loss in the network. Moreover, the power loss in the system due to the power exchange of agents j and i is estimated by Eq. (14.15). It is noteworthy that the system operator considers that the power loss in the energy grid would be addressed by purchasing power from the upper network in order to determine the costs associated with the power loss in the energy grid. Moreover, $TE_{O\&M}$ is considered as a constant cost that the consumer agents have to pay regarding the purchasing power from other agents to address the O&M costs in the energy grid. It is noteworthy that increasing the transactive control signal associated with the O&M costs would incentivize the agents to develop a more sustainable system, which would enable the agents to supply their demands utilizing their own local resources.

As a result, the optimization model conducted by consumer agents could be reformulated as shown in Eq. (14.16) to consider the transactive control signals associated with the power loss and O&M costs in the grid:

$$\text{Min } Total\_Cost_i^{Consumer\_Agent} = Operational\_Cost_i^{Consumer\_Agent} + \sum\nolimits_{m \in \omega^{Producer\_agent}} \left[\left(P_{m,i,t}^{Producer\_Agent} \right) \cdot \left(TE_{O\&M} + TE_{m,i}^{loss} \right) \right] \Big|_{t=t'} \tag{14.16}$$

Subject to Eqs. (14.2)–(14.9).

It is noteworthy that, after the convergence of the P2P energy scheduling, the system operator would check the costs associated with the power loss of each agent. In this regard, in the case that the cost of the power loss has changed significantly, the system operator would update the transactive control signal associated with P2P power transactions utilizing the formulation presented in Eqs. (14.14)–(14.15).

This section expanded the preliminary P2P energy scheduling scheme to address the costs of active power loss and O&M in the energy system. Nevertheless, the system operator should be able to exploit the P2P energy scheduling of agents in order to alleviate the operational issues in the grid. In this regard, the next section would discuss activating the operational service in the P2P management system to address the network congestion, which could occur in the system with high integration of DERs.

4 Extending the developed P2P power market to address the congestion issue in the energy grid

The high integration of DERs in local energy systems and the development of decentralized structures could lead to operational issues in energy systems. In this structure, system operators should be able to exploit the scheduling of independently operated local resources to address the operational issues in the energy grid. In other words, system operators should be able to activate operational services from independent agents to ensure the reliable operation of the grid. Consequently, this section is devoted to discussing the activation of operational services from local resources to address the congestion issue in multiagent systems. In this regard, the system operator should be able to exploit the scheduling of independent agents in the developed P2P energy management scheme to alleviate the congestion issue in the grid.

Local energy grids are conventionally designed based on the fit-and-forget concept to ensure that the energy grid could supply the estimated load in the system. Nevertheless, the high integration of DERs as well as the limited resources for expansion investment in local energy grids could result in potential congestion issues in the system. Specifically, the overpower consumption by load demands as well as the overpower generation by RESs could cause congestion in the grid. In this regard, the system operator would rely on local flexible resources operated by independent agents to address the congestion issue in the grid. In other words, the system operator as the entity responsible for the reliable operation of the grid has to employ an approach to alleviate the congestion issue in the grid, while addressing the decentralized operation of the system in a P2P energy management scheme. As a result, the transactive energy concept is employed in this section to enable the system operator to exploit the scheduling of agents participating in the P2P energy management scheme.

Transactive energy facilitates the development of decentralized energy management frameworks in the local energy system. This approach would facilitate the privatization and flexible operation of the system without compromising the security of local resources operated by independent agents. Transactive energy provides agents with monetary control signals in order to incentivize their contribution in providing the required operational service and exploit their resource scheduling in local power markets. In this context, the announced transactive control signals, as well as the operational characteristics of local resources, would be modeled in the operational optimization conducted by independent agents. Accordingly, transactive control signals should effectively link the operational issue in the grid with the operational optimization of agents in order to ensure the contribution of agents in alleviating the operational issue in the grid. In this regard, the transactive control signal should be determined in a way that models the economic value associated with activating the operational service from independent agents. In other words, the main concept of exchanging operational services in the transactive energy framework is to define the optimal point where all the entities come to an agreement regarding the value of the service transaction. Note that the process of defining the value of the service exchange in the system could be developed based on various approaches such as local markets, tariffs, and bilateral contracts. However, the developed process should comply with the decentralized structure of the multiagent system and the P2P energy management scheme applied for resource scheduling in the system.

Based on the above discussions, the system operator should develop a transactive control signal associated with the congestion issue in the system in order to decrease the power flow and finally alleviate the congestion issue in the grid. On the other hand, independent agents of the system would have to pay the cost proportional to the received transactive control signal from the system. In this regard, agents would be incentivized to minimize their power transactions through the network line with the congestion issue. This methodology would enable the system operator to revise the output of the P2P energy management scheme with the aim of alleviating the congestion issue in the system.

The primary characteristic of developing the transactive-based congestion alleviation scheme is to determine the transactive control signal, which enables the system operator to exploit the power transactions between peers in the system. That is why the power flow through the line with the overloading aims to be decreased in order to alleviate the congestion issue. The system operator would strive to decrease the power transactions between system agents that could cause overpower loading in the designated line by allocating the cost to these P2P power transactions. The cost associated with the power transactions caused by the congestion issue should be compensated by the producer and consumer agents. However, as discussed in the previous section, it is assumed that the consumer agents would be responsible to pay the cost to the system operator. Consequently, consumer agents would be incentivized to purchase power from other producer agents, which would finally result in alleviating the congestion issue in the grid. Note that the system operator would marginally increase the power injection by the producer agent as well as the power consumption by the consumer agent in order to determine the changes in the loading of the respective line with the congestion issue. This procedure is called the sensitivity analysis and enables the system operator to determine the share of each P2P power transaction on the overloading of the network lines with the congestion issue. As a result, the system operator would update the transactive control signal associated with the line with the congestion issue and finally allocate the share of P2P power transactions on the overloading in the respective line. The mathematical formulation of determining the transactive control signal associated with the congestion issue in the grid is formulated as follows:

$$TE_{line}^{Congestion} = TE_{line}^{Congestion-old} + \rho_{line}^{Congestion} \cdot \left(loading_{line} - loading_{line}^{Permissible}\right) \tag{14.17}$$

$$TE_{j,i}^{Congestion} = TE_{line}^{Congestion} \cdot SF_{j,i}^{line} \tag{14.18}$$

where $SF_{j,i}^{line}$ models the sensitivity factor and $TE_{j,i}^{Congestion}$ shows the transactive control signal associated with the P2P power transaction between the consumer agent i and producer agent j. In the developed formulation, Eq. (14.17) models the updating procedure of the transactive signal associated with the overloading of the *line*. Furthermore, the allocated transactive control signal to agent i based on the P2P power transaction between the consumer agent i and producer agent j is shown in Eq. (14.18). This tariff cost would incentivize the agents to decrease the P2P power transactions that cause overloading in the respective line of the grid. Note that the system operator would update the transactive control signals associated with the congestion issue in the grid based on the obtained results in the P2P energy management scheme.

As a result, the optimization model conducted by consumer agents could be reformulated as follows to consider the transactive control signals associated with the congestion issue in the grid:

$$\text{Min } Total\_Cost_i^{Consumer\_Agent} = Operational\_Cost_i^{Consumer\_Agent} + \sum\nolimits_{m \in \omega^{Producer\_agent}} \left[\left(P_{m,i,t}^{Producer\_Agent} \right) \cdot \left(TE_{m,i}^{congestion} \right) \right] \Big|_{t=t'} \tag{14.19}$$

Subject to Eqs. (14.2)–(14.9).

Based on the developed transactive-based model, the system operator would check the loading of network lines after convergence of the P2P energy management scheme. In this regard, in the case that the congestion issue is not alleviated, the system operator would update the transactive control signals associated with the congestion issue in the grid utilizing the formulation presented in Eqs. (14.17) and (14.18). The stepwise algorithm conducted for alleviating the congestion issue in a multiagent system is presented in Fig. 14.3.

The proposed scheme for congestion management is taken into account to study its effectiveness in managing the congestion issue in a multiagent test system. In this regard, the test system shown in Fig. 14.4 is considered, where 36 agents are defined based on each node of the test system. Each agent of the system independently operates its local resources while participating in the P2P energy management scheme. Note that the operational characteristics of the network and resources operated by each agent are adapted from Refs. [20–22].

It is assumed that the preliminary P2P energy management scheme at hour 12 causes the congestion issue in the line connecting the agents 21 and 22. In this regard, the system operator utilizes the developed transactive-based mechanism to alleviate the

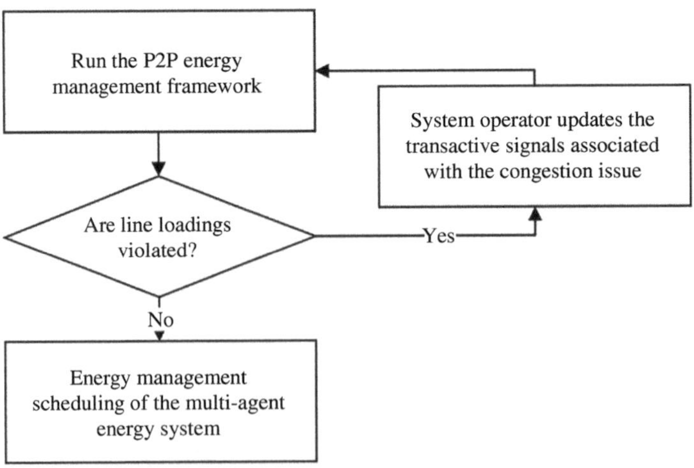

Fig. 14.3 The stepwise algorithm associated with congestion alleviation in a multiagent energy system.

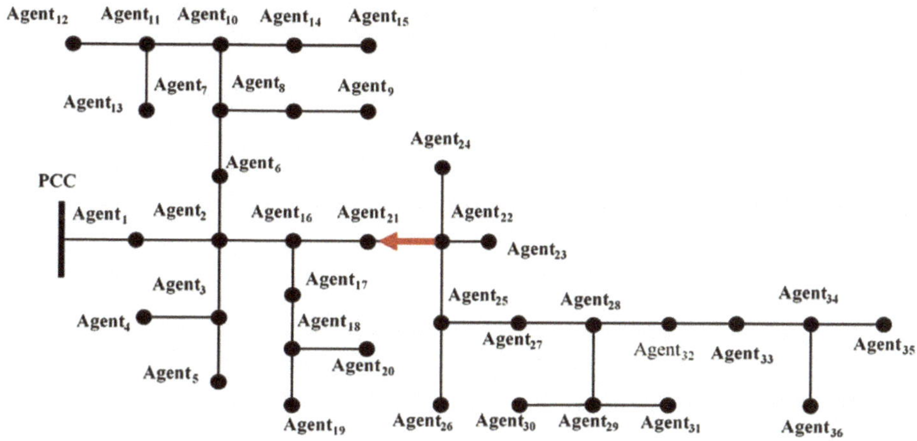

Fig. 14.4 The multiagent test system.

congestion issue in the system. After implementing the stepwise algorithm shown in Fig. 14.3, the finalized selling prices associated with producer agents are presented in Fig. 14.5. It is noteworthy that agents 3, 9, 22, 23, 24, 25, 28, 30, 31, 32, 33, 35, and 36 are producer agents at the considered operational condition, while other agents of the system act as consumer agents. Based on the obtained results, the system confronts the congestion issue in the line connecting agents 21 and 22 due to the overpower generation by RESs. In this regard, the announced transactive control signals to consumer agents have incentivized them to decrease their purchasing power from producer agents 22, 23, 24, 25, 28, 30, 31, 32, 33, 35, and 36 in order to alleviate the congestion issue in the grid. On the other hand, these producer agents have decreased their selling prices in the P2P energy management scheme to minimize their potential RESs curtailment by selling power to the consumer agents. This procedure has eventually

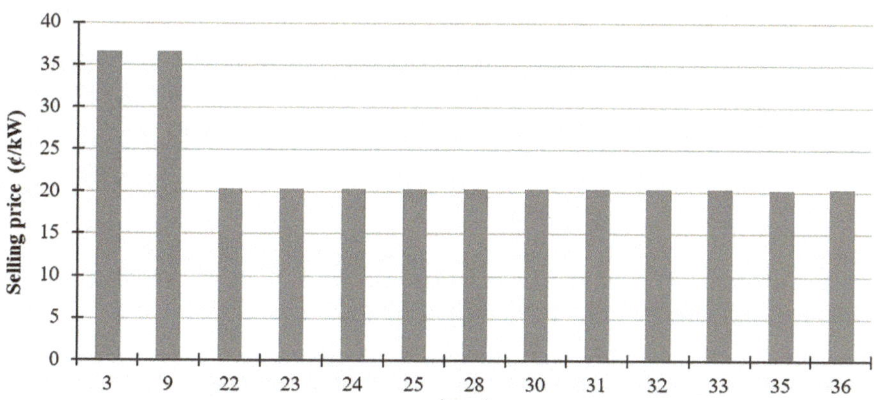

Fig. 14.5 The converged selling prices associated with the producer agents in the multiagent test system.

resulted in the selling prices shown in Fig. 14.5. Note that the selling prices of producer agents 3 and 9 are close to 36.65 ¢/kW, which is the selling price of retail entity to consumer agents in the system. As a result, the obtained results and the above discussions show the effectiveness of the developed framework to address the congestion issue in energy systems with decentralized structures. In other words, the proposed procedure enables the system operator to exploit the energy management of independent agents participating in the P2P power market in order to address the congestion issue in the network.

5 Further operational points associated with modeling the P2P energy management framework

- In order to improve the convergence procedure of the P2P energy management scheme, several conditions could be taken into account by the market operator as follows:
  - The market operator could enforce limitations over the potential changes in the purchasing power by consumer agents in consecutive iterations of running the P2P energy management scheme. As a result, this limitation would lead to a smooth change in the operational point of the system during the running of the P2P energy management scheme, which would also improve the convergence of the developed framework. This operational limitation over the purchasing power from agent m by agent i at iteration l could be mathematically formulated as follows:

$$(1-\sigma)\cdot P_{m,i}^{Producer_{Agent}}\Big|_{l-1} \leq P_{m,i}^{Producer_{Agent}}\Big|_{l} \leq (1+\sigma)\cdot P_{m,i}^{Producer\_Agent}\Big|_{l-1} \tag{14.20}$$

In this formulation, σ is a confining factor that enforces limitations over the changes in the P2P power transaction from agent m to agent i.

  - In a similar manner, the market operator could impose limitations over the selling prices announced by the producer agents in order to ensure the smooth convergence of the developed P2P energy management scheme. In this regard, the limitation over the changing of the selling prices in two consecutive iterations of running the P2P energy management scheme could be formulated as

$$(1-\tau)\cdot \lambda_{i}^{Producer\_Agent}\Big|_{l-1} \leq \lambda_{i}^{Producer\_Agent}\Big|_{l} \leq (1+\tau)\cdot \lambda_{i}^{Producer\_Agent}\Big|_{l-1} \tag{14.21}$$

In this formulation, τ is a confining factor that enforces limitations over the changes in the selling price announced by producer agent i.

- The operational optimization conducted by each independent agent requires forecasting of the power production/consumption by DERs as well as the operational condition of the system, which could impact the profits of the agents participating in the P2P energy management scheme. In this context, efficient techniques should be taken into account to model the uncertainty associated with decision parameters.

- A stochastic programming technique could be taken into account in the operational optimization modeling of each agent to improve the precision of the obtained results and eventually decrease the economic risk associated with the uncertainties of decision parameters [23]. As a result, a set of the operational scenarios associated with each decision parameter as well as their incidence probability would be employed to model the optimization problem based on stochastic programming [21].
- A robust optimization approach, which aims to define the worst-case realization of uncertain parameters, could be taken into account to model the operational problem of each agent. In comparison with stochastic programming that requires the distribution probability of uncertain parameters, utilizing the robust optimization technique could be conducted with limited information.
- Besides stochastic and robust optimization approaches, the distributionally robust optimization technique [24], data-driven robust optimization [25], chance-constrained optimization [26], and information gap decision theory (IGDT) [27] could be taken into account by agents participating in the P2P energy management scheme to model the uncertainty associated with the operational optimization of DERs.

6 Conclusions

The conventional hierarchical structure of power systems is undergoing a substantial transition by the introduction of privatization and restructuring in local energy systems. In this new operational environment, multiagent structures are developed to facilitate the integration of independently operated DERs into the energy network. Nevertheless, this decentralized structure would impede the application of conventional centralized approaches in the management of local energy systems. As a result, the application of the P2P concept in secure energy management of multiagent energy systems is investigated in this chapter. The developed scheme enables the P2P power transactions among system agents while ensuring the reliable operation of the energy grid.

References

[1] S. Fattaheian-Dehkordi, et al., An incentive-based mechanism to alleviate active power congestion in a multi-agent distribution system, IEEE Trans. Smart Grid 12 (3) (2020) 1978–1988, https://doi.org/10.1109/TSG.2020.3037560.

[2] A. Rajaei, et al., Developing a distributed robust energy management framework for active distribution systems, IEEE Trans. Sustain. Energy 12 (4) (2021) 1891–1902, https://doi.org/10.1109/TSTE.2021.3070316.

[3] T. Sousa, et al., Peer-to-peer and community-based markets: a comprehensive review, Renew. Sustain. Energy Rev. 104 (2019) 367–378.

[4] S. Fattaheian-Dehkordi, A. Abbaspour, M. Lehtonen, Electric vehicles and electric storage systems participation in provision of flexible ramp service, in: Energy Storage in Energy Markets, Academic Press, 2021.

[5] F. Kamrani, et al., Flexibility-based operational management of a microgrid considering interaction with gas grid, IET Gener. Transm. Distrib. 15 (19) (2021) 2673–2683, https://doi.org/10.1049/gtd2.12155.

[6] F. Kamrani, et al., A two-stage flexibility-oriented stochastic energy management strategy for multi-microgrids considering interaction with gas-grid, IEEE Trans. Eng. Manage. (2021) 1–11, https://doi.org/10.1109/TEM.2021.3093472.
[7] A. Rajaei, et al., Decentralized transactive energy management of multi-microgrid distribution systems based on ADMM, Int. J. Electr. Power Energy Syst. 132 (2021) 107126.
[8] Council, T, Gridwise Transactive Energy Framework Version 1.0, The GridWise Architecture Council, 2015. Tech. Rep.
[9] O. Abrishambaf, et al., Towards transactive energy systems: an analysis on current trends, Energy Strat. Rev. 26 (2019) 100418.
[10] S. Fattaheian-Dehkordi, et al., Transactive-based control algorithm for real-time energy imbalance minimisation in microgrids, Open Access Proc. J. 2020 (2020) 549–552.
[11] S. Fattaheian-Dehkordi, et al., Incentive-based ramp-up minimization in multi-microgrid distribution systems, in: 2020 IEEE PES Innovative Smart Grid Technologies Europe (ISGT-Europe), 2020.
[12] S. Fattaheian-Dehkordi, et al., Distribution grid flexibility-ramp minimization using local resources, in: 2019 IEEE PES Innovative Smart Grid Technologies Europe (ISGT-Europe), 2019.
[13] S. Cui, Y. Wang, J. Xiao, Peer-to-peer energy sharing among smart energy buildings by distributed transaction, IEEE Trans. Smart Grid 10 (6) (2019) 6491–6501.
[14] A. Paudel, et al., Peer-to-peer energy trading in a prosumer-based community microgrid: a game-theoretic model, IEEE Trans. Ind. Electron. 66 (8) (2019) 6087–6097.
[15] E. Sorin, L. Bobo, P. Pinson, Consensus-based approach to peer-to-peer electricity markets with product differentiation, IEEE Trans. Power Syst. 34 (2) (2019) 994–1004.
[16] Y. Zhou, J. Wu, C. Long, Evaluation of peer-to-peer energy sharing mechanisms based on a multiagent simulation framework, Appl. Energy 222 (2018) 993–1022.
[17] C. Zhang, et al., Review of existing peer-to-peer energy trading projects, Energy Procedia 105 (2017) 2563–2568.
[18] E. Mengelkamp, et al., Designing microgrid energy markets: a case study: the Brooklyn microgrid, Appl. Energy 210 (2018) 870–880.
[19] M.L.D. Silvestre, et al., A technical approach to the energy blockchain in microgrids, IEEE Trans. Ind. Inf. 14 (11) (2018) 4792–4803.
[20] S. Fattaheian-Dehkordi, et al., The role of transactive energy on management of flexible resources, in: Flexible Resources for Smart Cities, Springer Nature, Switzerland AG, 2021.
[21] S. Fattaheian-Dehkordi, et al., Distributed transactive framework for congestion management of multiple-microgrid distribution systems, IEEE Trans. Smart Grid (2021) 1–11, https://doi.org/10.1109/TSG.2021.3135139.
[22] S. Fattaheian-Dehkordi, et al., A new framework for mitigating voltage regulation issue in active distribution systems considering local responsive resources, IEEE Access 9 (2021) 152585–152594, https://doi.org/10.1109/ACCESS.2021.3127240.
[23] H. Gharibpour, F. Aminifar, Multi-stage equilibrium in electricity pool with flexible ramp market, Int. J. Electr. Power Energy Syst. 109 (2019) 661–671.
[24] E. Guevara, et al., A machine learning and distributionally robust optimization framework for strategic energy planning under uncertainty, Appl. Energy 271 (2020) 115005.
[25] F. Shen, et al., Large-scale industrial energy systems optimization under uncertainty: a data-driven robust optimization approach, Appl. Energy 259 (2020) 114199.

[26] J.D. Vergara-Dietrich, et al., Advanced chance-constrained predictive control for the efficient energy management of renewable power systems, J. Process Control 74 (2019) 120–132.
[27] M.A. Mirzaei, et al., Integration of emerging resources in IGDT-based robust scheduling of combined power and natural gas systems considering flexible ramping products, Energy 189 (2019) 116195.

False data injection attacks on distributed demand response 15

I'm paying less: A targeted false data injection attack against distributed device scheduling

Thusitha Thilina Dayaratne, Carsten Rudolph, Ariel Liebman, and Mahsa Salehi
Faculty of IT, Monash University, Clayton, VIC, Australia

Nomenclature

$\mathbf{a_{h,i}}$	ith appliance of household h
$\mathbf{A_h}$	Number of appliances in hth household
$\mathbf{ad_p}$	Aggregated demand value of pth pricing slot
$\mathbf{d_{h,i}}$	Demand value of ith appliance of household h
$\mathbf{es_{h,i}}$	Earliest start scheduling index of ith appliance of household h, $es_{h,i}\in\{1,2,\ldots,S\}$
$\mathbf{H}$	Number of households, $H\in\mathbb{N}^+$
$\mathbf{k}$	Number of scheduling slots per single pricing slot, $k\in\mathbb{N}^+$
$\mathbf{lf_{h,i}}$	Latest finish scheduling index of ith appliance of household h, $lf_{h,i}\in\{1,2,\ldots,S\}$
$\mathbf{os_{h,i}}$	Optimized start scheduling index of ith appliance of household h, $os_{h,i}\in\{1,2,\ldots,S\}$
$\mathbf{P}$	Number of pricing slots per day, $P\in\mathbb{N}^+$
$\mathbf{p}$	Pricing slot index
$\mathbf{pf_{h,i}}$	Penalty costs of ith appliance of household; indicates the costs per pricing slot that an appliance is moved from the desired slot h.
$\mathbf{ps_{h,i}}$	Preferred start scheduling index of ith appliance of household h, $ps_{h,i}\in\{1,2,\ldots,S\}$
$\mathbf{R_{s(h,i)}}$	Indicates whether ith appliance of household h is running on sth scheduling slot, $R_{s(h,i)}\in\{0,1\}$
$\mathbf{rs_{h,i}}$	Real start scheduling index of ith appliance of household h, $rs_{h,i}\in\{1,2,\ldots,S\}$
$\mathbf{rt_{h,i}}$	Number of scheduling slots ith appliance of household h runs, $rt_{h,i}\in\{1,2,\ldots,S\}$
$\mathbf{s}$	Scheduling slot index
$\mathbf{S}$	Number of scheduling slots per day, $S\in\mathbb{N}^+$
$\mathbf{up_p}$	Unit price value of pth pricing slot

An ever-increasing population and the use of electricity throughout businesses induce the growth of peak demands. Traditional power grids struggle to maintain stability [1] as the efficiency of the grids requires a reduction in superfluous baseload provided by traditional (e.g., coal-powered) methods. The formation of smart grids and its descendant concepts such as demand response (DR) schemes and transactive energy markets try to ensure that grids are capable of serving the needs of a highly advanced society while reducing emissions. Smart grid concept enhances traditional power grids with the use of information and communication technology and enables a range of

Decentralized Frameworks for Future Power Systems. https://doi.org/10.1016/B978-0-323-91698-1.00001-7

improved operational functionalities, including more convenient and efficient energy management. Many countries are augmenting existing power systems with advanced metering infrastructure (AMI) that enables two-way communication between utility providers and consumers as well as increase the overall system visibility with the use of smart meters. The United Kingdom is planning to replace existing meters with smart meters by 2024 [2] while smart meter penetration in North America has forecasted to reach 81% by 2024 [3]. AMI can be used to enhance the efficiency of DR schemes, which are programs designed by utility companies that drive consumers to change their consumption patterns and better manage their demand with financial incentives. Commonly considered financial incentives include real-time pricing (RTP), day-ahead pricing (DAP), and time of use (ToU) [4, 5]. Utility providers expected to implement DR schemes to increase the efficiency of power grids in the retail market [6, 7]. These DR programs help utility companies to maintain grid reliability with lower levels of reserve capacity by effectively providing more accurate demand forecast through DR mechanisms and at the same time providing improved cost management capabilities for customers. DR schemes also allow customers to make more informed decisions about their usages ultimately help utility companies in managing peak demands [8].

In general, DR schemes can be classified as either centralized or decentralized/distributed depend on the control mechanism that the scheme utilize [9]. However, scalability issues with the ever-increasing user and devices size and user privacy concerns are causing centralized DR schemes to be less favorable over decentralized schemes. Hence, researchers have focused more on distributed/decentralized DR schemes. Moreover, the advancements in IoT-related technologies enable the use of distributed/decentralized DR schemes in the residential energy market. For example, a combination of smart meter and home energy management systems (HEMS) can be used to schedule a set of smart appliances in a household (or in a community) by making intelligent decisions based on the particular DR program price signal or incentives and consumer convenience [10–13]. Thus, device scheduling-based residential DR scheme are gaining significant attention in recent years.

Data integrity is an indispensable criterion in any DR scheme. Having complete, accurate, and timely data can guarantee that DR schemes generate optimal decisions, where all users gain expected benefits. However, it is challenging to preserve the data integrity given the complexity of energy networks and the large attack surfaces enabled by cyber-physical integration. Moreover, the increase in user interactions with the adaptation of more decentralized and distributed schemes opens more opportunities for adversaries [14–16]. In particular, false data injection attacks (FDIAs) where adversaries inject or manipulate system data can induce incorrect or suboptimal power flows, excessive or insufficient generation, financial losses, and inconvenience for the users by impacting the integrity. Manipulated forecasts can further trigger a crippling impact on power grids as transactive energy markets allow energy providers to use demand forecasts in determining economic dispatching processes in power grids that involve the generation, transmission, and distribution. FDIAs on residential DR schemes require less effort compared to the attacks on state estimation as adversaries can use edge devices such as smart meters, HEMSs, home/neighborhood networks, and smart devices to mount attacks. The attack motives can be either to obtain benefits (e.g., cheaper energy) that would not be possible in normal circumstances or to cause harm to other users or a utility provider.

Although extensive research has been carried out on FDIA on state estimation and power flows, FDIAs on residential DR schemes such as device scheduling are still nascent. In our previous work [15], we have shown that an adversary could gain benefits by injecting false demand to a DR scheme where devices are fully flexible. However, given the real-world context, the majority of demands are uncontrollable and, controllable demands have constrained flexibilities. Further, in a previous study, we did not analyze how the attack impact varies with the community size, flexibility, or with the number of adversaries. Thus, in this work, we extend our previous work and do a thorough analysis of an FDIA on device scheduling distributed DR scheme with a more realistic dataset from an attacker's perspective.

The remainder of the chapter is organized as follows. In Section 1, we discuss the current state of the art in FDIA. Section 2 discusses the system model of the DR scheme, and in Section 3, we present the attack model. We explain the experiment and the results in Sections 4 and 5, respectively. Section 6 opens up a discussion about the results and we conclude the chapter in Section 7.

1 Literature review

Motivated by Liu et al. [17], an extensive amount of work has been carried out on the FDIA against state estimation [18–23]. Several works analyzed the impact of FDIAs on economic dispatch/optimal power flows [19, 24, 25]. Dong et al. [26] showed that the existence of a denial-of-service (DoS) attack on communication channels and false price signal of an RTP system could lead to a suboptimal solution of the deferrable load scheduling optimization algorithm rather than the global optimal solution. Focusing more on DoS attacks, Wood et al. [27] showed that strategic adversaries could increase their revenue by 69% in an RTP system.

Barreto et al. [7] modeled two strategic attackers (defraud and malicious) with two different motivations in RTP systems. They have considered both dynamic pricing and direct load control (DLC) systems and concluded that the dynamic pricing systems are more resilient than DLC systems in terms of data integrity attacks. Focusing on malicious adversaries, Abdallah et al. [28] showed how an attacker can manipulate demand profiles to increase the cost of generation for an operator. Zhang et al. introduced a novel DR scheme where they divided the users into demand users, supply users, and traditional power plant. They have considered both renewable and traditional power generation in their scheme [11]. Using experimental and performance evaluation results, they have shown that FDIA on their scheme can cause a significant impact on the welfare of the affected users. Yang et al. [29] showed that increasing the traditional power consumption factor of a set of demand users leads to a state, where the power plant receives an advantage while demand users lose their welfare due to the higher traditional power rates. Besides, they showed that normal demand users could obtain a welfare gain while compromised demand users and power plan suffer a loss in their welfare if the adversary manipulates the broadcast traditional power price signal.

Liu et al. showed that a malicious adversary can increase the peak to average ratio of a DR system by 35.7% and a fraud attacker can gain about 34.3% of electricity bill reduction while increasing the normal users' bills by another significant percentage [30, 31]. They analyzed how an adversary can reduce their bills using pricing cyber attack. Their results are based on the assumption that an attacker is capable of manipulating the broadcast pricing signal, which is sent by the utility company (UC). Goitom et al. investigated how FDIA can lead their decentralized DR scheme to a sub-optimal solution or force it not to converge [32]. Liu et al. proposed a coordinated attack [33] where the adversary manipulates both the price signal and demand consumption value in a set of smart meters. The adversary can obtain about 32% price reduction using the coordinated attack according to their simulation results. Though all these works are significant steps toward understanding the effect of different FDIAs, all these attacks require to manipulate communication or to breach one or more entities in the system. However, with the existence of lightweight efficient encryption mechanisms, protecting price signals and measurement data can be effectively realized using standard security mechanisms. Moreover, FDIAs on device scheduling-based DR schemes are not thoroughly explored.

2 System model

A typical household contains both controllable (flexible) and uncontrollable (inflexible) demands. Compared to controllable demands (refrigerator, kitchen appliances, lights, TV), uncontrollable demands are not instantaneous where operating times can be flexible (washing machines, dishwashers, pool pump, clothes dryer). These uncontrollable devices can be optimized using DR systems to reduce both the electricity generation cost for utility companies and the consumption cost for consumers. In particular, scheduling devices optimally under these DR programs, such as programs that use RTP, can effectively maximize the benefits by lowering energy costs [34, 35]. The commonly adopted high-level communication flow of a typical distributed DR scheme for device scheduling [12, 13, 36] is depicted in Fig. 15.1. Note that the idea

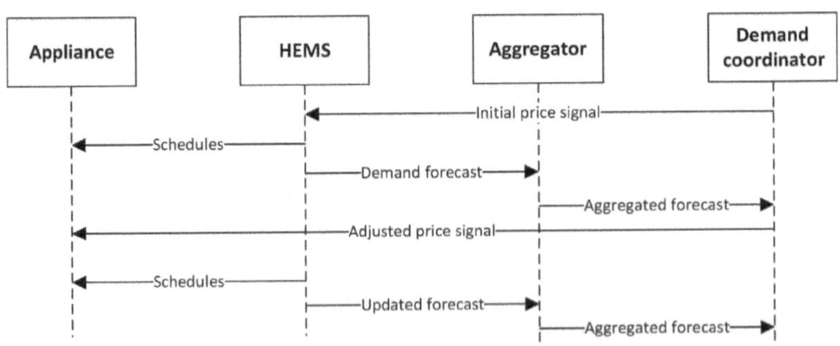

Fig. 15.1 Sequence of communication flow of a distributed DR scheme in general.

of a device scheduling-based DR scheme is to decide optimal starting time for devices rather curtailing any demands. Thus, in general, distributed device scheduling DR schemes leverage multiple iterations until the system converges.

For the following discussion, we assume a distributed device scheduling DR scheme, which equally divides a day into P pricing slots, where each pricing slot is subdivided into $k(\geq 1)$ scheduling slots (Eq. 15.1). The goal of the device scheduling DR scheme is to decide the optimal starting time of connected devices to achieve the optimization goal of minimizing the total cost.

$$kP = S \tag{15.1}$$

Before the start of each day, the UC/demand coordinator sends a guideline pricing signal for the following day to all the connected smart meters. The price signal consists of P values, where each value correspondents to the unit price for one pricing slot for some initially assumed demand. The unit price for the k scheduling slots (that belongs to the pth pricing slot) remains same. We indexed unit price values from 1 to p and scheduling slots from 1 to s. Each household has several devices that need to be scheduled. The house owner configures the earliest start scheduling index, the latest finish scheduling index, preferred start scheduling index, device demand (power rate), running duration and the penalty factor for each device. Setting up these configurations can be considered as an one-time operation unless the preference changes or users want to add new devices. The HEMS locally schedules devices between the configured earliest start scheduling index and the latest finish scheduling index. If a device does not start at its preferred start index, there will be an associated cost (inconvenience/penalty) which is calculated based on the penalty factor. A device may operate and finish within a single scheduling slot or operates for multiple slots.

In this work, we assume that the goal of optimization is to achieve the lowest overall costs without curtailing any demand. Thus, the optimization needs to flatten out peaks and reconstructs the demand distribution, given the flexibility and penalty constrain. For the generic discussion, we assume a perfect optimization oracle that achieves an ideal distribution of demand given the user constrains.

We define the unit price of the pth pricing slot as a function of the aggregated demand of the same slot, where f is monotonically increasing. Unit price values of pricing slots can be different. Nevertheless, the unit price monotonically increases, either step wise, linear of exponential with the demand.

$$f: ad_p \rightarrow up_p$$

For the formal definition of optimal demand distribution, we need to consider the following values.

The sum of the demand of all appliances scheduled to run during scheduling period s depicts the aggregated demand value of scheduling period s.

$$ad_s = \sum_{h=1}^{H}\sum_{i=1}^{A_h} d_{h,i} \times R_{s(h,i)} \tag{15.2}$$

As UC requires to maintain the maximum demand among the k scheduling slots of pth pricing slot, the required demand of pth pricing slot is the maximum among k scheduling slots:

$$ad_p = \max\left(ad_{s=kp-k+1}, ad_{s=kp-k+2}, \ldots, ad_{s=kp}\right) \tag{15.3}$$

Total electricity bill of the community for the day (P slots) is calculated using Eq. (15.4), which is the summation of the product between each pricing slot's aggregated demand ad_p and its correspondent unit price up_p.

$$Total\,bill = \sum_{p=1}^{P} ad_p \times up_p \tag{15.4}$$

Depending on the price signal and the flexibility of the devices, HEMSs select optimized start times for each appliance. The difference between preferred start index $ps_{h,i}$ and the optimized start index $os_{h,i}$, multiplied by the appliance's penalty costs $pf_{h,i}$ conceives the inconvenience to a user of a given schedule for a given appliance. Sum of all the inconvenience conceives the total penalty costs for the community, which is

$$Total\,penalty = \sum_{h=1}^{H}\sum_{i=1}^{A_h} \left|os_{h,i} - ps_{h,i}\right| \times pf_{h,i} \tag{15.5}$$

The optimization induces HEMSs to change $os_{h,i}$ of the devices with the ultimate goal to move the system toward minimizing the total cost:

$$\min\ Total\,cost = Total\,bill + Total\,penalty \tag{15.6}$$

Finding the minimum cost can be modeled as a convex optimization problem and can solve either using centralized or distributed approaches. Given the privacy issues and scalability factors, decentralized methods are favorable over centralized solutions. However, for this discussion, the mode of optimization is irrelevant. Instead, we assume a perfect optimization oracle as depicted in Fig. 15.2, which produces optimal demand distribution to minimize the total costs for a given community given initial demand distribution for P number of slots and the set of penalty costs.

2.1 Optimization

This section explains in more detail the result of the demand schedule optimization. For simplicity, we restrict the explanation to the following case:

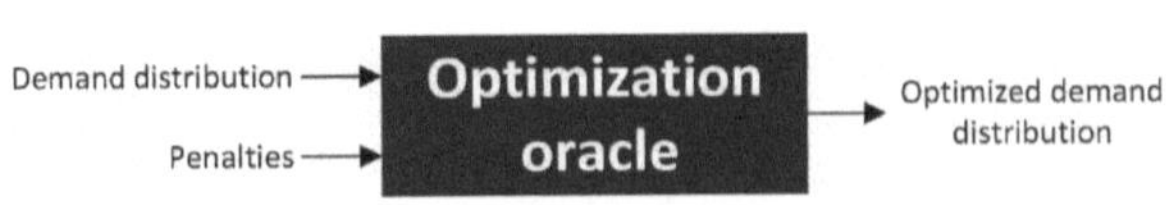

Fig. 15.2 Optimization oracle.

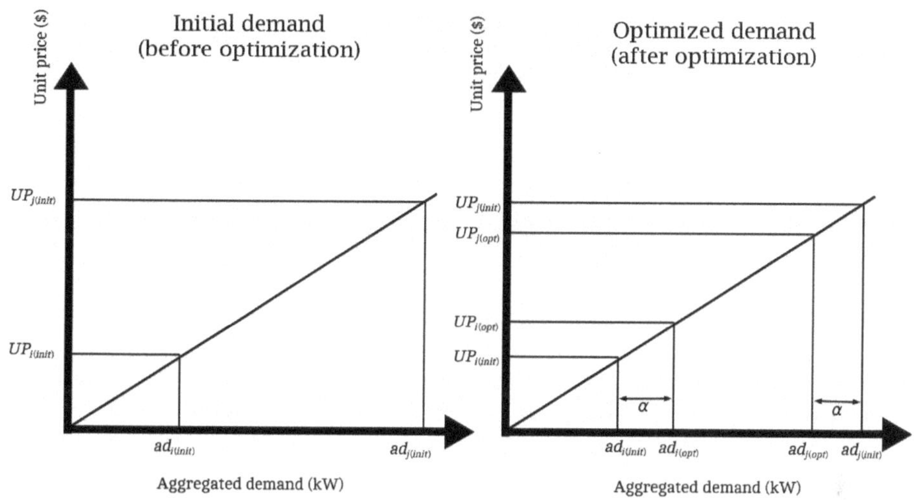

Fig. 15.3 Unit price variation with aggregated demand before and after the optimization.

1. We identify two pricing slots p_i and p_j with $i \neq j$.
2. Some appliances may run during either one of the two pricing slots or in both pricing slots.

We denote $ad_{i(init)}$, $ad_{j(init)}$, $ad_{i(opt)}$, and $ad_{j(opt)}$ as the initial aggregated demand of the ith pricing slot, the initial aggregated demand of the jth pricing slot, the optimized aggregated demand of the ith pricing slot, and the optimized aggregated demand of the jth pricing slot, respectively. Similarly, we denote respective unit price values as $up_{i(init)}$, $up_{j(init)}$, $up_{i(opt)}$, and $up_{j(opt)}$. Left-side chart in Fig. 15.3 depicts the initial unit price value variation with the aggregated demand.

Before the optimization, all the appliances are scheduled to run on their preferred start times. Thus, the *Total cost* is equal to the *Total bill*, as no penalty needs to be applied.

$$Total\ cost = Total\ bill = ad_{i(init)} \times up_{i(init)} + ad_{j(init)} \times up_{j(init)}$$

The goal of the optimization is to minimize the *Total cost* by adjusting the starting time of the devices. Thus, the total demand of the community remains the same. Now, as an example assume that the optimization only moves load between pricing slots p_i and p_j without curtailments. Then,

$$ad_{i(init)} + ad_{j(init)} = ad_{i(opt)} + ad_{j(opt)} \tag{15.7}$$

The optimization oracle is capable of calculating the aggregated demand values $ad_{i(opt)}$ and $ad_{j(opt)}$ such that *Total cost* is minimized. We assume that, given feasible, the oracle can find optimal values for $ad_{i(opt)}$ and $ad_{j(opt)}$ by moving α: $(\alpha > 0)$ amount of demand from the second pricing slot p_j to the first pricing slot during the optimization as depicted in the right-side graph in Fig. 15.3.

Then, the followings hold for the optimization.

$$ad_{j(opt)} = ad_{j(init)} - \alpha$$
$$ad_{i(opt)} = ad_{i(init)} + \alpha \quad \because \text{Eq. (15.7)}$$
$$\therefore \quad ad_{i(opt)} > ad_{i(init)} \quad \text{and} \quad ad_{j(opt)} < ad_{j(init)}$$

Due to the monotonic growth of the price, we can conclude the following for the unit prices.

$$up_{i(opt)} > up_{i(init)} \quad \text{and} \quad up_{j(opt)} < up_{j(init)}$$

And *Total penalty* > 0 as a set of devices' start times differ from their preferred start times.

Now let us consider the influence of this change to the total costs, as expressed by the following expression:

$$\begin{aligned}
\Delta cost &= Total\,cost_{(init)} - Total\,cost_{(opt)} \\
\Delta cost &= Total\,cost_{(init)} - (Total\,bill_{(opt)} + Total\,penalty) \\
&= ad_{i(init)} \times up_{i(init)} + ad_{j(init)} \times up_{j(init)} \\
&\quad - (ad_{i(opt)} \times up_{i(opt)} + ad_{j(opt)} \times up_{j(opt)} + Total\,penalty)
\end{aligned}$$

Replacing the optimized demand with the initial demand $\pm\alpha$ leads to this equivalent expression.

$$\begin{aligned}
&\Rightarrow ad_{i(init)} \times up_{i(init)} + ad_{j(init)} \times up_{j(init)} \\
&\quad - ((ad_{i(init)} + \alpha) \times up_{i(opt)} + (ad_{j(init)} - \alpha) \times up_{j(opt)} + Total\,penalty)
\end{aligned}$$

This is equal to

$$\begin{aligned}
&\Rightarrow ad_{i(init)} \times (up_{i(init)} - up_{i(opt)}) + ad_{j(init)} \times (up_{j(init)} - up_{j(opt)}) \\
&\quad + \alpha \times (up_{j(opt)} - up_{i(opt)}) - Total\,penalty
\end{aligned}$$

The initial assumption was that the optimization oracle will lead to the perfect result. Thus, it needs to produce the optimal values for $ad_{i(opt)}$, $ad_{j(opt)}$, and α for given unit prices $up_{i(opt)}$, $up_{j(opt)}$ so that the $\Delta cost > 0$. Thus,

$$\begin{aligned}
&ad_{i(init)} \times (up_{i(init)} - up_{i(opt)}) + ad_{j(init)} \times (up_{j(init)} - up_{j(opt)}) \\
&\quad + \alpha \times (up_{j(opt)} - up_{i(opt)}) - Total\,penalty > 0
\end{aligned}$$

Likewise, for the general case, we assume that the Oracle is capable of finding the optimal demand distribution for any given P pricing slots, which results as follows:

$$\sum_{p=1}^{P} ad_{p(init)} \times up_{p(init)} - \left(\sum_{p=1}^{P} ad_{p(opt)} \times up_{p(opt)} + Total\,penalty \right) > 0$$

s.t.

$$\sum_{p=1}^{P} ad_{p(init)} = \sum_{p=1}^{P} ad_{p(opt)}$$

3 Attack model

3.1 Attack motivation

In this work, we assume that the adversary's motive is to consume a significant amount of energy during a particular timeslot without paying a higher bill (or lower the consumption cost compared to a normal scenario).

3.2 Preliminaries

In this section, we compare two scenarios, where the DR scheme executes with and without FDIAs from an adversary's perspective. We assume that there is a perfect DR optimization oracle as we discussed in the previous section and, all the communications channels are reliable and secure. Further, we assume that all data are securely encrypted throughout the network to ensure that adversaries cannot manipulate or intercept the data. There can be adversaries who want to gain some financial advantages in this system. However, since the communication is secure, adversaries cannot intercept (and alter) the price signals or other demand profiles. Thus, traditional FDIA approaches cannot impact the DR scheme. Nevertheless, given that adversaries are also participants in the system, they can falsify data that they input into the DR scheme via smart devices, smart meters, HEMSs, or in combinations. These can be either their systems within their direct control or compromised systems located at another household.

Let us assume that our perfect Oracle uses either a linear or a quadratic unit price function. It should be noted that a linear increase models a quite modest price increase. Demand peaks induce high costs for production and distribution as both need to be prepared to react quickly to demand changes and need to ensure that sufficient energy reserves are available. However, though the attack works for the linear, quadratic, and step-wise cases, we only focus on the linear case where demand and price are proportional.

$$up_p \propto ad_p$$

For the optimization as well as the attacks to work, we need a few more assumptions on the flexibility of demand (i.e., can sufficient demand be moved to other pricing slots) and on potential penalty costs (i.e., is the gain in a unit price higher than the penalty). Note that these assumptions constitute some of the primary preconditions for DR schemes to make a difference for the efficiency of energy distribution. More formally, we assume the following. Let δ_{init} be the initial demand difference between two pricing slots i and j. We assume that there is a threshold demand value γ such that if $\delta_{init} > \gamma$, the optimization oracle will move to demand such that the optimized difference β_{opt} is now smaller than γ. As the optimization oracle minimizes overall costs, this also expresses that the penalty for moving demand from the preferred pricing slot is not higher than the gain in energy costs through the lower unit prices.

3.3 Attack-free scenario

Let us denote $ad_{p(init)(attack\text{-}free)}$, $ad_{p(opt)(attack\text{-}free)}$, and $ad_{real(attack\text{-}free)}$ as the initial aggregated demand of the pth pricing slot before the optimization, the optimized aggregated demand value for the pth pricing slot after the optimization, and the real demand consumption of the pth pricing slot, respectively. Similarly, we denote unit price values $up_{p(init)(attack\text{-}free)}$, $up_{p(opt)(attack\text{-}free)}$, and $up_{real(attack\text{-}free)}$. Further, $Total\ bill_{(init)(attack\text{-}free)}$, $Total\ bill_{(opt)(attack\text{-}free)}$, and $Total\ bill_{(real)(attack\text{-}free)}$ denote the initial total bill before the optimization, optimized total bill after the optimization, and the real total bill, respectively.

Similar to the optimization example, let assume that $P = \{2|i, j\}$ and the optimization oracle moves α amount of demand from the jth pricing slot to ith pricing slot. Then, once the oracle converges to a solution, the optimized aggregated demand value of the jth pricing slot would be computed as

$$ad_{j(opt)(attack-free)} = ad_{j(init)(attack-free)} - \alpha \tag{15.8}$$

If the system can preserve the integrity of data and all the users follow their schedules, optimized aggregated demand value and unit price value of the jth pricing slot should be equal to the consumption values of the subsequent day.

$$\begin{aligned} ad_{j(opt)(attack-free)} &= ad_{j(real)(attack-free)} \\ \Rightarrow up_{j(opt)(attack-free)} &= up_{j(real)(attack-free)} \end{aligned}$$

Similarly, $up_{i(opt)(attack\text{-}free)} = up_{i(real)(attack\text{-}free)}$ since $ad_{i(opt)(attack\text{-}free)} = ad_{i(real)(attack\text{-}free)}$. Hence, in an attack-free scenario users are able to receive the expected return.

3.4 Attack scenario

In this work, we focus on a more practical and strategic attack scenario where compromising communication channels are not a necessity to execute the attack. As adversaries are a part of the community, they can manipulate their data that get inputted into the device scheduling DR scheme via HEMSs or IoT devices in their controls. However, being able to compromise a set of other HEMSs is an added advantage for adversaries as it hinders the detection.

Let us assume that the adversary wants to operate their demand on jth slot. Thus, they inject some false demand into the jth slot such that

$$ad_{j(init)(with-attack)} = ad_{j(init)(attack-free)} + \phi \tag{15.9}$$

where $\phi(\phi > 0)$ is the amount of false demand, and everything else remains identical to the attack-free scenario.

Then

$$\begin{aligned} ad_{j(init)(with-attack)} &> ad_{j(init)(attack-free)} \quad \because \phi > 0 \\ \Rightarrow up_{j(init)(with-attack)} &> up_{j(init)(attack-free)} \end{aligned}$$

Since $ad_{j(init)(with\text{-}attack)} > ad_{j(init)(attack\text{-}free)}$,

$$\sum_{\forall P} ad_{p(with-attack)} > \sum_{\forall P} ad_{p(attack-free)}$$

Thus,

$$\therefore \; Total\,bill_{(opt)(with-attack)} > Total\,bill_{(opt)(attack-free)}$$

However, the optimization oracle tries to minimize the $Total\ bill_{(opt)(with\text{-}attack)}$ as much as possible. If the adversary keeps maintaining the false demand ϕ in the jth slot, the optimization must move some of the appliances that are scheduled in the jth pricing slot to slots with lower demand (this would be ith slot according to our case) to reduce the $ad_{j(opt)(with\text{-}attack)} \times up_{j(opt)(with\text{-}attack)}$, which results in a minimized $Total\ bill_{(opt)(with\text{-}attack)}$. Note that, as explained previously, we assume the penalty for flexibility is smaller than the benefit in unit price. If this was not the case, the overall costs could not be used as a target for optimization, as high penalties would remove the incentive to reschedule demand.

Let us assume the oracle moves β amount of demand from jth pricing slot to ith slot. Therefore, once the oracle optimized the distribution,

$$ad_{j(opt)(with-attack)} = ad_{j(init)(with-attack)} - \beta \tag{15.10}$$

From Eqs. (15.9), (15.10)

$$ad_{j(opt)(with-attack)} = ad_{j(init)(attack-free)} + \phi - \beta \tag{15.11}$$

However, ϕ amount of demand will not be consumed on the real run as that is the injected fake demand. Hence, $ad_{j(opt)(with\text{-}attack)}$ can be deviate from $ad_{j(opt)(attack\text{-}free)}$. This deviation produces results in favor of the adversary where the demand of attacked slot is lower than the attack-free scenario, given that majority of the houses are following their predicted schedules. We have proved and shown in detail that attackers could potentially achieve lower bill independent from the optimization algorithm in our previous work [37].

4 Experiment

We adopted the algorithm introduced by He et al. [36] to analyze the impacts of the depicted FDIAs on distributed device scheduling DR systems, which uses the abstract model that we explained in Section 2. Further, the selected DR scheme optimally

Time of the day	0000–0029 hours			0030–0059 hours			...	2330–2359 hours		
Pricing slot	1			2			...	48		
Scheduling slot	1	2	3	4	5	6	...	142	143	144

Fig. 15.4 Mapping of individual slots.

schedules devices in a highly efficient and scalable manner with limited information exchange between the households and the UC reducing trivial privacy concerns. Moreover, the code of this algorithm is publicly available. However, the attack is not specific to this particular algorithm. Any distributed DR system with the goal of optimal demand distribution is vulnerable against FDIAs [37].

The adopted scheme divides a day into 48 thirty-minute pricing slots ($P = 48$), where each pricing slot consists of 3 ten-minute scheduling periods ($k = 3$), which produces 144 scheduling slots ($S = 144$) that are mapped as shown in Fig. 15.4. The scheme is divided into a local optimization and a master problem. Each HEMS solves the local optimization problem of minimizing household bill while the UC solves the master problem of minimizing overall community cost. HEMSs decide optimal start time for each device depends on the price and the penalty, to minimize the total cost. Constructed demand profiles are sent to the aggregator to accumulate and produce an aggregated profile. UC solves the master optimization problem of minimizing community cost to derive the optimal price signal for the give aggregated demand distribution. The calculated price signal is received back by HEMSs for rescheduling. This process iterates until the system converges to the global optimal where HEMSs cannot move devices further to reduce their total costs. Further details of this scheme are discussed in [36].

As explained in the abstract DR model, both optimization and scheduling use day-ahead pricing. However, real-consumption values or RTP is used for billing purposes instead of predicted or optimized demands. We assume that households follow their schedules once the system converges. The adversary can operate their devices at any timeslot of the day. However, the adversary prefers to use their appliance(s) at a peak time. Thus, to reduce costs, the attacker aims to get a lower unit price during the selected/attacked period. We rationalized this assumption as adversaries able to lower their cost by operating devices during off-peak times even without any attacks as off-peak unit prices are lower than of peak-time prices.

4.1 Dataset

Although there are real-world datasets that contain energy consumption data such as Pecanstreet dataset, there is no single real-world device scheduling dataset as decentralized device scheduling DR algorithms are its inception. Therefore, methods have been introduced to generate synthetic datasets [13, 38] that can be used in device scheduling experiments. Thus, in this work, we generated a realistic dataset with the combination of the synthetic dataset generation approach mentioned in [36] with real consumption data from the Pecanstreet dataset [39].

We have extracted the energy consumption data of 168 houses in Austin, Texas. The extracted data comprises minute-wise energy consumption of various appliances

Algorithm 15.1 Consumption to demand mapping for flexible appliances.

```
Input: Minute-wise flexible devices consumption
Output: Appliance demand per each house on each day
for each customer do
    for each day do
        devicesPerCustomerPerDay ← set()
        for each 30 minutes period do
            for each controllable device do
                PLevel ← mean(consumption of 30 minutes)
                if pLevel > 0 then
                    add pLevel to devicesPerCustomerPerDay
                end
            end
        end
        deviceDemands[customer][day] ← devicesPerCustomerPerDay
    end
end
```

operating on each house for 3 months period, between June and August (summer) 2017. Electric cars, washers, combined washer and dryers, electric dryers, gas dryers, dishwashers, jacuzzi, pool 1, pool 2, and pool pump were considered as controllable devices and while all other appliances/consumption were considered uncontrollable. However, Pecan street dataset does not contain any information about the appliance demand. Thus, we need to extract demand values from energy consumption data. Algorithm 15.1 depicts the demand extraction procedure. Initially, we divided the dataset into individual customers. Then, we extracted the average half-hourly consumption values of each controllable appliance as the demand value of that appliance. As an appliance can operate in several demand levels, we have extracted all distinct demand levels from our procedure. The subsequent step is to extract the average controllable and uncontrollable demand value in each 30 min duration of each day per each customer, using Algorithm 15.2. Then the following steps were followed to create datasets for 168 houses for 92-day period.

1. A probability distribution was derived based on the controllable demand profiles obtained from Algorithm 15.2 of each house to generate preferred start times of devices.
2. Devices' duration were sampled using a Rayleigh distribution.
3. Device demands were the values extracted using Algorithm 15.1.
4. Early start time and least finish time were randomly selected following a uniform distribution.
5. Penalty factors were randomly selected with a uniform distribution.

Algorithm 15.2 Extracting controllable/uncontrollable demands.

```
Input: Aggregated minute-wise flexible/inflexible consumption
Output: Flexible/inflexible demand per each house on each day
controllableDemand ← {}
uncontrollableDemand ← {}
for each customer do
    index ← 0
    for each day do
        for each 30 minutes period do
            pLevel ← max(consumption of 30 minutes)
            if controllable then
                controllableDemand[day][index] ← pLevel
            end
            else
                uncontrollableDemand[day][index] ← pLevel
            end
            index ← index + 1
        end
    end
end
```

Uncontrollable demands of each customer per each day were incorporated via hypothetical devices, which are assumed to be running for each 30 min duration (48 uncontrollable devices per customer per day). Fig. 15.5 compares the real and synthetic demand values for few days. As can be seen from the figure, the synthetic data that we generated are closely aligned with the corresponding real-demand distribution. Thus, our dataset is more realistic. The price table was derived using the National Energy Market wholesale spot prices [36], where 30 distinct unit price values were available in each price slot.

4.2 Process

We choose one household and the controllable device with the highest demand of the selected house as the adversary house and the appliance that the adversary wants to operate, respectively. We assume that the adversary operates the chosen appliance during a peak scheduling slot (peak time falls between 3 and 7 p.m. every day during the summer of Texas) for a previously determined duration. However, it is worth noting that the exact appliance is irrelevant to the experiment. We set the preferred and early start time of the device to the peak scheduling slot of the day and the latest finish

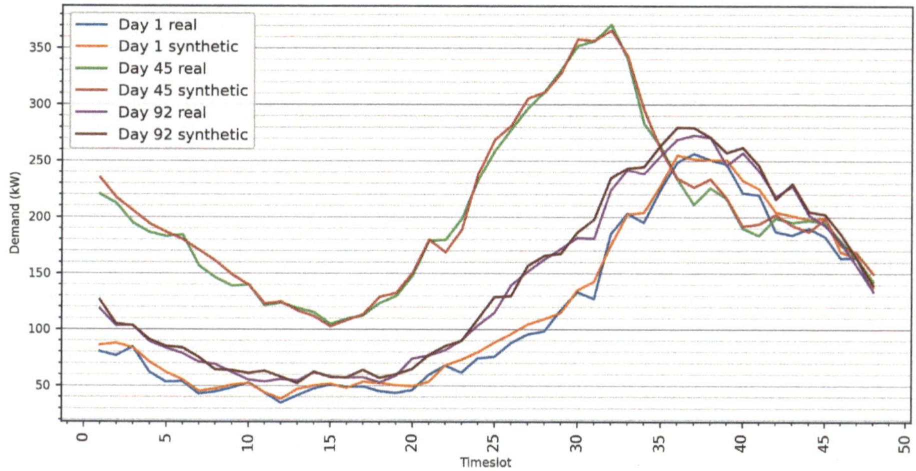

Fig. 15.5 Comparison between real and synthetic demand.

time to map with the operating duration of the appliance. Then we let the optimization to schedule appliances and, calculated the adversary's cost of the usage for the selected device, the total bill for the adversary (considering both controllable and uncontrollable appliances), the total bill for the community, and the total penalty for the community once the optimization converged to an optimal solution. The obtained results depict the standard scenario where no FDIAs in the system. The optimal demand distribution in the attack-free scenario and the calculated values were used as the base values to compare the results of different attack scenarios.

Next, we inject 0.1% (of the total demand of the community) of false demand into the system by increasing the selected device's demand. We obtained the optimized demand distribution and deducted the injected fake demand from the optimized distribution to get the real distribution once the system converged. The obtained demand distribution was used to calculate the price signal. Attacker's device bill, full bill, the total bill, and the total penalty for the community were calculated using the new price signal. These values depict the outcome of an attack scenario. Subsequently, the false demand percentage was gradually increased (up to 5%) in each attack round and repeated the same steps to calculate the attack impact. We used the optimal/real solutions in all calculations. However, due to the random choice of households, actual schedules might slightly differ from the optimal solution. Thus, we repeated the steps with 10 randomized schedules with each attack scenario and averaged those results to obtain a more realistic view of the attack impact. We executed 1104 FDIAs (12 attacks per each day with different injection percentages). The same set of steps were repeated with the assumption that the attacker is operating their appliance on nonpeak timeslots. FDIAs on nonpeak timeslots were used to obtain a more generalized view of the attack impact.

Both FDAIs on peak and nonpeak timeslot experiments were repeated with a different number of households (50 and 100 houses) in the community to analyze the impact of an attack with the community size. Further, the full-flexible household count was gradually increased between 5% and 100% and, the adversary's bill reduction percentages were recalculated using the same steps. The resulted bill values were used to analyze the impact of demand flexibility on FDIAs. The impact of PV battery penetration with FDIA was also analyzed. We assumed that 1% of the community (1.6 houses which we used as 1) is having a fully charged 12 kW PV battery in their households. The HEMSs can decide the optimal discharging schedule for battery depending on the price signal through the optimization. We calculated the adversary's device bill once the optimization converges. Consequently, we injected fake demand into the system, similar to other attack scenarios and recalculated the cost values. The PV battery percentage of the community was gradually increased up to 100% and, the device bill was recalculated with each attack scenario.

We have also analyzed how an adversary's bill reduction impacted when multiple adversaries execute FDIAs without coordination using the experiments with an increasing number of adversaries. We set the attacked timeslot of uncoordinated adversaries to be on a timeslot overlapping (one scheduling slot earlier) with the initial adversary's attack, adjacent to the initial adversary's attack (10 scheduling slot earlier) and apart from the attacked slot (50 scheduling slot earlier).

5 Results

In Fig. 15.6, the blue color solid line and the green color dashed line depict the initial demand distributions (before optimizing) of a no attack and with a 0.3% of false demand injection on 36th timeslot (attack executed on the 110th scheduling slot, which is mapped to 37th pricing period and the device runs for one scheduling slots), respectively. The optimized demand distribution of the no-attack scenario is shown in orange color, which is also equivalent to the actual demand distribution as all households follow their forecasts. The optimal solution induced with the FDIA is shown in the red color. The actual demand is less than the predicted demand in attack scenario as fake demand will not be consumed. Thus, the real demand distribution shown in the red color was obtained by subtracting the injected false demand from the optimized demand curve. The actual consumption is lower than the predicted value during 37th pricing slots. However, predicted and actual consumption values remain the same for

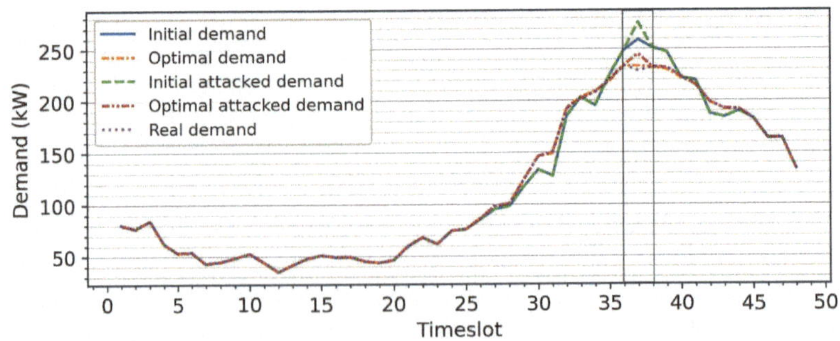

Fig. 15.6 Demand variation without attack, during attack, and actual scenarios.

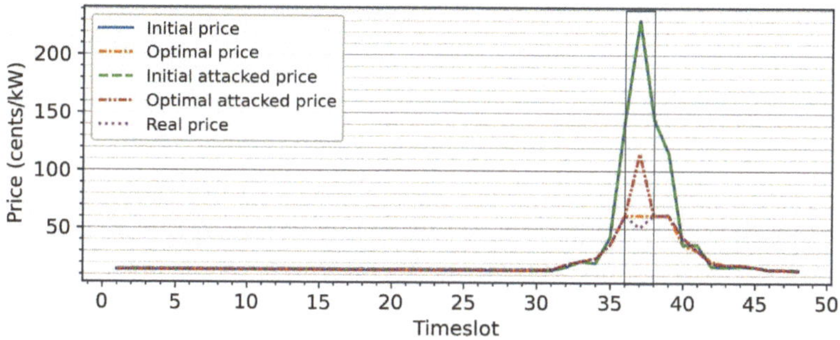

Fig. 15.7 Unit price variation without attack, during attack, and actual scenarios.

all other timeslots since other users follow their predicted schedules. Fig. 15.7 shows the price signal variation for the same attack. The optimization changes the initial price signal shown in blue color solid line to the orange color optimal price signal. The FDIA resulted in the optimized price signal shown in the red color line. However, the actual price signal shown by the purple color line is lower during that 37th slot as the real consumption is lower than the expected due to the FDIA. We calculated the actual price signal based on the actual consumption using the pricing table. Based on the price signals, we calculated the electricity cost for the adversary in each scenario.

Table 15.1 depicts the changes in demand and unit price values of the 37th timeslot due to an FDIA. Without any false demand, the expected optimal demand value was 233.239 kW. Injection of 0.3% of false demand into the system caused the optimized demand value for the same slot to be 244.887 kW. However, the real demand for the timeslot was 228.379 kW as the fake demand was not consumed. Consequently, the real/billed unit price value for the slot was dropped from $1.15 to 24.5 cents. The adversary received a 17.37% reduction in the cost of using the selected devices. Further increase in fake demand could not result in more reduction in the adversary's bill as the remaining demand on the timeslot is inflexible. Thus, the adversary is able

Table 15.1 Impact on demand and price values.

	Value
Initial demand (no attack)	258.823 kW
Initial demand (with attack)	275.331 kW
Optimal demand (no attack)	233.239 kW
Optimal demand (with attack)	244.887 kW
Actual demand	228.379 kW
Initial unit price (no attack)	230.4 cents/kW
Initial unit price (with attack)	230.4 cents/kW
Optimal unit price (no attack)	61.6 cents/kW
Optimal unit price (with attack)	115.5 cents/kW
Actual unit price	50.9 cents/kW

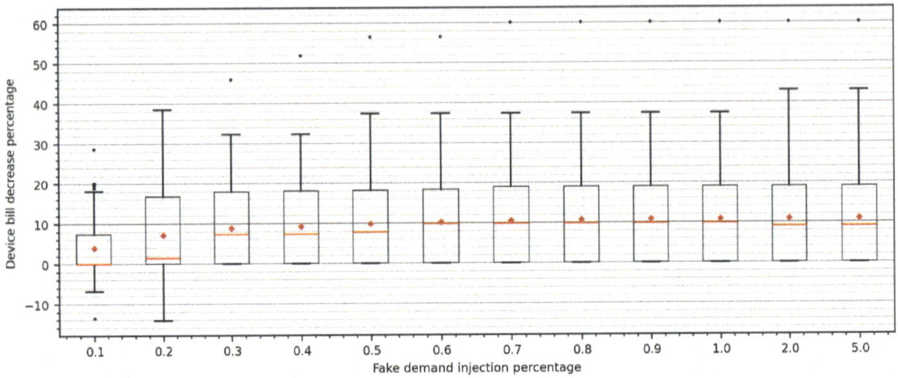

Fig. 15.8 Device bill reduction variation with different false demand percentage during 3-month period (peak).

to obtain the minimum bill for the device with 0.3% of the fake demand injection in this particular attack instance. The adversary's bill reduction percentage variation when the attack executed on a peak timeslot with different false demand injection percentages during 3-month period is depicted in Fig. 15.8. It can be observed that the adversary's bill reduction percentage (on average) increases with the overall false demand percentage and saturated after a certain injection percentage. Comparatively, Fig. 15.9 depicts the adversary's bill reduction percentage variation with the same fake demand injection percentages during 3 months period when attacks are executed on nonpeak timeslots. The adversary was able to achieve more reduction in their bill compared to attacking a peak timeslot. In particular, the average bill reduction percentage among 1104 FDIAs on nonpeak timeslots was 20.07% compared with 9.43% when the attack was on a peak slot, which was more than a 100% increase.

A fake demand injection of below 1% of the overall demand could produce maximum device bill reduction for the adversary in 59 days out of 92 days period on

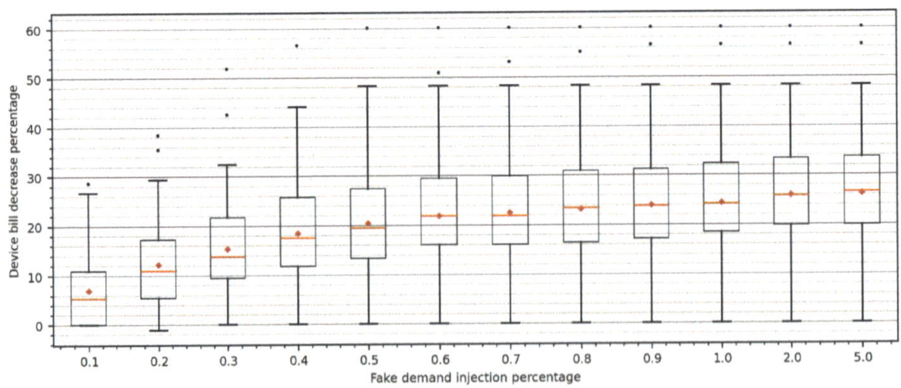

Fig. 15.9 Device bill reduction variation with different false demand percentage during 3-month period (nonpeak).

Table 15.2 Minimum false demand percentage to achieve maximum device bill reduction.

False demand percentage	0–0.5	0.51–1	1.01–2	2.01–5
No. of days—on peak	46	13	6	
No. of days—on nonpeak	23	27	20	19

instances where the attack was executed on a peak timeslot. The corresponding number was 50 when the attack is executed on a nonpeak timeslot. Table 15.2 depicts the number of days that the adversary could obtain the maximum device bill reduction with different amount of false demand injection percentages. There were 27 and 3 days where the adversary could not achieve any bill reduction even up to 5% of fake demand injection on peak timeslots and nonpeak timeslots, respectively. The price signal was at its minimum possible value in those occurrences, given the aggregated demand on attacked timeslot(s) cannot be further moved. It is worth noting that the attacker could have achieved some bill reduction on those instances as well if there were sufficient controllable demand that could be moved.

Figs. 15.10 depicts how the adversary's device bill reduction percentage impacted with the number of household in the community. The left-side graph shows the average bill reduction percentage variation for 3-month period when the attack is executed on a peak timeslot. The right-side graph shows the corresponding results when the attack executed on a no-peak timeslot. Results were calculated using the average device bill reduction for 92-day period under each demand injection percentage. We could observe that the adversary can gain a significant device bill reduction independent to the number of households in the community. In particular, the adversary's gain is higher when the attack is on nonpeak timeslots. Nonpeak timeslots are having less base demand compared to peak timeslots. Thus, the injected fake demand on peaks slots can move more demand away from the timeslots to produce higher reductions. Interestingly, having more number of households in the community can induce the adversary's gain as can be seen from Fig. 15.10. The amount of controllable

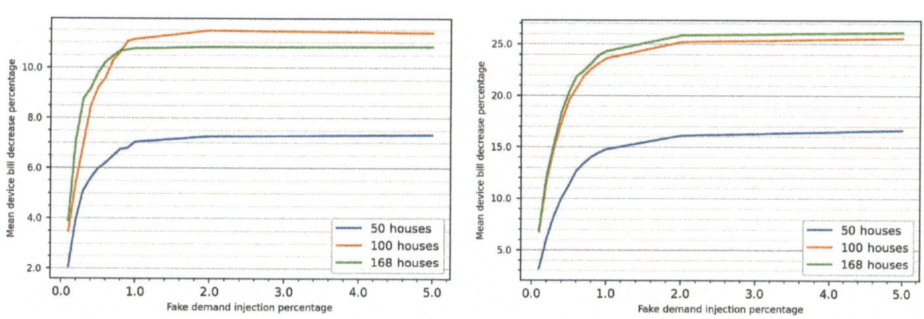

Fig. 15.10 Mean device bill reduction variation with different false demand percentage and with different number of households during 3-month period for FDIAs on peak and nonpeak timeslots.

demand on the attacked slot can be increased with the number of households as they bring more devices into the system. Thus, the attack can move more demand away from the attacked slot. However, having more households will not always beneficial for the adversary, as they can also increase the base demand.

Fig. 15.11 depicts the adversary's average bill reduction percentage variation with the demand flexibility of the community. In the box plot, each box summarizes device cost reduction percentages for 1012 attack scenarios for 92-day period. Under normal circumstances, where the full-flexible household percentage is 0% (all devices' of each household are constrained with some starting and finishing times), the adversary's average cost reduction percentage was 9.43% and, the median was 7.23%. However, when 5% of the community (8 houses) have full-flexible demand, the mean and median cost reduction percentages were 28.66% and 31.05%, respectively. The values further increased up to 56.77% and 60.8% when all households are full flexible. Having a higher amount of flexible demand in the system allows the optimization to move more devices away from the timeslots where the unit price is high. Thus, the unit price spike triggered by injected demand induces the optimization to schedule more devices away from the attacked slots to reduce the overall cost. However, the adversary cannot lower their bill over a certain threshold (depend on base demand and unit price) as minimum unit price value is always positive. The detailed summary of the attack impact with both injection percentages and flexibility is shown in Fig. 15.12. It can be observed that having higher flexibility in the community can significantly increase the adversary's gain for higher injection percentages. In particular, the mean and median reduction percentages were 52.81% and 53.36%, respectively, for 1% of fake demand injection.

The adversary's average bill reduction percentage variation over PV battery penetration in the community is shown in Fig. 15.13. The adversary could achieve more bill reduction when more households are having PV batteries similar to increase in

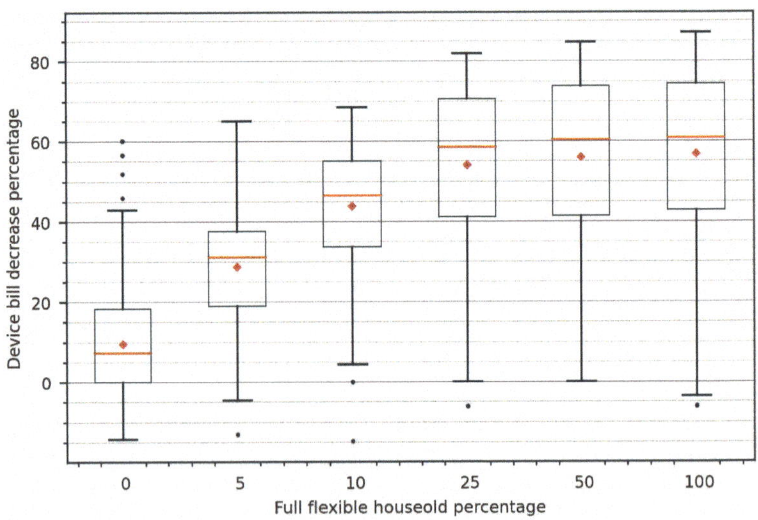

Fig. 15.11 Average device bill reduction variation with the flexibility of the community.

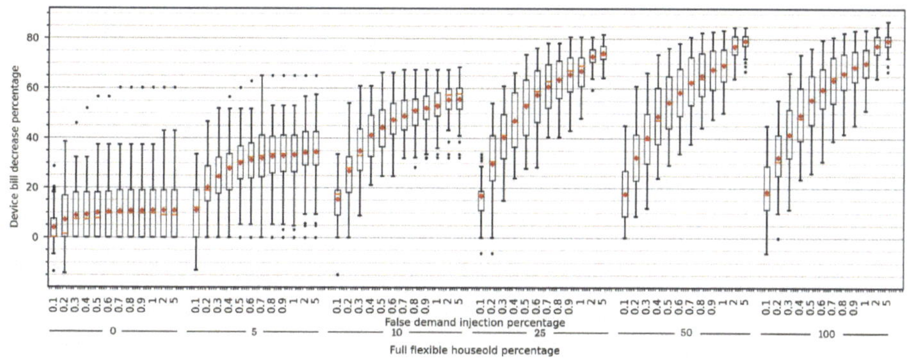

Fig. 15.12 Device bill reduction variation with different false demand percentage during 3-month period with the flexibility of the community.

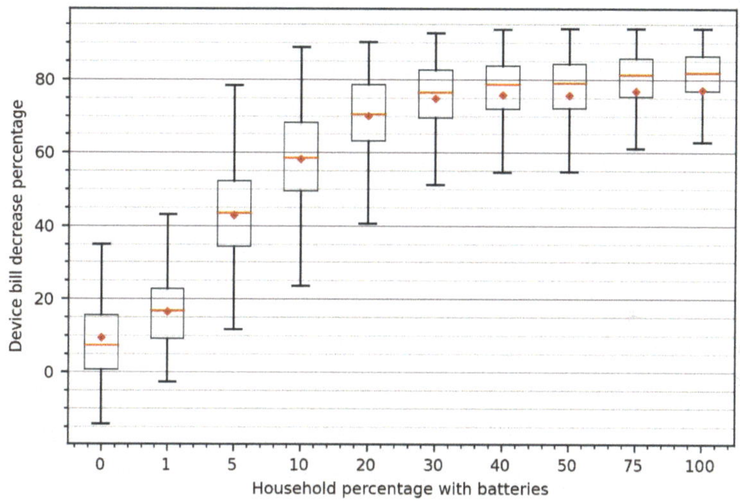

Fig. 15.13 Device bill reduction variation with the percentage of PV batteries.

gain with higher flexibility in the community. HEMSs use PV batteries during attacked timeslots as the projected unit price is high. The demand on the central grid decreases when more households use their batteries during attack instead of depending on central supply. Thus, the real unit price is lower for the attacker.

Fig. 15.14 depicts how the adversary's device bill reduction percentage varies when multiple attackers are in the system who are executing FDAIs in overlapping timeslots with the initial attack. Starting from the left, the adversary injects no fake demand, 0.1% and 1% of fake demand into the system, respectively. In each scenario, the number of attackers and their individual fake demand injection percentages are shown on the *X*-axis. The *Y*-axis shows the initial adversary's device bill reduction percentage where negative values represent an increase in the bill. As can be seen from the left-side box plot in Fig. 15.14, the initial adversary's device bill decreases with the increase in fake demand when a single adversary executes an FDIA. However, the

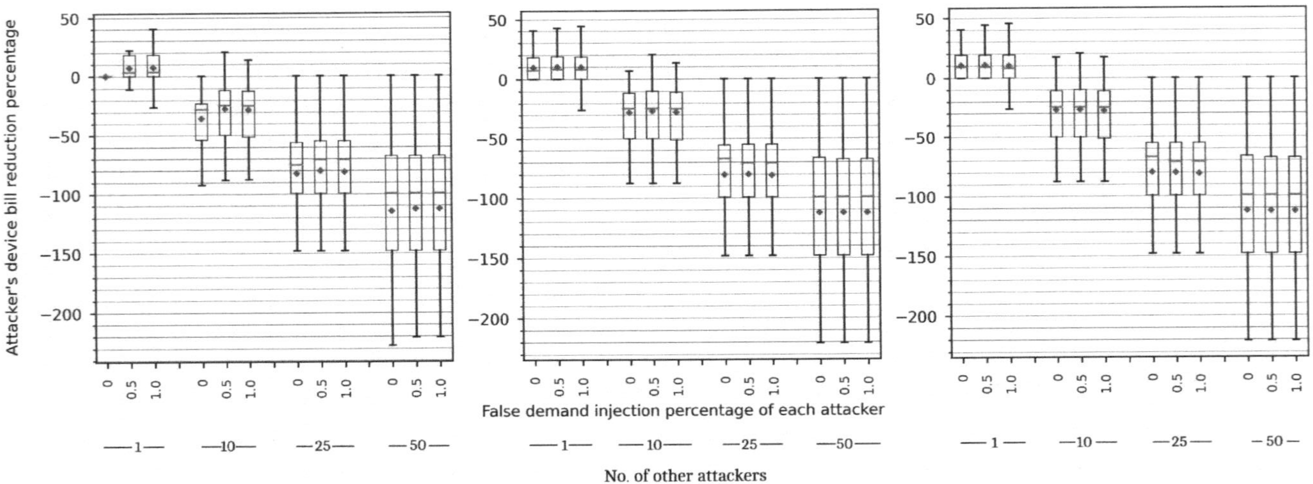

Fig. 15.14 Device bill reduction variation with multiple attackers when executing FDIAs in overlapping timeslots.

initial adversary's bill increases when 10 other adversaries execute FDIAs in overlapping timeslots, though the increase in bill marginally decreases with the fake demand injection percentages. On average, the adversary's bill almost doubled when there are 25 uncoordinated attackers in the system and, it further increases with the number of adversaries. The similar observation can be made with the increase in initial adversary's fake demand percentage as shown in the middle and the right-side box plots of Fig. 15.14. The increase in the initial adversary's device cost can be observed even without any fake demand in the system as the higher number of attackers consolidate more uncontrollable load in the attacked timeslot making the demand of the timeslot significantly high.

Unlike in overlapping uncoordinated attackers scenarios, the initial adversary did not receive any significant bill reduction when a single adversary injects fake demand close to the originally attacked timeslot. However, as shown in the left-side box plot in Fig. 15.15, the adversary's bill increases when more attackers execute FDIAs close to the original timeslot. The increase in the demand of secondary attacked slots (timeslots where other attackers execute FDIAs), moves some of the flexible load (depend on their flexibility) toward the initial attack slot. Thus, the initial adversary's bill increases. Nevertheless, the initial adversary is able to achieve a decrease in their cost with an increasing amount of fake demand injection, as shown in the middle and the right-side box plots of the figure.

In contrast, the initial adversary's bill decreases with the number of adversaries when more adversaries execute FDIAs away from the original timeslot, as shown in the left-side box plot of Fig. 15.16. The demand in the originally attacked slot decreases as attackers move their loads (that are scheduled to run in the original attacked slot) away from the timeslot. Thus, the unit price decreases even if the attacker does not inject any fake demand. Similar to the previous scenario, the initial adversary able to further reduces their cost with an increasing amount of fake demand injection, as shown in the middle and the right-side box plots of the figure.

6 Discussion

Table 15.3 depicts a portion of one of a pricing lookup table that used in our experiments. As shown in the table, a small increase in the aggregated demand can cause a significant increase in the unit price due to the quadratic characteristics of the unit price function. When an attacker injects false demand into the system, the aggregated demand of the attacked timeslot(s) increases. Once the demand exceeds the boundary of a given demand level, the new unit price can become very high. Consequently, the costs of the households that have scheduled devices in attacked timeslots increases. This cost increase triggers HEMSs to move some of the devices (that have been scheduled in attacked timeslots) to other slots, to minimize the cost. However, the actual demand of the attacked timeslot(s) will always be lower than the expected optimized demand value due to the movement of devices away from the attacked slot(s). Therefore, the actual unit price(s) for the attacked timeslot(s) will be lower than the expected unit price. Hence, the attacker will gain a lower usage cost and less inconvenience by

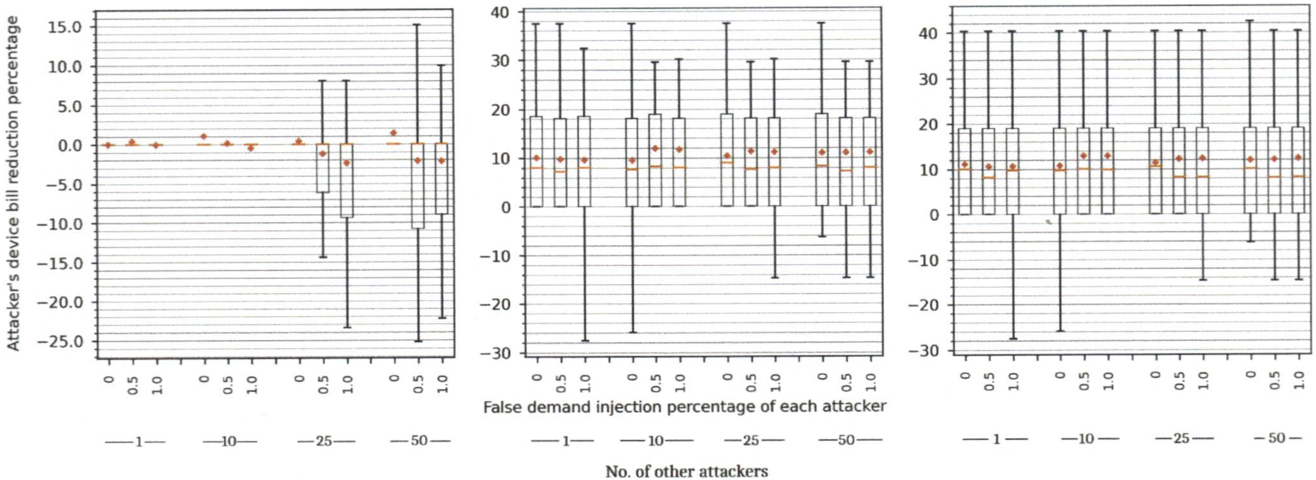

Fig. 15.15 Device bill reduction variation with multiple attackers when executing FDIAs near the original timeslot.

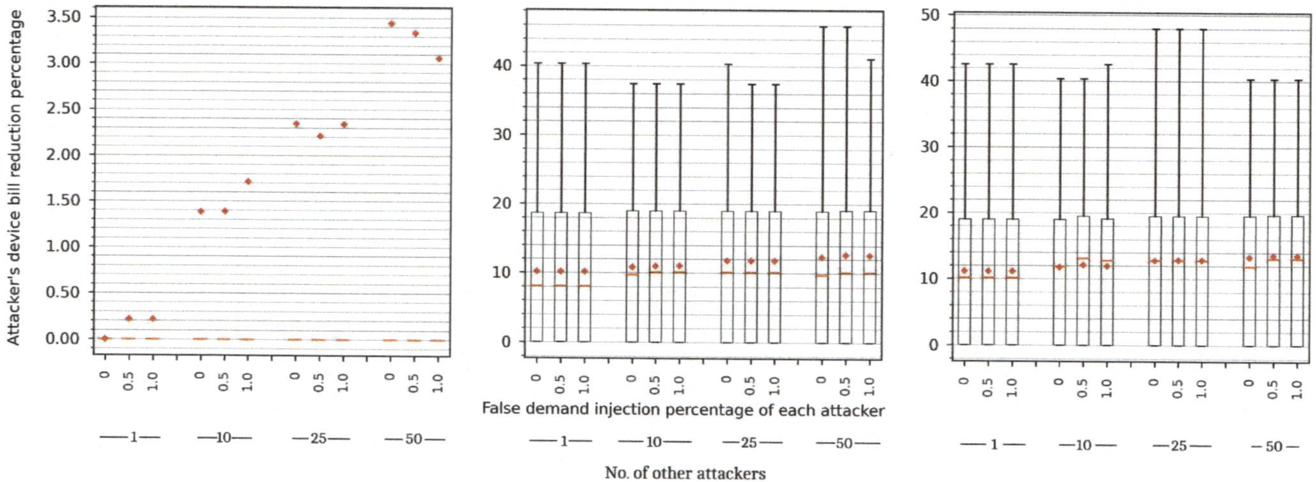

Fig. 15.16 Device bill reduction variation with multiple attackers when executing FDIAs in away from the original timeslot.

Table 15.3 Sample price table

Generation	Unit price (cents/kW)		
	Timeslot 1	**Timeslot 2**	...
0–224.39 kW	14.1	14.1	...
224.4–228.79 kW	14.2	14.2	...
...	...	...	...
...	...	...	...
396–400.39 kW	75.2	75.2	...
400.4–409.19 kW	92.8	92.8	...
...	...	...	...
426.8–431.19 kW	182.1	182.1	...
431.2 kW or more	230.4	230.4	...

running their devices in the attacked timeslot(s). However, it is worth noting that having less number of distinct price levels can impact both the adversary's gain and the overall performance of the DR scheme, compared to having more distinct price levels, as small demand fluctuations cannot cause alterations in price levels with fewer levels. Adversaries can gain a bill reduction irrespective of the number of users, the flexibility of demand, the use of PV batteries or attacked slots, as they are capable of injecting a sufficient amount of false demand into the scheme. Nevertheless, adversaries incur an increase in device bill when multiple adversaries execute similar attacks in an uncoordinated manner in overlapping timeslots as real demand increases with the number of adversaries. However, adversaries are still able to obtain a bill reduction, if the number of adversaries is not significantly high and the attacking timeslots are not overlapped with each other.

There is a highly inverse correlation between the false demand percentage and device consumption cost for the adversary. However, a similar correlation could not be observed in the total bill, since the adversary operating other controllable and uncontrollable appliances during different timeslots. Nevertheless, the adversary could operate more appliances during the attacked timeslot(s) to reduce the total bill. In the dataset, the daily average total controllable appliance demand percentage was 7.3% among the 168 houses during 92-day period, which is a very low value, as shown in Fig. 15.17. Having low controllable demand affects the adversary's gain as FDIA cannot move the inflexible demand. However, with the penetration of the smart house concept, more appliances will be used as controllable appliances in the future. A higher percentage of controllable demand yield to higher bill reduction with lower false demand injection for adversaries.

Any user can execute this FDIA without much effort as the strategy does not involve any tampering or compromise of the communication channel. Malicious users can easily manipulate HEMS' data since HEMSs reside within their premises. Further, additional false demand can be inputted to the system in many forms, other than

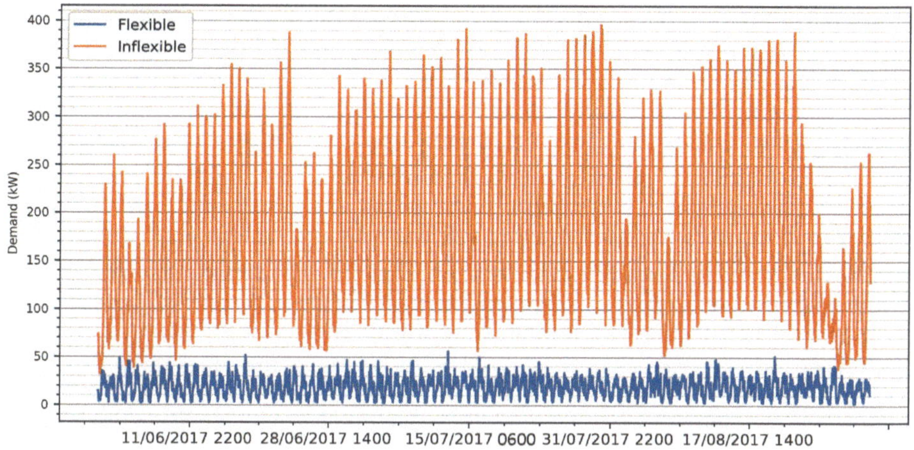

Fig. 15.17 Controllable and uncontrollable demand of each half-hour period of 92 days.

changing the demand value of a selected device. Nevertheless, as no actual security breach is needed, that is the easiest option for the attacker. Other attack possibilities with minimum effort are

1. Increase the demand value of single or multiple device(s) in the HEMS.
2. Add one or more fake home appliance(s) to the HEMS.
3. Add one or more real home appliance(s) to the HEMS but disconnect them once the system converges.
4. Collaborate with a set of other households and distribute the false demand partially among those set of households.

As depicted by the abstract DR model, on high level, each household adjusts their demand prediction for the next 24 h to minimize their cost based on the receiving price signal, which is sent by the optimization to minimize the overall cost. Accordingly, any optimization algorithm that produces an optimal demand distribution will remove peaks and produce similar results. The outcome of the optimization mainly depends on how flexible demands are, characteristics of the predicted price and the total cost. Thus, if an adversary can input false demand into the DR system, they are capable of pushing demand into other timeslots and getting cost benefits from the attack independent from the optimization algorithm [37].

The assumption that proper authentication, security protocols, security controls, or sealed/controlled devices can provide security against FDIAs represents a common misconception concerning the security of SGs. The boundaries of these systems are not rigidly defined and, users are in total control over their preferences in HEMS, which is a part of home automation. Utility providers have no control over user preference. As the false data originates from the systems, which are outside the control of the energy provider, utility companies are unable to provide reliable security protocols to remediate these attacks unlike typical pricing cyber attacks, where utility companies can secure the broadcast signals (price, demand) from tampering. Further, there is no guarantee that user preferences remain consistent due to the inherently chaotic nature of energy usage, which makes accommodating typical machine learning-based detection models hard. Further, low false demand percentage makes detection much harder since fluctuations are expected to exist in a typical grid environment due to various reasons. Additionally, the price reduction that a typical user receives via the DR scheme is comparatively higher compared to the impact of a single attack. Therefore, the probability of someone noticing the attack impact is low. However, the attacker can gain a significant amount of collective bill reduction, if the attacker is capable of executing the attack throughout the time without getting detected.

Injecting false demand via a single HEMS might be easier for the utilities to identify such outliers if the individual demand profiles can be isolated. Thus a competent attacker can exploit weaknesses in a set of HEMSs to mount the attack through several houses in parallel where they can distribute the false demand among compromised HEMSs. Additionally, a set of attackers can execute this attack as a coordinated attack with collaboration. Therefore, applying penalties over outliers will not be an effective solution if the false prediction comes from a compromised device/HEMSs. It allows attackers to receive benefits, while the compromised users pay the penalty. Optimally schedule local energy storage mechanisms such as PV batteries can be utilized to alleviate the inconvenience and an increase in cost for affected users when there is an increase in the predicted price signal. We expect that a more resilient solution would need to combine optimization with suitable FDIA detection mechanisms and potential penalties for deviations between the predicted demand and the actual demand. The latter could minimize the cost incentive for the attack. Detailed information on detection and mitigation strategies are discussed in our recent work [40].

7 Conclusions

Using a realistic dataset, we have demonstrated that FDIA can produce a significant cost reduction to the adversary without any technical skills. Compared to some recent work, a significantly low percentage of controllable appliances were used to represents a more realistic real-world scenario. The impact of the attack was analyzed under various circumstances. The adversary was able to gain a significant bill reduction even with a small portion of controllable appliances in the community. Further, we showed that an adversary's gain decreases when the number of uncoordinated attackers executes similar attacks in overlapping timeslots. As small percentages of demand fluctuations are to be expected in most of the scenarios, having reliable and accurate false data detection mechanism is necessary to identify this type of attacks. Reliable FDIA detection mechanisms are inevitable if the DR system shall be beneficial to the community and to achieve the goal to reduce overall costs and improve the efficiency of the grids. Future works need to analyze more generic attack models that can compromise various parameters and processes of a smart grid. Further, a suitable combination of attack detection, correction, and prevention needs to be explored to increase the resilience of DR schemes against FDIAs.

References

[1] US Department of Energy, The SMART GRID: an introduction, 2010, https://doi.org/10.1016/B978-1-59749-570-7.00011-X. Technical Report.

[2] Delivering a smart system consultation on a smart meter policy framework post 2020, 2019. https://assets.publishing.service.gov.uk/government/uploads/system/uploads/attachment_data/file/831734/smart-meter-policy-framework-post-2020-consultation.pdf. Technical Report.

[3] T. Ryberg, L. Östling, B. Insight, Smart metering in North America and Asia-Pacific, Technical Report, Berg Insight, *www.berginsight.com*.

[4] M.H. Albadi, E.F. El-Saadany, Demand response in electricity markets: an overview, in: Power Engineering Society General Meeting, 2007, IEEE, 2007, pp. 1–5, https://doi.org/10.1109/PES.2007.385728.

[5] C. Eid, E. Koliou, M. Valles, J. Reneses, R. Hakvoort, Time-based pricing and electricity demand response: existing barriers and next steps, Util. Policy 40 (2016) 15–25, https://doi.org/10.1016/j.jup.2016.04.001.

[6] J. Giraldo, A. Cardenas, N. Quijano, Integrity attacks on real-time pricing in smart grids: impact and countermeasures, IEEE Trans. Smart Grid 8 (5) (2017) 2249–2257, https://doi.org/10.1109/TSG.2016.2521339.

[7] C. Barreto, A.A. Cárdenas, N. Quijano, E. Mojica-Nava, CPS: market analysis of attacks against demand response in the smart grid, in: ACM International Conference Proceeding Series, vol. 2014, 2014, https://doi.org/10.1145/2664243.2664284.

[8] B.P. Esther, K.S. Kumar, A survey on residential demand side management architecture, approaches, optimization models and methods, Renew. Sust. Energ. Rev. 59 (2016) 342–351, https://doi.org/10.1016/j.rser.2015.12.282.

[9] J.S. Vardakas, N. Zorba, C.V. Verikoukis, A survey on demand response programs in smart grids: pricing methods and optimization algorithms, IEEE Commun. Surv. Tutor. 17 (1) (2015) 152–178, https://doi.org/10.1109/COMST.2014.2341586.

[10] S. He, A. Liebman, A. Rendl, M. Wallace, C. Wilson, Modelling RTP-based residential load scheduling for demand response in smart grids, in: International Workshop on Constraint Modelling and Reformulation, 2014, pp. 36–51.

[11] X. Zhang, X. Yang, J. Lin, G. Xu, W. Yu, On data integrity attacks against real-time pricing in energy-based cyber-physical systems, IEEE Trans. Parallel Distrib. Syst. 28 (1) (2017) 170–187, https://doi.org/10.1109/TPDS.2016.2546259.

[12] N. Li, L. Chen, S.H. Low, Optimal demand response based on utility maximization in power networks, in: Power and Energy Society General Meeting, 2011, IEEE, 2011, pp. 1–8, https://doi.org/10.1109/PES.2011.6039082.

[13] S. Mhanna, A.C. Chapman, G. Verbic, A fast distributed algorithm for large-scale demand response aggregation, CoRR abs/1603.00149 (2016).

[14] A. Hansen, J. Staggs, S. Shenoi, Security analysis of an advanced metering infrastructure, Int. J. Crit. Infrastruct. Prot. (2017), https://doi.org/10.1016/j.ijcip.2017.03.004.

[15] T. Dayaratne, C. Rudolph, A. Liebman, M. Salehi, S. He, High impact false data injection attack against real-time pricing in smart grids, in: Proceedings of the 2019 IEEE PES Innovative Smart Grid Technologies Conference Europe (ISGT-Europe), 2019.

[16] C. Barreto, H. Neema, X. Koutsoukos, Attacking electricity markets through IoT devices, Computer 53 (5) (2020) 55–62.

[17] Y. Liu, P. Ning, M.K. Reiter, False data injection attacks against state estimation in electric power grids, ACM Trans. Info. Syst. Sec. 14 (1) (2009) 1–33, https://doi.org/10.1145/1952982.1952995.

[18] O. Kosut, L. Jia, R.J. Thomas, L. Tong, Malicious data attacks on the smart grid, IEEE Trans. Smart Grid 2 (4) (2011) 645–658, https://doi.org/10.1109/TSG.2011.2163807.
[19] L. Xie, Y. Mo, B. Sinopoli, Integrity data attacks in power market operations, IEEE Trans. Smart Grid 2 (4) (2011) 659–666, https://doi.org/10.1109/TSG.2011.2161892.
[20] L. Jia, R.J. Thomas, L. Tong, Impacts of malicious data on real-time price of electricity market operations, in: Proceedings of the Annual Hawaii International Conference on System Sciences, 2012, pp. 1907–1914, https://doi.org/10.1109/HICSS.2012.313.
[21] S. Amini, F. Pasqualetti, H. Mohsenian-Rad, Dynamic load altering attacks against power system stability: attack models and protection schemes, IEEE Trans. Smart Grid (2016) 1, https://doi.org/10.1109/TSG.2016.2622686.
[22] J. Kim, L. Tong, R.J. Thomas, Dynamic attacks on power systems economic dispatch, in: 48th Asilomar Conference on Signals, Systems and Computers, 2014, pp. 345–349, https://doi.org/10.1109/ACSSC.2014.7094460.
[23] G. Liang, J. Zhao, F. Luo, S.R. Weller, Z.Y. Dong, A review of false data injection attacks against modern power systems, IEEE Trans. Smart Grid 8 (4) (2017) 1630–1638, https://doi.org/10.1109/TSG.2015.2495133.
[24] J. Lin, W. Yu, X. Yang, G. Xu, W. Zhao, On false data injection attacks against distributed energy routing in smart grid, in: Proceedings—2012 IEEE/ACM 3rd International Conference on Cyber-Physical Systems, ICCPS, 2012, pp. 183–192, https://doi.org/10.1109/ICCPS.2012.26.
[25] J. Lin, W. Yu, X. Yang, On false data injection attack against multistep electricity price in electricity market in smart grid, in: GLOBECOM—IEEE Global Telecommunications Conference, 2013, pp. 760–765, https://doi.org/10.1109/GLOCOM.2013.6831164.
[26] Q. Dong, D. Niyato, P. Wang, Z. Han, Deferrable load scheduling optimization under power price information attacks in smart grid, in: IEEE Wireless Communications and Networking Conference, WCNC, 2013, pp. 4683–4688, https://doi.org/10.1109/WCNC.2013.6555333.
[27] P. Wood, S. Bagchi, A. Hussain, Profiting from attacks on real-time price communications in smart grids, in: 2017 9th International Conference on Communication Systems and Networks, COMSNETS 2017, 2017, pp. 158–165, https://doi.org/10.1109/COMSNETS.2017.7945372.
[28] Y. Abdallah, Z. Zheng, N.B. Shroff, H.E. Gamal, T.M. El-Fouly, The impact of stealthy attacks on smart grid performance: tradeoffs and implications, IEEE Trans. Control Netw. Syst. 4 (4) (2017) 886–898, https://doi.org/10.1109/TCNS.2016.2615158.
[29] X. Yang, X. Zhang, J. Lin, W. Yu, X. Fu, W. Zhao, Data integrity attacks against the distributed real-time pricing in the smart grid, in: 2016 IEEE 35th International Performance Computing and Communications Conference (IPCCC), 2016, pp. 1–8, https://doi.org/10.1109/PCCC.2016.7820657.
[30] Y. Liu, S. Hu, T.Y. Ho, Vulnerability assessment and defense technology for smart home cybersecurity considering pricing cyberattacks, in: IEEE/ACM International Conference on Computer-Aided Design, Digest of Technical Papers, ICCAD, vol. 2015, 2015, pp. 183–190, https://doi.org/10.1109/ICCAD.2014.7001350.
[31] L. Yang, H. Shiyan, H.T. Yi, Leveraging strategic detection techniques for smart home pricing cyberattacks, IEEE Trans. Dependable Secure Comput. 13 (2) (2016) 220–235, https://doi.org/10.1109/TDSC.2015.2427841.
[32] G.K. Weldehawaryat, P.L. Ambassa, A.M.C. Marufu, S.D. Wolthusen, A.V.D.M. Kayem, Decentralised scheduling of power consumption in micro-grids: optimisation and security, in: Lecture Notes in Computer Science (including subseries Lecture Notes in Artificial Intelligence and Lecture Notes in Bioinformatics), vol. 10166, LNCS, 2017, pp. 69–86, https://doi.org/10.1007/978-3-319-61437-3_5.

[33] Y. Liu, Y. Zhou, S. Hu, Combating coordinated pricing cyberattack and energy theft in smart home cyber-physical systems, IEEE Trans. Comput. Aided Des. Integr. Circuits Syst. (2017), https://doi.org/10.1109/TCAD.2017.2717781.

[34] C. Joe-Wong, S. Sen, S. Ha, M. Chiang, Optimized day-ahead pricing for smart grids with device-specific scheduling flexibility, IEEE J. Sel. Areas Commun. 30 (6) (2012) 1075–1085.

[35] Y. Ozturk, D. Senthilkumar, S. Kumar, G. Lee, An intelligent home energy management system to improve demand response, IEEE Trans. Smart Grid 4 (2) (2013) 694–701.

[36] S. He, M. Wallace, G. Gange, A. Liebman, C. Wilson, A fast and scalable algorithm for scheduling large numbers of devices under real-time pricing, in: J. Hooker (Ed.), Principles and Practice of Constraint Programming, Springer International Publishing, Cham, 2018, pp. 649–666.

[37] T. Dayaratne, C. Rudolph, A. Liebman, M. Salehi, Inherent vulnerability of demand response optimisation against false data injection attacks in smart grids, in: NOMS 2020–2020 IEEE/IFIP Network Operations and Management Symposium, IEEE, 2020, pp. 1–9.

[38] M. Van Den Briel, P. Scott, S. Thiébaux, Randomized load control: a simple distributed approach for scheduling smart appliances, in: Proceedings of the Twenty-Third International Joint Conference on Artificial Intelligence, AAAI Press, 2013, pp. 2915–2922.

[39] Pecan Street Inc., Dataport, 2019, https://dataport.cloud/.

[40] T. Dayaratne, C. Rudolph, A. Liebman, M. Salehi, Robust demand response for device scheduling under false data injection attacks in smart grids, in: Proceedings of the 2020 IEEE PES Innovative Smart Grid Technologies Conference Europe (ISGT—Europe), 2020.

Toward building decentralized resilience frameworks for future power grids

16

Yaser Al Mtawi[a] *and Anwar Haqu*[b]
[a]Department of Applied Computer Science, The University of Winnipeg, Winnipeg, MB, Canada, [b]Department of Computer Science, Western University, London, ON, Canada

1 Introduction

The ever-increasing electrical load consumption poses a burden on power grids that, along with natural disasters, causes resiliency issues that impact the power distribution to residential and business customers. In recent years, the number of power fault events has increased in many parts of the world. The growth of the total number of incidents over 8 years has a compound annual growth rate (CAGR) of 13% [1]. Power fault events can significantly impact the power grid infrastructures. For example, hurricane Sandy hit the United States in 2012, causing economic damage of around $50 billion; 97 people died, and thousands were displaced from their homes in the great New York area [2]. It affected the power grid leaving more than 8 million customers without electricity [3]. Therefore, it is critical to develop modernized resilient power systems that would seamlessly cope with failures. Power failure incidents cause power disruption and distort operational power paths that deliver the power. These incidents could disconnect several components of the power grid. The whole area will blackout if the existing techniques and procedures cannot reroute bulk power through new alternative paths.

Introducing resiliency in a current power grid is one of the core mandates of next-generation smart grid architecture. Resiliency in a power grid enables electrical distribution companies to maintain their service level objectives. It balances the supply–demand power and overcomes fault incidents with minimal impact. The existing monitoring systems lack agility as they are centralized and require local data to be collected and aggregated from all power grid components. In this regard, a new decentralized and scalable resiliency framework is highly appealing. The proposed framework focuses on two features: (a) real-time monitoring systems based on a cyber-physical network that fully utilizes smart technologies and (b) decentralized distributed control functions for a dynamic autonomous self-healing grid. In doing so, the power grid is divided into cells or blocks (i.e., microgrids). Microgrid application is most suitable to the distribution system, the last mile located close to consumers, the end beneficiaries. Each microgrid should maintain its power demands using renewable energy resources, local generators, and power storage.

Decentralized Frameworks for Future Power Systems. https://doi.org/10.1016/B978-0-323-91698-1.00010-8

The macro- and microlevel of the smart grid will have their own controllers. As you may intuitively guess, a controller at the grid's macrolevel, i.e., macrolevel controller, will play the orchestration over components of the smart grid as a whole and a microlevel controller for each microgrid. A microlevel controller will enable a local solution of all load balancing and voltage issues at an intra-microgrid level. The macrolevel controller should have the whole picture of the smart grid. Consequently, macrolevel controllers must keep the macrolevel controller updated about the state local macrogrid in real-time. Thus, the macrolevel controller will be able to identify microgrids with maximum power generation more than its demand and share them with other microgrids with a load that exceeds the supply.

The first part of this study discusses incorporating smart electrical components such as reclosers, switches, and relays in real-time monitoring of the power grid and isolating the faults when they occur. The second part proposes two-phase resiliency frameworks for microgrids. The first phase aims to assess the resiliency of a power grid. This phase provides important information regarding the weaknesses and the strengths of components in a microgrid. Phase two will process the input of phase one to increase resiliency by employing redundant grid elements such as power lines and poles to back up/strengthen the vulnerable components. Strengthening the structure of microgrids needs considerable budget and time to be accomplished. However, phase two will only be triggered in the case of critical weaknesses. This preventive framework increases the inter-microgrid resiliency and provides an efficient self-recovery policy upon any possible power failure. To achieve this goal, we utilize graph-theoretic metrics, mainly distortion and clustering coefficient metrics, to implement our proposed framework. The proposed framework is implemented, and its effectiveness with link failure events is demonstrated through three IEEE bus systems: 14, 30, and 118.

There are quite a few proposals of schemes to assess and boost the resiliency level of power grids. However, many of them are either not applicable outside the context in which they are formulated, implemented under certain conditions, or focused on proactive islanding of power segments of higher failure probability under specific weather parameters [1].

Corrective actions can be deployed during and postnatural disasters to enhance power grid resiliency [1]. This approach will physically strengthen the power grid structure and increase resiliency. Examples include elevating substations and maintaining underground power lines. Thus, this resiliency approach is limited because it provides a reactive policy without an insightful analysis of the structural characteristics of the power grid.

A highlight on the current status of shifts and changes of the power grid toward sustainability, resiliency, and affordability was addressed in Ref. [2]. A proactive procedure was proposed in Ref. [3] to assess and boost grid resiliency. However, it did not provide any structural analysis and no procedure to harden vulnerable segments. Other proposals include a multi-microgrid system that provides resiliency by diversifying power sources [4–6].

The security perspective of the smart grid is also another direction for resiliency enhancement [7]. Critical infrastructures of a smart power grid can be modeled from a security perspective. Accordingly, an adaptive islanding scheme to achieve the self-healing concept will be triggered upon a cyberattack. This is similar to the formation of microgrids. However, this work did not assume the existence of distributed generation (DG); they depended on the characteristics of the graph representing the power network to fragment the network into connected components. Unlike the works presented in Refs. [5,6], the work in Ref. [7] used a graph-theoretic heuristic to partition the network and isolate the fault to avoid an impending blackout.

A new CC-based metric was proposed in Ref. [8] for identifying influential spreaders in complex networks. The proposed algorithm has near-linear time complexity and can avoid selecting node spreaders that are too close to each other. On the other hand, Chen et al. [5] proposed mixed-integer linear programming (MILP) to obtain a resilient distribution system that maximizes the power delivered to different nodes after a natural disaster. Self-healing systems based on microgrids were proposed by many research works such as in Refs. [6, 7, 9].

Although there is a rich body of the literature that addresses power grid resiliency, yet no work has studied resiliency analysis, evaluation, and enhancement proactively (i.e., prior to fault incidents) in the context of decentralized perspective using structural metrics such as distortion and clustering-coefficient metrics under link failure probability distribution (i.e., operational resiliency). There is an urgent need to provide detailed resiliency characteristics using sophisticated tools that use topological and functional features of the smart grid. In this sense, link failure probability needs to be employed to study the operational lifetime of a power grid.

In this chapter, we propose a decentralized approach where a macrolevel controller will be able to enhance resilience at the microgrid level. To this end, we address two main parts: incorporating smart devices to collect electricity-related data from all over the power grid and implementing the two-level decentralized resiliency framework. Note that partial technical materials from our two recent articles [10,11] will contribute to this chapter. The contributions are as follows:

(1) Reinforce communications over smart grids by incorporating smart devices in all grid components. This will enable forming a clear picture of the status of each microgrid and a smart grid as a whole, thus, effectively orchestrating a macrolevel controller.
(2) Proposes novel two-phase frameworks to assess and enhance the resiliency in the locality of each microgrid.
- The resiliency assessment: The assessment of resiliency to identify resiliency-related weaknesses in a microgrid.
- The resiliency enhancement: The weaknesses identified in phase one will be backed up by placing extra/redundant elements.

(3) Provides an in-depth analysis of resiliency metrics with a particular focus on distortion and clustering coefficient metrics.
(4) Study detailed distortion and clustering coefficient-based resiliency characteristics with link failure that follows a Nakagami probability distribution function (PDF) [12].
(5) Provide several algorithms to assess microgrid resiliency and enhance it based on the analyses of distortion clustering coefficient metrics.
(6) Comprehensively evaluate and compare the overall power grid resiliency assessment before and after applying our proposed framework. We conduct the experiments on IEEE 14-, 30-, and 118-bus.

The organization of the proposed chapter is as follows. Section 1 is devoted to the Introduction in which the motivations behind the decentralized resiliency framework will be established. Power network modeling will be addressed in Section 2; each microgrid is modeled as a graph, where nodes are step-up and step-down transformers, power poles, switches, relays, reclosers, and links are power lines. The problem formulation will be presented in Section 3. In Section 4, we provide our decentralized framework, which comprises two phases. The first phase will assess the current resiliency of each microgrid. This phase is critical for electrification planning, which will include reliability in its costing model. In contrast, the second phase will propose two solutions to elevate resiliency: distortion-based and clustering-coefficient solutions. Experimental results, including validation and effectiveness of our frameworks, will be provided in Section 5. We conclude the chapter in Section 6.

2 Power grid modeling

A power grid can be mathematically modeled as a graph $G(V,E)$, where V is the set of vertices/nodes that represents the network's elements and E is the set of edges/links that describe the connectivity between its nodes. If two nodes v_i and v_j are connected through a physical power line, there is an edge $e_{i,j}=(v_i,v_j)$ in E. Let $N=|V|$ and $M=|E|$ be the number of nodes and links, respectively. V could include power poles, power plants, or power consumers (i.e., homes, schools, hospitals, etc.). In this paper, we focus mainly on power poles as elements of V and power lines as elements of E. A path between two nodes u and v is defined as a set of adjacent links that connect u with v. Each link $e_{i,j}$ in a power grid has a failure probability Pr($e_{i,j}$ *is failed*) and, consequently, any path p_k in a power grid is operational with probability Pr(p_k *is operational*), which depends on the failure probability of its links.

The assumptions made in this chapter are as follows. (1) A power grid is represented as a simple undirected graph in which there is no more than one link between any pair of nodes. (2) We assume that the power grid is optimally partitioned into microgrids according to the maximum power generation of available DGs and the local required load. (3) The terminal nodes (i.e., power plants or power consumers) are assumed to be perfect (i.e., do not fail). (4) Although our approach can be applied on nodes too, in this paper, we allow only links to fail independently from each other with known probability. (5) The failed/removed link is not restored and remains down.

3 Problem formulation

Given a graph G that represents a power grid. Let $\mathcal{B}$ denote a budget limit. The goal of this study is twofold:

(1) Assess the redundancy of each microgrid power grid by evaluating some significant graph measures such as distortion and clustering coefficient metrics.

(2) While there are critical lines that weaken the grid resiliency, we *utilized* distortion metrics to determine in which cases extra standby power elements (i.e., poles and lines) should be placed such *as* resiliency is maximized subject to distortion evaluation of each link, and *budget* $\mathcal{B}$.

Resiliency refers to the probability of a power grid recovering from a failure to any of its elements. That is, it measures the robustness of the power grid upon any damage. We represent any path in a graph G from source s to destination d by a pair (s,d), where s, $d \in V$. Let $\aleph = \{p_1, p_2, \ldots, p_r\}$ denote the set of all possible paths p_k with various lengths that connect the pair (s,d); p_k is sorted by a length in ascending order such as $p_1 \leq p_2 \leq \ldots \leq p_r$, where $1 \leq r \leq \sum_{k=0}^{N-2} \binom{N-2}{k} k!$. The upper bound of r represents the size of set $\aleph$ when G is a complete graph of N vertices. The goal is to assess to what extent a link failure will distort operational paths that connect (s,d). The output of this assessment suggests which power link requires a backup to improve its resiliency and to minimize the distortion it causes to the alternative operational path. That is, after distortion-based assessment, if a backup is required to a critical/vulnerable link, a set of grid elements will be deployed such that the distortion is minimized and the alternative paths have a high probability to be used within a budget $\mathcal{B}$. Similar handling will be followed for the clustering coefficient metric.

4 Part one: Incorporating smart devices

Incorporation of smart devices in the power grid contributes to better resiliency. These smart devices can collect a comprehensive picture of the grid condition and send it to a macrolevel controller for effective decision-making. As the complexity of the next-generation power grid increases, this approach of adopting sensors is necessary to protect microgrids and the grid as a whole.

Power transmission is the first subsystem to transport bulk power from a power generation subsystem to the distribution subsystem. The transmission system includes many fundamental components: step-up and step-down transformers, power lines, towers, switches, relays, and reclosers. As in conventional methods, analog electromechanical protective relays, placed at both ends of a transmission line, sense the fault immediately and isolate the faulted line by opening the associated circuit breakers [7]. Faults may be temporary (i.e., the fault is cleared after breaker reclosing) or permanent (i.e., the fault is not cleared even after several reclosing attempts). Restoring a power service after a permanent fault requires accurate localization of the faulty spot so that the technical team can quickly repair the faulted line section. One severe drawback of conventional methods is that they do not provide accurate localization of the fault. In addition, identification of the type of fault is difficult using relays.

The core idea of a smart grid is the communication of information between different grid subsystems to provide stable, sustainable, and reliable power services. In this sense, the smart grid enables intelligence, efficiency, and dynamicity to the electrical system to close any gap between supply and demand and be more resilient to faults. Smart grid techniques, on the other hand, overcome these drawbacks by providing digital adaptive protective relays. Communication networks comprise smart electronic devices that measure various physical and electrical parameters such as current, voltage, temperature, weather, etc. These smart electronic devices have different versions to apply to both transmission subsystem and distribution subsystem adequately. Smart grids leverage various communication networks to exchange information between various components as follows.

- Global system for mobile communications (GSM)

GSM is a telecommunication standard developed by the European Telecommunication Standard Institute (ETSI). It is considered the second generation (2G) of cellular networks used by mobile devices [13].

GSM has various applications, including fault analysis of power transmission lines [14,15]. For instance, in Ref. [15], a technique has been proposed to detect and classify a faulty transmission line. The proposed system consists of protective equipment such as relays, switch breakers, reclosers, voltage sensors, microcontrollers, GSM modules, and an LED display. The sensed fault signal is delivered to the microcontroller to analyze the characteristic condition of the signal in terms of current and voltage readings. The microcontroller will detect and classify the fault once it occurs. GSM will be used to send a message to the person in charge for any existing fault. Furthermore, a signal will be sent to relays and switch breakers to isolate the faulty section.

- Wireless sensor networks (WSNs)

Sensed data is collected by sensor nodes, aggregated at BS, and then transmitted to the end-user via a communication medium such as TCP/IP protocols or GPRS.

Since WSN is one of the leading IoT enablers, researchers often use the term IoT to indicate WSN [16,17]. WSNs have been applied in smart buildings, smart vehicles, health care, environmental studies, security, tracking objects, and agriculture [18]. WSNs can play an essential role in monitoring the status of transmission lines in a real-time manner [19,20]. For example, in Ref. [19], the authors provide a WSN-based system to monitor power transmission lines. The proposed system is divided into two subsystems: monitoring and operation. In the monitoring subsystem, heterogeneous sensors with different vision and magnetic induction capabilities are deployed on transmission towers. These sensors collect data on a real-time basis and send it to the operation subsystem. The collected data is transmitted wirelessly through the transmission line themselves with up to 138 kV of power. In this case, there is no need for any kind of cellular or optical communication to deliver the sensed data to the operating system to analyze it. The operation subsystem stores the collected sensed data for analysis. It shows the analytical results via a graphical user interface display—the installation and testing of the prototype offer real-time monitoring and fault detection.

5 Part two: The proposed decentralized resiliency framework

Resiliency can be defined as the system's ability to cope with external stresses [21]. As there are many external stresses (i.e., fault incidents), in this chapter, we opt not to identify faults, the reasons behind their presence, or their cascading behavior. This

Table 16.1 A comparison between conventional grid and next-gen smart grid in terms of resiliency parameters.

Parameter	Conventional grid	Next-gen smart grid
Sensors usage	No	Yes, used heavily
Communication networks	None or one-way nonreal-time communication	Multiple communication paradigms: GSM, LTE, WSNs, fiber optic backhaul
Cloud computing	No	Extensive
Protective devices	Analog electromechanical	Digital electronic
Power generation	Centralized	Both centralized and distributed
Mean time to repair (MTTR)	High due to inaccurate fault localization and manual restoration	Low due to accurate localization, self-recovering, or manual fixing
Frequency of failure	High as none or minimum coordination between the different components	Low as maximum real-time coordination is maintained through a sophisticated controller

will require several chapters on its own, if not an entire book. Table 16.1 shows the differences of resiliency related parameters between conventional grids and next-generation smart grids.

A resilient power system requires a self-healing/restoration method. However, this method is constrained by the existing topological structure and the physical limitations/capacities of its elements such as power lines, poles, switches, and relays. Our proposed resiliency framework will be generic and will include two phases: resiliency assessment and enhancement. Our approach will change microgrid topologies as new electrical and communication/control elements will be added in a decentralized manner for each microgrid to enhance its resiliency against faults.

5.1 Phase one: The resiliency assessment

Quantifying resiliency has attracted much discussion within the risk analysis community; yet, there is no standard method to do so [21]. In this phase, we utilize graph-theoretic measures to assess the redundancy in a power grid. This assessment enables us to identify design weaknesses in a power grid and avoid component failures that may lead to a complete system failure. Many graph-theoretic metrics can be used to assess the resiliency of the power grid. However, from the point of view of computational complexity, we seek to find a small set of these topological measures so that the given set of metrics can still characterize any power network uniquely.

Graph metrics such as average nodal degree [22], clustering coefficient [23], and cut links [24] provide precious information about the graph's structure. Yet, it is macrolevel information of the graph as there are no detailed characteristics about the graph. Graph-theoretic metrics are not totally independent. Indeed, the distortion metric [25], for example, is correlated to other metrics. A link failure that leaves a network disconnected (i.e., edge cut) makes a distortion value approach infinity to indicate the disconnectivity. Furthermore, high nodal degree and cluster coefficient contribute to lower distortion metric values. Next, we present two approaches for resiliency assessment: distortion-based resiliency assessment and clustering coefficient-based resiliency assessment.

5.1.1 Distortion-based resiliency assessment

Distortion: Let T be a spanning tree of a graph G. The distortion metric measures how extra distance is required to go from node v_i to node v_j if the spanning tree T is used instead of G, where v_i and $v_j \in V(G)$. The distortion D is defined in Ref. [25] to be the smallest such average distance over all possible Ts.

Calculating distortion in a graph G is NP-hard [25]. In this paper, we redefine the distortion metric to refer to the extra distance needed to traverse any path, in G, that includes $e_{i,j}$ if $e_{i,j}$ is failed. Therefore, instead of focusing on all possible spanning trees, we locally determine all possible paths that include $e_{i,j}$. In this study, we adopt the number of hops as a distance measure. We formulate the new definition as follows.

Definition 1. The failure of link $e_{i,j}$ in network G distorts the distances of all paths that include $e_{i,j}$ as follows.

(a) $D_{i,j} = D(G - \{e_{i,j}\}) = \infty$, if $e_{i,j}$ is a cut-edge
(b) $\min_{G-\{e_{ij}\}} \left(d\left(v_i, v_j\right)\right) \leq D_{i,j} \leq \max_{G-\{e_{ij}\}} \left(d\left(v_i, v_j\right)\right)$, otherwise

where $d(v_i, v_j)$ denotes the distance between nodes v_i and v_j, and $G - \{e_{i,j}\}$ is the graph obtained from removing the link $e_{i,j}$ from G. The maximum distance between these two nodes is upper bounded by $N-1$.

Let $S = \{(v_i, v_j)$: i and $j \in \{1, 2, .., M\}, i \neq j\}$ be the set of all directly connected pair nodes in G (i.e., set of links E). Let $P_{i,j} = \{P^2_{i,j}, ..., P^l_{i,j}\}$, where $P^h_{i,j}$ denotes the set of paths of length h that connect (v_i, v_j), where $2 \leq l \leq N-1$. We sort all paths in $P_{i,j}$ such that $|p_1| \leq |p_2| \leq, ..., \leq |p_r|$, where $|p_k|$ denotes the number of links/hops that form p_k, and ties are broken according to the highest operational probability, and $1 \leq r \leq \sum_{k=0}^{N-2} \binom{N-2}{k} k!$. Note that $P_{i,j} = \bigcup_{k=1}^{r} p_k$. Clearly, the path p_k is operational if all its links are up/present. In other words,

$$\Pr(p_k \; is \; operational) = \prod_{e_{i,j} \in p_k} \Pr\left(e_{i,j} \; is \; up\right) \tag{16.1}$$

The lowest possible distortion upon the removal of the link (v_i, v_j) can be obtained by selecting the shortest operational path that connects v_i and v_j. The probability that p_k is the shortest path to be used to connect v_i and v_j is given by

$$\Pr(p_k \; is \; used) = \prod_{z=2}^{k-1} (\Pr(p_z \; is \; failed)).\Pr(p_k \; is \; operational) \tag{16.2}$$

$$\Pr(p_z\, is\, failed) = \prod_{z=2}^{k-1}\left(1 - \prod_{e_{i,j}\in p_z}\Pr\left(e_{i,j}\, is\, up\right)\right) \tag{16.3}$$

The probability that the failed link is a cut-edge is given by

$$\Pr\left(\left(v_i, v_j\right)\, is\, a\, cut-edge\right) = 1 - \prod_{\substack{k=2 \\ p_k\in P_{i,j}}}^{r}\Pr(p_k\, is\, operational) \tag{16.4}$$

Regarding distortion, we are interested in keeping distortion associated with the link (v_i, v_j) within a specific threshold θ_d.

$$\Pr\left(D_{i,j} < \theta_d\right) = \sum_{\substack{k=2 \\ P_{i,j}^{\theta_d}\in P_{i,j}}}^{r'}\Pr(p_k\, is\, used) \tag{16.5}$$

where $r' = \sum_{w=2}^{\theta_d}|P_{i,j}^{w}|$. The event that the removal of any link will keep the network's distortion, D, within a threshold θ_d has the following probability.

$$\Pr(D < \theta_d) = \prod_{e_{i,j}\in E}\Pr\left(D_{i,j} < \theta_d\right) \tag{16.6}$$

We are ready to present an algorithm to calculate distortion based on the above analysis.

Algorithm 16.1: Calculation of distortion and its probability distribution (CDPD)

Input: Network graph G, the maximum number of paths r
Output: Distortion and its probability distribution

```
For each (v_i, v_j) ∈ E do
    S_{i,j} ← {p_1, p_2, ..., p_r} // Set of shortest paths with |p_k| > 1
    P(i,j) ← ∅
    D(i,j) ← ∅
    For each p_k ∈ S_{i,j} do
        D(i,j) ← append(Length(p_k) − 1, D(i,j))
        Pr(k) ← Pr(p_k is used)
        P(i,j) ← append(Pr(k), P(i,j))
```

Proposition 1. *The CDPD algorithm calculates distortion and its probability distribution of a network G in $O(M^2 + rMN)$ time complexity.*

Proof. The time complexity of the CDPD algorithm can be calculated as follows. The outer loop iterates a number of times equal to the size of set E (i.e., M), while the number of inner loop iterations is upper bounded by r. Finding each path in step 2 requires a time of $O(M)$. The calculation of step 7 requires $O(N)$. The total time complexity is $O(M^2+rMN)$. ▪ □

We use the link failure rate of the IEEE 14-bus in Ref. [26]. An IEEE 14-bus graph is depicted in Fig. 16.1.

Using MATLAB, we found that the link failure distribution of the IEEE 14 network follows the Nakagami probability distribution function (PDF) with the two parameters $\mu=0.28015$ and $\Omega=0.86472$. We use the same PDF parameters to generate the link failure distributions for IEEE 30- and 118-bus systems.

Table 16.2 shows the average probability (overall links) of a path of length K to be used as an alternative to a failed link for IEEE 14-, 30-, and 118-bus systems. The results show that the probability of using a path of length greater than 8 is almost zero for all buses. Therefore, there is no need to explore all possible paths and, hence, 8 is a suitable upper bound for K. Hence drops the computational complexity.

5.1.2 Clustering-coefficient-based resiliency assessment

Clustering coefficient: The clustering coefficient (CC) of a node v_i in a graph G measures how close the subgraph induced by v_i and its neighboring nodes form a fully connected subgraph [23]. The clustering coefficient of a node v_i in a graph G is given by the following formula.

$$CC_i = \frac{x_i}{\binom{\deg(v_i)}{2}} \tag{16.7}$$

where x_i is the total actual number of edges between v_i's neighbors. The clustering coefficients for all nodes is denoted by $CC(G)$.

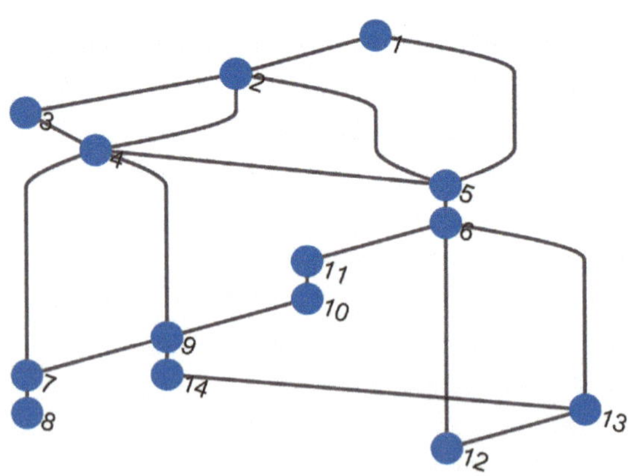

Fig. 16.1 IEEE 14-bus system.

Table 16.2 Evaluation of the probability that a path of length K is used in each IEEE system averaged overall links' failure.

K/ System	**$K=2$**	**$K=3$**	**$K=4$**	...	**$K=7$**	**$K=8$**	**$K=9$ (10^{-5})**	**$K=10$ (10^{-6})**
IEEE 14	0.89	0.07	0.01		0.003	0.0004	26	74
IEEE 30	0.88	0.70	0.07		0.21	0.02	171	77
IEEE 118	0.89	0.59	0.35		0.03	0.02	2	88

The average clustering coefficient is an important indicator of a network's structural connectivity. The higher the clustering coefficient of a node, the more connected its neighbors are and, hence, increasing the alternative local routes around the node. The clustering coefficient demonstrates an excellent key precision indicator for the prediction of network resilience [27].

Similar to distortion's algorithm 16.1, we propose an algorithm called "Calculation of CC with reliable microbased alternative paths (CCRMP)" to assess the criticality of a power grid based on CC and the probability of link failure distributions.

Algorithm 16.2: Calculation of CC with reliable microbased alternative paths (CCRMP)

Input: Grid graph G , the maximum number of paths r
Output: CC, Paths with their probability distribution

1 $P \leftarrow \emptyset; CC \leftarrow \emptyset$
2 For each $(v_i, v_j) \in E$ do
3 $\quad S_{i,j} \leftarrow \{p_1, p_2, \ldots, p_r\}$ /* *Set of shortest paths with* $|p_k| > 1$ */
4 $\quad G' \leftarrow rmedge(G, (v_i, v_j))$ /* *remove link*
5 $\quad (v_i, v_j)$ *from* G */
6 $\quad CC_{i,j} \leftarrow CC(G')$
7 $\quad Append(CC,\ CC_{i,j})$
8 $\quad$ For each $p_k \in S_{i,j}$ do
9 $\quad\quad \Pr(k) \leftarrow \Pr(p_k)$
10 $\quad\quad Append(P, \{p_k, \Pr(k)\})$

Proposition 1. *CCRMP Algorithm calculates Reliable Micro-based Alternative Paths of a grid G in O(rMN(M + NlogN)) time complexity.*

Proof. The time complexity of the CCRMP algorithm can be calculated as follows. The outer loop iterates a number of times equal to the size of set E (i.e., M), while the number of inner loop iterations is upper bounded by r. Finding paths in step 3 can be done using Yen's algorithm, which requires a time of $O(rN(M+NlogN))$ [28]. The calculations of steps 6 and 9 require $O(N)$ each. The total time complexity is $O(rMN(M+NlogN))$. ▪ □

5.2 Phase two: The resiliency enhancement

In this phase, we use the previous distortion analysis to enhance resiliency. Our goal is to minimize the distortion of overall vulnerable links. The deployment of extra grid elements will depend on the failure probability of network links and the distortion values upon their failure. Therefore, deploying some grid elements may be required to increase redundancy and, consequently, resiliency upon line failure. We should also consider the probability of the operational paths that will be used to overcome this failure.

Next, we propose an algorithm to show the link criticality and when extra elements should be added to maintain distortion within a specific threshold θ_d.

Algorithm 16.3: Identifying links' criticality (ILC)

Input: Network G, distortion, failure probability distribution, the probability distribution of alternative paths used, Budget $\mathcal{B}$.

Output: Sorted list of links to be backed up

1 $E' \leftarrow Sort(E)$// sort edges in descending order by //their failure rate

2 For each $e_{i,j} \in E'$ do

3 $P_{i,j} \leftarrow \{P_{i,j}^2, \ldots, P_{i,j}^l\}$// sort paths in ascending //order by their length

4 While $\mathcal{B} > 0$

5 *If* $D_{i,j}(P_{i,j}(1)) > \theta_d$ *OR* $Pr(P_{i,j}(1)\ is\ used) < \theta_{used}$

// $e_{i,j}$ is critical and must be backed

// up to make distortion goes below/over

// θ_d / θ_{used}, respectively.

6 *add_ elements*$(P_{i,j}(1), e_{i,j})$

7 $\mathcal{B} = \mathcal{B} - \text{cost}$

The ILC algorithm provides extra redundancy based on distortion analysis of each link in a graph. It gives priority to back up links of the highest failure rates. Thus, the links are sorted in descending order according to their failure probability, as shown in line 1. In line 2, the algorithm iterates over each of the sorted links $e_{i,j}$. All possible paths that connect the end-nodes of $e_{i,j}$ are sorted in ascending order by their lengths (line 3). While there is still budget $\mathcal{B}$, the algorithm checks for more links that need extra redundant elements. In line 5, ILC tests whether the distortion value of the first shortest alternative path, i.e., $P_{i,j}(1)$, is greater than a threshold θ_d or its probability to be used is less than θ_{used}; then, the $e_{i,j}$ is considered a critical link and should be backed up and, hence, invoke the function *add_elements(*$P_{i,j}(1)$, $e_{i,j}$*)* and subtract the cost of this addition from the budget (lines 6 and 7). Moreover, the function *add_elements(*$P_{i,j}(1)$, $e_{i,j}$*)* added nodes and links to the path $P_{i,j}(1)$ such as the adjusted new path will have both low distortion and high probability to be used. That is, $D_{i,j}(P_{i,\ j}(1)) \approx \theta_d$ and $\Pr(P_{i,j}(1)$ *is used*$) \approx \theta_{used}$.

A similar algorithm can be developed to consider the output of CCRMP as follows.

Algorithm 16.4: Critical links support (CLS)

Input: Grid G, CC, failure probability distribution, budget $\mathcal{B}$.
Output: Sorted list of links to be backed up

```
1  For each e_{i,j} ∈ E' do
2      if (1 − pr(e_{i,j})) > f_thresh /* failure threshold */
3      if var(CC_{i,j}) > cc_thresh /* check the variance of CC associated with e_{i,j} > cc threshold */
4          if B > 0
5              cost ← harden_line(e_{i,j})
6              B = B − cost
```

In line 1, the algorithm iterates over each of the links $e_{i,j}$. The link $e_{i,j}$ is supported with redundant power lines only if three conditions hold: the failure rate of $e_{i,j}$ is greater than a failure threshold, *f_thresh* (line 2); the variance of CC difference vector associated with $e_{i,j}$ is greater than CC threshold, *cc_thresh* (line 3); lastly, there should be an available budget to strengthen $e_{i,j}$ (line 4). Once these conditions are met, then $e_{i,j}$ is considered a critical link and should be backed up and, the function *harden_line*$(e_{i,j})$ should be hardened and the cost of this addition from the budget (lines 5 and 6) subtracted. Moreover, the function *harden_line*$(e_{i,j})$ added links to connect at least two neighbors of $e_{i,j}$. Note that *var*$(CC_{i,j})$ calculates the variance of the difference vector of $CC(G)$ and $CC(G-e_{i,j})$, where $CC(G)$ is the CC vector for all nodes, and $G-e_{i\ j}$ is the graph resulting from removing the link $e_{i,\ j}$.

ILC and CLS algorithms consider the cost constraint within the given budget $\mathcal{B}$. The optimal placement of redundant elements should also consider the surrounding ecosystem and the growth of residential areas to make the model adaptable to dynamic environmental and social changes. However, the placement strategy is out of the scope of this paper.

6 Experimental results

In this section, we validate our resiliency frameworks and show their effectiveness. We conduct our experiments using IEEE 14, 30, and 118 bus systems. Applying our framework to a bus system G will generate a modified bus system G'. To this end, we apply a failure process on both G and G' to remove a link (or multiple links) every time. G and G' are then compared to each other against three performance evaluation metrics that represent the quality of resiliency. The metrics are the "size of the large component" (SoLC), the "number of components" (NoC), and the "percentage of power delivery" (PoPD). The main experimental parameter is time. We assume that the failure process follows an exponential distribution. That is, $Pr(e_{i,j}$ *is removed at* *time t*) = 1 − *exp*(−$\lambda_{i,j}$ * *t*), where $\lambda_{i,j}$ is the failure rate of the link $e_{i,j}$. We split our experimental results into two sets. One for the distortion-based framework, while the other for the clustering coefficient-based framework.

6.1 Set 1: Distortion-based experimental results

We set the following values to our framework-related parameters: the normal link failure threshold $\epsilon = 1\%$. Thus, any link with a higher failure rate than ϵ is considered vulnerable; $\theta_{used} = 70\%$, and $\theta_d = 3$. In order to test our framework's full potential, we assume no cap on budget $\mathcal{B}$ in all experiments. In our experiments, we limit the number of paths to the first 8 shortest paths. Finally, the failure rate of the newly deployed links, as a result of phase 2, will be set to the minimum failure rate in the system. This is obvious because they are new, strongly established, and should be placed in such a way as to avoid disastrous incidents.

We divided our experiment into "validation" and "effectiveness." In the former, we demonstrate the "validation" of our proposed framework. We choose IEEE 14-bus for this experiment as it has a relatively smaller number of nodes and edges, making the demonstration easier. In the latter, we show the "effectiveness" of our framework by comparing each IEEE system and its modified distortion-based version against the abovementioned performance evaluation metrics.

6.1.1 Validation

The link failure distribution of IEEE 14 bus systems alongside the probability of the shortest path being used is shown in Fig. 16.2. This figure is important for cross-referencing with Fig. 16.3. For example, it shows that the failure probability of the

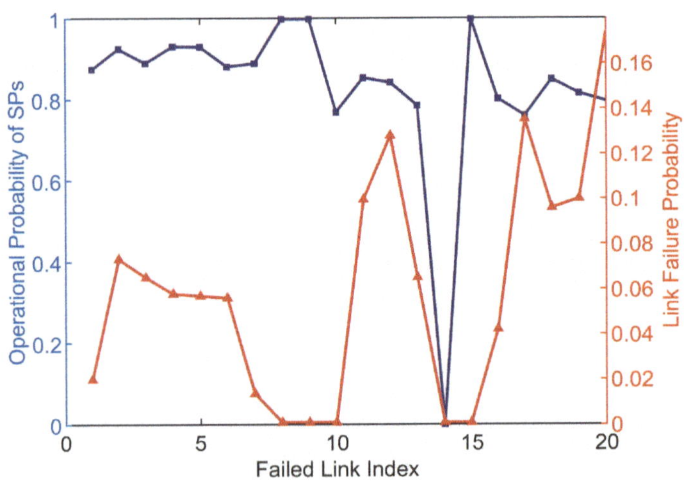

Fig. 16.2 IEEE 14-bus system: The link failure probability vs probability of the shortest path being used.

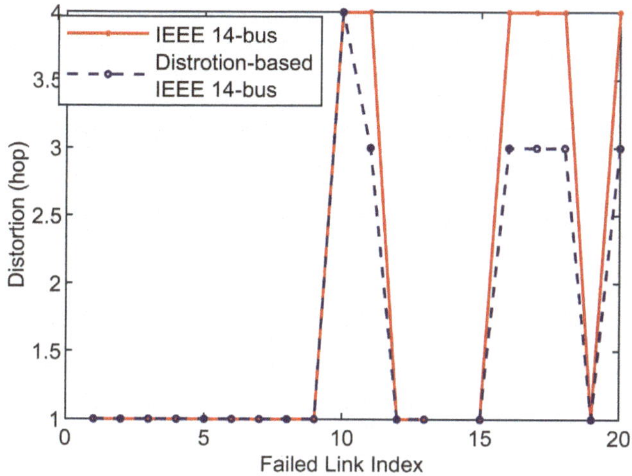

Fig. 16.3 The modified IEEE 14-bus system: distortion and probability of the shortest path being used vs link index.

link with index 14 (the edge-cut link) is very close to zero, which means that it is not vulnerable and will not have a priority in the enhancement phase of our proposed framework.

Next, we present the result of applying our framework to the IEEE 14-bus system. We recall that $\epsilon = 1\%$, $\theta_d = 3$, $\theta_{used} = 70\%$, and $\mathcal{B}$ has no cap. The results are shown in Fig. 16.3. Comparing Figs. 16.2 and 16.3, we notice some changes that occurred on the original network after applying our framework. For example, links 15–20 have been backed up in the modified network and, hence, their distortion reached the preferred threshold. The probability of the shortest path used was already above 70% except for link 14, which was not selected for enhancement in phase 2 because it has a failure probability under the threshold of 1%. Like link 14, link 10 has a failure probability of less than the threshold and, consequently, was not backed up despite distortion over the threshold.

6.1.2 *Effectiveness*

After we showed the validation of our framework in the previous experiment, we now show its effectiveness. We compare each IEEE power system with its modified distortion-based version. We use the SoLC, the NoC, and the PoPD as performance metrics with time as a parameter. To conduct this comparison, we use the link failure process mentioned at the beginning of this section. The failure process follows an exponential distribution to down/remove links. Fig. 16.4A and B show the comparison of both power systems as the failure process continues over time. Fig. 16.4A shows that the modified system has a main component with a larger size (i.e., number of nodes) than the original IEEE 14-bus system. It also shows that the modified system is more stable and robust over failure. It maintained a large size of about 80% of its

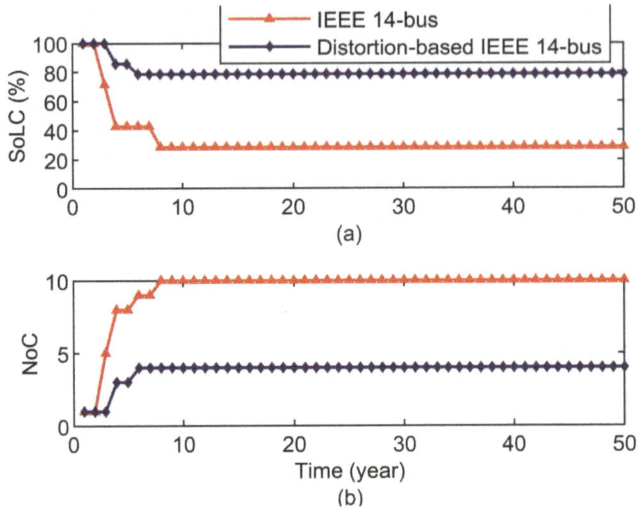

Fig. 16.4 (A) The SoLC and (B) the NoC.

initial size (i.e., before the failure process). On the other hand, Fig. 16.4B demonstrates a significant improvement in the number of components of the modified system over the original one. The figure shows that the original system has double the number of components most of the time. The interpretation of these results stems from the structural changes that occurred through hardening some critical links, namely the five links shown in Fig. 16.3. This enhancement contributed to more routes to deliver the bulk power and form a better structural resistance against failure.

In the next experiment, we test the power load interruption for the three IEEE bus systems. This experiment tends to compare each IEEE bus system with its modified distortion-based version by evaluating to what extent the failure process impacts their power load delivery over the operational paths that connect each of the pair nodes. A successful power delivery between any pair of nodes is valued by one (i.e., high resiliency); otherwise, it is zero. If one pair had a successful delivery at time t while the other pair failed, then the percentage of power delivery (PoPD) is 50%, i.e., the average between the results of both pairs. We create a list of pair nodes for each IEEE bus system as follows. {(8,6), (2,10)}, {(29,18), (26, 4)}, and {(100,19), (11, 49)} for IEEE 14-, 30-, and 118-bus systems, respectively. The results are shown in Fig. 16.5.

The results of all performance metrics for all IEEE systems are tabulated in Table 16.3. In this table, the horizontal arrow (→) denotes the average value of the corresponding metric/system before and after the system modification, while the vertical arrow (↑, ↓) shows the increase and decrease, respectively, in the

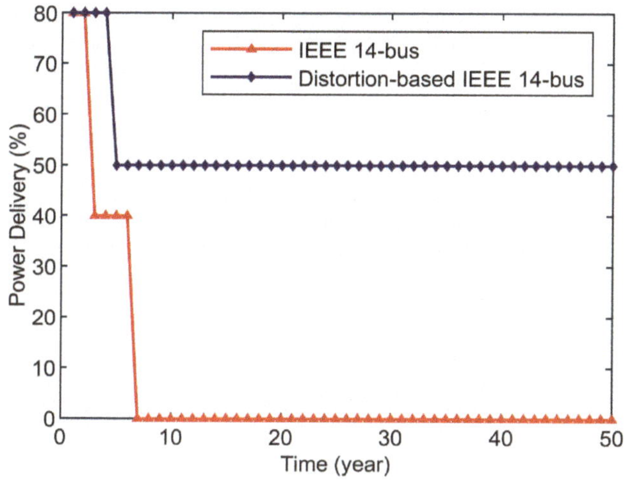

Fig. 16.5 The PoPD in IEEE 14-buses and its modified version under link failure over time.

Table 16.3 Summary of the performance evaluation metrics for all IEEE bus systems. Each value is the average over 50 years.

Performance metric/system	Size of the largest component	Number of components	Percentage of power delivery via pair nodes
IEEE 14	4.7%→15.2% (2.23-fold ↑)	9.42→4.74 (0.49-fold ↓)	8%→52% (5.5-fold ↑)
IEEE 30	4.3%→20.9% (3.9-fold ↑)	24.5→13.4 (0.45-fold ↓)	5%→18% (2.6-fold ↑)
IEEE 118	13.7%→37.7% (1.8-fold ↑)	82→61.8 (0.33-fold ↓)	4%→18% (3.5-fold ↑)

corresponding metric values. The parenthesis shows the ratio of the improvement.

Table 16.3 shows significant improvement in all performance metrics and for all modified distortion-based IEEE bus systems. The highest improvement peak was at the PoPD of IEEE 14-bus with a 5.5-fold improvement of the modified system over the original system. Furthermore, the lowest improvement was at the number of components for IEEE 118-bus. The modified 118 system has a number of components less by a third than the original system.

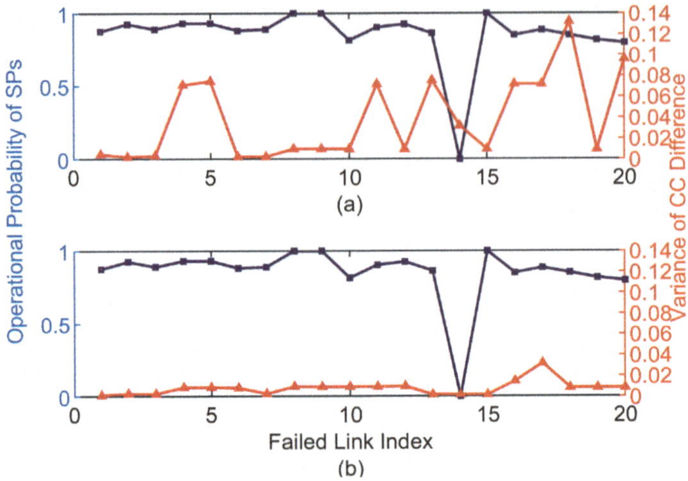

Fig. 16.6 (A) IEEE 14-bus system, and (B) the modified IEEE 14-bus system. The variance of cluster coefficient difference and the operational probability of the shortest paths (SPs) were used to connect the two end nodes of the failed link.

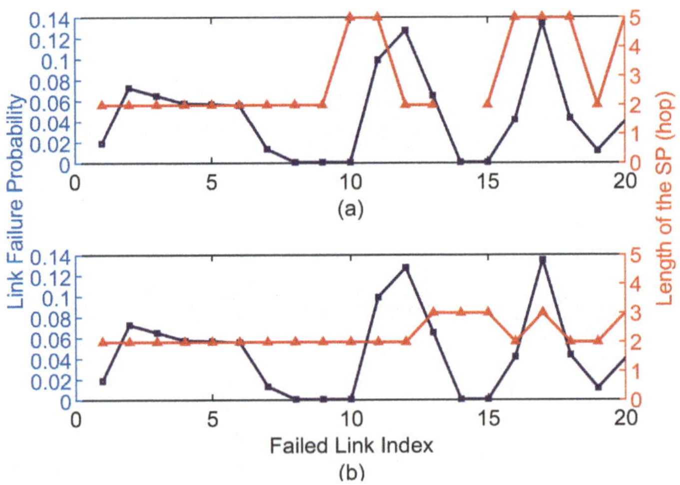

Fig. 16.7 (A) IEEE 14-bus system, and (B) the modified IEEE 14-bus system. The link failure probability and length of the shortest paths are used to connect the two end nodes of the failed link vs failed link index.

6.2 Set 2: Clustering coefficient-based experimental results

6.2.1 Validation

Fig. 16.6 shows the IEEE 14-bus system and its modified version, respectively. They show the variance of CC difference and the operational probability of the shortest path being used, both for all links. In Fig. 16.6A, some of the CC variances have values larger than *cc_thresh*, which indicates a disruption in CC upon a failure of the corresponding link. The existence of edge-cut at link 14 is obvious because it has a probability of zero to connect the end-nodes of this link. In Fig. 16.6B, the CC variance values in the enhanced system became more stable with minimal disruption.

The link failure distribution and the length of the shortest path used are shown in Fig. 16.7A and B for the IEEE 14 bus system and its modified version. The cross-referencing between these two subplots shows that some links adjacent to the edge-cut link with index 14 are suitable for enhancement, resulting in a new path connecting the end nodes of link 14. We also found that the alternative routes became shorter after the enhancement, as expected.

Next, we show the effectiveness of our framework on the IEEE 14-bus system.

6.2.2 Effectiveness

We compare the IEEE power system with its modified CC-based version. Similar to Set 1, we use the SoLC, the NoC, and the percentage of PD as performance metrics. Fig. 16.8 shows the performance evaluation of both power systems against the SoLC and NoC metrics.

Fig. 16.8A shows that the CC-based system is more stable and robust over failure. It maintained an SoLS of about 70% of its initial size. On the other hand, Fig. 16.8B shows both systems' performance against the NoC metric. The results demonstrate a significant outperformance of the CC-based system over the original one, with around 50% less fragmentation over time.

The next experiment tends to show how resilient a power grid is under the same failure process. We follow a similar setting as in Set 1. The results are shown in

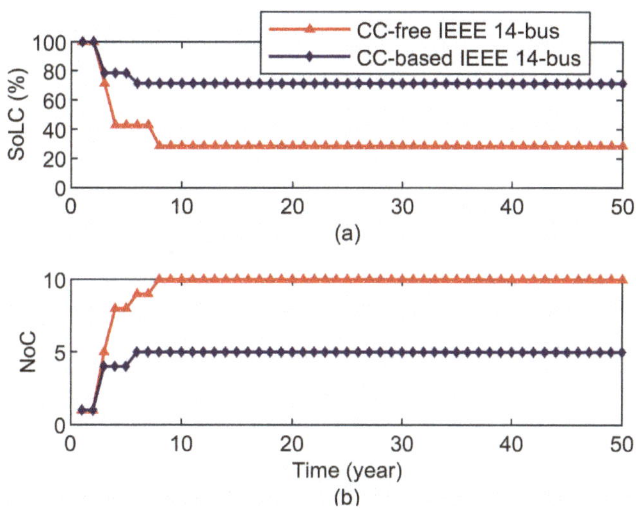

Fig. 16.8 The performance of the CC-based IEEE 14-bus and CC_free (original) IEEE 14-bu under link failure over time and against the following performance metrics: (A) The percentage of the SoLC, and (B) the NoC.

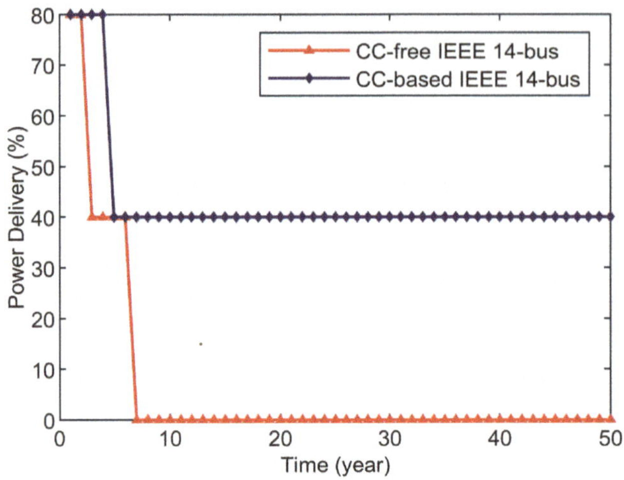

Fig. 16.9 The PoPD in IEEE 14-buses and its updated version under link failure over time.

Fig. 16.9. It shows that the modified system (CC-based system) maintains higher resiliency over time with almost 40% more power delivery than the CC-free system.

6.3 Discussion

This experimental results section showed the validation and effectiveness of our distortion and clustering coefficient metrics. Our proposed distortion-based approach evaluates the vulnerability and utilizes the distortion analysis to install extra power elements to strengthen the grid. For each link failure, we test IEEE bus system resiliency and distortion changes. Then, to enhance resiliency, we detect the critical links that weaken the grid and install power elements according to a given budget. Distortion metric impacts the whole grid because a failure of one link may distort the bulk power of many delivering paths. This is different from what we propose using the clustering coefficient metric. In clustering coefficient, a failed link will impact only neighboring nodes. Higher CC, which exhibits a higher resiliency, does not require installation of extra nodes (power poles). This is important from the costing model's perspective. However, from the simulation, we can see that the distortion-based approach is about 10% more effective in enhancing the reliability of microgrids. The interpretation is that having extra nodes deployed using the distortion approach will provide new alterative paths to reroute the bulk power upon contingencies. This enhancement hardens the structure and makes it more resistant to failures.

As the energy industry continues to expand and renewable generation sources have emerged, our frameworks are great tools for operators of future power grid. They provide high reliability to smart grids through a decentralized approach for enhancing the resiliency of the microgrids. There is still a need for an effective costing model to auction the power surplus. Indeed, some microgrids will have excess power resources, and their owners seek the highest possible prices. On the other hand, consumers from other microgrids are looking for power with minimum expenditures. A decentralized approach is possible in this case through bidding in auctions initiated by the seller. Accordingly, there will be a price curve for power depending on the supply–demand formula. An optimization formulation should be conducted according to each party's optimization criteria.

There are numerous research opportunities to cover the knowledge gap in the area of the decentralized power grids. Some examples are the optimal placement of power generators within microgrids so that the cost and time of power interruption are minimal. Also, fog and edge computing can be utilized to have the best communication performance between microgrid controllers and macrolevel controllers at higher levels. This is important to reduce the latency and overhead of communication and promote the grid to scale up.

7 Conclusions

This chapter highlights futuristic insights on the decentralized smart power grid away from the legacy grid built around the assumption of centralized resources. The future power grid should be autonomous and have the self-adaptive and recovery mechanisms to overcome any faulty event in a timely manner. In this regard, we provide two-phase novel frameworks that combine both topological characteristics and operational features of a power grid to assess the microgrid resiliency, identify the criticalities, and to enhance the resiliency. This will minimize the impact of faults and reduce fault propagation. Metrics based on graph theory are used, namely, distortion and clustering coefficient. Then numerical results from three benchmark grids demonstrate the validation and effectiveness of our frameworks. This study provides a microlevel of a grid to address the probability distributions of link failure and operational paths used as alternatives to connect the end nodes upon a link failure. We anticipate that this study will highly benefit power grid operators to start migrating their grids in a more decentralized manner for further effective and reliable power services. We expect a surge in decentralized smart power grid research to mitigate the impact of power faults on the economy and people's lives, especially faults that hit a transmission system.

Future research should focus on the correlation between resiliency metrics. These metrics are not independent; the higher the correlation between two metrics, the less the information retrieved, and the less chance to characterize a given grid. Identifying the correlation among topological metrics is not a straightforward task as it requires a deep understanding of the metrical space. However, figuring out such correlation will help to develop a comprehensive framework for reliability smart power grid.

References

[1] Y. Wang, C. Chen, J. Wang, R. Baldick, Research on resilience of power systems under natural disasters—a review, IEEE Trans. Power Syst. 31 (2) (2016) 1604–1613, https://doi.org/10.1109/TPWRS.2015.2429656.

[2] P. Cicilio, et al., Resilience in an evolving electrical grid, Energies 14 (3) (2021), https://doi.org/10.3390/en14030694.
[3] M. Panteli, D.N. Trakas, P. Mancarella, N.D. Hatziargyriou, Boosting the power grid resilience to extreme weather events using defensive islanding, IEEE Trans. Smart Grid 7 (6) (2016) 2913–2922, https://doi.org/10.1109/TSG.2016.2535228.
[4] H. Farzin, M. Fotuhi-Firuzabad, M. Moeini-Aghtaie, Enhancing power system resilience through hierarchical outage management in multi-microgrids, IEEE Trans. Smart Grid 7 (6) (2016) 2869–2879, https://doi.org/10.1109/TSG.2016.2558628.
[5] C. Chen, J. Wang, F. Qiu, D. Zhao, Resilient distribution system by microgrids formation after natural disasters, IEEE Trans. Smart Grid 7 (2) (2016) 958–966, https://doi.org/10.1109/TSG.2015.2429653.
[6] Z. Wang, J. Wang, Self-healing resilient distribution systems based on sectionalization into microgrids, IEEE Trans. Power Syst. 30 (6) (2015) 3139–3149, https://doi.org/10.1109/TPWRS.2015.2389753.
[7] P. Chopade, M. Bikdash, Modeling for survivability of Smart Power Grid when subject to severe emergencies and vulnerability, in: 2012 Proceedings of IEEE Southeastcon, 2012, pp. 1–6, https://doi.org/10.1109/SECon.2012.6196924.
[8] D.J. Watts, S.H. Strogatz, Collective dynamics of 'small-world9 networks, Nature 393 (6684) (1998) 440–442, https://doi.org/10.1038/30918.
[9] F. Teymouri, T. Amraee, H. Saberi, F. Capitanescu, Towards controlled islanding for enhancing power grid resilience considering frequency stability constraints, IEEE Trans. Smart Grid (2017) 1, https://doi.org/10.1109/TSG.2017.2777142.
[10] Y. Al Mtawa, A. Haque, Clustering-coefficient based resiliency approach for smart grid, in: Presented at the 2021 IEEE International Wireless Communications and Mobile Computing Conference (IWCMC), July, 2021, pp. 1569–1574, https://doi.org/10.1109/IWCMC51323.2021.9498658.
[11] Y. Al Mtawa, A. Haque, H. Lutfiyya, Towards a customized resiliency framework for smart grid, in: 2019 IEEE Global Communications Conference (GLOBECOM), Dec. 2019, pp. 1–6, https://doi.org/10.1109/GLOBECOM38437.2019.9013960.
[12] F. Gianaroli, F. Pancaldi, G.M. Vitetta, The impact of statistical noise modeling on the error-rate performance of OFDM power-line communications, IEEE Trans. Power Deliv. 29 (6) (2014) 2622–2630, https://doi.org/10.1109/TPWRD.2014.2311167.
[13] P.S. Emi, R. Sivasankari, P.R. Kumar, R. Prabha, J. Jayageetha, A. Karhikeyan, Fault detection in transformer using GSM technology, 2019 5th International Conference on Advanced Computing Communication Systems (ICACCS), March, 2019, pp. 868–873, https://doi.org/10.1109/ICACCS.2019.8728502.
[14] M.S. Sujatha, M.V. Kumar, On-line monitoring and analysis of faults in transmission and distribution lines using GSM, J. Theor. Appl. Inf. Technol. 33 (2) (2011) 258–265.
[15] P.A. Gulbhile, J.R. Rana, B.T. Deshmukh, Overhead line fault detection using GSM technology, in: 2017 International Conference on Innovative Mechanisms for Industry Applications (ICIMIA), February, 2017, pp. 46–49, https://doi.org/10.1109/ICIMIA.2017.7975510.
[16] P.P. Namdev, M.P. Sudhir, R.D. Shantaram, M.R. Goutam, S.V. Shankar, S.A. Palake, Transmission line fault detection with mini weather station using IoT, Asian J. Converg. Technol. 7 (1) (2021), https://doi.org/10.33130/AJCT.2021v07i01.026.
[17] Y. Al Mtawa, H.S. Hassanein, N. Nasser, Identifying bounds on sensing coverage holes in IoT deployments, in: 2015 IEEE Global Communications Conference (GLOBECOM), December, 2015, pp. 1–6, https://doi.org/10.1109/GLOCOM.2015.7417792.

[18] Y. Al Mtawa, H. Hassanein, N. Nasser, The impact of anchor misplacement on sensing coverage, in: 2016 IEEE Wireless Communications and Networking Conference, 2016, pp. 1–7, https://doi.org/10.1109/WCNC.2016.7564860.
[19] J.L. Leoni, et al., Real-time monitoring of transmission lines using wireless sensor networks, in: 2014 IEEE PES Transmission Distribution Conference and Exposition—Latin America (PES T D-LA), September, 2014, pp. 1–6, https://doi.org/10.1109/TDC-LA.2014.6955193.
[20] Y. Yang, G. Xie, X. Xu, Y. Jiang, A monitoring system design in transmission lines based on wireless sensor networks, Energy Procedia 12 (suppl C) (2011) 192–199, https://doi.org/10.1016/j.egypro.2011.10.027.
[21] H. Thorisson, J.H. Lambert, J.J. Cardenas, I. Linkov, Resilience analytics with application to power grid of a developing region, Risk Anal. 37 (7) (2017) 1268–1286, https://doi.org/10.1111/risa.12711.
[22] A.T. Le, R. Sankar, Complex network approach for power grids vulnerability and large area blackout, in: Computational Social Networks, 2016, pp. 206–213.
[23] M. Rosas-Casals, S. Valverde, R. Solé, A simple spatiotemporal evolution model of a transmission power grid, IEEE Syst. J. 12 (4) (2018) 3747–3754, https://doi.org/10.1109/JSYST.2018.2800766.
[24] S. Arora, S. Rao, U. Vazirani, Expander flows, geometric embeddings and graph partitioning, J. ACM 56 (2) (2009) 1–37, https://doi.org/10.1145/1502793.1502794.
[25] H. Tangmunarunkit, et al., Network topology generators, in: Proceedings of the 2002 Conference on Applications, Technologies, Architectures, and Protocols for Computer Communications—SIGCOMM '02, vol. 32, 2002, p. 147, https://doi.org/10.1145/633025.633040. no. 4.
[26] E. Zio, R. Piccinelli, M. Delfanti, V. Olivieri, M. Pozzi, Application of the load flow and random flow models for the analysis of power transmission networks, Reliab. Eng. Syst. Saf. 103 (2012) 102–109, https://doi.org/10.1016/j.ress.2012.02.005.
[27] A. Albarakati, M. Bikdash, Analyzing transmission lines robustness and resilience prediction accuracy with line graph based model, in: Conference Proceedings—IEEE SOUTHEASTCON, vol. 2018-April, Oct, 2018, https://doi.org/10.1109/SECON.2018.8479232.
[28] J.B. Ahuja, R.K. Magnanti, T.L. Orlin, Network Flows-Theory, Algorithms and Applications, 1993.

Modeling and evaluation of power system vulnerability against the hurricane

17

Amirhossein Nasri[a], *Amir Abdollahi*[a], *Masoud Rashidinejad*[a], *and Wei Peng*[b]
[a]Department of Electrical Engineering, Shahid Bahonar University of Kerman, Kerman, Iran,
[b]Faculty of Engineering and Applied Science, University of Regina, Regina, SK, Canada

1 Introduction

High impact-low probability (HILP) incidents such as hurricanes gravely and usually damage power systems. Hurricanes begin as tropical disturbances in warm ocean waters with surface temperatures of at least 26.5°C (80°F). These low-pressure systems are fed by energy from warm seas. Hurricanes are enormous heat engines that deliver energy on a staggering scale. They draw heat from the warm, moist ocean air and release it through the condensation of water vapor in thunderstorms. Hurricanes spin around a low-pressure center known as the eye. Sinking air makes this 32- to 64-km (20- to 40-mile)-wide area notoriously calm. But the eye is surrounded by a circular "eyewall" that contains the storm's strongest rain and winds. Climate change may be driving more frequent, more intense extreme weather, and that includes hurricanes. The 2018 hurricane season was one of the most active on record, with 22 major hurricanes in the Northern Hemisphere in under 3 months, and 2017 also saw seriously devastating Atlantic storms. While many factors determine a hurricane's impact and strength, warmer temperatures in certain locations play an important role. In the Atlantic, warming in the Arctic could drive future hurricane tracks farther west, making a US landfall more likely. Hurricanes produce significant, widespread, and often prolonged electrical power outages. For example, Hurricane Irene caused more than 500,000 Long Island Power Authority customers to lose power and it took 8 days to achieve 99% customer restoration. Businesses and individuals are heavily dependent on a continuous supply of electricity. Given this strong dependence on reliable electricity, private industries and individuals are increasingly putting collective pressure on regulators to require system hardening by utilities. Hence, different references study the effect of a hurricane on a power grid and the proactive efforts against this event. Khodaei [1] introduced a resiliency-oriented microgrid (MG) optimal scheduling structure to calculate the enhancement of power grid resiliency by the local supply of loads and curtailment reduction in a hurricane situation. An MG preventive management structure is presented in Ref. [2] to cope with the adverse effects of extreme hurricanes. According to the alerts received for the anticipated windstorm, the structure finds a conservative operation scheme of MG with the minimum number of vulnerable lines in service while

Decentralized Frameworks for Future Power Systems. https://doi.org/10.1016/B978-0-323-91698-1.00013-3

a total load is supplied. Amirioun et al. [3] introduced a quantitative model for calculating the MG resilience against HILP hurricanes. The presented model jointly applies the windstorm profile and fragility curves of distribution lines to evaluate the degradation in the MG operation. Huang et al. [4] proposed an integrated resilience response structure, which not only links conditional awareness with resilience enhancement but also suggests impressive and efficient strategies in both emergency and restrictive cases. In the suggested framework, a two-level robust mixed-integer optimization framework is illustrated as the main part. A two-stage resilience enhancement planning for power distribution grids against hurricanes is depicted in Ref. [5]. The decisions on distributed generation placement, line hardening, mobile emergency generators allocation, and tie switch placement are considered in the first step to improving the system resilience. In the second step, the operation costs of the system pertaining to the distribution network planners power purchase from the upstream grid, forced load shedding, and (distribution generation) DG power production in emergency situations are optimized to reach a techno-economic compromise of investment costs and enhanced operation/resilience advantages over both operation and planning scales. Amirioun et al. [6] suggested a resilience-oriented proactive model to enhance the preparedness of multiple energy carrier MGs against a windstorm. Moreover, due to improvement of a power system resiliency against HILP incidents, events should be evaluated and modeled as well as possible. Hurricanes as natural disasters are generally highly uncertain events that are difficult to predict, estimate, and model. A lot of efforts have been made to increase our awareness of natural disasters according to history. The forecasting of a natural disaster is often based on statistical models or simulation models. However, a hurricane as an atmospheric system is a complex nonlinear dynamic system, in which small changes of atmospheric state may lead to a dramatic variation on the subsequent atmospheric evolution. Therefore, on consideration of the atmospheric dynamical system, a combined methodology based upon the least-squares support-vector machine (LS-SVM), chaos theory, and associated data is employed to predict the hurricane speed. Then a novel optimization model is suggested to determine the hurricane effects on the distribution network. The objective function of the proposed hurricane happening model considers the lines importance according to its downstream connected loads, lines fragility function, and the lines' difference angle with the hurricane direction. The multiperiod and multizone limitations for the damaged equipment budget provide the hurricane model constraints.

2 Temporal and spatial dynamics of hurricanes

As revealed by the hurricane forecast improvement program [7], a hurricane often follows a path that consists of several associated geographic regions and multiple periods according to Fig. 17.1. Moreover, the velocity of a hurricane reduces once the storm lands and drifts away from the sustaining heat and moisture provided by the ocean or gulf waters. As shown in Fig. 17.2 hurricane velocity decays quickly over time after landfall. The hurricane velocity decays geographically along its path as illustrated in Fig. 17.3 according to the inland wind model [8].

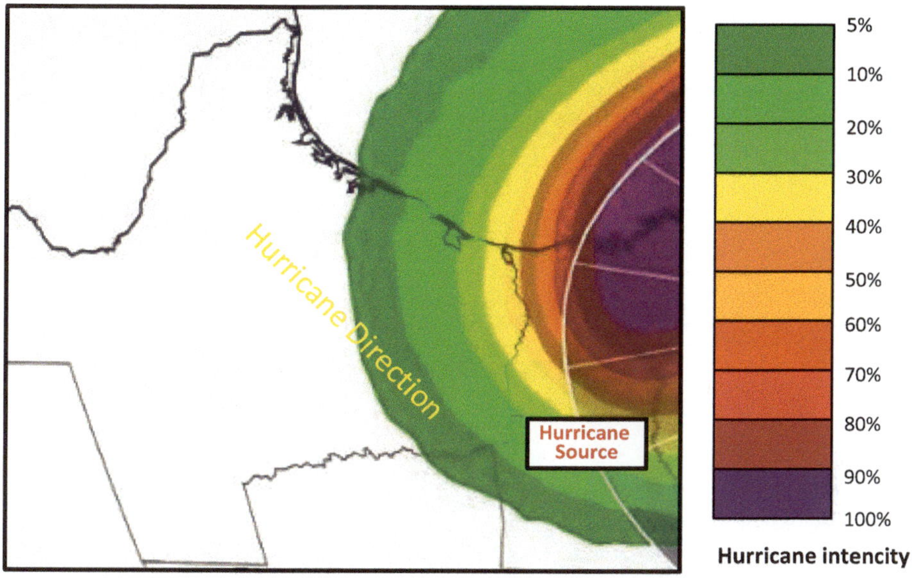

Fig. 17.1 A typical evolution of hurricane and its decay procedure [9].

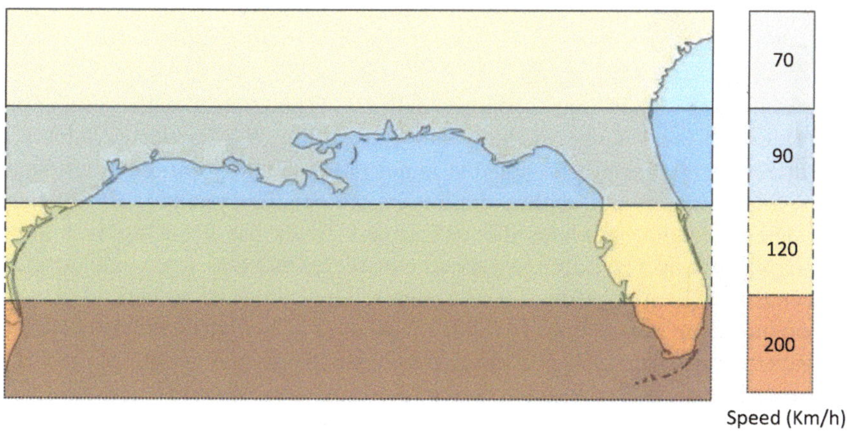

Fig. 17.2 Decay of hurricane attack.

3 Hurricane velocity anticipation based on the chaos theory and LS-SVM

Hurricanes are uncertain events that are complex to model and forecast. Many studies have been made to enhance our awareness of natural events according to the usual anticipation methods and historical data. The prediction of a natural incident is often based upon simulation models or statistical ones as discussed in Ref. [10]. Predefined

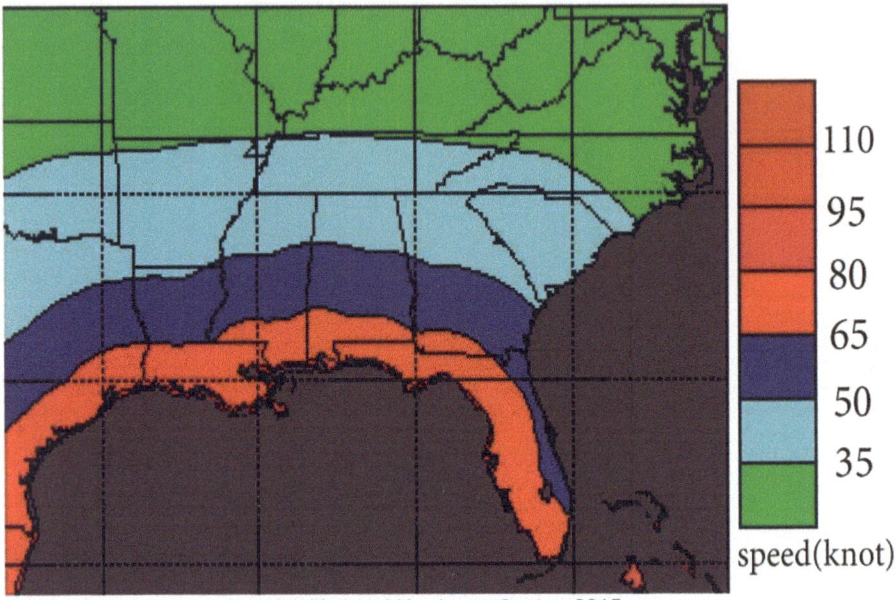

Source: The inland wind model, National Hurricane Center, 2015

Fig. 17.3 The extent of inland winds from category 3 hurricanes with forwarding velocity 20 knots.

natural phenomena scenarios are supposed in Ref. [11] and treated with a similar probability. Due to a low change in the primary condition of a hurricane quickly and greatly affecting its future behavior that is not predictable by common anticipation methods, a hurricane is a complex, nonlinear, and chaotic system [12] as a chaotic system. The utilization of chaos theory in many fields has been highly improved by the Lorenz system. The Lorenz system is considered as a classical model to address the essential specification of nonlinear systems. The Lorenz disturbance has a great impact on hurricane forecasting, which is revealed by using hurricane data. Different from conventional researches working on enhancing the numerical approaches for hurricane anticipation, dynamic features of the hurricane system are discussed here. To model the nonlinear dynamics and chaotic behavior of hurricanes, this study employs the Lorenz system as chaos theory combined with the LS-SVM method that is named LS-SVM^{CH} as the hurricane prediction model. Therefore, the proposed model applies the perturbation structure for hurricane prediction by the notion of the Lorenz comprehensive disturbance flow to properly consider the disturbance of the aerial system in the hurricane velocity anticipation to enhance its exactitude. More clarification for the suggested model is presented as:

- Implementation of the LS-SVM method

The support-vector machine (SVM) approach employs a nonlinear kernel function to map the input sample space to high dimension linear feature space due to make possible the pattern detection and function assessment of greatly nonlinear systems. The significant preponderance of employing this approach is a model-free description and no requirement of the complete system model. Therefore, the abovementioned approach is proper for nonlinear systems with uncertainties such as hurricane behavior. LS-SVM is the upgraded form of SVM, which transforms the conventional SVM inequality limitations into equality ones and takes the error squared and the loss function as the training set experiences loss. Hence, solving the programming is converted into solving a linear problem, which promotes the hurricane velocity and convergence of the solution [13]. Supposing the training data set as $\{(t_k, V_k^H) | k = 1, 2, \ldots, N\}$, where N describes the total number of samples, $t_k \in R^n$ depicts the input variables, $V_k^H \in \mathrm{R}$ indicates the output variables, n is the dimension of t_k. According to the SVM mapping theory, the samples from the original space R^n are transformed to the feature space $R^{n_h}$by a nonlinear mapping $\varnothing(.)$. The real-valued function $V^H(t)$ in a feature space is determined as [14]

$$V^H(t) = \omega^T.\varnothing(t) + b\,\omega \in R^n, b \in R \tag{17.1}$$

Eq. (17.2) is applied to evaluate the regression error variables for the LS-SVM fitting as follows:

$$e_k = \omega^T.\varnothing(t) + b - V_k^H\ k = 1,2,\ldots,N \tag{17.2}$$

$\{t_k, V_k^H\}_{k=1}^N$ as a training set cooperated with Eq. (17.2) as N limitations, are employed to construct the following objective function that is considered in the primal weight space.

$$\begin{aligned} &\mathrm{P}: \min_{\omega, e} J_P(\omega, e) \\ &J_P(\omega, e) = \frac{1}{2}\omega^T.\omega + \frac{1}{2}\gamma.\sum_{k=1}^{N} e_k^2 \end{aligned} \tag{17.3}$$

Eq. (17.3) consists of a compromise between the summation of squared errors and an objective function managed by Υ (i.e., The penalty parameter, which controls the degree of deviation beyond the error of the sample). The $\frac{1}{2}\omega^T\omega$ term of regression formulation specifies the smoothness of the model instead of hyperplane separation. Similar to the ridge regression problem formulated in the feature space, the parameter Υhas the same role of smoothing the resultant model in LS-SVM formalism. The dual Lagrangian-based formulation is presented as follows:

$$\begin{aligned} &D: \max_{a} L(\omega, b, e; a) \\ &L(\omega, b, e, a) = J_P(\omega, e) - \sum_{k=1}^{N} a_k.\left(\omega^T.\phi(t_k) + b + e_k - V_k^H\right) \end{aligned} \tag{17.4}$$

where a_k is the Lagrange multiplier. The optimal solutions should satisfy the following equations.

$$\begin{cases} \dfrac{\partial L}{\partial \omega} = 0 \rightarrow \omega = \sum_{k=1}^{N} a_k.\varnothing(t_k) & k = 1,2,\ldots,N \\ \dfrac{\partial L}{\partial b} = 0 \rightarrow \sum_{k=1}^{N} a_k = 0 & k = 1,2,\ldots,N \\ \dfrac{\partial L}{\partial e_k} = 0 \rightarrow a_k = \Upsilon e_k & k = 1,2,\ldots,N \\ \dfrac{\partial L}{\partial a_k} = 0 \rightarrow \omega^T.\varnothing(t_k) + b + e_k - V_k^H = 0 & k = 1,2,\ldots,N \end{cases} \tag{17.5}$$

After omission of the variables ω and e, the optimization problem leads to a linear structure as follows:

$$\begin{bmatrix} 0 & 1^T \\ 1 & K + r^{-1} \end{bmatrix} \begin{bmatrix} b \\ a \end{bmatrix} = \begin{bmatrix} 0 \\ V^H \end{bmatrix} \tag{17.6}$$

where $V^H = [V_1^H; V_2^H; \ldots; V_N^H]$, $1 = [1_1; 1_2, \ldots; 1_N]$, and $a = [a_1; a_2; \ldots; a_N]$. The kernel trick is employed here as

$$K_{kq} = \varnothing(t_k)^T.\varnothing\left(t_q\right) = K\left(t_k, t_q\right), k, q = 1,2,\ldots,N \tag{17.7}$$

Therefore, the LS-SVM model for hurricane velocity estimation is obtained as follows [15]:

$$V_{LS-SVM}^H(t) = \sum_{k=1}^{N} a_k.K(t, t_k) + b \tag{17.8}$$

where b and a_k are the solution to the linear system of Eq. (17.8). Moreover, b, a_k, and $K(t, t_k)$ sequentially depict the bias parameter, weight factor, and kernel mapping function between the training sample t and support vector t_k. The radial basis function networks kernel function utilized for LS-SVM is presented as follows [15]:

$$K(t, t_k) = \exp\left(-\frac{|t - t_k|^2}{2\sigma^2}\right) \tag{17.9}$$

- Implementation of the Lorenz system

The Lorenz system is a three-variable atmospheric convection model that describes the same motion state [16]. Eq. (17.10) illustrates the simplest notation of the chaotic state of a nonlinear system as the Lorenz equations.

$$\dot{\omega}_\Lambda^H = \begin{cases} \sigma^L \times \left(\omega_y^H - \omega_\Lambda^H\right) \omega_\Lambda^H = x \\ \omega_x^H \times \left(r^R - \omega_z^H\right) - \omega_\Lambda^H \omega_\Lambda^H = y \\ \omega_x^H \times \omega_y^H - b^L \times \omega_\Lambda^H \omega_\Lambda^H = z \end{cases} \tag{17.10}$$

In this equation, $\omega_x^H$is attributed to convection severity, $\omega_y^H$is attributed to the warmth degree deviation between rising and declining streams, and ω_z^H is attributed to the warmth degree departure from linearity. σ^L, r^R, and b^L are nondimensional constants that are sequentially defined as the Prandtl number, the Rayleigh number, and the area of microclimate. Since the solution of the Lorenz equations is three-dimensional, to unify the effect of the Lorenz system solution dimensions, the results have been standardized as

$$\acute{\omega}_{\Lambda,\mathrm{t}}^H = \frac{\omega_{\Lambda,\mathrm{t}}^H - \omega_{\Lambda,\mathrm{Min}}^H}{\omega_{\Lambda,\mathrm{Max}}^H - \omega_{\Lambda,\mathrm{Min}}^H} \forall \Lambda \in \{x, y, z\}, \forall t \in \{1, 2, \ldots, n\} \tag{17.11}$$

The standard form of the Lorenz system result is indicated as $\acute{\omega}_{\Lambda,\mathrm{t}}^H$ for $t=1, 2, \ldots, n$. Moreover, $\omega_{\Lambda,\,\mathrm{Min}}^H$ and $\omega_{\Lambda,\,Max}^H$ sequentially demonstrate the minimum and maximum amounts of ω_{Λ}^H. To unify the hurricane velocity and three dimensions of the Lorenz system solution, the Euclidean distance (ED) is employed to decrease the mentioned dimensions. Therefore, Eq. (17.12) should be utilized as the ED formula [17].

$$d\left(C_t^H - C_0^H\right) = \sqrt{\left(\acute{\omega}_{\mathrm{x,t}}^H - \acute{\omega}_{\mathrm{x,0}}^H\right)^2 + \left(\acute{\omega}_{\mathrm{y,t}}^H - \acute{\omega}_{\mathrm{y,0}}^H\right)^2 + \left(\acute{\omega}_{\mathrm{z,t}}^H - \acute{\omega}_{\mathrm{z,0}}^H\right)^2} \tag{17.12}$$

where C_t^H signifies $C_t^H\left(\acute{\omega}_{\mathrm{x,t}}^H, \acute{\omega}_{\mathrm{y,t}}^H, \acute{\omega}_{\mathrm{z,t}}^H\right)$ and C_0^H determines $C_0^H\left(\acute{\omega}_{\mathrm{x,0}}^H, \acute{\omega}_{\mathrm{y,0}}^H, \acute{\omega}_{\mathrm{z,0}}^H\right)$. Moreover, $L^H$is set as the length of vectors in the phase space according to the comprehensive perturbation, which indicates the deviance from a balanced state $C_0^H(0,0,0)$ to an absolute movement situation in the disturbed system as follows [17]:

$$L^H = \left\|C_t^H\right\| = \sqrt{\left(\omega_{x,t}^H\right)^2 + \left(\omega_{y,t}^H\right)^2 + \left(\omega_{z,t}^H\right)^2} \tag{17.13}$$

Finally, due to considering the chaotic behavior of hurricanes and their historical data simultaneously, the results of LS-SVM method and Lorenz equations are implemented coordinately as follows [18]:

$$\begin{aligned} &S^H\left(v_1^H, v_1^H, \ldots, v_f^H\right) = \\ &V_{LS-SVM}^H\left(v_1^H, v_2^H, \ldots, v_f^H\right) + \xi . L_{C-T}^H\left(l_1^H, l_2^H, \ldots, l_f^H\right) \end{aligned} \tag{17.14}$$

In Eq. (17.14), $S^H(v_1^H, v_2^H, \ldots, v_f^H)$is the hurricane velocity anticipation series according to the effect of LS-SVM and the Lorenz system solution. The LS-SVM method yields $V_{LS-SVM}^H(v_1^H, v_2^H, \ldots, v_f^H)$ as the hurricane velocity anticipation series. The Lorenz system results in $L_{C-T}^H(l_1^H, l_2^H, \ldots, l_f^H)$ as the atmospheric disturbance series. The effective Lorenz system coefficient is presented by ξ, and f indicates the forecasted specimen number. To evaluate the effectiveness of the suggested hurricane anticipation method, the performances of several approaches such as RBF, SVR, and LS-SVM are compared with the LS-SVM^{CH} (i.e., suggested approach). Hence, several error metrics such as mean absolute error (MAE), mean squared error (MSE), and mean absolute percentage error (MAPE) are applied to illustrate the accuracy of the mentioned anticipation methods and compared. The abovementioned anticipation error metrics are depicted as follows [19]:

$$MAE = \frac{1}{M}\sum_{t=1}^{M}|y(t) - f(t)| \tag{17.15}$$

$$MSE = \frac{1}{M}\sum_{t=1}^{M}(y(t) - f(t))^2 \tag{17.16}$$

$$MAPE = \frac{1}{M}\sum_{t=1}^{M}\left|\frac{y(t) - f(t)}{y(t)}\right| \tag{17.17}$$

wherein $y(t)$ and $f(t)$ present the observations and forecasts of hurricane speed at time t, respectively. Moreover, M depicts the sample size.

4 Vulnerability of lines and poles against the hurricane

In an electric grid, poles and overhead lines are the most vulnerable to confront an excessive hurricane. The fragility curves of overhead lines and poles indicate their failure probability by the maximum predicted hurricane velocity as shown in Fig. 17.4.

The phenomenon happening model uses the fragility curve as its input. To predict damaged infrastructures, a novel and multilevel model is presented here. It should be noted that the exact anticipation of outage lines and poles is impossible, similar to other power system variables anticipations that have errors. In this study, a hurricane prediction and optimization framework is simultaneously employed to anticipate damaged infrastructures against the hurricane. The proposed optimization framework is organized so that the devices with high vulnerability from the hurricane (i.e., according to the technical issues and hurricane intensity prediction) and their importance (i.e., according to the downstream load) will be determined. Therefore, the proposed damaged equipment as the optimal solution is more probable and more crucial that failure of them has significant effects on the network performance. Hence, the determination of outage devices is crisis-based as well as hurricane prediction-based. Although the suggested outage equipment (i.e., optimal solution) may not happen completely, the preventive scheduling for this situation will be more capable of improving the performance of the network against the hurricane than the other ones. The outage infrastructures can be determined by solving the optimization problem

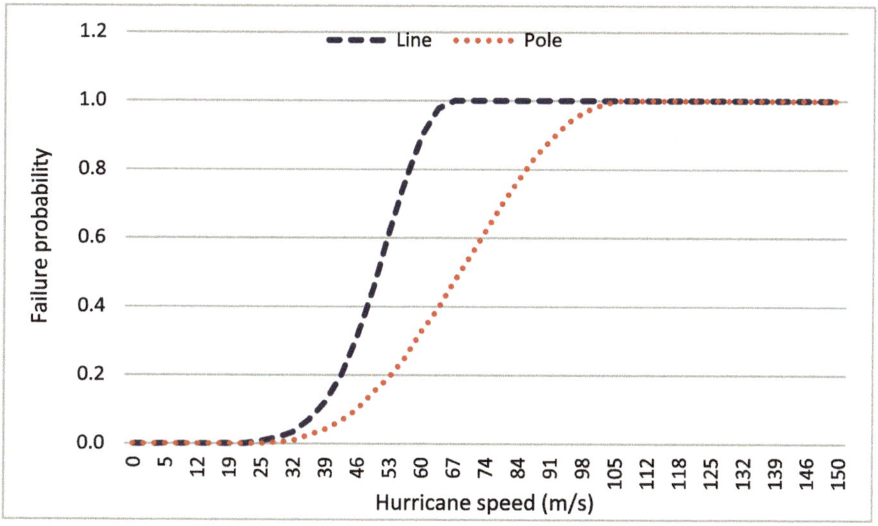

Fig. 17.4 Typical wind-affected fragility curves were applied for a pole and overhead branch.

to specify the most probable and crucial devices. The objective function of the proposed hurricane happening model is named the vulnerability index (VI) that is presented as

$$\psi^H = VI\left(u^t_{ij,\wp}, \omega_H, \theta^{\circ}_{ij}, \theta_H{}^{\circ}\right)$$
$$= \text{Max} \sum_{t=1}^{T} \sum_{\wp=1}^{N_Z} \sum_{(i,j)\in Z_\wp} \left(\frac{\left|\angle\theta_{ij}{}^{\circ} - \angle\theta_H{}^{\circ}\right|}{\angle 90^{\circ}} + \frac{\int_0^{\omega_H^{Max}} f_{ij}(\omega_H).d_\omega}{\int_0^{\omega_H^{Max}} 1.d_\omega} + \frac{\sum_{n\in\mathfrak{J}_{ij}} \vartheta_n.P_n}{\text{Max}\left(\vartheta_n.P_n\right)} \right).\left(1 - u^t_{ij,\wp}\right) \quad (17.18)$$

In the above equation, the whole damaged devices are defined as the set ψ^H. θ°_{ij} and θ°_H are the sequentially depicted deviation angle of the line ij and its poles and a hurricane from the north–south direction. $u^t_{ij,\ \wp}$ points to the line ij and its poles' status in the zone $\wp$ at period t (i.e., 0 is damaged and 1 is not damaged). P_n represents the load of node n that the whole of the downstream nodes of line ij create set $\mathfrak{J}_{ij}$. $\vartheta_n$is the load importance coefficient of node $n.f_{ij}$is the fragility function of line ij *or* its poles. Moreover, Eq. (17.18) is constructed from three terms. The first term of the proposed objective function calculates the deviation angle line with the hurricane direction. If the hurricane direction is more perpendicular to a line, the line will be more vulnerable. Furthermore, due to the elimination of the angle unit, the calculated deviation should be divided into 90°as the maximum deviation angle to make possible the summation of different terms. In addition to the angle of difference of lines with the direction of the storm, the hurricane speed is considered as the other crucial factor of a hurricane in the VI. The VI employs the hurricane speed as the input of its second term that considers the lines and poles fragility curves for the relation of the hurricane velocity with the line or pole failure probability. How much the below area of the fragility curve (i.e., Integral value of fragility function) is higher, the lines or poles are more vulnerable. To make dimensionless the second term, the evaluated value should be divided into the maximum amount of the below area of the fragility curve. Equipment importance is presented as the last term of Eq. (17.18). The quantity and priority of loads downstream of each device determine its importance. Due to normalization of the calculated value, it should be divided into the maximum multiplication of loads quantity and priority. This study considers hurricane provisional and spatial dynamics behavior. When a hurricane arrives in a region, it will descend on the area that is near the beach. The hurricane strength and intensity will affect the network equipment (i.e., lines and poles) in the mentioned area, while the outlying network lines and poles will remain spotless. In the impressed zone from the hurricane, Eq. (17.19) is utilized to consider the number of destroyed devices limitation based upon the effect of the provisional and spatial behavior of the hurricane. In Eq. (17.19), $B_{\wp}$ is the maximum number of destroyed network equipment in the region $\wp$ that is named the fundamental budget. In the suggested hurricane happening model, the number of outage infrastructures in each region, as the quantity of the optimization problem solution, should not be more than its $B_{\wp}$. Hence, different values of $B_{\wp}$ lead to a different number of damaged devices in the region $\wp$. When the hurricane lands in the inward area, it will impress the area from one district to another one and its intensity reduces zone by zone. Therefore, the value of $B_{\wp}$ should be determined according to the region $\wp$ distance from the hurricane source. The $B_{\wp}$ value of the farther zones from the hurricane origin should not be more than the closer regions. The network operator determines the fundamental budget $B_{\wp}$ for the region $\wp$ according to an event intensity, network restoration equipment, and expected resilient level by sensitivity analysis. It means that $B_{\wp}$ should not be too high that all nodes are disconnected or too low that the significant effects of the hurricane will be ignored.

$$\sum_{(i,j)\in Z_{\wp}} \left(1 - u^{t}_{ij,\wp}\right) \leq B_{\wp} \tag{17.19}$$

$$u^{t}_{ij,\wp} = \begin{cases} 0 \text{ if device}(i,j)\text{damaged} \\ 1 \text{ if device}(i,j)\text{not damaged} \end{cases} \forall\ (i,j)\in\{Z_{\wp}\}, \tag{17.20}$$

$$\forall\ \wp = 1,2,\ldots,N_Z, t = 1,2,\ldots,T$$

$$u^{t}_{ij,\wp-1} = u^{t-1}_{ij,\wp-1}$$
$$\forall\ (i,j)\in\{Z_{\wp-1}\},\ \forall\ \wp = 2,3,\ldots,N_Z, t = 2,3,\ldots,T \tag{17.21}$$

$$u^{t}_{ij,\wp+1} = 1$$
$$\forall\ (i,j)\in\{Z_{\wp+1}\},\ \forall\wp = 1,2,\ldots,N_Z - 1, t = 1,2,\ldots,T-1 \tag{17.22}$$

When the hurricane lands in the zone $\wp$ at duration t, Eq. (17.20) indicates the device's status limitation of this zone. In this condition, the equipment status of the zone $\wp-1$ at duration t is the same as their status at duration $t-1$ according to Eq. (17.21). Eq. (17.22) determines the situations of the infrastructures of zone $\wp+1$ that are not affected (i.e., $u_{ij,\wp+1}^{t}=1$) while the hurricane arrives in the zone $\wp$.

Furthermore, most work on electricity generation vulnerability to wind has focused on wind turbines. Wind turbines are vulnerable to hurricanes because the maximum wind speeds in those storms can exceed the design limits of wind turbines. Failure modes can include loss of blades and buckling of the supporting tower [20]. In many areas being actively considered by developers of offshore wind farms, for instance, nearly half the turbines in a given farm are likely to be destroyed in a 20-year period [20]. The distribution of the number of wind turbine towers buckled by hurricanes can be modeled using parameters estimated as the combined probabilities of multiple single wind turbine towers buckling along with the probability of an area receiving winds of hurricane magnitude. The probability that a single wind turbine tower is buckled by a 10-min sustained hub-height hurricane velocity u is modeled using a log-logistic function with a scale parameter α and a shape parameter ß is presented as follows:

$$D(u)=\frac{(u/\alpha)^{\beta}}{1+(u/\alpha)^{\beta}} \tag{17.23}$$

Values for α and β are given for two cases in which the turbine is pointed into the wind and the turbine is pointed perpendicular to the wind. The number of turbine towers buckled by a single hurricane in a wind farm of n turbines is modeled as a beta-binomial distribution with parameters α_B and β_B. The probability that X, the number of turbine towers that buckle out of n total, equals a particular value x is given by

$$\Pr(X=x)=\binom{N}{K}\frac{B(x+\alpha_B,n-x+\beta_B)}{B(\alpha_B,\beta_B)} \tag{17.24}$$

where $B()$ is the beta function.

5 Scheduling of a network in a normal/hurricane condition

The charging/discharging situation variables of energy storage systems (ESSs) and their related powers scheduling are presented here. The proposed charging/discharging and the stored energy (E) profile of storage should consider the following limitations [21].

$$0\le P_{s,h}^{ch}\le\delta_{s,h}^{ch}P_{s,h}^{ch,\max} \qquad \delta_{s,h}^{ch}\in\{0,1\},\forall s,h \tag{17.25}$$

$$0\le P_{s,h}^{dch}\le\delta_{s,h}^{dch}P_{s,h}^{dch,\max} \qquad \delta_{s,h}^{dch}\in\{0,1\},\forall s,h \tag{17.26}$$

$$E_s^{\min}\le E_{s,h}\le E_s^{\max} \qquad \forall s,h \tag{17.27}$$

$$\delta_{s,h}^{ch} + \delta_{s,h}^{dch} = 1 \qquad \forall\, s,h \tag{17.28}$$

$$E_{s,h+1} = E_{s,h} + \left(\delta_{s,h}^{ch} P_{s,h}^{ch} \eta_s^{ch} \Delta t - \delta_{s,h}^{dch} P_{s,h}^{dch} \Delta t / \eta_s^{dch}\right) \quad \forall\, s,h \tag{17.29}$$

The charging/discharging power constraints are indicated as Eqs. (17.25) and (17.26). Limitation Eq. (17.27) relates to the state-of-charge (SOC) constraints. Concurrent charging and discharging avoidance is considered as Eq. (17.28) and the charging/discharging and SOC relevance are presented in Eq. (17.29).

In the normal operating condition without considering the hurricane occurrence, the scheduling formulation for minimizing the operation cost of the system can be employed as Eq. (17.30):

$$\begin{aligned} OC^N = Min\ & \underbrace{\sum_{l=1}^{NL}\sum_{h=1}^{NH}\sum_{g=1}^{NG}(I_{l,g,h} \times F_g(p_{l,g,h}))}_{\text{MTs generation cost}} + \\ & \underbrace{\sum_{l=1}^{NL}\sum_{h=1}^{NH}(\sum_{r=1}^{NR}(p_{l,r,h}C_r) + p_{l,u,h}U_h}_{\text{DERs and Upstream network power}} + \\ & \underbrace{\sum_{l=1}^{NL}\sum_{h=1}^{NH}\sum_{s=1}^{Ns}(\delta_{l,s,h}^{ch} p_{l,s,h}^{ch} \eta_s^{ch} - \delta_{l,s,h}^{dch} p_{l,s,h}^{dch} / \eta_s^{dch}) \times U_h}_{\text{ESSs operation cost}} + \\ & \underbrace{\sum_{l=1}^{NL}\sum_{h=1}^{NH}\sum_{g=1}^{NG} I_{l,g,h} \times (SU_{l,g,h} + SD_{l,g,h})}_{\text{Start-up and shut-down costs of MTs}} \end{aligned} \tag{17.30}$$

In this condition, the system constraints should be applied as follows:

$$\begin{aligned} \sum_{g=1}^{NG} I_{l,g,h} p_{l,g,h} + \sum\nolimits_{r=1}^{NR} p_{l,r,h} & \\ + \sum\nolimits_{s=1}^{NS} & \left(\delta_{l,s,h}^{dch} p_{l,s,h}^{dch} / \eta_s^{dch} - \delta_{l,s,h}^{ch} p_{l,s,h}^{ch} \eta_s^{ch}\right) + p_{l,u,h} - p_{l,h} \\ = \sum\nolimits_{j=1}^{NL} & p_{lj,h} \qquad \forall h,l \end{aligned} \tag{17.31}$$

$$q_{l,g,h} - q_{l,h} = \sum_{j=1}^{NL} q_{lj,h} \qquad \forall\, g,h,l \tag{17.32}$$

$$p_{l,g}^{\min} I_{l,g,h} \le p_{l,g,h} \le p_{l,g}^{\max} I_{l,g,h} \qquad \forall l,g,h \tag{17.33}$$

$$q_{l,g,h}^{Min} I_{l,g,h} \le q_{l,g,h} \le q_{l,g,h}^{Max} I_{l,g,h} \qquad \forall l,g,h \tag{17.34}$$

$$p_{l,g,h} - p_{l,g,h-1} \le UR_g \qquad \forall l,g,h \tag{17.35}$$

$$p_{l,g,h-1} - p_{l,g,h} \le DR_g \qquad \forall l,g,h \tag{17.36}$$

$$SU_g \ge CU_g\left(I_{l,g,h} - I_{l,g,h-1}\right), SU_g \ge 0 \qquad \forall l,g,h \tag{17.37}$$

$$SD_g \ge CD_g\left(I_{l,g,h-1} - I_{l,g,h}\right), SD_g \ge 0 \qquad \forall l,g,h \tag{17.38}$$

$$0 \le p_{l,u,h} \le p_{u,h}^{\max} \qquad \forall l,h \qquad (17.39)$$

$$0 \le p_{l,r,h} \le p_{r,h}^{\max} \qquad \forall l,r,h \qquad (17.40)$$

Active and reactive power balance limitations, i.e., Eqs. (17.31) and (17.32), should be satisfied at each hour. The output power limitations of MTs are indicated by Eqs. (17.33)–(17.38). The other limitations present active/reactive injected power limits (17.33), (17.34), ramping up/down constraints Eqs. (17.35)and (17.36), and start-up/shutdown cost limitations Eqs. (17.37) and (17.38). The distribution network limitation for the power exchange with the upstream network is presented in Eq. (17.39). Ultimately, the utmost output power of renewable energy resources at each hour is considered by Eq. (17.40).

In order to schedule the system in the hurricane condition, the following equations should be considered as follows:

$$\begin{aligned} OC^H = Min \; & \underbrace{\sum_{l=1}^{NL}\sum_{h=1}^{NH}(C_{loss}\, p_{l,h}^{shed})}_{\text{Load shedding cost}} + \\ & \underbrace{\sum_{l=1}^{NL}\sum_{h=1}^{NH}\left(\sum_{g=1}^{NG} I_{l,g,h} F_g(p_{l,g,h}) + \sum_{r=1}^{NR}(p_{l,u,h}C_r) + p_{l,u,h}U_h\right)}_{\text{MTs, RESs generation and upstream network power cost}} \\ & \underbrace{\sum_{l=1}^{NL}\sum_{h=1}^{NH}\sum_{s=1}^{NS}(\delta_{l,s,h}^{ch} p_{l,s,h}^{ch}\eta_s^{ch} - \delta_{l,s,h}^{dch} p_{l,s,h}^{dch}/\eta_s^{dch}) \times U_h}_{\text{ESSs operation cost}} + \\ & \underbrace{\sum_{l=1}^{NL}\sum_{h=1}^{NH}\sum_{g=1}^{NG} I_{l,g,h} \times (SU_{l,g,h} + SD_{l,g,h})}_{\text{Start-up and shut-down costs of MTs}} \end{aligned} \qquad (17.41)$$

$$\begin{aligned} & \sum\nolimits_{g=1}^{NG} I_{l,g,h}p_{l,g,h} + \sum\nolimits_{r=1}^{NR} p_{l,r,h} \\ & \quad + \sum\nolimits_{s=1}^{NS} \left(\delta_{l,s,h}^{dch} p_{l,s,h}^{dch}/\eta_s^{dch} - \delta_{l,s,h}^{ch} p_{l,s,h}^{ch}\eta_s^{ch}\right) + p_{l,u,h} + p_{l,h}^{shed} - p_{l,h} \\ & \quad = \sum\nolimits_{j=1}^{NL} p_{lj,h} \qquad \forall h,l \end{aligned} \qquad (17.42)$$

$$\sum_{g=1}^{NG} I_{l,g,h}q_{l,g,h} + q_{l,h}^{shed} - q_{l,h} = \sum_{j=1}^{NL} q_{lj,h} \qquad \forall\, h,l \qquad (17.43)$$

Eq. (17.41) is proposed to minimize the operation cost and load shedding cost, simultaneously, when the hurricane phenomenon appears to improve the distribution system resiliency and provide the most economical solution. Moreover, Eqs. (17.42) and (17.43) consider the active and reactive power equilibrium at each bus. Furthermore, Eqs. (17.33)–(17.40) should be satisfied in this condition.

6 Test system and main assumptions

The suggested model is performed on the IEEE 33-bus distribution network as a part of a power system [22], which has been developed by considering the multiple renewable energy resources, storage, and reserve lines, as indicated in Fig. 17.5 [21]. It is

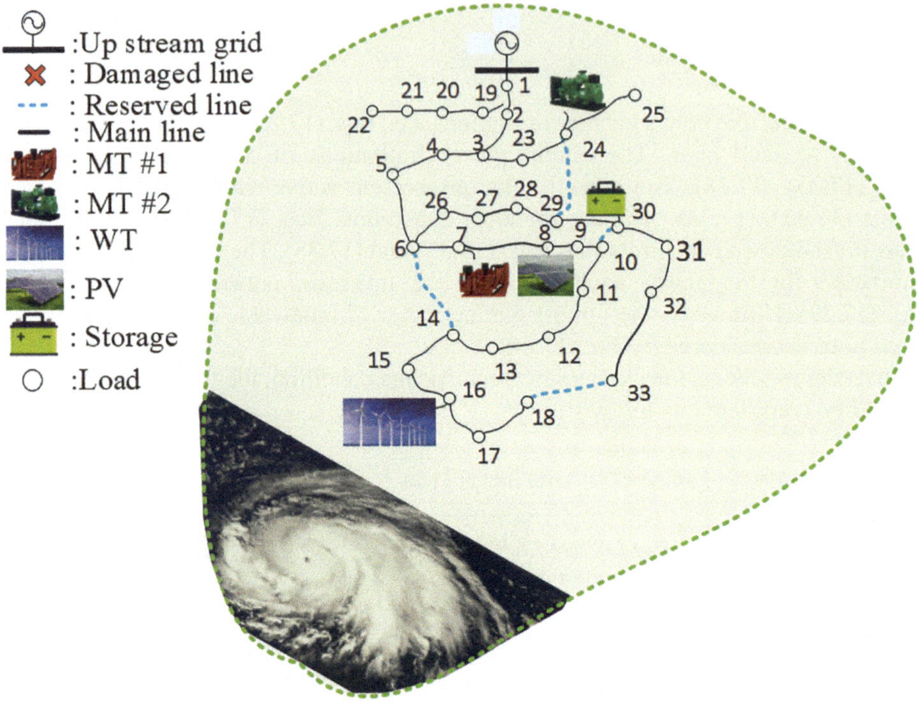

Fig. 17.5 The 33-bus network.

supposed that the reserve lines are out of service in the normal condition. The system loads have multifarious priorities according to their type (e.g., *residential, commercial, industrial, military, and medical*) as presented in Table 17.1. The specification of ESS and MT units are presented in Tables 17.2 and 17.3, respectively [21]. Load shedding costs and other associated constants are considered according to [23,24]. Moreover, the ESS is assumed to be completely discharged at 0'clock and the operation cost of renewable energy resources units is ignored. The network load and the

Table 17.1 Loads priority list.

Load no #	Priority	Load No #	Priority	Load No#	Priority	Load no #	Priority
2	5	10	1	18	1	26	1
3	5	11	2	19	4	27	1
4	3	12	3	20	4	28	2
5	2	13	1	21	2	29	2
6	4	14	2	22	2	30	3
7	3	15	1	23	2	31	3
8	3	16	3	24	5	32	3
9	1	17	1	25	5	33	2

Table 17.2 Characteristics of the ESS

E_S^{min} (MWh)	E_S^{max} (MWh)	$P_S^{ch,\max}$ (MW)	$P_S^{dch,\max}$ (MW)	η_S^{ch}	η_S^{dch}
0	0.7	0.175	0.175	0.88	0.88

Table 17.3 Characteristics of dispatchable DG units.

Unit #	Operation cost ($/MWh)	Min-max capacity (MW)	Ramp up/down rate (MW/h)
1	51.86	0–0.6	0.3
2	85	0–0.9	0.45

electric market price are forecasted by the Monte Carlo method based on the PDF and historical data as [21] that is illustrated in Table 17.4. Furthermore, Table 17.4 indicates the PV output power profile forecasted by the Monte Carlo method according to Beta PDF and historical data [25]. According to Table 17.4, most solar radiation occurs at 12 o'clock and sun radiation continues from hours 6 to 19.

In order to anticipate the hurricane's strength, historical and proper data should be employed. Here the abovementioned data is taken from Refs. [26, 27] that is related to the Gulf and East Coasts of the United States as shown in Fig. 17.6. This data consists of hurricane speeds in 16 specified directions, beginning with NNE and moving clockwise to N.

As mentioned before, this study employs Lorenz equations to consider the hurricane's chaotic behavior. According to reference [19], the initial Lorenz parameters

Table 17.4 Forecasted profile for price, load, and PV output power.

Time (h)	Price ($/MW)	Load (MW)	PV (MW)	Time (h)	Price ($/MW)	Load (MW)	PV (MW)
1	18	3.45	0	13	59	4.23	0.575
2	15	2.98	0	14	63	4.90	0.45
3	13	2.44	0	15	76	5.40	0.3
4	13	2.44	0	16	85	5.76	0.2
5	10	2.21	0	17	90	6.05	0.1
6	25	3.24	0.025	18	100	6.56	0.05
7	20	3.2	0.1	19	110	6.85	0.02
8	22	3.94	0.2	20	120	7.21	0
9	24	4.02	0.3	21	105	6.63	0
10	35	4.12	0.45	22	92	6.13	0
11	40	4.28	0.575	23	75	5.19	0
12	65	4.50	0.6	24	60	4.7586	0

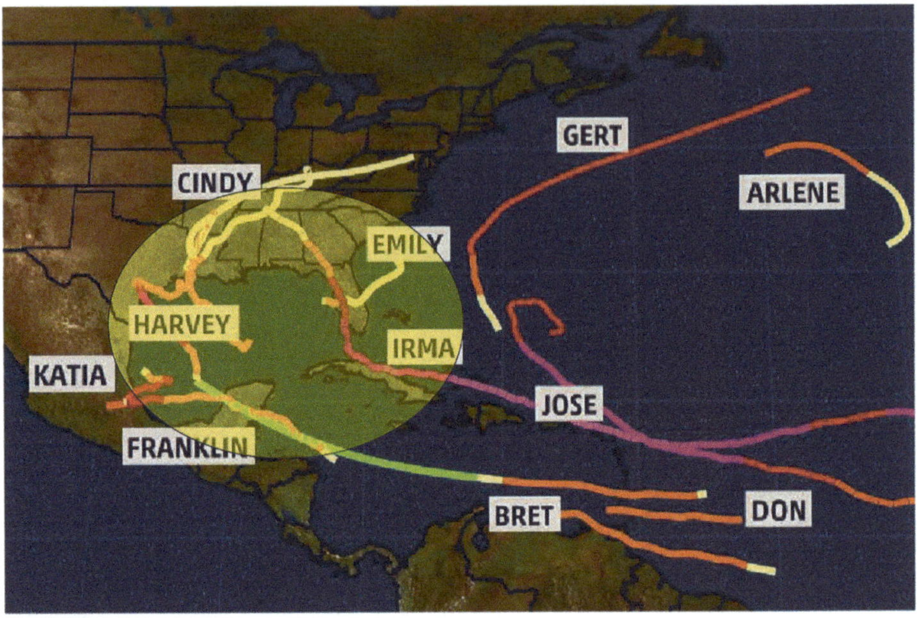

Fig. 17.6 Location of the employed hurricane data.

for the airflows perturbation as b, r, and σ^L were sequentially chosen equal to 8/3, 10, and 28. The other Lorenz parameters are determined according to [19]. Fig. 17.7 illustrates the Lorenz attractor in the chaotic situation for the abovementioned initial conditions.

The hurricane speed and direction of the abovementioned zones are demonstrated in Fig. 17.8. As shown in Fig. 17.8, the main hurricane direction is north. Furthermore, the hurricane strength steps down by going on each zone.

The errors of several approaches for the hurricane anticipation speed are compared with the suggested method in Table 17.5. Most values of the mentioned error metrics belong to RBF. MSE, MAE, and MAPE indicators of the SVR approach in comparison with RBF, 13.4%, 1.2%, and 8.4% are enhanced sequentially. 34.6%, 15.3%, and 4.1%

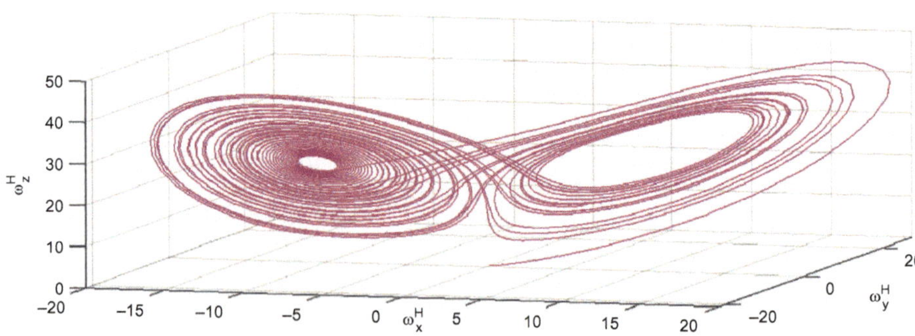

Fig. 17.7 The Lorenz attractor.

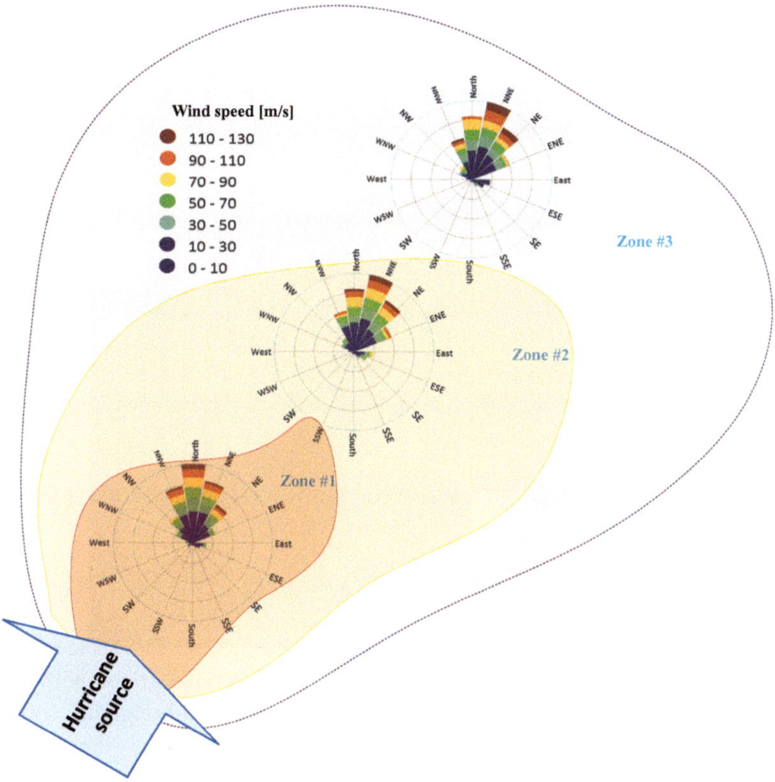

Fig. 17.8 The hurricane speed and direction.

Table 17.5 Error metrics of hurricane anticipation approaches.

Methods	Error		
	MSE (m^2/s^2)	MAE (m/s)	MAPE (%)
RBF	49.6	6.08	0.075
SVR	42.9	6	0.069
LS-SVM	28.1	5.08	0.066
LS-SVM^{CH}	4.0	1.88	0.021

reduction of MSE, MAE, and MAPE metrics, respectively, for LV-SVM in comparison with SVR, depicts that LV-SVM is more accurate than SVR and RBF. At last, the best accuracy occurs in the LS-SVM^{CH} method as the suggested approach. The MSE, MAE, and MAPE metrics for the LS-SVM^{CH} method are 4, 1.88, and 0.02, sequentially.

The hurricane velocity anticipation results for the RBF, SVR, LS-SVM, and LS-SVM^{CH} approaches are illustrated in Fig. 17.9. The RBF anticipation profile

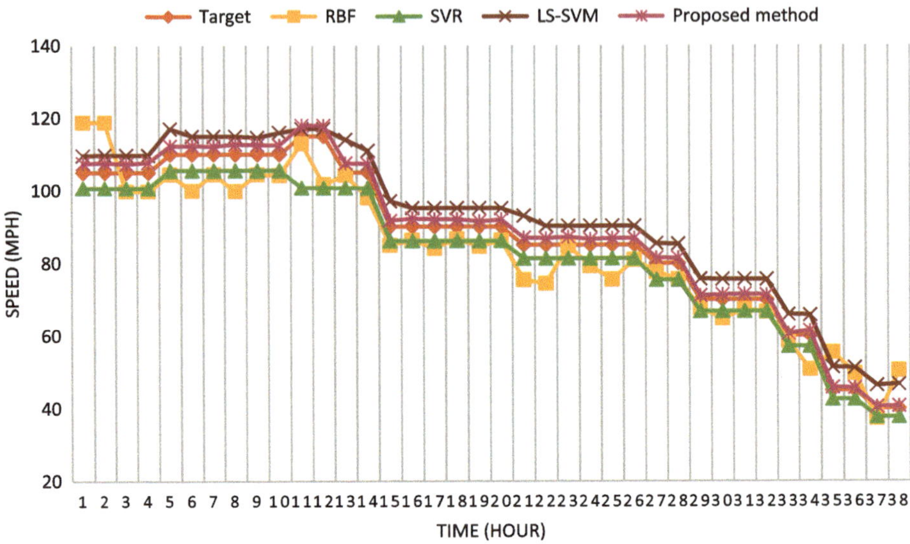

Fig. 17.9 Hurricane velocity anticipation profiles.

has the most deviation from the real hurricane velocity profile. The SVR anticipation outcome is more accurate than the RBF one. However, in comparison with LS-SVM and LS-SVM^{CH}, SVR is not near the target velocity profile. Although the anticipation results of the LS-SVM and LS-SVM^{CH} approaches are similar to the actual storm velocity series, the anticipation result of the LS-SVM^{CH} approach is closer to the real hurricane velocity series in more hours.

7 Results and analysis of the proposed model

The hurricane effect model that is proposed in this study is based on the multistage, multiperiod, and multizone model, which is given from an optimization procedure. The hurricane effect model for the modified IEEE 33-node distribution system is constructed on the hurricane speed prediction by the LS-SVM method coordinated with the chaos theory and the suggested optimization structure. The progress procedure of the hurricane on the abovementioned electric distribution network consists of three periods and three zones based on the route of the hurricane motion and the regions of the distribution network lines. Zone #1 is supposed to be close to the beach. Therefore, the distribution network lines of zone #1 will be damaged severely from the hurricane effects in the first period. Then the hurricane will land on the inland region and cause damage to the distribution network lines in zone #2. Finally, the hurricane will reach region #3 and cause damage, mostly on a lesser scale for the attributed distribution network lines. By executing the proposed model, Fig. 17.10 indicates the worst-case hurricane scheme effects based on the predefined budget (i.e., $B_1 = 2$, $B_2 = 2$, $B_3 = 1$).

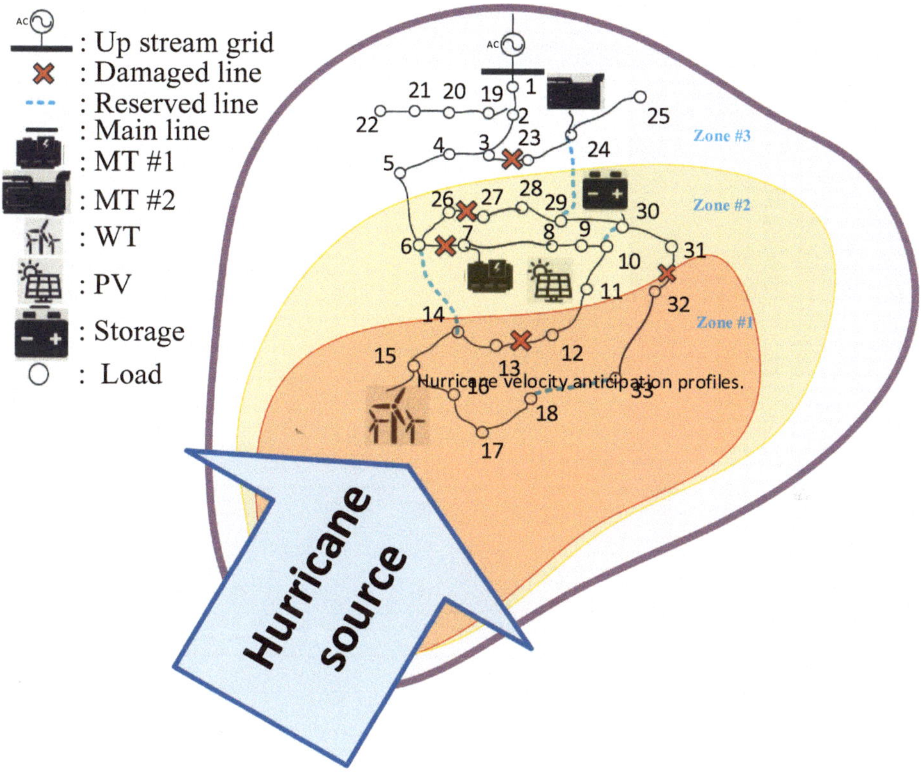

Fig. 17.10 The hurricane effect on the understudy network.

In order to evaluate the implementation of the proposed model, two scenarios are considered as follows:

✓Scenario #1: Optimal scheduling of the distribution network in normal condition.
✓Scenario #2: Optimal and proactive scheduling of the distribution network against the hurricane as an HILP event.

Table 17.6 depicts the MTs 1 and 2 output power profiles in scenarios #1 and #2. In scenario #1, MTs have been employed when the market price is more than the MTs' operation cost. However, in scenario #2, in addition to the difference between the market price and MTs' operation cost, isolation of zone #2 caused MT #1 to supply the disconnected area from hours 7 to 24. Moreover, MT #2 is utilized from hours 17 till 24 to supply the isolated zone #3. Furthermore, Table 17.6 depicts WT output power in scenarios #1 and #2. In scenario #1, the output power of the WT is related to the wind speed in all hours that is forecasted by the Monte Carlo method, the Rayleigh PDF, and historical data [28]. In scenario #2, at hours 1–8 AM, due to the high intensity of the hurricane, the WT protection mechanism has stopped it. Therefore, WT output power has been equal to zero. In the other time intervals, the output power is according to the wind speed fluctuations. At hours 9–24, due to the hurricane condition, the wind speed

Table 17.6 MTs and WT output power and the exchange power of distribution network with the upstream network profiles.

Time (h)	MT #1 output power (MW)		MT #2 output power (MW)		WT output power (MW)		Upstream grid exchanged power (MW)	
	Scenario #1	Scenario #2	Scenario #1	Scenario #2	Scenario #1	Scenario #2	Scenario #1	Scenario #2
1	0	0	0	0	0.24	0	3.21	2.88
2	0	0	0	0	0.255	0	2.73	2.52
3	0	0	0	0	0.27	0	2.17	2.10
4	0	0	0	0	0.3	0	2.14	2.10
5	0	0	0	0	0.3	0	1.91	1.96
6	0	0	0	0	0.3	0	2.78	2.82
7	0	0	0	0	0.3	0	2.83	2.71
8	0	0.3	0	0	0.285	0	3.46	3.36
9	0	0.6	0	0	0.27	0.38	3.45	2.20
10	0	0.6	0	0	0.24	0.36	3.33	2.20
11	0.3	0.6	0	0	0.255	0.34	3.15	2.24
12	0.6	0.6	0	0	0.249	0.332	3.05	2.34
13	0.6	0.6	0	0	0.24	0.32	2.82	2.28
14	0.6	0.6	0	0	0.24	0.32	3.61	2.61
15	0.6	0.6	0	0	0.21	0.28	4.30	2.84
16	0.6	0.6	0.45	0.45	0.18	0.24	4.34	3.01
17	0.6	0.6	0.9	0.9	0.165	0.22	4.29	1.23
18	0.6	0.6	0.9	0.9	0.15	0.2	4.86	1.33
19	0.6	0.6	0.9	0.9	0.18	0.24	4.97	1.39
20	0.6	0.6	0.9	0.9	0.15	0.2	5.38	1.46
21	0.6	0.6	0.9	0.9	0.18	0.24	4.78	1.34
22	0.6	0.6	0.9	0.9	0.21	0.28	4.33	1.24
23	0.6	0.6	0	0.9	0.21	0.28	4.38	1.05
24	0.6	0.6	0	0.9	0.24	0.32	3.92	0.96

is more in scenario #2 than in scenario #1 and WT output power follows its speed. Moreover, both scenarios for exchange power with the upstream network profiles are determined in Table 17.6. It is observed that until 8 o'clock, due to the effect of zone #1 from the hurricane, less load shedding is provided in cooperation with the other zones. Here the exchange power profiles have an almost identical pattern in scenarios #1 and #2. The maximum difference between the two scenarios belongs to hours 17–24. In this period, the delivered power is more than for the other periods for scenario #1. However, in scenario #2, due to disconnection of the major part of the distribution network, this profile is the most minimum at hours 17–24.

The reserve lines status alterations in the three zones are illustrated in Fig. 17.11. It is observed that the hurricane affected zones #1, #2, and #3 when it arrived there. Here the most vulnerable lines will be disconnected. After that, the reserved lines should be managed to feed the most prioritized loads by DGs and storage. Line #34 has been connected for the whole horizon scheduling time, in order to unify the disconnected areas of zone #1. From 1 a.m. to 8 a.m. (i.e., *not affected zone #2 from the hurricane*), line #33 (i.e., *the line between zone #2 and zone #1*) has been connected to transmit electric power from zone #2 to zone #1, in limitation of line power capacity. Furthermore, from hours 11 to 14, due to injection of the surplus power generated from zone #2 into zone #1, line #33 could be connected again. When the hurricane disconnected the network of zone #2 from the main grid and damaged it, in order to adjoin the segregated electric network sections of zone #2, line #35 has been utilized. Furthermore, line #36 has been employed from hours 9 to 16 (i.e., *not damaged zone #3 from the hurricane*) to transmit electric power from zone #3 to zone #2.

The charging, discharging, and SOC profiles of storage at both scenarios are shown in Fig. 17.12.A–C, respectively. In scenario #1, due to the lowest price market, the storage should be charged and SOC grows up from hours 1 to 6. At 7 o'clock, owing to market price increment the storage has been discharged and the SOC has been

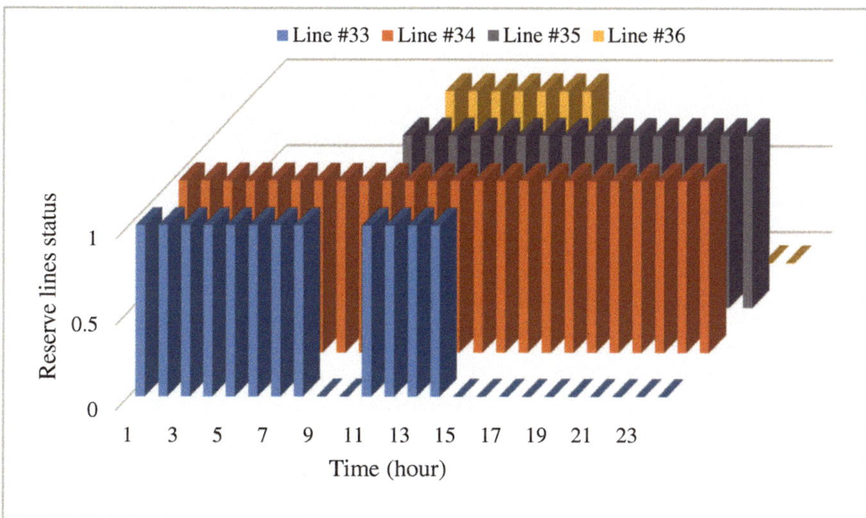

Fig. 17.11 Reserve lines status alterations against the hurricane.

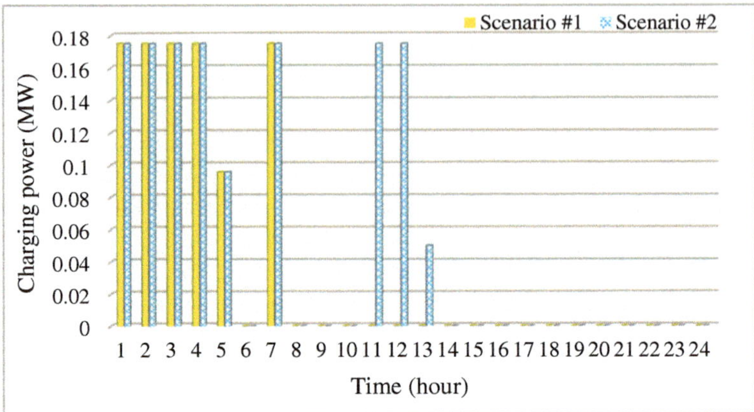

(a). Charging profile

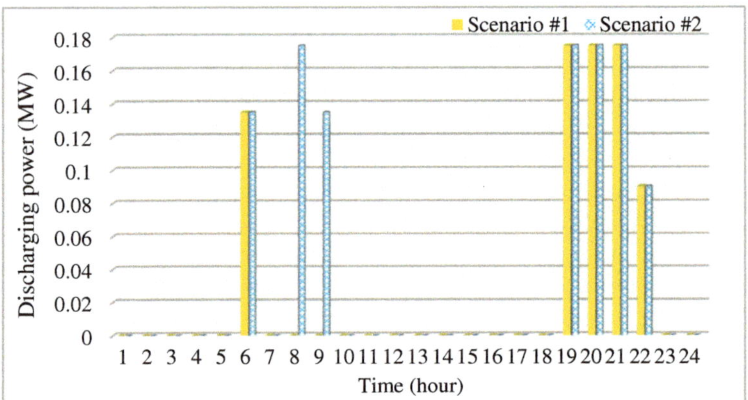

(b). Discharging profile

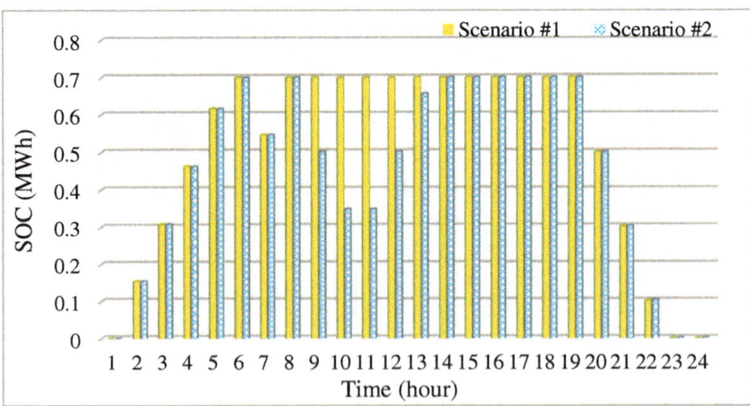

(c). SOC profile

Fig. 17.12 Scheduling of storage: (A) charging profile, (B) discharging profile, and (C) SOC profile.

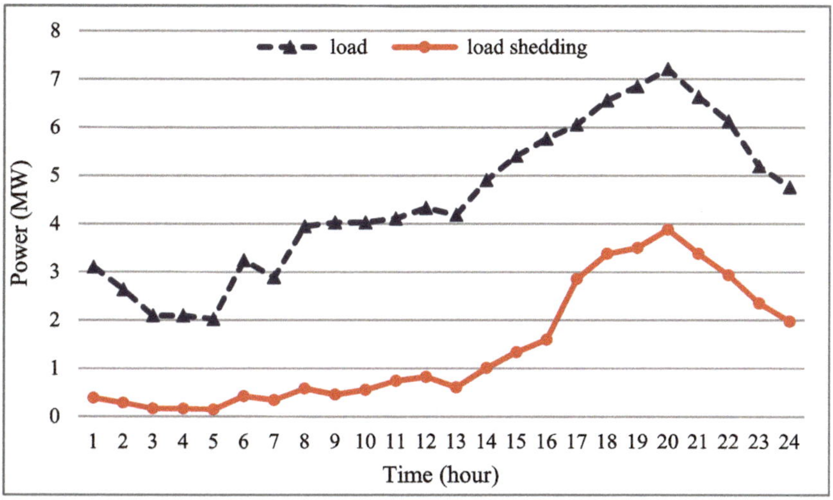

Fig. 17.13 The total load and load shedding profiles for scenario #2.

stepped down. From hours 8 to 19, SOC is preserved at its maximum capacity. After that, due to the great increment in market price, SOC was stepped down until it is empty at hour 23. In scenario #2, the storage charging, discharging, and SOC profiles have the manner approximately like those in scenario #1. However, from hours 9 to 16, due to the arrival of the hurricane in zone #2, its electric network will be damaged and isolated. Nevertheless, due to the existence of PV, storage, and DG in the peak solar radiation time, supply energy will be above the local demand. Therefore, before this period, the storage is discharged from hours 8 until 9; then it is charged at the peak radiation time, i.e., *from 11 o'clock until 13*. Therefore, the storage SOC will be full after hour 13.

The total load and load shedding profiles in scenario #2 are presented in Fig. 17.13. Both the mentioned profiles approximately have the same behavior. Most load shedding occurs at the peak load time. The cumulative load shedding and load are equal to 34 and 108 MWh, respectively, for 24 h in hurricane conditions.

Fig. 17.14 emphasizes that the supplied power according to the proposed model against the hurricane is more than one in normal scheduling. The supplied power difference in both conditions is 3.91 MW. In other words, the proposed model has protected the distribution network operator company from 62,400 $ financial loss for 24 h against the hurricane.

8 Conclusions

In this chapter, a hurricane happening model is suggested that consists of velocity prediction and damaged devices determination. The LS-SVM method coordinated with the chaos theory is applied for hurricane velocity prediction. The hurricane multiperiod and multizone dynamic, as well as a novel optimization model, which consists of the *VI* metric as an objective function, and the destroyed equipment budget

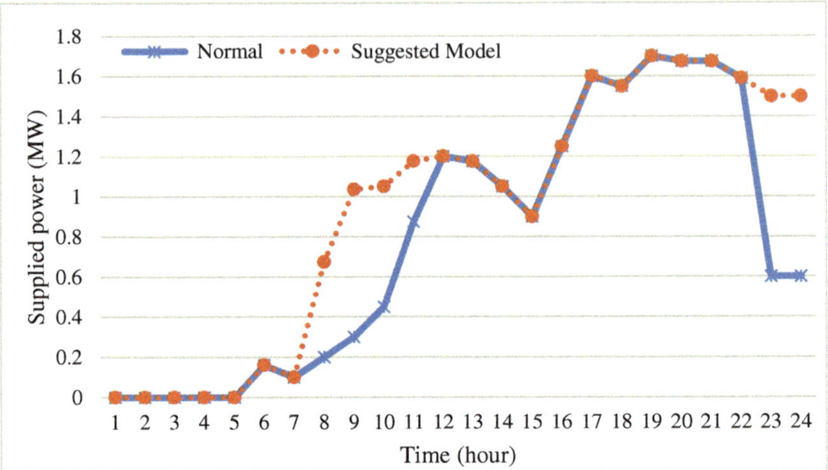

Fig. 17.14 Supplied power against the hurricane damages.

constraints are employed to specify the damaged topology. Executing the proposed hurricane model determines the most effective and vulnerable topology for damaged infrastructures. Moreover, a resilience-based operation strategy is employed here to reduce the effects of hurricanes on a power system. This strategy is based on storage and dispatchable resources scheduling and network reconfiguration. Finally, from the authors' point of view, the proposed future work for this study includes the environmental damage against the hurricane such as breaking trees and their effects on the power system performance. Future work also includes the aging of equipment such as poles and their effects on the vulnerability of a power system against a hurricane.

References

[1] A. Khodaei, Resiliency-oriented microgrid optimal scheduling, IEEE Trans. Smart Grid 5 (4) (2014) 1584–1591.

[2] M.H. Amirioun, F. Aminifar, H. Lesani, M. Shahidehpour, Metrics and quantitative framework for assessing microgrid resilience against windstorms, Int. J. Electr. Power Energy Syst. 104 (2019) 716–723.

[3] M.H. Amirioun, F. Aminifar, H. Lesani, Resilience-oriented proactive management of microgrids against windstorms, IEEE Trans. Power Syst. 33 (4) (2018) 4275–4284.

[4] G. Huang, J. Wang, C. Chen, J. Qi, C. Guo, Integration of preventive and emergency responses for power grid resilience enhancement, IEEE Trans. Power Syst. 32 (6) (2017) 4451–4463.

[5] M. Ghasemi, A. Kazemi, E. Bompard, F. Aminifar, A two-stage resilience improvement planning for power distribution systems against hurricanes, Int. J. Electr. Power Energy Syst. 132 (2021) 107214.

[6] M.H. Amirioun, F. Aminifar, M. Shahidehpour, Resilience-promoting proactive scheduling against hurricanes in multiple energy carrier microgrids, IEEE Trans. Power Syst. 34 (3) (2019) 2160–2168.

[7] National Hurricane Center, Hurricane Decay: Demise of a Hurricane, 2015, [Online]. Available *http://www.hurricanescience.org/science/science/hurricanedecay/*.

[8] J. Kaplan, M. DeMaria, A simple empirical model for predicting the decay of tropical cyclone winds after landfall, J. Appl. Meteorol. 34 (11) (2002) 2499–2512.
[9] . *https://www.nhc.noaa.gov*.
[10] K. Wu, X. Zhang, Global dynamics of the generalized Lorenz systems having invariant algebraic surfaces, Physica D 244 (1) (2013) 25–35.
[11] A.B. Coffin, E.N. Lorenz, The essence of chaos, Rocky Mt. Rev. Lang. Lit. 49 (1) (1995) 87.
[12] M. Gan, M. Ding, Y.Z. Huang, X.P. Dong, The effect of different state sizes on Mycielski approach for wind speed prediction, J. Wind Eng. Ind. Aerodyn. 109 (2012) 89–93.
[13] Y. Zhang, P. Wang, T. Ni, P. Cheng, S. Lei, Wind power prediction based on LS-SVM model with error correction, Adv. Electr. Comput. Eng. 17 (1) (2017) 3–8.
[14] G. Sharma, A. Panwar, I. Nasiruddin, R.C. Bansal, Non-linear LS-SVM with RBF-kernel-based approach for AGC of multi-area energy systems, IET Gener. Transm. Distrib. 12 (14) (2018) 3510–3517.
[15] Z.B. Shi, Y. Li, T. Yu, Short-term load forecasting based on LS-SVM optimized by bacterial colony chemotaxis algorithm, in: 2009 International Conference on Information and Multimedia Technology. ICIMT 2009, 2009, pp. 306–309.
[16] D. Viswanath, The fractal property of the Lorenz attractor, Physica D 190 (1–2) (2004) 115–128.
[17] Y. Zhang, J. Yang, K. Wang, Z. Wang, Y. Wang, Improved wind prediction based on the Lorenz system, Renew. Energy 81 (2015) 219–226.
[18] A. Nasri, A. Abdollahi, M. Rashidinejad, Probabilistic–proactive distribution network scheduling against a hurricane as a high impact–low probability event considering chaos theory, IET Gener. Transm. Distrib. 15 (2) (2021) 194–213.
[19] Y. Zhang, P. Wang, C. Zhang, S. Lei, Wind energy prediction with LS-SVM based on Lorenz perturbation, J. Eng. 2017 (13) (2017) 1724–1727.
[20] S. Rose, P. Jaramillo, M.J. Small, I. Grossmann, J. Apt, Quantifying the hurricane risk to offshore wind turbines, Proc. Natl. Acad. Sci. U. S. A. 109 (9) (2012) 3247–3252.
[21] H. Farzin, M. Fotuhi-Firuzabad, M. Moeini-Aghtaie, A stochastic multi-objective framework for optimal scheduling of energy storage systems in microgrids, IEEE Trans. Smart Grid 8 (1) (2017) 117–127.
[22] M.E. Baran, F.F. Wu, Network reconfiguration in distribution systems for loss reduction and load balancing, IEEE Trans. Power Deliv. 4 (2) (1989) 1401–1407.
[23] X. Wang, Z. Li, M. Shahidehpour, C. Jiang, Robust line hardening strategies for improving the resilience of distribution systems with variable renewable resources, IEEE Trans. Sustain. Energy 10 (1) (2019) 386–395.
[24] H. Farzin, M. Fotuhi-Firuzabad, M. Moeini-Aghtaie, Enhancing power system resilience through hierarchical outage management in multi-microgrids, IEEE Trans. Smart Grid 7 (6) (2016) 2869–2879.
[25] A.M. Eltamaly, Performance of MPPT techniques of photovoltaic systems under normal and partial shading conditions, Adv. Renew. Energies Power Technol. 1 (2018) 115–161.
[26] Hurricane Irma, 2014. *https://www.wunderground.com/hurricane/atlantic/2017/Tropical-Depression-Irma*.
[27] National Hurricane Center, "https://www.nhc.noaa.gov/data/," 2018.
[28] D. Lew, Development of Regional Wind Resource and Wind Plant Output Datasets, no. March, 2010, pp. 1–29.

Index

Note: Page numbers followed by *f* indicate figures, *t* indicate tables, and *b* indicate boxes.

F

G

Q

R

FSC
www.fsc.org
MIX
Papier | Fördert gute Waldnutzung
FSC® C083411

Kolibri 360 GmbH
Ferdinand-Jühlke-Straße 7
99095 Erfurt, Deutschland
produktsicherheit@kolibri360.de